MATHEMATICS FOR OPERATIONS RESEARCH

MATHEMATICS FOR OPERATIONS RESEARCH

W. H. MARLOW
The George Washington University
Washington, D.C.

DOVER PUBLICATIONS, INC.
New York

Copyright

Published in Canada by General Publishing Company, Ltd., 30 Lesmill Road, Don Mills, Toronto, Ontario.

Bibliographical Note

This Dover edition, first published in 1993, is an unabridged, corrected republication of the work first published by John Wiley & Sons, Inc. ("A Wiley-Interscience Publication"), New York, 1978.

Library of Congress Cataloging-in-Publication Data

Marlow, W. H., 1924–
Mathematics for operations research / W. H. Marlow.
p. cm.
Reprint. Originally published: New York : Wiley, 1978.
Includes bibliographical references and index.
ISBN 0-486-67723-0
1. Operations research—Mathematics. I. Title.
T57.6.M345 1993
510—dc20 93-24974
CIP

Manufactured in the United States of America
Dover Publications, Inc., 31 East 2nd Street, Mineola, N.Y. 11501

To Delphine E. Marlow

Preface

Operations research is the application of scientific methods, and especially mathematical and statistical ones, to problems of making decisions. Many illustrations are given in Wagner (1975) where particular attention is devoted to problems of management. Other textbooks that introduce methods for operations research are Gaver and Thompson (1973), Hillier and Lieberman (1974), and—from members of engineering faculties—Phillips, Ravindran, and Solberg (1976). Among more specialized works are: Dantzig (1963), which makes essential uses of mathematics; Gross and Harris (1974), which depends on probability theory; and Mann, Schafer, and Singpurwalla (1974), which shows how statistical methods can be used. The contributors to Marlow (1976) give many problems, methods, and solutions for logistics. Finally, in scientific journals there are ever increasing numbers of methods and reports of significant applications.

The present book is devoted to parts of mathematics that are used in operations research. Probability and statistics are two of the most prominent, but I have omitted them because they are well covered elsewhere. For example, from among the many books that could be studied as an introduction to probability and statistics for any reason, I would cite Hogg and Tanis (1977). The parts of mathematics that I have brought together are of equal importance, and they need to be presented in a single book whose only prerequisite is calculus.

I explain origins and organization in the section on *Notes on the Use of This Book*. Here I would like to explain my objective in another way. This is a book on how to do things. The things to be done are mathematical tasks that frequently arise, not only in operations research, but also in engineering, systems sciences, statistics, economics, and the many other fields where optimization and linear systems figure prominently. Examples of such tasks are

Resolving linear independence and finding null spaces, factors of matrices, and so on, in Section 3.3.

Determining existence of restricted solutions to linear equations and inequalities, in Chapter 4.

Resolving definiteness of Hermitian and real symmetric matrices by Gaussian pivoting, in Section 5.3.

Diagonalizing, or "nearly" diagonalizing, square matrices, in Section 5.4.

Differentiating vectors and matrices by the chain rule, in Section 6.3.

Establishing convexity of functions of several variables, in Section 6.5.

Establishing criteria for constrained minima or maxima, in Chapter 7.

Solving linear systems of differential equations, in Chapter 9.

Solving difference equations, and differential–difference equations, in Chapter 10.

My objective is to make readily available a set of effective procedures for performing such tasks. To this end I aim to

Formulate systematic procedures,
Guarantee the results, and
Develop associated techniques.

Representative examples of procedures are

Resolving Linear Independence	(3.22)
Resolving Definiteness	(5.28)
Differentiating Composite Transformations	(6.87)
Solving Initial Value Problems	(9.12)

Guarantees are possible because procedures are established as theorems. Techniques are illustrated in most of the examples, and nearly all of the tables display reference material for procedures. Many of the 1300 problems are also devoted to matters of technique.

It is a pleasure for me to acknowledge the support of the Logistics Program in the Office of Naval Research under Contract N00014-75-C-0729 and earlier contracts.

I am grateful to many students from various departments of The George Washington University; their comments on my notes for underlying courses have been most helpful for writing this book. Professors James E.

Falk, Anthony V. Fiacco, and Garth P. McCormick have had considerable influence on my outlook toward optimization theory. I also have gained much from my long association in logistics research with Professor S. Zacks of Case Western Reserve University. My original colleagues in founding the Department of Operations Research at The George Washington University—Professors Donald Gross, Carl M. Harris, Charles E. Pinkus, and Nozer D. Singpurwalla, and Henrietta Jones—have done much for which I am grateful. Above all, I wish to acknowledge the continuing patience and strong encouragement of my wife, Delphine, and our children.

W. H. MARLOW

Arlington, Virginia
February 1978

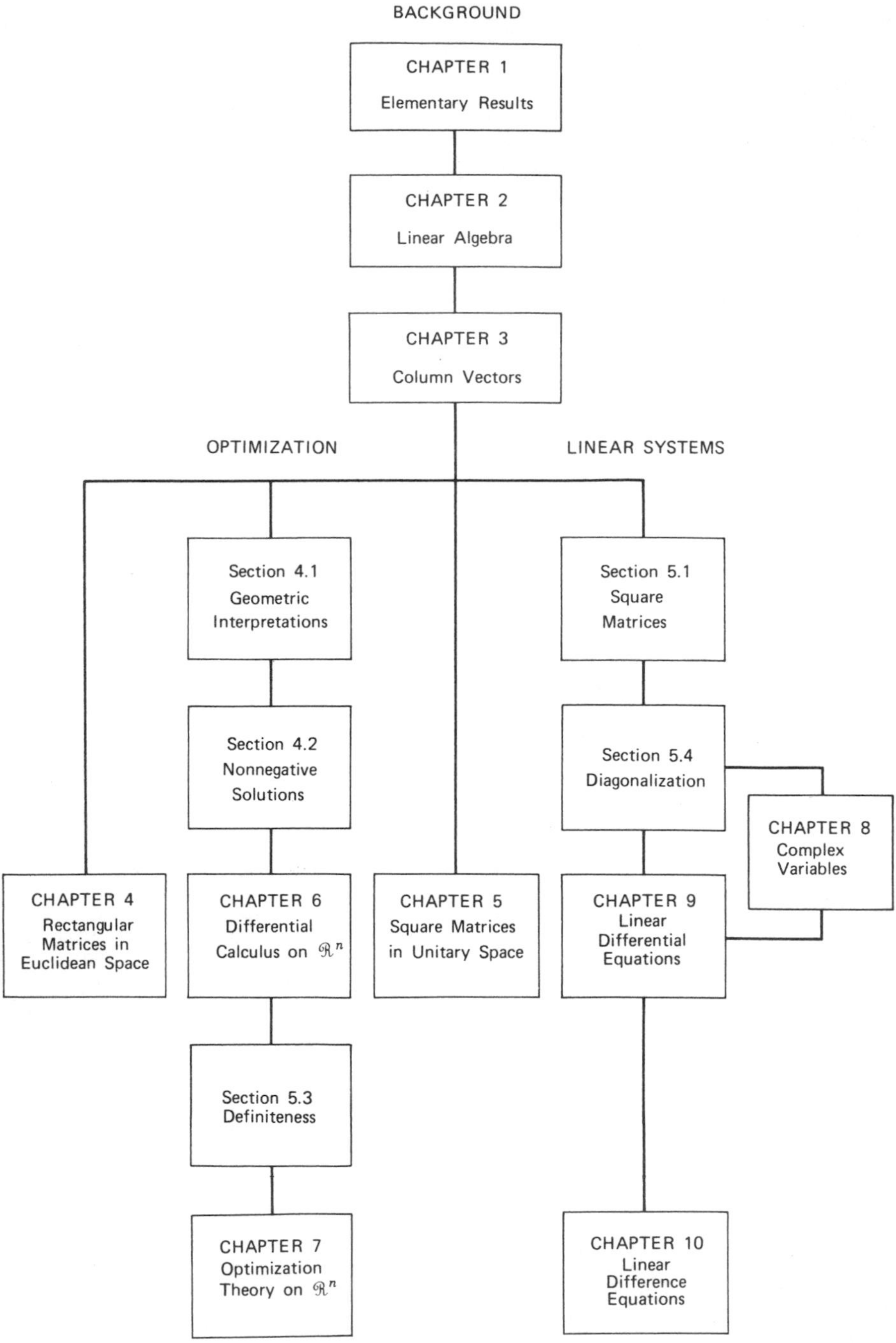

INTERDEPENDENCE OF CONTENTS

Notes on the Use of this Book

This book is based on notes distributed for two graduate courses aimed at removing mathematics—except for probability and statistics which were covered by another course—as a barrier to graduate programs in operations research. Students had completed calculus and usually at least one more advanced course in mathematics. Their graduate programs were drawn from 25 courses in operations research and other courses from departments in Engineering and Applied Science, Arts and Sciences, and Government and Business Administration. As a result of experience during 1971–1977, *optimization*—referring to minima and maxima of functions—was chosen for the first course, and *linear systems*—referring to systems of linear differential and difference equations—for the second.

The diagram *Interdependence of Contents* reveals the plan for the two graduate courses and the organization of this book. Early chapters furnish background and Chapter 5 contributes to both optimization and linear systems. No prerequisites are out of place if Chapters 1–10 are considered in order. But the book can also be entered at other places as indicated by the diagram. If Chapter 6 is entered following Section 4.2, then the reader can proceed to Chapter 7 as shown (and find no gaps except in Example 7.4 where Section 4.3 is used). More than this, relatively abstract parts such as Section 6.4 can be omitted at first reading, Section 7.6 can be omitted, and so on, according to directions in the text. Chapter 9 depends on Section 5.4, in ways that are explained in the text, but Chapter 8 is optional.

This book has also been designed for individual courses as follows.

Matrices: Chapters 1–5
Mathematics for Optimization: Chapter 4, Section 5.3, Chapters 6, 7
Mathematics for Linear Systems: Sections 5.1, 5.4, Chapters 9, 10
Advanced Calculus I: Chapters 1–6
Advanced Calculus II: Chapters 7–10

Numbered examples from throughout this book would be accessible in each such course and applications in particular subject areas could be drawn from works on operations research such as those cited in the Preface.

Displays are labeled in the format $(x.y)$ where x is the number of the chapter and y is the consecutive number within chapter; for brevity, part (z) of display $(x.y)$ is cited as $x.y(z)$. The same format $x.y$ is used for sections, examples, tables, and figures. Problems are collected, titled, and numbered in sets $a.b.c$ under Problems for Solution where a denotes the chapter, b the section, and c the consecutive number within section. Parenthetical citations $(a.b.c)$ are to be read "see problem set $a.b.c$." Items in the index referring to problems also refer to appropriate material in Answers, Hints, and Comments. In the latter are entries for each problem set, and "Drill" means that it is a matter of following, or slightly changing, something in the text; no entry for an individual problem is to be interpreted in the same way. Works in the references are cited using surnames and years of publication as, for example, Abramowitz and Stegun (1964). All special symbols and abbreviations appear in the List of Special Notation.

Contents

MATHEMATICS FOR OPERATIONS RESEARCH

1
ELEMENTARY RESULTS

Differential and integral calculus are the prerequisites for this book. In this chapter we introduce terminology, notation, and a few results for working with sets, functions, and real or complex numbers. Most of the Problems for Solution (supported by answers) furnish review and drill material for the reader with minimal mathematical background.

1.1 SETS AND FUNCTIONS

Here we introduce words and symbols that we rely on throughout the book.

Let us first note some matters connected with theorems in general. A *statement* is a declarative sentence that is *true* or *false* but not both. A *theorem* is a statement about two statements, an *hypothesis* p and a *conclusion* q, namely

$$\text{If } p \text{ is true then } q \text{ is true.}$$

In other words, the case p true and q false cannot occur. We generally abbreviate a theorem to "if p then q" or "p *implies* q." Most of the numbered displays in this book are theorems but we include the label "theorem" only for the results we consider to be most important. An *equivalence theorem* is a statement

$$p \text{ is true iff } q \text{ is true}$$

where "iff" is an abbreviation for "if and only if." We also use RHS and RH for *right-hand side* and *right-hand*, respectively, in referring to members of equations, inequalities, and so on, and we use LHS and LH for objects on the *left*. The symbol ■ is read "this completes the proof" to mark the end of proof for a theorem.

Let us next review some notions of sets in mathematics. A *set* is a collection of objects called its *elements*. We generally denote sets by capital letters such as A, B, C, and elements by lowercase a, b, c,.... Then $a \in A$ is read "a is an element of A" or "a belongs to A." For many symbols we use the shilling / to denote negation as in the familiar $\neq$ for "not equal". Thus $a \notin A$ is read "a is not an element of A." We use abbreviated notation such as $a, b, c \in A$ to specify elements of a set. Let us presuppose the existence of some *universal set*, say $\mathfrak{U}$, so that all elements of A and B are understood to be elements of $\mathfrak{U}$. If every element of A is an element of B, then A is a *subset* of B and we write $A \subset B$; if in this case $A \neq B$ then A is a *proper subset* of B. The *empty set* or *null set* $\varnothing$ is the set that contains no elements. All elements of $\mathfrak{U}$ that are not elements of A compose A^c the *complement* of A; similarly, elements of B that are not elements of A make up the *relative difference* $B-A$. The *intersection* of A and B, $A \cap B$, is the set of elements common to A and B; when $A \cap B = \varnothing$ we call A and B *disjoint* and otherwise we say that each of A and B *intersects* the other. The *union* of A and B, $A \cup B$, is the set consisting of all elements that are elements either of A or of B or of both. Suppose we have a statement $p(x)$ once we have specified any $x \in \mathfrak{U}$. Then

$$\{x : p(x)\}$$

denotes the set of elements $x \in \mathfrak{U}$ for which resulting statements $p(x)$ are true. We also use brackets { } to specify a set by enumerating its elements; for example,

$$\{1,2,3,\ldots,10\}$$

denotes the set consisting of the first 10 positive integers.

If X and Y are sets, the *Cartesian product* $X \times Y$ consists of all *ordered pairs* (x,y) where $x \in X$ and $y \in Y$. The significance of the modifier "ordered" is that it underscores the necessity of distinguishing the *first coordinate* x from the *second coordinate* y, in the ordered pair (x,y): By definition

$$(x_1,y_1) = (x_2,y_2) \quad \text{iff} \quad x_1 = x_2 \text{ and } y_1 = y_2$$

Clearly the concept applies to any finite collection of sets.

A *relation R from X to Y* is a subset of $X \times Y$. If $(x,y) \in R$ we write xRy and say that x *satisfies* the relation R with y. If $Y \subset X$ then R is *reflexive* if xRx for each $x \in X$; it is *symmetric* if xRy implies yRx; it is *transitive* if

$$xRy \text{ and } yRz \text{ imply } xRz$$

If R possesses all three properties it is called an *equivalence relation*; then each $x \in X$ determines a subset of X

$$C(x) = \{y : yRx\}$$

called the *equivalence class* determined by x. Such relations generalize ordinary equality in important ways (1.1.5).

A *function* f *on* a set X *to* a set Y is denoted by

$$f : X \to Y$$

and is commonly described as a rule that assigns to each $x \in X$ a unique element $f(x) \in Y$ called the *value* of f at x. What we call a function was in former years called a *single-valued function* and its definition is: A relation f from X to Y is a *function* if $(x,y) \in f$ and $(x,z) \in f$ imply $y = z$. In other words, f is a collection of ordered pairs (x,y) where each $x \in X$ appears exactly once and no special restriction is placed on $y \in Y$. Sometimes we write $f(\cdot)$ for the function and it is customary to refer informally to the function as $y = f(x)$. But in place of using, for example, "the function $y = x^2$" we often write

$$x \mapsto x^2$$

in *symbol-free functional notation* which can be used in particular cases without assignment of a letter such as f for the function. If $A \subset X$ then $f : A \to Y$ is the function called the *restriction* of f to A.

The *domain* of $f : X \to Y$ is X and the *set of images*, or *image*, of $A \subset X$ is

$$f(A) = \{f(a) : a \in A\}$$

It is also convenient to have notation for the set of *inverse images*

$$f^{-1}(B) = \{x : f(x) \in B\}$$

where $B \subset Y$ (and where the present f^{-1} is not to be confused with the function f^{-1} appearing in (1.1)). The set of all images $f(X)$ is called the *range* of f. If each $y \in Y$ appears at least once as an image, that is, if $f(X) = Y$, then f is said to be *onto* Y. If each image appears exactly once, that is, if

$$x_1 \neq x_2 \quad \text{implies} \quad f(x_1) \neq f(x_2)$$

then f is called *one-to-one*.

We use the name *one-to-one correspondence*, and notation such as $x \leftrightarrow y$, for functions that are one-to-one and onto. The empty set $\varnothing$ is a *finite set* with $n=0$ elements. A nonempty set is called a *finite set* with n elements if it can be put into one-to-one correspondence with

$$\{1,2,\ldots,n\}$$

An *infinite set* is a set that is not finite. A set that can be put into one-to-one correspondence with the set of all positive integers is called *countably infinite*

Consider $f: X \to Y$. If the ordered pairs

$$\{(y,x): y \in Y \quad \text{and} \quad x \in X\}$$

which define a relation from Y to X, also define a function then this function is the *inverse* of f and it is denoted by $f^{-1}: Y \to X$. The following result is fundamental.

A function $f: X \to Y$ has an inverse iff it is one-to-one and onto. (1.1)

Proof is an exercise (1.1.7).

Two functions f and g are *equal* if they have the same domain, say X, and if $f(x)=g(x)$ for all $x \in X$. On this basis we can say that f^{-1} in (1.1) is determined uniquely (1.1.7). Suppose next that $f: X \to Y$ and $g: X \to Y$ are functions to Y where sums y_1+y_2 and products $y_1 y_2$ are defined. The *sum* of f and g is then the function $f+g$ with values $f(x)+g(x)$ and the *product* fg has values $f(x)g(x)$. Let $f: X \to Y$ and $g: Y \to Z$ be given; then $h: X \to Z$ with values $h(x)=g(f(x))$ is called a *composite function.*

PROBLEMS FOR SOLUTION

1.1.1 Statements. Give examples of declarative sentences that are (1) statements even though we cannot be sure which of true or false applies, and (2) not statements.

1.1.2 Negations. Write the negation by taking "not" as far to the right as practicable. (1) All angles can be trisected using compass and straightedge alone. (2) There exist two nonintersecting lines. (3) All quintic equations cannot be solved in terms of radicals. (4) One and only one of Problems I and II has a solution. (5) To each x there corresponds a quantity y such that the statement $p(x,y,z)$ holds for all z. (6) The equations can be solved for any two of x_1, x_2, x_3, x_4, and x_5 in terms of the remaining three.

1.1.3 Theorems. Give the following for "p implies q": (1) Restatements. (2) Contrapositive. (3) Converse. (4) A more general theorem. (5) A stronger theorem. (6) Restatements of "p iff q."

1.1.4 Proofs. Describe the following. (1) Direct proof. (2) Indirect proof. (3) Proof by counterexample. (4) Weak induction. (5) Strong induction. (6) Backward induction.

1.1.5 Equivalence Relations. Establish: (1) xRy iff $C(x)=C(y)$. (2) Equivalence classes *partition* X in that each $x \in X$ belongs to one and only one equivalence class.

1.1.6 Examples of Functions. Let $N=\{0,1,2,\ldots\}$ and find simple examples where $f: N \to N$ is (1) one-to-one and onto, (2) one-to-one but not onto, (3) not one-to-one but onto, (4) not one-to-one and not onto.

1.1.7 Theorem (1.1). Establish: (1) Proof. (2) Uniqueness of f^{-1}.

1.1.8 Deducing Functional Expressions. Find $f(x)$ given (1) $f((x-1)^{-1}) = x^{-1}$, (2) $2f((x-1)^{-1})=x^{-1}+x^{-2}$, (3) $f(x+1)=(x-1)^{-1}$, (4) $f(\exp(x)) = 2x+1$, (5) $f(x(x-1)^{-1})=(x+1)^{-1}$, (6) $f((x+1)x^{-1})=(x^2+3x+1)\times (x+1)^{-2}$.

1.2 GROUPS AND FIELDS

We need the definition of a group so that we can verify that certain sets of objects form vector spaces (2.1.1, 2.3.2). We need the definition of a field so that we can be sure, especially in Chapters 2 and 3, of the properties that real and complex numbers have in common. But most of this section is devoted to a review of special properties not shared by these two fields.

A group is determined by a set $\mathcal{G}$ and a function from $\mathcal{G} \times \mathcal{G}$ onto $\mathcal{G}$ satisfying several conditions. Following custom we refer to the function as a *binary operation* and denote it by a symbol such as *. Then we say $\mathcal{G}$ is *closed* under * to mean that $a,b \in \mathcal{G}$ implies $a*b \in \mathcal{G}$. The operation is *associative* provided

$$a,b,c \in \mathcal{G} \text{ implies } (a*b)*c = a*(b*c)$$

An *identity* $e \in \mathcal{G}$ satisfies $a*e=a=e*a$ for all $a \in \mathcal{G}$. An element $a^{-1} \in \mathcal{G}$ satisfying

$$a*a^{-1}=e=a^{-1}*a$$

is called an *inverse* of a. The operation is *commutative* provided

$$a,b \in G \text{ implies } a*b=b*a$$

With these concepts we have the

Definition. A *group* is a set $\mathcal{G}$ and a binary operation * satisfying these conditions:

(1) The set $\mathcal{G}$ is closed under *.
(2) The operation * is associative.
(3) There is an identity $e \in \mathcal{G}$.
(4) Each $a \in \mathcal{G}$ has an inverse $a^{-1} \in \mathcal{G}$. (1.2)

If, moreover, the operation * is commutative, then $\mathcal{G}$ is called a *commutative group*.

If we consider the definition of a group together with recollections from elementary mathematics we find that both the real number system and the complex number system form commutative groups under ordinary addition. Furthermore, if we remove the element zero, which is the only element having no multiplicative inverse, then these systems are also commutative groups under multiplication. In each it is also true that the two different group operations interact under the following *distributive properties*: For any three elements a, b, c, and using the familiar $a+b$ for addition and ab for multiplication

$$a(b+c)=ab+ac, \quad (a+b)c=ac+bc$$

In words, *multiplication is distributive with respect to addition*. This discussion motivates the

Definition. A *field* is a set $\mathcal{F}$ and two binary operations—addition and multiplication—satisfying these conditions:

(1) Under addition, $\mathcal{F}$ is a commutative group with identity 0 called *zero*.
(2) Under multiplication, the elements that are not zero form a commutative group with identity 1 called *unity*.
(3) Multiplication is distributive with respect to addition. (1.3)

We need to work with the field $\mathcal{R}$ of real numbers and the field $\mathcal{C}$ of complex numbers.

The special properties of $\mathcal{R}$ we first recall involve the familiar *order* relation $>$ read "greater than" and the associated $\geqq$, $<$, and $\leqq$ read "greater than or equal," " less than," and "less than or equal," respectively.

ALGEBRA OF INEQUALITIES. The following properties hold for a, b, $c \in \mathcal{R}$:

(1) Three Possibilities. Exactly one of the following holds: $a<b$, $a=b$, $a>b$.

(2) Transitive Property. If $a>b$ and $b>c$ then $a>c$.
(3) Addition. If $a>b$ then $a+c>b+c$ for any c; if also $c>d$ then $a+c>b+d$.
(4) Multiplication. If $a>b$ and $c>0$ then $ca>cb$ and $(a/c)>(b/c)$; but if $c<0$ then $ca<cb$ and $(a/c)<(b/c)$.
(5) Division. If $a>b>0$ and $c>d>0$ then $ac>bd$ but $(a/d)>(b/c)$.
(6) Powers. If $a>b>0$, and if p and q are positive integers, then $a^{p/q}>b^{p/q}$. (1.4)

The preceding are used to establish particular inequalities, to find real numbers that satisfy inequalities, and so on. Often conditions on *absolute values* $|x|$ are involved where, by definition for $x\in\mathfrak{R}$

$$|x|=x, \qquad x\geqq 0$$

$$=-x, \qquad x<0$$

See 1.2.3 for a number of exercises.

Let us also recall the following terminology for $\mathfrak{R}$. A set $A\subset\mathfrak{R}$ is *bounded above* if there exists an element that is greater than or equal to all $x\in A$. Such an element is called an *upper bound* of A and if the set of upper bounds of A has a least element it is called the *supremum* of A and written $\sup A$. Now $\sup A$ may or may not belong to A; if $\sup A\in A$, then A has a largest element called the *maximum* of A, written $\max A$, and $\max A=\sup A$. A characteristic property of $\mathfrak{R}$ is

Every nonempty subset of $\mathfrak{R}$ that has an upper bound has a supremum in $\mathfrak{R}$. (1.5)

The concepts *bounded below*, *lower bound*, *greatest lower bound* or *infimum* of A, written $\inf A$, and the *minimum* of A, written $\min A$, are defined in straightforward fashion by interchanging "greater" and "less," and so on, in this paragraph. Finally, a set is *bounded* if it is both bounded below and bounded above (1.2.4).

It is convenient to use geometric notions in working with real numbers. The most fundamental is the *axis of reals*: On a given straight line we select an arbitrary point as origin, an arbitrary direction as positive, and an arbitrary unit of length. In this way we obtain the familiar one-to-one correspondence between $\mathfrak{R}$ and the so-called *real line*. Elements of $\mathfrak{R}$ are called *points*, $a<b$ means that b lies in the positive direction from a, and so on. This makes it easy to visualize the subsets of $\mathfrak{R}$ that are called *intervals*. Given points $a<b$ there are four *finite intervals* with descriptors

and notation as follows.

Closed:	$[a,b]=\{x : a\leqq x\leqq b\}$
Open:	$(a,b)=\{x : a<x<b\}$
Half open:	$[a,b)=\{x : a\leqq x<b\}$
Half open:	$(a,b]=\{x : a<x\leqq b\}$

The points a and b are called *endpoints*. Given $a\in\mathcal{R}$ there are four *semiinfinite intervals*.

Closed:	$[a,\infty)=\{x : a\leqq x\}$
Open:	$(a,\infty)=\{x : a<x\}$
Closed:	$(-\infty,a]=\{x : x\leqq a\}$
Open:	$(-\infty,a)=\{x : x<a\}$

The conventional symbol ∞, read "infinity," does not denote an element of $\mathcal{R}$ but simply appears in notation for unbounded subsets in $\mathcal{R}$.

A *linear function* on $\mathcal{R}$ to $\mathcal{R}$ is any function of the form $x\mapsto ax$. The name for a function of the form $x\mapsto ax+b$ is *affine function*.

Three distinctive properties of the complex field $\mathcal{C}$ can be recalled informally as follows. First, $\mathcal{C}$ is more general than $\mathcal{R}$ (in a way that is explained at the start of Section 3.4). Second, $\mathcal{C}$ contains a zero of the polynomial z^2+1, namely the *imaginary unit* i satisfying

$$i=\sqrt{-1} \text{ and } i^2=-1$$

Third, there is no possible order relation $>$ under which $\mathcal{C}$ can be ordered to produce a counterpart to (1.4). See 1.2.5.

The field $\mathcal{C}$ is composed of the familiar numbers

$$z=x+iy$$

where $x=\operatorname{Re} z$, the *real part* of z, and $y=\operatorname{Im} z$, the *imaginary part* of z are real numbers and where i is the imaginary unit previously introduced. If $\operatorname{Re} z=0$ then $z=iy$ is called a *pure imaginary*. Complex numbers can be represented as ordered pairs of real numbers and

$$z_1=z_2 \text{ iff } \operatorname{Re} z_1=\operatorname{Re} z_2 \text{ and } \operatorname{Im} z_1=\operatorname{Im} z_2$$

The *complex conjugate* of z is

$$\bar{z}=\operatorname{Re} z-i\operatorname{Im} z$$

and the *absolute value* of z is $|z|\in\mathcal{R}$ which is the nonnegative square root of

$$z\bar{z}=|z|^2$$

Fundamental properties are collected in 1.2.6 and 1.2.7.

It is natural to work with the Cartesian product $\mathcal{R}\times\mathcal{R}$ as the *complex plane* $\mathcal{C}$ wherein the x axis is called the *real axis* and the y axis is the *imaginary axis*. The *polar* or *trigonometric* representation of $z=x+iy$ is

$$z=\rho(\cos\theta+i\sin\theta)$$

where $\rho=(x^2+y^2)^{1/2}$ and θ is determined from $\cos\theta=x/\rho$ and $\sin\theta=y/\rho$. We see that $\rho=|z|$ but $\theta=\arg z$, the *argument* (or *amplitude*) of z, is not determined uniquely. In case we need a unique argument we assign $\theta=0$ for $z=0$ and otherwise restrict θ to $-\pi<\theta\leqq\pi$ and call this the *principal value of the argument* (1.2.8). Finally, let us note that *Euler's formula*

$$e^{i\theta}=\cos\theta+i\sin\theta \tag{1.6}$$

is used in the concise representation $z=\rho e^{i\theta}$.

We follow the common practice of using the term *polynomial* to refer to the specific expression defining a *polynomial function*. If $a_0,a_1,\ldots,a_n\in\mathcal{C}$ and $a_n\neq 0$ then

$$a(z)=a_nz^n+a_{n-1}z^{n-1}+\cdots+a_1z+a_0 \tag{1.7}$$

is the value at $z\in\mathcal{C}$ of the nth *degree polynomial function* $a:\mathcal{C}\to\mathcal{C}$ defined by the *coefficients* $a_0,a_1,\ldots,a_n$. The coefficient a_n is the *leading coefficient* and a_nz^n is called the *leading term*. When $a_n=1$ the polynomial (1.7) is *monic*. The nonnegative integer n is called the *degree*. A *real polynomial* has all $a_i\in\mathcal{R}$.

Any real or complex number z_1 such that $a(z_1)=0$ is called a *zero* of $a(z)$ or equivalently a *root* of the polynomial equation $a(z)=0$, and the same terminology applies to any function. We recall the

FUNDAMENTAL THEOREM OF ALGEBRA. Every polynomial of degree one or higher has at least one zero. (1.8)

A polynomial of positive degree n can be factored

$$a(z)=a_n(z-z_1)(z-z_2)\ldots(z-z_n)$$

and this expression is unique once the n labels have been assigned to the n (not necessarily distinct) zeros. If the same complex number appears more than once, say m times, in the preceding expression it is called a zero of *multiplicity* m or briefly a *multiple*, or *repeated, zero*. Otherwise a zero appears but once and it is called a *simple*, or *nonrepeated*, zero of $a(z)$.

The concept of the quotient of two polynomials is so important that the special name *rational function* is assigned. Such a function is defined except at the finite set of points where its denominator vanishes.

PROBLEMS FOR SOLUTION

1.2.1 A Finite Field. Show that $\{0,1,x,x^2\}$ forms a field where 0 is the identity for addition and 1 is the multiplicative identity, and where $1+1=0$, $x^2=x+1$.

1.2.2 Divisors of Zero. Whenever $ab=0$ with nonzero a and b, the elements a and b are called *divisors of zero*. Show that there are no divisors of zero in a field: if $a,\ b\in\mathfrak{F}$, then $ab=0$ iff $a=0$ or $b=0$.

1.2.3 Elementary Inequalities. Use (1.4) to establish, or solve, the following in $\mathfrak{R}$.

(1) If $u=1-x-y$ where $0<y<z$, then $1-x-z<u<1-x$.
(2) If $a,\ b>0$ and $t\in[0,1]$ then $[ta+(1-t)b]^{-1}\leqq ta^{-1}+(1-t)b^{-1}$ and equality holds if $t=0$, 1, and when $t\in(0,1)$, iff $a=b$.
(3) Find $\{x:(x-8)/(x-5)>2\}$.
(4) If $k>0$ then $|x-a|<k$ iff $a-k<x<a+k$.
(5) $|x|<|v|$ iff $x<|v|$ and $-x<|v|$ where $v\neq0$.
(6) Find $\{x:|1/3-a/2x^2|<1\}$ where $a>0$.
(7) Find $\{x:|ax+b|<k\}$ where $a,\ k>0$.
(8) $\min\{a,b\}=(a+b-|a-b|)/2$.
(9) $\max\{a,b\}=(a+b+|a-b|)/2$.

1.2.4 Suprema and Infima. Furnish proofs.

(1) We have $s=\sup A$ iff two conditions hold: (i) $a\leqq s$ for all $a\in A$, and (ii) to any $\varepsilon>0$ there corresponds at least one $a'\in A$ satisfying $s-\varepsilon<a'$. (2) Condition (ii) in the preceding can be replaced by (iii) if $b<s$ there exists at least one $s''\in S$ satisfying $b<s''$. (3) If A has a supremum then it is unique. (4) Suppose $A\subset\mathfrak{R}$ is nonempty and define $B=\{x:-x\in A\}$; then $\inf A=-\sup B$. (5) If $A\subset\mathfrak{R}$ is bounded below then it has an infimum in $\mathfrak{R}$.

1.2.5 Impossibility of Ordering $\mathcal{C}$. Show that there can be no order relation $<$ in $\mathcal{C}$ for which (1.4) holds.

1.2.6 Complex Conjugates. Establish:
(1) $\bar{\bar{z}}=z$.
(2) $\overline{z_1 \pm z_2}=\overline{z_1} \pm \overline{z_2}$.
(3) $\overline{z_1 z_2}=\overline{z_1} \cdot \overline{z_2}$.
(4) If $z_2 \neq 0$ then $\overline{z_1/z_2}=\overline{z_1}/\overline{z_2}$.
(5) $\operatorname{Im} z=0$ iff $z=\bar{z}$.
(6) $\operatorname{Re} z=0$ iff $z=-\bar{z}$.

1.2.7 Inequalities for Absolute Values. Establish:
(1) $|z| \geqq 0$ for all z, and $|z|=0$ iff $z=0$.
(2) $-|z| \leqq \operatorname{Re} z \leqq |z|$, $-|z| \leqq \operatorname{Im} z \leqq |z|$.
(3) $|z_1 z_2|=|z_1| \cdot |z_2|$ and if $z_2 \neq 0$ $|z_1/z_2|=|z_1|/|z_2|$.
(4) $|z_1+z_2| \leqq |z_1|+|z_2|$.
(5) $||z_1|-|z_2|| \leqq |z_1-z_2|$.

1.2.8 Principal Value of Argument. Establish: (1) The argument θ in the polar form is determined only to within a multiple of 2π. (2) Restricting θ to $-\pi<\theta \leqq \pi$ and assigning $\theta=0$ for $z=0$ produces a function $\arg: \mathcal{C} \to \mathcal{R}$. (3) $\arg z$ does not vary continuously for $z \in \mathcal{C}$.

1.2.9 Rational Functions. Show that any rational function f can be uniquely expressed as $f(x)=r(x)+p(x)/q(x)$ where $r(x)$, $p(x)$ and $q(x)$ are polynomials such that (1) the degree of $p(x)$ is strictly less than that of $q(x)$, and (2) $p(x)$ and $q(x)$ have no zeros in common.

1.3 CALCULUS

In this section we review the parts of elementary calculus that we take as given for use in Chapter 6 and following.

We presuppose familiarity with the theory of limits for *sequences* of real numbers, that is, for $f: D \to \mathcal{R}$ where D can be put into one-to-one correspondence with the positive integers, and for *subsequences*, that is, for restrictions of f to infinite subsets of D. We also need the basic theory of limits, continuity, and differentiability of $f: I \to \mathcal{R}$, where I is an interval, and where one-sided objects are used for endpoints. For example, we recall the definition: a function $f:[a,b] \to \mathcal{R}$ is *piecewise continuous* on $[a,b]$ if (1) there are at most a finite number of points $x_1<x_2<\cdots<x_{m-1}$ where f is discontinuous and (2) one-sided limits exist and are finite at points of discontinuity. Thus, if f is discontinuous at x_1, where $a<x_1<b$, then both $f(x_1-)$ and $f(x_1+)$ must exist as finite limits.

It is convenient to use the *large O small o* notation for brief, yet precise, specification of phenomena connected with limiting operations. Let u and v denote real functions on a common domain where a variable is understood to tend to a limit; for example, let $n \to \infty$ for two sequences, and let $h \to 0$ or $t \to \infty$ be understood for two real functions. Then we write $u = O(v)$, which is read "large oh of v", to mean

$$\frac{u}{|v|} \text{ remains bounded} \tag{1.9}$$

Thus $u_n = O(1)$ means that u_n is the nth term of a bounded sequence: $|u_n| \leqq M$ for all n. Similarly $f(h) = O(h^p)$ means $|h|^{-p} f(h)$ is bounded as $h \to 0$. If $f(t) = O[\exp(x_0 t)]$ for some finite x_0, the function f is said to be of *exponential order*; by definition this means that for some K, $|f(t)| \leqq K\exp(-x_0 t)$ as $t \to \infty$. Provided that the limiting operation is understood, we use simply $O(h)$, for example, to denote an arbitrary function u with the boundedness property (1.9).

The *small o notation* $u = o(v^p)$, where p is a positive integer, means

$$\frac{u}{|v|^p} \to 0 \tag{1.10}$$

Most commonly we are concerned with $h \to 0$ in which case the preceding means

$$\lim_{h \to 0} \frac{u(h)}{|v(h)|^p} = 0$$

For example, $f(h) = o(h)$ means $|h|^{-1} f(h) \to 0$ so that intuitively $f(h) \to 0$ faster than $h \to 0$. As noted above for $O(h)$, we also write simply $o(h)$, $o(h^2)$, and so on, for arbitrary functions u satisfying (1.10). For example, we can write

$$(1+h)^3 = 1 + 3h + o(h)$$

as a more precise expression of

$$(1+h)^3 \approx 1 + 3h$$

where $\approx$ is read "is approximated by."

Let us also recall the concepts of *global* and *local extrema* (minima or maxima that are special cases of definitions in Chapter 7) and the familiar

A necessary condition for a differentiable function $f: (a,b) \to \mathfrak{R}$ to have a local extremum at $x_0 \in (a,b)$ is $f'(x_0) = 0$. (1.11)

Of even more importance is the elementary

MEAN VALUE THEOREM. If $f:[a,b]\to\mathfrak{R}$ is differentiable on (a,b) and continuous on $[a,b]$, then for some $u\in(a,b)$

$$f(b)=f(a)+(b-a)f'(u) \tag{1.12}$$

This result is often called the *law of the mean* and it is the first, $n=1$, case of

TAYLOR'S THEOREM. Suppose $f:[a,b]\to\mathfrak{R}$ has a finite nth derivative $f^{(n)}$ on (a,b) and $f^{(n-1)}$ is continuous on $[a,b]$. Then for some $u\in(a,b)$,

$$\begin{aligned} f(b)&=f(a)+f'(a)(b-a)+\cdots \\ &\quad+\frac{f^{(n-1)}(a)}{(n-1)!}(b-a)^{n-1}+\frac{f^{(n)}(u)}{n!}(b-a)^n \end{aligned} \tag{1.13}$$

In Section 6.5 we work with *convex functions* and obtain many extensions of elementary results such as

Let $f:I\to\mathfrak{R}$ be twice differentiable on the interval I. Then f is convex iff $f''\geqq 0$ on I; a sufficient condition for strict convexity is $f''>0$ except at a finite number of points in I. (1.14)

PROBLEMS FOR SOLUTION

1.3.1 Continuity. Identify the subsets of $\mathfrak{R}$ on which the following are continuous. (1) $x\mapsto x^n$ where n is a positive integer. (2) Polynomials. (3) Rational functions. (4) $x\mapsto|x-a|$ where $a\in\mathfrak{R}$ is fixed. (5) The *greatest integer function* $x\mapsto[x]$ whose value at x is the algebraically largest integer that does not exceed x. (6) The roots of $x^2+2bx+1=0$ as functions of b. (7) The roots of $x^2+2x+c=0$ as functions of c.

1.3.2 Extrema. Recall the definitive result on extrema of functions $f:I\to\mathfrak{R}$ where f is differentiable except at possibly a finite number of points on the interval I.

1.3.3 Taylor's Theorem. Rewrite for $a=x_0$ and $b-a=h$.

1.3.4 L'Hôpital's Rule. Let f and g be differentiable functions on (a,b) to $\mathfrak{R}$ and suppose that the quotient of $\lim_{x\to a}f(x)$ and $\lim_{x\to a}g(x)$ is *indeterminate* of the form ∞/∞, $-\infty/\infty$, $\infty/-\infty$ or else $0/0$ where here

$g' \neq 0$ on (a,b). Then the rule is

$$\lim_{x\to a}\frac{f(x)}{g(x)}=\lim_{x\to a}\frac{f'(x)}{g'(x)}$$

provided the RHS exists or equals one of $\pm\infty$. Clearly we understand limits as $x\to a+$ as necessary and the rule is equally applicable to $x\to b-$. Apply the rule to

(1) $\lim_{x\to 1}(\log x)/(x-1)$.

(2) $\lim_{x\to 0+}(1-e^x)/(x\sin x)$.

(3) $\lim_{x\to 0+} x\log x$.

(4) $\lim_{x\to\infty} x^n e^{-x}$ for $n>0$.

(5) $\lim_{x\to 0} x^3(\sin x - x)^{-1}$.

(6) $\lim_{x\to\infty} e^x x^{-4}$.

1.3.5 Extensions. Are there other types of indeterminate forms to which the rule applies?

1.3.6 Leibniz's Rule. Let a and b be differentiable functions of $\alpha\in A=[\alpha_0,\alpha_1]$. Let $f(\cdot,\cdot)$ and $(\partial/\partial\alpha)f(\cdot,\cdot)$ be continuous functions of two variables in the rectangle determined by $[a,b]$ and A. Then

$$\alpha\mapsto F(\alpha)=\int_{a(\alpha)}^{b(\alpha)} f(x,\alpha)\,dx$$

is differentiable on A and

$$F'(\alpha)=\int_{a(\alpha)}^{b(\alpha)}\frac{\partial f(x,\alpha)}{\partial\alpha}\,dx - f(a(\alpha),\alpha)\frac{da(\alpha)}{d\alpha}+f(b(\alpha),\alpha)\frac{db(\alpha)}{d\alpha}$$

Find the derivatives in two ways: First, integrate and then differentiate; second, apply the subject rule.

(1) $\int_0^\alpha e^x\,dx$. (2) $\int_0^1 e^{\alpha x}\,dx$. (3) $\int_\alpha^1 e^{\alpha x}\,dx$. (4) $\int_\alpha^{10\alpha} x\exp(\alpha^2 x)\,dx$.

2

LINEAR ALGEBRA

The subject is the study of vector spaces and linear transformations. Vector spaces are used as domains of functions in this book and all derivatives are linear transformations. Here we develop the fundamentals from scratch.

2.1 VECTOR SPACES

In this section we define vector spaces, give several illustrations, and show why we work with familiar spaces of column vectors.

Let us consider two sets: a field $\mathcal{F}$ of scalars and a commutative group $\mathcal{V}$ (under addition) of vectors. It suffices to require that $\mathcal{F}$ be $\mathcal{R}$ or $\mathcal{C}$ and it is furthermore effective (until Section 3.4) to establish results that hold for both $\mathcal{R}$ and $\mathcal{C}$; the latter endeavor mainly requires that we not use order in $\mathcal{R}$ or zeros of polynomials in $\mathcal{C}$.

Definition. A *vector space* $\mathcal{V}$ over a field $\mathcal{F}$, which is $\mathcal{R}$ or $\mathcal{C}$, of elements $a, b, \dots$ called *scalars* is a set of elements $\mathbf{x}, \mathbf{y}, \dots$ called *vectors*, such that

(1) The set $\mathcal{V}$ is a commutative group under addition.
(2) There exists a function $(a, \mathbf{x}) \mapsto a\mathbf{x} \in \mathcal{V}$ called *multiplication by a scalar* satisfying the *distributive* properties for all $a, b \in \mathcal{F}$ and $\mathbf{x}, \mathbf{y} \in \mathcal{V}$

$$a(\mathbf{x}+\mathbf{y}) = a\mathbf{x} + a\mathbf{y}, \qquad (a+b)\mathbf{x} = a\mathbf{x} + b\mathbf{x}$$

and the *associative* properties for all $a, b \in \mathcal{F}$ and $\mathbf{x} \in \mathcal{V}$

$$(ab)\mathbf{x} = a(b\mathbf{x}), \; 1\mathbf{x} = \mathbf{x}$$

where $1 \in \mathcal{F}$ is the multiplicative identity. (2.1)

We use boldface lowercase letters $\mathbf{a}, \mathbf{b}, \ldots, \mathbf{x}, \mathbf{y}, \ldots$ for vectors and lowercase $a, b, \ldots, x, y, \ldots$ for scalars. A good procedure for handwriting is to use the printer's convention of underlining boldface with a wavy line: for example, $\underset{\sim}{x}$ denotes a vector and x does not. (Such distinctions become increasingly helpful as we progress.) The *zero vector* $\mathbf{0}$ is the identity for addition. Functional values $a\mathbf{x}$ are called *scalar multiples* and at times it is convenient to denote particular ones such as $b^{-1}\mathbf{x}$ by $\mathbf{x}/b$ where b^{-1} is the multiplicative inverse of nonzero $b \in \mathcal{F}$, and always $(-1)\,\mathbf{x} = -\mathbf{x}$.

Before we indicate the wide range of possibilities for realization of (2.1), let us introduce the special case that is most important in this book.

Definition. Let $\mathcal{F}$ be $\mathcal{R}$ or $\mathcal{C}$ and let n be a positive integer. Then $\mathcal{V}^n$ is the *column vector space* of all n-component *column vectors*

$$\mathbf{x} = \begin{bmatrix} x_1 \\ x_2 \\ \vdots \\ x_n \end{bmatrix}$$

with *components* $x_i \in \mathcal{F}$. If $\mathbf{y} \in \mathcal{V}^n$ has components $y_i \in \mathcal{F}$, then

$$\mathbf{x} + \mathbf{y} = \begin{bmatrix} x_1 + y_1 \\ x_2 + y_2 \\ \vdots \\ x_n + y_n \end{bmatrix}$$

and the result of multiplication of $\mathbf{x}$ by the scalar $a \in \mathcal{F}$ is

$$a\mathbf{x} = \begin{bmatrix} ax_1 \\ ax_2 \\ \vdots \\ ax_n \end{bmatrix} \tag{2.2}$$

Given an element in $\mathcal{V}^n$ denoted by a particular letter, we generally use the same letter together with subscripts for its components. We also use the *transpose* of $\mathbf{x} \in \mathcal{V}^n$

$$\mathbf{x}^T = [x_1, x_2, \ldots, x_n]$$

and $(\mathbf{x}^T)^T = \mathbf{x}$ written as

$$[x_1, x_2, \ldots, x_n]^T = \mathbf{x}$$

for economy of space in writing. We have $\mathbf{x} = \mathbf{y}$ iff $x_i = y_i$ for $i = 1, 2, \ldots, n$.

EXAMPLE 2.1. Particular Vector Spaces. It is easy in most of the following cases to verify that requirements of (2.1) are satisfied. See 2.1.1. (Further cases appear in Example 2.2.)

(1) The set of column vectors $\mathcal{V}^n$ forms a vector space with zero vector

$$\mathbf{0} = [0, 0, \ldots, 0]^T$$

where 0 is the additive identity in $\mathcal{F}$.

(2) The vector space $\mathcal{R}^n$ is obtained by specifying the real field $\mathcal{R}$ in (2.2). Later in (4.1) we associate $\mathcal{R}^n$ with *Euclidean space*.

(3) The vector space $\mathcal{C}^n$ is obtained by specifying the complex field $\mathcal{C}$ in (2.2). Later in (3.42) we associate $\mathcal{C}^n$ with *unitary space*.

(4) The set of all "free" vectors in the real plane forms a vector space where two quantities determine a vector: a *length* and a *direction*. These are the familiar vectors from physics which are drawn as arrows; a vector is represented by an arrow of the given length pointing in the given direction and it is "free" because it may originate at any point in the plane. Given two such vectors, their sum is found by the familiar parallelogram law. The scalar multiple, where a is the scalar, has the same or opposite direction as the vector involved and its length is $|a|$ times the given length.

(5) The set of all "position" vectors in the real plane forms a vector space where a single point determines a vector: the vector from the fixed *origin* to the given point. Here the sum of two vectors is determined as in (4) so that it also emanates from the origin, and so also does any scalar multiple. This example coincides with the *Cartesian plane* $\mathcal{R} \times \mathcal{R}$ setting for analytic geometry. (For this latter reason we often use the customary notation (x_1, x_2) for points in the plane in place of the cumbersome $[x_1, x_2]^T \in \mathcal{R}^2$. For larger n, we use vector notation, for example, $[x_1, x_2, x_3]^T$ to stress that we work with column vectors.)

(6) The complex plane $\mathcal{C}$ is itself a nontrivial instance of a vector space (and (2.1.4) has "dimension" 1).

The following sets also form vector spaces given appropriate operations.

(7) All real polynomials of degree less than m.
(8) All polynomials.

(9) All twice-differentiable functions $f:(-1,1)\to\mathcal{R}$.
(10) All solutions to the differential equation.

$$y''(t)+\alpha_1 y'(t)+\alpha_0 y(t)=0$$

considered in Example 9.2.
(11) All solutions to the difference equation

$$y(k+2)+\alpha_1 y(k+1)+\alpha_0 y(k)=0$$

considered in Example 10.4.

Let us now introduce a number of concepts and work toward our main result (2.11) which assures us that we can concentrate on spaces $\mathcal{V}^n$ of column vectors where, in fact, $\mathcal{V}^n$ can be read "$\mathcal{R}^n$ or $\mathcal{C}^n$." Suppose $\{\mathbf{a}_1,\mathbf{a}_2,\ldots,\mathbf{a}_k\}$ is a nonempty set of vectors in a vector space $\mathcal{V}$. Then

$$\mathbf{x}=c_1\mathbf{a}_1+c_2\mathbf{a}_2+\cdots+c_k\mathbf{a}_k,\qquad \text{all } c_j\in\mathcal{F}$$

is a vector in $\mathcal{V}$ called the *linear combination*, abbreviated LC, with *coefficients* $c_1,c_2,\ldots,c_k$, of the given set of vectors. If all $c_j=0$ the LC is *trivial*; otherwise at least one $c_j\neq 0$ and the LC is *nontrivial*. The set of vectors is *linearly dependent*, abbreviated LD, if there exists a nontrivial LC equalling $\mathbf{0}$; otherwise it is *linearly independent*, abbreviated LI. For brevity we sometimes write "the vectors are LD" rather than "the set of vectors is LD" but it must always be realized that LD and LI are concepts associated with sets of vectors. The given set is LD if it contains $\mathbf{0}$ or if one of the vectors is a LC of the remaining ones, and it is LI iff

$$c_1\mathbf{a}_1+c_2\mathbf{a}_2+\cdots+c_k\mathbf{a}_k=\mathbf{0} \tag{2.3}$$

implies $c_1=c_2=\cdots=c_k=0$ (2.1.2).

Often we need to work with the members of a set of vectors in some special order: a first vector, a second, and so on. Then we use *ordered sets* of vectors which are merely sets in which the vectors are arranged in the definite order shown in the enumeration within { }. For such sets we use

The ordered set of nonzero vectors $\{\mathbf{x}_1,\mathbf{x}_2,\ldots,\mathbf{x}_k\}$ is LD iff some $\mathbf{x}_p$, $p\geqq 2$, is a LC of the preceding vectors. (2.4)

PROOF. If the set is LD and (2.3) holds there is a largest subscript p for which $c_p\neq 0$; furthermore $p\neq 1$ because $p=1$ would imply $\mathbf{x}_1=\mathbf{0}$. Thus, we can "solve" (2.3) for $\mathbf{x}_p$ to furnish the required LC. The condition is therefore necessary. Conversely, if

$$\mathbf{x}_p = d_1\mathbf{x}_1 + d_2\mathbf{x}_2 + \cdots + d_{p-1}\mathbf{x}_{p-1}$$

then by adding $-\mathbf{x}_p$ to both sides we obtain $\mathbf{0}$ on the LHS, and we have a nontrivial LC on the RHS because -1 is the coefficient of $\mathbf{x}_p$. ■

A subset of a vector space $\mathcal{V}$ that is itself a vector space, with respect to addition and multiplication by scalars as defined in $\mathcal{V}$, is a *subspace* of $\mathcal{V}$.

Each of the following is necessary and sufficient for $S \subset \mathcal{V}$ to be a subspace of $\mathcal{V}$.
(1) S is closed under addition and multiplication by scalars.
(2) S contains all LCs of its members. (2.5)

Evidently $\mathbf{0}$ must belong to every subspace and $\{\mathbf{0}\}$ constitutes a trivial subspace as does $\mathcal{V}$ itself. Subspaces different from these two are naturally called *proper subspaces* of $\mathcal{V}$. By (2) just above, the set of LCs of a given set of vectors forms a subspace; it is called the subspace *spanned* by the given vectors and we say that they *span* the subspace. It is readily verified that it is the smallest subspace containing the given vectors (2.1.3).

A *basis* for $\mathcal{V}$ is a set of vectors from $\mathcal{V}$ meeting two independent requirements: (1) it is LI, and (2) it spans $\mathcal{V}$. Certainly no basis can contain $\mathbf{0}$. A vector space is *finite dimensional* if it has a basis consisting of a finite number of vectors, that is, if it has a *finite basis*. Clearly any finite basis can be converted into an ordered set and called an *ordered basis*. If $\mathcal{V}$ has no finite basis it is called *infinite dimensional*. The *dimension* of a finite-dimensional vector space $\mathcal{V}$, written $\dim \mathcal{V}$, is defined as follows. If $\mathcal{V} = \{\mathbf{0}\}$ then $\dim \mathcal{V} = 0$ and otherwise, $\dim \mathcal{V} = n$ where there is a basis containing n vectors.

The space $\mathcal{V}^n$ has dimension n because it has the *standard basis*

$$E = \{\mathbf{e}_1, \mathbf{e}_2, \ldots, \mathbf{e}_n\} \tag{2.6}$$

where the ith component of $\mathbf{e}_i$ is the multiplicative identity 1 and all other components are zero. In order to complete the justification of the definition of dimension we establish

Let $\mathcal{V}$ be a finite-dimensional vector space. If $\mathcal{V}$ contains a single element then $\dim \mathcal{V} = 0$. In all other cases each basis for $\mathcal{V}$ contains exactly $n = \dim \mathcal{V}$ vectors. (2.7)

PROOF. If there is but one element in $\mathcal{V}$ it must be $\mathbf{0}$ which by (2.5) must belong to every space and subspace. Otherwise, we have nontrivial

LCs in $\mathcal{V}$ and we suppose that $\{\mathbf{x}_1,\mathbf{x}_2,\ldots,\mathbf{x}_n\}$ is an ordered basis. Suppose $\{\mathbf{y}_1,\mathbf{y}_2,\ldots,\mathbf{y}_k\}$ is any LI ordered set in $\mathcal{V}$; we first show $n \geqq k$. Now $\mathbf{y}_1$ is a LC of the $\mathbf{x}_i$ and

$$\{\mathbf{y}_1,\mathbf{x}_1,\ldots,\mathbf{x}_n\}$$

is LD and spans $\mathcal{V}$. By (2.4) we can remove one $\mathbf{x}_i$, relabel those remaining, and obtain

$$\{\mathbf{y}_1,\mathbf{x}'_1,\ldots,\mathbf{x}'_{n-1}\}$$

which is also LI and spans $\mathcal{V}$. We apply this argument a total of k times until we reach

$$\{\mathbf{y}_1,\mathbf{y}_2,\ldots,\mathbf{y}_k,\mathbf{x}_1^{(k)},\ldots,\mathbf{x}_{n-k}^{(k)}\}$$

and conclude $n-k \geqq 0$ because $\{\mathbf{y}_1,\mathbf{y}_2,\ldots,\mathbf{y}_k\}$ cannot be LD and so only an $\mathbf{x}$ can be removed each time. Since the roles of the two LI sets can be reversed, we can also show $k-n \geqq 0$ and hence $n=k$. ■

Not only does (2.7) justify the definition of dimension, but it can be used to establish a number of fundamental properties of finite-dimensional vector spaces. A frequently used property is (2.1.5).

Any set of $k<n$ LI vectors in an n-dimensional vector space $\mathcal{V}$ can be augmented by $n-k$ vectors so that the resulting set is a basis for $\mathcal{V}$. (2.8)

Bases are also called *coordinate systems* on account of

An ordered set $A=\{\mathbf{a}_1,\mathbf{a}_2,\ldots,\mathbf{a}_n\}$ is a basis for $\mathcal{V}$ iff each $\mathbf{v}\in\mathcal{V}$ can be expressed as a unique LC

$$\mathbf{v}=x_1\mathbf{a}_1+x_2\mathbf{a}_2+\cdots x_n\mathbf{a}_n \tag{2.9}$$

PROOF. If A is a basis there is one such LC. If there were a second then by subtraction we would find A LD. Conversely, if in particular $\mathbf{0}\in\mathcal{V}$ is uniquely expressible then all $x_i=0$ and A is LI. ■

The scalars $x_1,x_2,\ldots,x_n$ in (2.9) are called the *coordinates* of $\mathbf{v}$ *with respect to the ordered basis A* and we write

$$(\mathbf{v})_A=[x_1,x_2,\ldots,x_n]^T \tag{2.10}$$

Now the RHS is a column vector in $\mathcal{V}^n$ whatever $\mathbf{v}$, A, and $\mathcal{V}$ may be and in order to deal effectively with the situation we have reached we need the

Definition. A one-to-one correspondence $\mathbf{v} \leftrightarrow \mathbf{v}'$ between the elements of two vector spaces $\mathcal{V}$ and $\mathcal{V}'$ over the same field $\mathcal{F}$ is an *isomorphism* if it preserves sums and scalar multiples.

$$(\mathbf{v}+\mathbf{w})' = \mathbf{v}' + \mathbf{w}', \ (a\mathbf{v})' = a\mathbf{v}'$$

for all $\mathbf{v}, \mathbf{w} \in \mathcal{V}$ and all $a \in \mathcal{F}$.

The idea is that any two isomorphic mathematical systems are abstractly the same, and we can choose to work with either one without loss of generality: it is merely a matter of notation whether we use the one or the other. Our main result is the

THEOREM. Any finite-dimensional vector space $\mathcal{V}$ is isomorphic to some one $\mathcal{V}^n$. (2.11)

PROOF. Of course $n = \dim \mathcal{V}$ and any given basis A can be made into an ordered basis so that (2.10) furnishes a one-to-one correspondence. Since $\mathcal{F}$ is a field, $\mathbf{v}+\mathbf{w} \in \mathcal{V}$ corresponds to the column vector whose components are the sums of corresponding ones for $\mathbf{v}$ and $\mathbf{w}$, and scalar multiples are similarly preserved. ■

In this book we work mainly with spaces $\mathcal{V}^n$ in Chapter 3 and following.

PROBLEMS FOR SOLUTION

2.1.1 Example 2.1. Verify that each of the cases satisfies the requirements (2.1).

2.1.2 Linear Dependence. Establish:

(1) The assertion for (2.3).
(2) Any set containing $\mathbf{0}$ is LD.
(3) If one vector is a multiple of another in the set then the set is LD.
(4) Repeat for LCs.
(5) Any set having a LD proper subset is LD.
(6) Any nonempty subset of a LI set is LI.
(7) A set is LI iff $\mathbf{0}$ is represented by a unique LC of the set.
(8) If $c \neq 0$ then $\{\mathbf{a}_1, \mathbf{a}_2, \ldots, \mathbf{a}_k\}$ is LD iff $\{c\mathbf{a}_1, c\mathbf{a}_2, \ldots, c\mathbf{a}_k\}$ is LD.

2.1.3 Subspaces. Establish where S_1 and S_2 are subspaces of $\mathcal{V}$:

(1) Criteria (2.5).
(2) Every subspace must contain $\mathbf{0}$.

(3) All LCs of the set $\{\mathbf{a}_1, \mathbf{a}_2, \ldots, \mathbf{a}_k\}$ form a subspace that is the smallest subspace containing the set.
(4) The *intersection* $S_1 \cap S_2$ is a subspace.
(5) Repeat for the *union* $S_1 \cup S_2$.
(6) Repeat for the *linear sum* $S_1 + S_2 = \{\mathbf{x}_1 + \mathbf{x}_2 : \mathbf{x}_1 \in S_1 \text{ and } \mathbf{x}_2 \in S_2\}$.
(7) Repeat for the *direct sum* defined when $S_1 \cap S_2 = \{\mathbf{0}\}$:

$$S_1 \oplus S_2 = \{\mathbf{x} : \mathbf{x} = \mathbf{s}_1 + \mathbf{s}_2 \text{ for some } \mathbf{s}_1 \in S_1,\ \mathbf{s}_2 \in S_2\}$$

2.1.4 Bases and Dimension. Establish:
(1) Several bases for $\mathcal{V}^n$.
(2) A necessary and sufficient condition that $\mathcal{V}$ be infinite-dimensional.
(3) If $\mathbf{v} \neq \mathbf{0}$ satisfies (2.10) then (i) if $x_i \neq 0$ then $\mathbf{a}_i$ may be replaced by $\mathbf{v}$ and the resulting set is a basis, and (ii) if $x_i = 0$ then the resulting set is not a basis.
(4) Consider $m=4$ in Example 2.1(7). Which of the following form a basis? $A = \{1+x, x+x^3, x^2, 1+x+x^3\}$, $B = \{1, 1+x, x^2, x^2+x^3\}$, $C = \{1, 1+x, x+x^2, x^2+x^3\}$.
(5) Solve the preceding as a problem in $\mathcal{R}^4$.
(6) The complex plane $\mathcal{C}$ has dimension 1 as a vector space.

2.1.5 Finite-Dimensional Space. Suppose $\dim \mathcal{V} = n \geqq 1$ and establish
(1) Any set of $n+1$ or more vectors is LD.
(2) Every proper subspace has dimension equal to one of $1, 2, \ldots, n-1$.
(3) A set of vectors is a basis iff it (i) is LI and (ii) has n members.
(4) Result (2.8)
(5) $S \oplus T = \mathcal{V}$ means there exists a basis for $\mathcal{V}$

$$\{\mathbf{s}_1, \mathbf{s}_2, \ldots, \mathbf{s}_k, \mathbf{t}_1, \mathbf{t}_2, \ldots, \mathbf{t}_{n-k}\}$$

where all $\mathbf{s}_i \in S$ and all $\mathbf{t}_i \in T$.
(6) If S_1 and S_2 are subspaces of $\mathcal{V}$ then $\dim S_1 + \dim S_2 = \dim(S_1 \cap S_2) + \dim(S_1 + S_2)$.

2.2 PARTITIONED MATRICES

Matrices are the most important tools used in this book. They furnish economical notation, they help in working with more abstract objects such as linear transformations, they help in finding "best" coordinate systems in which solutions are "easiest" to find, they suggest ways to generalize one-dimensional results, and so on. Everything that can be said in praise of matrices goes at least double for partitioned matrices.

Let us first recall simple properties from elementary mathematics where matrices are used to solve linear equations. A *matrix* is a rectangular array of elements from a field. The basic rule for describing a matrix is: *rows first, columns second.* Thus we say that the *rectangular matrix*

$$\mathbf{A}=\begin{bmatrix} a_{11} & a_{12} & \cdots & a_{1n} \\ a_{21} & a_{22} & \cdots & a_{2n} \\ \vdots & \vdots & & \vdots \\ a_{m1} & a_{m2} & \cdots & a_{mn} \end{bmatrix}$$

has m *rows* and n *columns*, and is of *order* $m\times n$, read m *by* n. If $m=n$ then $\mathbf{A}$ is a *square matrix of order* n. We use boldface capitals $\mathbf{A},\mathbf{B},\ldots$ for matrices and we continue to use boldface lowercase $\mathbf{a},\mathbf{b},\ldots$ for vectors from $\mathcal{V}$ and column vectors from $\mathcal{V}^n$. We recall that we can say

$$\mathbf{A}=[a_{ij}] \text{ is } m\times n$$

to denote the matrix $\mathbf{A}$ displayed above. The object a_{ij} is variously called the (i,j)th *element*, *component*, or *entry* of $\mathbf{A}$ and we suppose a_{ij} belongs to a field $\mathcal{F}$ which is $\mathcal{R}$ or $\mathcal{C}$. Thus column vectors $\mathbf{x}\in\mathcal{V}^n$ can be treated as $n\times 1$ matrices and we repeatedly do so. Matrices of order $1\times n$ are sometimes called "row vectors" but not by us: All vectors in this book come from either $\mathcal{V}$ or some $\mathcal{V}^n$ and transposes of $\mathbf{x}\in\mathcal{V}^n$ are $1\times n$ matrices and not vectors.

We say row i_1 *precedes* row i_2 if $i_1<i_2$, and equivalently, row i_2 *follows* row i_1. Terminology involving *between*, *above*, *below*, *to the right*, and so on, is used for rows and columns in obvious fashion. If every entry in a row is zero it is called a *zero row* and similarly for *zero column*.

Two matrices $\mathbf{A}=[a_{ij}]$ and $\mathbf{B}=[b_{ij}]$ are *equal* if they are of the same order and $a_{ij}=b_{ij}$ for all i,j. If they are of the same order, say $m\times n$, then their *sum* is the $m\times n$ matrix $[a_{ij}+b_{ij}]$. If $k\in\mathcal{F}$ the *scalar multiple* of $\mathbf{A}$ is $k\mathbf{A}=[ka_{ij}]$, also of order $m\times n$, and we write $-\mathbf{A}$ for $(-1)\mathbf{A}$. Of course addition is commutative, and the elementary associative and distributive properties hold for scalar multiples because of properties of $\mathcal{F}$.

The matrix *product* **AB** is defined only when the number of columns of $\mathbf{A}$ equals the number of rows of $\mathbf{B}$; when this happens we say $\mathbf{A}$ *and* $\mathbf{B}$ *are conformable*. If $\mathbf{A}$ is $m\times n$ and $\mathbf{B}$ is $n\times p$ then

$$\mathbf{AB}=\left[\sum_{k=1}^{n} a_{ik}b_{kj}\right] \text{ is } m\times p$$

is the familiar Σ-notation form of the product. Matrix multiplication is *associative* and *distributive over addition* because of properties of $\mathcal{F}$ but it is not generally *commutative* (2.2.1).

If $\mathbf{A}$ is $m \times n$, its *transpose* $\mathbf{A}^T$ is the $n \times m$ matrix formed by interchanging the rows and columns of $\mathbf{A}$. Of course $(\mathbf{A}^T)^T = \mathbf{A}$ and also

$$(\mathbf{A}+\mathbf{B})^T = \mathbf{A}^T + \mathbf{B}^T, \; (\mathbf{BC})^T = \mathbf{C}^T\mathbf{B}^T$$

whenever $\mathbf{A}+\mathbf{B}$ and $\mathbf{BC}$ are defined.

If $\mathbf{D} = [d_{ij}]$ is a square matrix of order n, its *principal* or *main, diagonal* consists of the elements in the n locations $(1,1),(2,2),\ldots,(n,n)$ where $i=j$. If $d_{ij}=0$ for $i \neq j$, $\mathbf{D}$ is called a *diagonal matrix* and we use the notation

$$\mathbf{D} = \mathbf{diag}(d_{11}, d_{22}, \ldots, d_{nn})$$

There is some additional terminology that is useful for square matrices. A *submatrix* of any matrix results from removing columns or rows or both from the matrix. A *principal submatrix* of a square matrix results when row i is removed iff column i is removed and this holds for all i. If the last $n-r$ rows and columns are removed from a matrix of order n we obtain a *leading principal submatrix* of order r. The latter can also be called the *upper LH corner* of the matrix since it consists of all elements in the first r rows and columns.

We write $\mathbf{O}$ for a *zero matrix* which is a matrix in which every element is $0 \in \mathcal{F}$. Its order can be denoted by writing $\mathbf{O}_{m,n}$ if it is $m \times n$ or $\mathbf{O}_n$ if it is square. Clearly, the concepts LC, LD, and LI apply to sets of matrices of fixed order, and also to sets of rows or sets of columns of a matrix.

We reserve the symbol $\mathbf{I}$ for *identity matrices*

$$\mathbf{I} = \mathbf{diag}(1, 1, \ldots, 1)$$

and we write $\mathbf{I}_n$ for $\mathbf{I}$ of order n. We note that $\mathbf{e}_1, \mathbf{e}_2, \ldots, \mathbf{e}_n$ introduced in (2.6) and also

$$\mathbf{e} = [1, 1, \ldots, 1]^T$$

are used as fixed notation throughout this book.

We *partition* a given matrix by inserting *cuts* between rows or columns, or both, as illustrated by the broken lines in

$$\mathbf{A} = \left[\begin{array}{cc:c} a_{11} & a_{12} & a_{13} \\ a_{21} & a_{22} & a_{23} \\ \hdashline a_{31} & a_{32} & a_{33} \end{array}\right]$$

Where originally $\mathbf{A}$ was 3×3 we now have a 2×2 representation as a *partitioned matrix*, say

$$\mathbf{A}=\begin{bmatrix}\mathbf{A}_{11} & \mathbf{A}_{12}\\ \mathbf{A}_{21} & \mathbf{A}_{22}\end{bmatrix}$$

where we omit broken lines and label resulting submatrices as matrices. Such submatrices are called *blocks* and they can be labeled in any convenient way, as for example in

$$\mathbf{A}=\begin{bmatrix}\mathbf{U} & \mathbf{X}\\ \mathbf{Y} & \mathbf{Z}\end{bmatrix}$$

It is permissible to partition only rows, or only columns, or both; however, the rule is that each individual cut must partition the rows, or the columns, into two disjoint sets. Rules whereby sums or products of partitioned matrices can be expressed in terms of those for individual blocks are readily formulated (2.2.5).

We use partitioned matrices throughout the book. They furnish convenient notation in the many situations where it is effective to work with blocks rather than elements. In fact, we work this way nearly everywhere in the book using, in particular, the material in the next two paragraphs.

Henceforth let us understand that enumerated sets of vectors such as

$$A=\{\mathbf{a}_1,\mathbf{a}_2,\ldots,\mathbf{a}_n\} \tag{2.12}$$

are ordered sets. Suppose each $\mathbf{a}_i$ is an m-component column vector, and let

$$\mathbf{A}=[\mathbf{a}_1,\mathbf{a}_2,\ldots,\mathbf{a}_n] \tag{2.13}$$

denote the $m\times n$ matrix $\mathbf{A}$ that corresponds to (2.12) with jth column $\mathbf{a}_j$ for $j=1,2,\ldots,n$. We call $\mathbf{A}$ the *matrix of the ordered set* A. We consistently use brackets [ ] to denote partitioned matrices associated with ordered sets of column vectors enumerated within { }; where possible we use pairs of capitals and the same letter in boldface lowercase as shown previously for A, $\mathbf{A}$, and

$$\mathbf{a}_j=[a_{1j},a_{2j},\ldots,a_{mj}]^T$$

for $j=1,2,\ldots,n$ so a_{ij} will stand in the ith row of (2.13) for $i=1,2,\ldots,m$ just as it does in the "$\mathbf{A}=[a_{ij}]$ is $m\times n$" notation.

Sets A and matrices $\mathbf{A}$ are in one-to-one correspondence and we can reverse the preceding process so that any $m\times n$ matrix with elements in $\mathcal{F}$

can be *specified by columns* in the form (2.13). We write

$$[\mathbf{A}, \mathbf{b}]$$

for the $m \times (n+1)$ matrix having $\mathbf{b} \in \mathcal{V}^m$ as its final column. If $\mathbf{x} \in \mathcal{V}^n$ then

$$\mathbf{A}\mathbf{x} = x_1\mathbf{a}_1 + x_2\mathbf{a}_2 + \cdots + x_n\mathbf{a}_n \tag{2.14}$$

and it is convenient to use $\mathbf{Ax}$ as notation for the LC on the RHS in place of the Σ-summation notation (2.2.2)

$$\left[\sum_{j=1}^{n} a_{ij} x_j\right]$$

Let us express this idea another way.

A vector $\mathbf{y} \in \mathcal{V}^m$ is a LC of the columns of a matrix $\mathbf{A}$ iff $\mathbf{y} = \mathbf{Ax}$ for some $\mathbf{x} \in \mathcal{V}^n$. (2.15)

If $\mathbf{Z}$ is $k \times m$ then

$$\mathbf{ZA} = [\mathbf{Za}_1, \mathbf{Za}_2, \ldots, \mathbf{Za}_n] \tag{2.16}$$

is a useful expression of the product. At times an $n \times p$ matrix, say $\mathbf{B}$, is *specified by rows* as

$$\mathbf{B}^T = [\mathbf{b}_1, \mathbf{b}_2, \ldots, \mathbf{b}_n] \tag{2.17}$$

where each $\mathbf{b}_i \in \mathcal{V}^p$. Then $\mathbf{b}_i^T$ is the ith row of $\mathbf{B}$ and

$$\mathbf{AB} = \mathbf{a}_1\mathbf{b}_1^T + \mathbf{a}_2\mathbf{b}_2^T + \cdots + \mathbf{a}_n\mathbf{b}_n^T \tag{2.18}$$

expresses the product as a sum of $m \times p$ *outer product* matrices.

A matrix $\mathbf{A}$ is *nonsingular* if there exists an *inverse matrix* $\mathbf{A}^{-1}$ satisfying

$$\mathbf{AA}^{-1} = \mathbf{A}^{-1}\mathbf{A} = \mathbf{I}$$

Consequently such matrices are necessarily square. A square matrix that has no inverse is *singular*. When $\mathbf{A}^{-1}$ exists it is unique, nonsingular and

$$(\mathbf{A}^{-1})^{-1} = \mathbf{A}, \ (\mathbf{A}^{-1})^T = (\mathbf{A}^T)^{-1} \tag{2.19}$$

When $\mathbf{A}$ and $\mathbf{B}$ are both nonsingular of order n, then so is $\mathbf{AB}$,

$$(\mathbf{AB})^{-1} = \mathbf{B}^{-1}\mathbf{A}^{-1} \tag{2.20}$$

and this extends to any finite number of nonsingular matrices of the same order (2.2.9). A fundamental property is

If $\mathbf{A}$ is nonsingular and $\mathbf{y}=\mathbf{A}\mathbf{x}$ then either (1) both $\mathbf{x}$ and $\mathbf{y}$ are zero vectors, or (2) both are nonzero. (2.21)

PROOF. The equation $\mathbf{x}=\mathbf{A}^{-1}\mathbf{y}$ also holds and from (2.15) we see that if either of $\mathbf{x}$ or $\mathbf{y}$ is $\mathbf{0}$ so also is the other. ■
Two characterizations of singular, and hence also of nonsingular, matrices are contained in

Each of the following is necessary and sufficient for a square matrix $\mathbf{A}$ to be singular:
(1) $\mathbf{A}\mathbf{x}=\mathbf{0}$ for some $\mathbf{x}\neq\mathbf{0}$.
(2) The columns of $\mathbf{A}$ are LD. (2.22)

PROOF. If $\mathbf{A}$ is singular then (1) results from indirect proof using (2.21). Clearly (1) implies (2), and when (2) holds, (2.15) shows $\mathbf{A}$ to be singular. ■

A particularly useful result is

Suppose $\mathbf{A}$ is of order n and $\mathbf{b}\in\mathcal{V}^n$. Then there is a unique $\mathbf{x}\in\mathcal{V}^n$ satisfying

$$\mathbf{A}\mathbf{x}=\mathbf{b}$$

iff $\mathbf{A}$ is nonsingular. (2.23)

PROOF. If $\mathbf{A}$ is singular there can be no unique solution because $\mathbf{A}\mathbf{x}=\mathbf{A}(\mathbf{x}+\mathbf{z})$ for some $\mathbf{z}\neq\mathbf{0}$ from 2.22(1). Conversely, if $\mathbf{A}$ is nonsingular, $\mathbf{x}=\mathbf{A}^{-1}\mathbf{b}$ is unique. ■

Let us close this section, which has been devoted to manipulative aspects of working with matrices, with a result that suggests their importance as objects connected with vector spaces.

THEOREM. A matrix of order n is nonsingular iff it is the matrix of a basis for $\mathcal{V}^n$. (2.24)

PROOF. Because of (2.15) the present theorem is a restatement of (2.23). ■

PROBLEMS FOR SOLUTION

2.2.1 Matrix Products. Prove or disprove given that sums and products are defined.

(1) $a\mathbf{X}=\mathbf{O}$ iff $(a=0)$ or $(\mathbf{X}=\mathbf{O})$.
(2) $\mathbf{A}\mathbf{B}=\mathbf{O}$ implies $(\mathbf{A}=\mathbf{O})$ or $(\mathbf{B}=\mathbf{O})$.

(3) $\mathbf{AB}=\mathbf{AC}$ implies $\mathbf{B}=\mathbf{C}$.
(4) $\mathbf{AB}=\mathbf{CB}$ implies $\mathbf{A}=\mathbf{C}$.
(5) $(\mathbf{A}+\mathbf{B})^2=\mathbf{A}^2+2\mathbf{AB}+\mathbf{B}^2$.
(6) $(\mathbf{A}-\mathbf{B})(\mathbf{A}+\mathbf{B})=\mathbf{A}^2-\mathbf{B}^2$.
(7) $(\mathbf{Ax}=\mathbf{Bx}$ for some $\mathbf{x})$ implies $(\mathbf{A}=\mathbf{B})$.

2.2.2 Vector-Matrix Notation. The subject notation can be used whether matrices are originally partitioned or not. The distinguishing feature is that Σ-summation signs generally do not appear because expressions such as $\mathbf{Ax}$ are used in place of $[\Sigma\cdot]$ as in (2.14), $\mathbf{x}^T\mathbf{Ax}$ is used instead of $[\Sigma\Sigma\cdot]$, and so on. Write in vector–matrix notation and specify the orders of all vectors and matrices.

(1) $\sum_{i=1}^{n} a_{ij}y_i=c_j,\ j=1,2,\ldots,n.$

(2) $\sum_{j=1}^{n} a_{ij}x_j=b_i,\ i=1,2,\ldots,m.$

(3) $\sum_{i=1}^{s}\sum_{j=1}^{r} a_{ji}x_ja_{ki},\ k=1,2,\ldots,r$ given $\mathbf{A}$ is $r\times s$.

(4) $\sum_{i=1}^{m}\sum_{j=1}^{n}\sum_{k=1}^{n} a_{ik}b_{kj}x_iy_j.$

2.2.3 Σ-summation Notation. Examples of the subject notation are given in the preceding problem. Suppose $\mathbf{A}$ is $m\times n$, $\mathbf{x}$ is $n\times 1$, $\mathbf{y}$ is $m\times 1$, $\mathbf{B}$ is $n\times p$, $\mathbf{z}$ is $p\times 1$, and $\mathbf{F}$, $\mathbf{G}$ are order n. Rewrite in Σ-summation notation. (1) $\mathbf{Ax}=\mathbf{u}$. (2) $\mathbf{y}^T\mathbf{A}=\mathbf{V}$. (3) $\mathbf{y}^T\mathbf{Ax}=b$. (4) $\mathbf{AB}=\mathbf{C}$. (5) $\mathbf{ABz}=\mathbf{u}$. (6) $\mathbf{y}^T\mathbf{AB}=\mathbf{V}$. (7) $\mathbf{y}^T\mathbf{ABz}=b$. (8) $\mathbf{AA}^T=\mathbf{C}$. (9) $\mathbf{AA}^T\mathbf{y}=\mathbf{u}$. (10) $\mathbf{A}^T\mathbf{A}=\mathbf{D}$. (11) $\mathbf{A}^T\mathbf{Ax}=\mathbf{w}$. (12) $\mathbf{F}^T\mathbf{G}=\mathbf{H}$. (13) $\mathbf{F}^T\mathbf{Gx}=\mathbf{u}$. (14) $\mathbf{FG}^T=\mathbf{K}$. (15) $\mathbf{x}^T\mathbf{FG}^T\mathbf{x}=q$.

2.2.4 Exercises. (1) Partition both $\mathbf{A}$ and $\mathbf{B}$ of order 3 by cuts between columns 2 and 3, rows 2 and 3. Verify that the correct product $\mathbf{AB}$ is obtained by multiplying blocks in the partitioned matrices. (2) Find $\mathbf{A}^T$ if $\mathbf{A}$ is partitioned into $[\mathbf{A}_{ij}]$ of order 2.

2.2.5 Rules for Partitioning. (1) When is $\mathbf{X}+\mathbf{Y}$ defined for partitioned $\mathbf{X}$ and $\mathbf{Y}$? (2) Suppose $\mathbf{A}$ is $m\times n$ and $\mathbf{B}$ is $n\times p$. Find necessary and sufficient conditions for conformability given (i) only $\mathbf{A}$ is partitioned, (ii) only $\mathbf{B}$ is partitioned, (iii) both $\mathbf{A}$ and $\mathbf{B}$ are partitioned.

2.2.6 Rows of Matrices. (1) Find another expression for (2.16) when $\mathbf{Z}$ is specified by rows. (2) Display the rows of $\mathbf{ZA}$.

2.2.7 Proof of (2.18). Prove using (2.12) and (2.17).

2.2.8 Empty Matrices. Some authors use *empty matrices* in working with partitioned matrices. Then there is one empty matrix of each of the orders: 0×0, $m\times 0$ for $m=1,2,\ldots$, and $0\times n$ for $n=1,2,\ldots$. For example, a given $\mathbf{B}$ of order $m\times n$ can be partitioned $\mathbf{B}=[\mathbf{B}_{11},\mathbf{B}_{12}]$, where $\mathbf{B}_{11}$ and $\mathbf{B}_{12}$ are of respective orders $m\times r$ and $m\times(n-r)$ for $r=1,2,\ldots,n$. (We circumvent this procedure by alternative definitions. In the preceding we define $\mathbf{B}=\mathbf{B}_{11}$ for $r=n$ and otherwise $\mathbf{B}=[\mathbf{B}_{11},\mathbf{B}_{12}]$.) Deduce the appropriate definitions of products so that the following hold for $\mathbf{B}_{12}$ and $\mathbf{C}_{21}$ empty: $\mathbf{A}[\mathbf{B}_{11},\mathbf{B}_{12}]=[\mathbf{AB}_{11},\mathbf{AB}_{12}]$,

$$\begin{bmatrix}\mathbf{C}_{11}\\ \mathbf{C}_{21}\end{bmatrix}\mathbf{D}=\begin{bmatrix}\mathbf{C}_{11}\mathbf{D}\\ \mathbf{C}_{21}\mathbf{D}\end{bmatrix}$$

$$[\mathbf{B}_{11},\mathbf{B}_{12}]\begin{bmatrix}\mathbf{C}_{11}\\ \mathbf{C}_{21}\end{bmatrix}=[\mathbf{B}_{11}\mathbf{C}_{11}+\mathbf{B}_{12}\mathbf{C}_{21}]$$

That is, answer (1) $m\times n$ times $n\times 0$ equals? (2) $0\times m$ times $m\times n$ equals? (3) $m\times 0$ times $0\times n$ equals?

2.2.9 Nonsingular Matrices. Establish: (1) Inverses are unique. (2) Inverses are nonsingular. (3) Properties (2.19). (4) Property (2.20). (5) Extension of (2.20) to any finite number of nonsingular matrices of the same order. (6) Necessary and sufficient conditions for nonsingularity. (7) If (2.13) is nonsingular then $\mathbf{A}^{-1}\mathbf{a}_i=\mathbf{e}_i$ for $i=1,2,\ldots,n$. (8) Let $\mathbf{X}$ and $\mathbf{Y}$ be $m\times n$. If $\mathbf{XA}=\mathbf{YA}$ where $\mathbf{A}$ is nonsingular then $\mathbf{X}=\mathbf{Y}$. (9) If any two of the matrices in $\mathbf{AB}=\mathbf{C}$ are nonsingular then so also is the third.

2.2.10 LCs of Rows and Columns. Given $\mathbf{XY}=\mathbf{Z}$ where no matrix is necessarily square, show: (1) The columns of $\mathbf{Z}$ are LCs of the columns of $\mathbf{X}$. If $\mathbf{Y}$ is nonsingular, then furthermore the columns of $\mathbf{X}$ are LCs of the columns of $\mathbf{Z}$. (2) The rows of $\mathbf{Z}$ are LCs of the rows of $\mathbf{Y}$. If $\mathbf{X}$ is nonsingular, then furthermore the rows of $\mathbf{Y}$ are LCs of the rows of $\mathbf{Z}$.

2.2.11 Zero Rows and Columns. Given $\mathbf{X}$, $\mathbf{Y}$, $\mathbf{Z}$ as previously, show: (1) If $\mathbf{Y}$ has a zero column j, then so also does $\mathbf{Z}$. (2) If $\mathbf{X}$ has a zero row i, then so also does $\mathbf{Z}$. (3) If $\mathbf{X}$ is nonsingular, then $\mathbf{Z}$ has a zero column j iff $\mathbf{Y}$ has a zero column j. (4) If $\mathbf{Y}$ is nonsingular, then $\mathbf{Z}$ has a zero row i iff $\mathbf{X}$ has a zero row i.

2.2.12 Diagonal Matrices. Given $\mathbf{D}=\mathbf{diag}(d_1,d_2,\ldots,d_n)$ is nonsingular, find (1) $\mathbf{D}^{-1}$, (2) $\mathbf{D}^p$ for p a positive integer, (3) $\mathbf{AD}$ where $\mathbf{A}$ is specified by columns, (4) $\mathbf{DB}$ where $\mathbf{B}$ is specified by rows.

2.2.13 Matrices of Bases. Establish for bases A and B in $\mathcal{V}^n$: (1) There exists a nonsingular matrix $\mathbf{P}$ satisfying $\mathbf{B}=\mathbf{AP}$. (2) The matrix $\mathbf{AQ}$ is the matrix of a basis iff it is nonsingular of order n. (3) If $(\mathbf{v})_A=\mathbf{x}$ and $(\mathbf{v})_B=\mathbf{y}$ then $\mathbf{x}=\mathbf{Py}$. (4) Generalizations for a proper subspace.

2.2.14 Linear Regression. The simplest example of the subject process is "fitting" a line to a set of more than two points in the plane "as well as possible." If the latter means that the total sum of squares of distances from points to the line is to be minimized over all lines in the plane, that is, "least squares," then the line of best fit is found by solving a pair of linear equations called the normal equations. Verify the form of the vector-matrix version of these equations for k points, as follows.

(1) Let $\mathbf{X}=[\mathbf{e},\mathbf{x}]$ where $\mathbf{e}=[1,1,\ldots,1]^T$, $\mathbf{x}\in\mathcal{R}^k$ and write the four elements of $\mathbf{X}^T\mathbf{X}$ in Σ-summation form.

(2) Repeat for the two elements of $\mathbf{X}^T\mathbf{y}$ where $\mathbf{y}\in\mathcal{R}^k$.

(3) Let $\mathbf{b}\in\mathcal{R}^2$ and write the *normal equations* $\mathbf{X}^T\mathbf{X}\mathbf{b}=\mathbf{X}^T\mathbf{y}$ as a pair of scalar equations in b_1 and b_2.

2.3 LINEAR TRANSFORMATIONS

In this section we introduce additional vector spaces and establish relationships between linear transformations and matrices.

It is conventional to use capital letters and the name *transformation* for functions $T:\mathcal{V}\to\mathcal{W}$ where $\mathcal{V}$ and $\mathcal{W}$ are vector spaces. We adhere to this usage and also to the traditional exceptions that lowercase letters and the name *function* are retained when $\dim\mathcal{W}=1$. Because we use column vectors in $\mathcal{V}^n$, we use *LH notation* $T\mathbf{x}, T\mathbf{y}$ for images of $\mathbf{x},\mathbf{y}\in\mathcal{V}$ where the *product* $T_1T_2\mathbf{x}$ means first $T_2\mathbf{x}$, and second $T_1(T_2\mathbf{x})$.

Definition. A transformation $T:\mathcal{V}\to\mathcal{W}$ is a *linear transformation* if $\mathcal{V}$ and $\mathcal{W}$ are vector spaces over a common field $\mathcal{F}$ and if

$$T(\mathbf{x}_1+\mathbf{x}_2)=T\mathbf{x}_1+T\mathbf{x}_2,\ T(a\mathbf{x})=aT\mathbf{x}$$

for all $\mathbf{x}_1$, $\mathbf{x}_2\in\mathcal{V}$ and all $a\in\mathcal{F}$.

The last requirement is that T preserve the operations of vector addition and multiplication by a scalar; it can be replaced by the single equation

$$T(a\mathbf{x}_1+b\mathbf{x}_2)=aT\mathbf{x}_1+bT\mathbf{x}_2 \tag{2.25}$$

for all $\mathbf{x}_1$, $\mathbf{x}_2\in\mathcal{V}$ and all a, $b\in\mathcal{F}$.

The *identity transformation* I, defined by $I\mathbf{x}=\mathbf{x}$ for all $\mathbf{x}$, is linear as is that given by $O\mathbf{x}=\mathbf{0}$. If a is any scalar then $\mathbf{x}\mapsto a\mathbf{x}$ is linear; but if $\mathbf{y}\neq\mathbf{0}$ then $\mathbf{x}\mapsto a\mathbf{x}+\mathbf{y}$ is not. If $\mathbf{D}$ is any $m\times n$ matrix, that is, if $\mathbf{D}$ is the matrix of any ordered set of n vectors in $\mathcal{V}^m$, then

$$D:\mathcal{V}^n\to\mathcal{V}^m \text{ defined by } \mathbf{x}\mapsto\mathbf{D}\mathbf{x}$$

is linear but the *affine transformation*

$$\mathbf{x}\mapsto\mathbf{D}\mathbf{x}+\mathbf{f}$$

is not unless $\mathbf{f}\in\mathcal{V}^m$ is $\mathbf{0}$.

Since transformations are functions we know how to tell if two are equal and we define *sums* and *scalar multiples* as follows. If T_1 and T_2 are linear transformations on $\mathcal{V}$ to $\mathcal{W}$ then for $\mathbf{v}\in\mathcal{V}$

$$T_1+T_2 \text{ has values } T_1\mathbf{v}+T_2\mathbf{v}$$

$$aT_1 \text{ has values } aT_1\mathbf{v}$$

where $a\in\mathcal{F}$. We readily verify that both of the preceding are linear transformations and, in fact, the set of all linear transformations on $\mathcal{V}$ to $\mathcal{W}$ forms a vector space $\mathcal{L}(\mathcal{V},\mathcal{W})$ over $\mathcal{F}$, and if $\mathcal{V}=\mathcal{W}$ we write $\mathcal{L}(\mathcal{V})$.

EXAMPLE 2.2. More Vector Spaces. The following supplement the list in Example 2.1. The several cases of spaces $\mathcal{L}(\mathcal{V},\mathcal{W})$ are of fundamental importance for differentiation in Chapter 6.

(1) The set of all solutions $\mathbf{x}\in\mathcal{V}^n$ to $\mathbf{A}\mathbf{x}=\mathbf{0}$, where $\mathbf{A}$ is $m\times n$, is an example of a vector space.

(2) A second is the set of all solutions to the linear differential system

$$\mathbf{y}'(t)=\mathbf{A}\mathbf{y}(t)$$

of Chapter 9.

(3) A third is the set of all solutions to the linear difference system

$$\mathbf{y}(k+1)=\mathbf{A}\mathbf{y}(k)$$

of Chapter 10.

(4) The set $\mathcal{M}(m,n)$ of all matrices of fixed order $m\times n$ with elements from a field $\mathcal{F}$ which is $\mathcal{R}$ or $\mathcal{C}$ forms a vector space. Sums and scalar multiples are formed according to the familiar rules. If $m=n$ we have the vector space $\mathcal{M}(n)$.

(5) The special case of $\mathcal{L}(\mathcal{V},\mathcal{W})$ for spaces of column vectors, namely $\mathcal{L}(\mathcal{V}^n,\mathcal{V}^m)$, is the space of linear transformations on $\mathcal{V}^n$ to $\mathcal{V}^m$.

(6) The particular set of linear transformations $\mathcal{L}(\mathcal{R}^n, \mathcal{R})$ is an important vector space. In Chapter 6 we find that the derivative of $f: D \subset \mathcal{R}^n \to \mathcal{R}$ is an element of this space. In works on linear algebra $\mathcal{L}(\mathcal{R}^n, \mathcal{R})$ is called the *dual space* of $\mathcal{R}^n$ and the dual of the dual is $\mathcal{R}^n$. Elements in the dual space correspond to $\mathbf{x}^T$ and are sometimes called *covectors*. Proceeding in this way gains the effect of introducing the concept of $\mathbf{v}^T$ for $\mathbf{v} \in \mathcal{V}$.

(7) The vector space $\mathcal{L}(\mathcal{R}^n, \mathcal{L}(\mathcal{R}^n, \mathcal{R}))$ has the second derivative as an element.

(8) The vector space $\mathcal{L}(\mathcal{R}^n, \mathcal{L}(\mathcal{R}^n, \mathcal{L}(\mathcal{R}^n, \mathcal{R})))$ has the third derivative as an element.

(9) When $\mathcal{F} = \mathcal{R}$ in (5) we have $\mathcal{L}(\mathcal{R}^n, \mathcal{R}^m)$ which generalizes (6). In Chapter 6 we find that the derivative of $f: D \subset \mathcal{R}^n \to \mathcal{R}^m$ is an element of this space.

(10) The vector space $\mathcal{L}(\mathcal{R}^n, \mathcal{L}(\mathcal{R}^n, \mathcal{R}^m))$ has the second derivative of $f: D \subset \mathcal{R}^n \to \mathcal{R}^m$ as an element.

The essential mathematical significance of matrices in this book is that they represent linear transformations provided we agree to use a fixed pair of bases as stated in the

THEOREM. Suppose $\dim \mathcal{V} = n$, $\dim \mathcal{W} = m$, and we fix bases in $\mathcal{V}$ and $\mathcal{W}$. Then the vector space of linear transformations $\mathcal{L}(\mathcal{V}, \mathcal{W})$ is isomorphic to the vector space of matrices $\mathcal{M}(m,n)$. (2.26)

PROOF. Let ordered bases $A = \{\mathbf{a}_1, \mathbf{a}_2, \ldots, \mathbf{a}_n\}$ in $\mathcal{V}$ and $C = \{\mathbf{c}_1, \mathbf{c}_2, \ldots, \mathbf{c}_m\}$ in $\mathcal{W}$ be fixed. First, to any $\mathbf{D} \in \mathcal{M}(m,n)$ there corresponds a unique $D \in \mathcal{L}(\mathcal{V}, \mathcal{W})$: If $\mathbf{v} \in \mathcal{V}$ and $(\mathbf{v})_A = \mathbf{x}$ then $(D\mathbf{v})_C = \mathbf{D}\mathbf{x}$. Second, given $T \in \mathcal{L}(\mathcal{V}, \mathcal{W})$, there corresponds a unique matrix $\mathbf{T}$ whose jth column is $(T\mathbf{a}_j)_C$ for $j = 1, 2, \ldots, n$. Suppose we start with a given

$$\mathbf{D} = [\mathbf{d}_1, \mathbf{d}_2, \ldots, \mathbf{d}_n]$$

Since $(\mathbf{a}_j)_A = \mathbf{e}_j$ we reach the linear transformation defined by $(D\mathbf{a}_j)_C = \mathbf{d}_j$ for $j = 1, 2, \ldots, n$. Then from the second correspondence, starting with this D we arrive back at the given $\mathbf{D}$. Similarly, we can start with any given T, reach $\mathbf{T}$, and then from the first correspondence arrive back at T. By (1.1) we have a one-to-one correspondence. Sums and scalar multiples are clearly preserved by the familiar rules for combining functions on the one hand, and matrices on the other. ■

See 2.3.3 for details of this proof.

If either basis is changed in (2.26) the matrix will in general also change. Concentrating on a particular matrix, we can introduce material needed in

Section 5.1 by asking the following question: How do we recognize two matrices of order n that represent the same linear transformation $A \in \mathcal{L}(\mathcal{V}^n)$ with respect to two different bases? The answer is phrased in terms of the following concepts. We call $\mathbf{A}$ and $\mathbf{Q}^{-1}\mathbf{AQ}$ *similar matrices* where $\mathbf{Q}$ is any nonsingular matrix of order n and we say that $\mathbf{Q}^{-1}\mathbf{AQ}$ results from $\mathbf{A}$ by a *similarity transformation*. Similarity is an equivalence relation and it suffices to prove the following for the case of change from the standard basis (2.3.4).

THEOREM. Two matrices in $\mathfrak{M}(n)$ represent the same linear transformation in $\mathcal{L}(\mathcal{V}^n)$ iff they are similar. (2.27)

PROOF. Let the linear transformation correspond to

$$\mathbf{v} = \mathbf{Au}$$

relative to the standard basis in $\mathcal{V}^n$. If we change to the basis with matrix $\mathbf{Q}$ then (2.2.13) $\mathbf{v} = \mathbf{Au}$ and $\mathbf{u} = \mathbf{Qx}$ where $\mathbf{y}$ and $\mathbf{x}$ are the "new" coordinates. If they are to be related as were the "old" then $\mathbf{v} = \mathbf{Qy} = \mathbf{AQx}$ and solving yields $\mathbf{y} = \mathbf{Q}^{-1}\mathbf{AQx}$. Thus changing to the basis with matrix $\mathbf{Q}$ changes $\mathbf{A}$ to the similar matrix $\mathbf{Q}^{-1}\mathbf{AQ}$. ■

Let us next introduce some subspaces determined by linear transformations. If $T: \mathcal{V} \to \mathcal{W}$ is a linear transformation, the set of images

$$\operatorname{Range} T = \{\mathbf{v} : \mathbf{v} = T\mathbf{x} \qquad \text{for some } \mathbf{x} \in \mathcal{V}\}$$

is a subspace of $\mathcal{W}$ called the *range space* or *range* of T; *rank* T is defined to be the dimension of this subspace. Also

$$\operatorname{Null} T = \{\mathbf{u} : T\mathbf{u} = \mathbf{0}\}$$

is a subspace of $\mathcal{V}$ called the *null space* of T and *nullity* T equals $\dim \operatorname{Null} T$. A fundamental result is

If $\mathcal{V}$ is finite-dimensional then

$$\operatorname{rank} T + \operatorname{nullity} T = \dim \mathcal{V} \tag{2.28}$$

PROOF. Suppose Null T has basis $\{\mathbf{a}_1, \mathbf{a}_2, \ldots, \mathbf{a}_s\}$. If $s = n = \dim \mathcal{V}$ then $T\mathbf{u} = \mathbf{0}$ for all $\mathbf{u} \in \mathcal{V}$ and $\operatorname{rank} T = 0$. If $s < n$ then by (2.8) $\mathcal{V}$ has basis

$$\{\mathbf{a}_1, \mathbf{a}_2, \ldots, \mathbf{a}_s, \mathbf{b}_1, \mathbf{b}_2, \ldots, \mathbf{b}_{n-s}\}$$

Since $T\mathbf{a}_i = \mathbf{0}$ for $i = 1, 2, \ldots, s$ the vectors

$$\{T\mathbf{b}_1, T\mathbf{b}_2, \ldots, T\mathbf{b}_{n-s}\}$$

span Range T. They are also LI since otherwise some nontrivial LC of the **b**s would belong to Null T. ■

A linear transformation is called *nonsingular* if it has an inverse T^{-1}. Such an inverse must be a unique linear transformation, whenever it exists, and

If $\mathcal{V}$ and $\mathcal{W}$ are finite-dimensional vector spaces then $T \in \mathcal{L}(\mathcal{V}, \mathcal{W})$ is nonsingular iff (1) $\dim \mathcal{V} = \dim \mathcal{W}$ and (2) relative to any pair of bases the associated matrix is nonsingular. (2.29)

See 2.3.6 for some additional properties.

PROBLEMS FOR SOLUTION

2.3.1 Elementary Properties. Establish for $\dim \mathcal{V} = n$ and $\dim \mathcal{W} = m$: (1) Condition (2.25) is necessary and sufficient for T to be linear. (2) The linear transformation T is completely determined if we know the images of any n LI vectors.

2.3.2 Example 2.2. (1) Verify that the cases (1)–(10) satisfy (2.1). (2) Can $\mathbf{Ax} = \mathbf{b}$ be used in case (1)? (3) Comment on the dimensions of all cases using (2.26) and other results in this section as appropriate.

2.3.3 Proof of (2.26). (1) Identify the spaces in which all vectors lie. (2) Supply details for use of (1.1). (3) Repeat for preservation of vector space operations.

2.3.4 Similarity. Establish: (1) Similarity is an equivalence relation. (2) $\mathbf{A}$ is similar to $\mathbf{B}$ iff $\mathbf{AC} = \mathbf{CB}$ for some nonsingular $\mathbf{C}$. (3) Show that the given proof suffices for (2.27) as stated. (4) Let the matrices of bases A and B in $\mathcal{V}^n$ be related through $\mathbf{B} = \mathbf{AP}$ and let bases C and D in $\mathcal{V}^m$ be related through $\mathbf{D} = \mathbf{CQ}$. Suppose the linear transformation $T: \mathcal{V}^n \rightarrow \mathcal{V}^m$ has matrix $\mathbf{T}$ relative to A and C, and $\mathbf{U}$ relative to B and D. Express $\mathbf{U}$ in terms of $\mathbf{T}$. (5) Deduce (2.27) from the preceding.

2.3.5 Rank and Nullity. Establish: (1) Range T and Null T are subspaces as stated. (2) $\text{rank}(T_1 + T_2) \leqq \text{rank}\, T_1 + \text{rank}\, T_2$.

2.3.6 Nonsingular Linear Transformations. Establish: (1) Inverses must be unique. (2) Repeat for linear. (3) (2.29). (4) Additional equivalent conditions for nonsingularity.

3

COLUMN VECTORS

In this chapter we show how to work with column vectors. We establish several systematic procedures and, at the end, we introduce inner products and norms.

3.1 STANDARD FORM

Henceforth we repeatedly work with ordered sets of column vectors, or what is the same, with $m \times n$ matrices **A**. To perform a number of tasks involving **A**—especially the ones in Section 3.3—it is effective to use a matrix **S** which is the standard form of **A**. In this section we show how to find **S** and we develop a few of its properties.

The procedure for finding **S** generalizes the familiar elimination process used to solve systems of linear equations: Solve for the first variable from the first equation and then eliminate this variable from all remaining equations, use the resulting equations to repeat for the second variable, and so on, until the penultimate variable has been eliminated so that solution values of all variables can be successively determined, from last to first. For efficiency we work with nonzero matrices rather than with systems of equations in

GAUSS–JORDAN PIVOTING. If $\mathbf{A}=[a_{ij}]$ is a nonzero $m \times n$ matrix with elements from $\mathcal{R}$ or $\mathcal{C}$, then a *Gauss–Jordan pivot operation in* **A** is performed as follows.

(1) Select a nonzero element a_{ij} as *pivot* and, equivalently, call row i the *pivot row* and column j the *pivot column*.
(2) Multiply the pivot row by $1/a_{ij}$.
(3) Answer: $m>1$? If "yes," reduce each element in the pivot column to zero, with the sole exception of the pivot, by adding appropriate multiples of the pivot row to the other rows, and stop. If "no," stop. (3.1)

The resulting matrix has elements a_{iq}/a_{ij} in row i, and unity at position (i,j) is the only nonzero element in column j. Elements at (p,q) in neither row i nor column j equal

$$a_{pq} - \frac{a_{pj}a_{iq}}{a_{ij}}$$

and we also say we have *pivoted on* a_{ij}, or *on position* (i,j) *in* **A**. For example, successive pivots are underlined in

$$\mathbf{A}=\begin{bmatrix} \underline{1} & 0 & 1 \\ 2 & 1 & 3 \\ 1 & 1 & 2 \end{bmatrix}, \begin{bmatrix} 1 & 0 & 1 \\ 0 & \underline{1} & 1 \\ 0 & 1 & 1 \end{bmatrix}, \begin{bmatrix} 1 & 0 & 1 \\ 0 & 1 & 1 \\ 0 & 0 & 0 \end{bmatrix} = \mathbf{S}$$

In the following general procedure we plan to pivot successively in columns $1,2,\ldots,n$ using rows $1,2,\ldots,m$ in order. We depart from this plan to avoid zeros but, by interchanging rows, we always move down one row, and to the right one or more columns, from one pivot to the next. We stop after a finite number r of pivots when there are no longer any nonzero elements below and to the right of where we have last pivoted.

GAUSS–JORDAN REDUCTION PROCEDURE. To start for $\mathbf{A}\neq\mathbf{O}$ of order $m\times n$, set $k=1$, designate $\mathbf{A}_1=\mathbf{A}$, and go to Step 1(1) in the following list of steps for $k=1,2,\ldots,r$ where $r\leqq \min\{m,n\}$ is to be determined.

Step 1(k). Answer: $\mathbf{A}_k=\mathbf{O}$? If "yes," go to Step 5(k). If "no," designate the m-component column containing the first nonzero column of $\mathbf{A}_k$ as the kth *pivot column*, assign the column number as the value of p_k, and go to Step 2(k).
Step 2(k). Answer: (k,p_k)th element zero? If "yes," interchange row k and the first of rows $k+1$, $k+2,\ldots,m$ that has a nonzero element in column p_k, and go to Step 3(k). If "no," go to Step 3(k).
Step 3(k). Designate the (k,p_k)th element as kth *pivot*, perform the Gauss–Jordan pivot operation in the $m\times n$ matrix, and go to Step 4(k).
Step 4(k). Answer: $k<m$ and $p_k<n$? If "no," go to Step 5(k). If "yes," designate the submatrix formed by the (i,j)th elements for $i>k$ and $j>p_k$ as $\mathbf{A}_{k+1}$, and go to Step 1($k+1$).
Step 5(k). Designate the number of pivots $r=\text{rank } \mathbf{A}$ and the resulting $m\times n$ matrix as **S** the *Gauss–Jordan Standard Form* (GJSF) of **A**, and stop.

In summary, input is **A** and output includes the column numbers $p_1,p_2,\ldots,p_r$ and the GJSF **S**. (3.2)

For example, the two pivoting operations displayed below (3.1) yield the GJSF.

GJSFs are easily found by hand for small m and n but in practice it is advantageous to use a computer program. Procedure (3.2) is readily validated as a well-defined procedure for any nonzero matrix of elements from $\mathcal{R}$ or $\mathcal{C}$: It is always possible to start, it is always certain what to do next, it always stops after r pivots, and **S** is uniquely determined as a function of **A**. We prove (in (3.8)) that **S** is uniquely determined in a much broader sense: Roughly speaking, we can omit rules for selecting successive pivots in (3.2) and still reach the same **S**. By the end of this section, we also prove that the preceding definition $r = \text{rank}\,\mathbf{A}$ determines the familiar integer from elementary algebra.

It is readily verified (3.1.1) that there are ten essentially different configurations for GJSFs. But we can assign $\mathbf{S}=\mathbf{A}$ for $\mathbf{A}=\mathbf{O}$ and then characterize GJSFs in precise terms as follows.

A zero matrix is its own GJSF. An $m\times n$ matrix $\mathbf{S}\neq\mathbf{O}$ is a GJSF iff

(1) Column vectors $\mathbf{e}_1,\mathbf{e}_2,\ldots,\mathbf{e}_r\in\mathcal{V}^m$ appear in pivot columns $p_1,p_2,\ldots,p_r$, respectively, where

$$1\leqq p_1<p_2<\cdots<p_r\leqq n,\ r\leqq \min\{m,n\}$$

and also

(2) If any column precedes column p_1 it must be a zero column; any column following column p_1 that is not a pivot column must be a LC of the pivot columns that precede it. (3.3)

PROOF. The specifications define the type of matrix produced by (3.2). Conversely, any such matrix is certainly its own GJSF. ■

(Note that (2) requires that p_i be the number of the first column in which $\mathbf{e}_i$ appears; this is significant because $\mathbf{e}_i$ can also appear to the right of column p_i.)

The GJSF **S** results from **A** in (3.2) following operations performed on the rows of **A**. In contrast to the rows, columns are never interchanged nor are they altered except through what is done to rows. If we analyze the operations on rows we are led to the following concepts.

A matrix **B** is said to be *row equivalent* to **A**, and we write

$$\mathbf{A}\xrightarrow{R}\mathbf{B}$$

if **B** results from a finite succession of the following three types of *elementary row operations*, briefly *row operations*, starting with **A**.

Type I. Interchange of two rows.

Type II. Replacement of a row by a nonzero multiple of that same row.

Type III. Replacement of a row by the sum of that same row and a multiple of a different row.

The key to working with row-equivalence is that the result of any single row operation on an $m \times n$ matrix $\mathbf{A}$ is duplicated by multiplying $\mathbf{A}$ on the left by the corresponding one of the three types of *elementary matrices* of order m.

Type I. $\mathbf{E}_{is}$ results from interchanging rows i and s in $\mathbf{I}_m$.
Type II. $\mathbf{E}_i(k)$ results from replacing row i of $\mathbf{I}_m$ by $k \neq 0$ times row i.
Type III. $\mathbf{E}_{is}(k)$ results from replacing row s in $\mathbf{I}_m$ by its sum with k times row i where $i \neq s$.

It is also readily verified that row-equivalence is an equivalence relation. Furthermore $\mathbf{E}_{is} = \mathbf{E}_{is}^{-1}$, $(\mathbf{E}_i(k))^{-1} = \mathbf{E}_i(k^{-1})$, and $(\mathbf{E}_{is}(k))^{-1} = \mathbf{E}_{is}(-k)$ so

Elementary matrices are nonsingular and the inverse of an elementary matrix is an elementary matrix. (3.4)

The preceding leads to three important results.

If $\mathbf{A}$ has GJSF $\mathbf{S}$ then $\mathbf{EA} = \mathbf{S}$ for some nonsingular $\mathbf{E}$. (3.5)

PROOF. If we perform successive row operations $1, 2, \ldots, q-1, q$ on $\mathbf{A}$ to obtain $\mathbf{S}$ then

$$\mathbf{E}_q \mathbf{E}_{q-1} \cdots \mathbf{E}_2 \mathbf{E}_1 \mathbf{A} = \mathbf{S}$$

where $\mathbf{E}_1, \mathbf{E}_2, \ldots, \mathbf{E}_{q-1}, \mathbf{E}_q$ are the corresponding elementary matrices (2.2.9). ■

A matrix is nonsingular iff its GJSF is an identity matrix. (3.6)

PROOF. If $\mathbf{A}$ is of order n and nonsingular then $\mathbf{EA} = \mathbf{S}$ from (3.5) shows $\mathbf{S}$ to be nonsingular. Then $\mathbf{S}$ has no dependent column by (2.22), there must be n pivots in (3.2), and $\mathbf{S} = \mathbf{I}_n$. Conversely, $\mathbf{EA} = \mathbf{I}_n$ shows $\mathbf{A}^{-1} = \mathbf{E}$. ■

THEOREM. Two matrices $\mathbf{A}$ and $\mathbf{B}$ are row-equivalent iff $\mathbf{EA} = \mathbf{B}$ for some nonsingular $\mathbf{E}$. (3.7)

PROOF. Given any succession of elementary row operations $1, 2, \ldots, q$ we arrive at $\mathbf{EA} = \mathbf{B}$ as in the proof of (3.5). Conversely, suppose $\mathbf{EA} = \mathbf{B}$ where we are given the single nonsingular $\mathbf{E}$. Then by (3.6) we can exhibit

$$\mathbf{B} = \mathbf{E}_1^{-1} \mathbf{E}_2^{-1} \cdots \mathbf{E}_t^{-1} \mathbf{A}$$

which shows we can go from $\mathbf{A}$ to $\mathbf{B}$ by a finite succession of elementary row operations. ■

UNIQUENESS THEOREM. A zero matrix is its own GJSF. Every succession of elementary row operations that starts with $\mathbf{A} \neq \mathbf{O}$ and produces a GJSF matrix (3.3) must produce the matrix $\mathbf{S}$ that results from the Gauss–Jordan reduction procedure (3.2). (3.8)

PROOF. Let us use the partitioned form (2.13) for all matrices. Suppose we can arrive at two GJSFs

$$\mathbf{S}=[\mathbf{s}_1,\mathbf{s}_2,\ldots,\mathbf{s}_n] \qquad \text{and} \qquad \mathbf{S}'=[\mathbf{s}'_1,\mathbf{s}'_2,\ldots,\mathbf{s}'_n]$$

Since $\mathbf{S}$ and $\mathbf{S}'$ are row-equivalent, $\mathbf{S}'=\mathbf{DS}$ for some nonsingular $\mathbf{D}=[\mathbf{d}_1,\mathbf{d}_2,\ldots,\mathbf{d}_m]$ by (3.7). By 2.2.11, $\mathbf{S}'$ has a zero column j iff $\mathbf{S}$ has one such; therefore we limit our attention to the case where neither has a zero column. By (3.3) we begin with $\mathbf{s}_1=\mathbf{s}'_1=\mathbf{e}_1$ and use (2.14) to conclude $\mathbf{d}_1=\mathbf{e}_1$. In the first case for two columns, $\mathbf{s}_1=a\mathbf{e}_1$, and

$$\mathbf{D}a\mathbf{e}_1=a\mathbf{e}_1=\mathbf{s}'_2$$

The second and remaining case is $\mathbf{s}_2=\mathbf{e}_2$ and here

$$\mathbf{De}_2=\mathbf{d}_2=\mathbf{s}'_2$$

We cannot have $\mathbf{s}'_2=b\mathbf{e}_1$ because then the first two columns of $\mathbf{D}$ would be LD and this would contradict 2.22(2). We conclude $\mathbf{s}'_2=\mathbf{e}_2$ in this case so $\mathbf{S}$ and $\mathbf{S}'$ agree through their first two columns in both cases. The matrix $\mathbf{D}$ is uniquely determined only in the second case so we proceed to $p+1$ columns by supposing we have established the following by the preceding procedure (3.1.6)

$$\mathbf{D}=[\mathbf{e}_1,\mathbf{e}_2,\ldots,\mathbf{e}_k,\ldots]$$

where all $\mathbf{e}_j\in\mathcal{V}^m$ and $\mathbf{s}_j=\mathbf{s}'_j$ for $j=1,2,\ldots,p$ where $p\geqq k$. Then if we are not finished showing $\mathbf{S}=\mathbf{S}'$, either

$$\mathbf{s}_{j+1}=a_1\mathbf{e}_1+\cdots+a_k\mathbf{e}_k$$

and we calculate $\mathbf{s}'_{j+1}=\mathbf{s}_{j+1}$, or else $\mathbf{s}_{j+1}=\mathbf{e}_{k+1}$. In the latter event, $\mathbf{d}_{k+1}=\mathbf{s}'_{j+1}$ cannot be a LC of the first k $\mathbf{e}$'s lest $\mathbf{D}$ be singular, so again $\mathbf{s}'_{j+1}=\mathbf{s}_{j+1}$. This procedure terminates when we reach the final column of $\mathbf{S}$ and we conclude $\mathbf{S}=\mathbf{S}'$. ■

Theorem (3.8) justifies use of *standard form* in GJSF. It also guarantees that any succession of elementary row operations can be used provided the end result satisfies (3.3). For example, we can interchange rows in any order at any time, we can pivot in any nonzero column at any time, and so on. Of course we must understand that results such as (3.8) neglect effects of using finite word lengths in computers where numbers generally must be rounded off. In this connection let us note that whereas (3.2) uses what can be called a *natural order strategy* for selecting pivots, experience indicates that a *maximum modulus strategy* generally works better. In the latter we plan to pivot successively in columns $1,2,\ldots,n$ using successive rows that provide pivots of greatest absolute value.

An important theoretical result that underlies many practical procedures is the

THEOREM. Two matrices are row-equivalent iff they have the same GJSF. (3.9)

PROOF. A matrix is row-equivalent to one and only one GJSF by (3.8). ■

Relative to standard bases in $\mathcal{V}^n$ and $\mathcal{V}^m$ there is a unique linear transformation A corresponding to $\mathbf{A}$ according to (2.26). We define Range $\mathbf{A}$ to be the subspace Range $A \subset \mathcal{V}^m$. Perhaps its most fundamental properties are listed in the

THEOREM. The following hold for any matrix $\mathbf{A}$.

(1) Range $\mathbf{A}$ consists of all LCs of columns of $\mathbf{A}$.
(2) Range $\mathbf{A}$ = Range $\mathbf{B}$ iff the GJSFs of $\mathbf{A}^T$ and $\mathbf{B}^T$ have the same submatrix of nonzero rows. (3.10)

PROOF. Property (1) is a restatement of (2.15). For "if" in (2), adjoin zero columns to $\mathbf{A}$ or $\mathbf{B}$ as necessary so both have the same number of columns. If the common GJSF is $\mathbf{DA}^T = \mathbf{GB}^T$ then

$$\mathbf{A} = \mathbf{B}(\mathbf{D}^{-1}\mathbf{G})^T$$

and the desired conclusion follows with 2.2.10. For "only if" it suffices to suppose that the columns of $\mathbf{A}$, and also those of $\mathbf{B}$, are LI. Again from 2.2.10, say $\mathbf{A} = \mathbf{BU}$, and $\mathbf{U}$ is nonsingular because if its columns were LD then so also would be those of $\mathbf{A}$. Thus $\mathbf{A}^T \overset{R}{\rightarrow} \mathbf{B}^T$ and they have the same GJSF by (3.9). ■

We similarly define

$$\text{Null}\,\mathbf{A} = \{\mathbf{x} : \mathbf{Ax} = \mathbf{0}\}$$

and nullity $\mathbf{A}$ = dim Null $\mathbf{A}$.

Let us complete the fundamental theory of GJSFs by justifying the definition of (Gauss–Jordan) rank $\mathbf{A}$ in (3.2). Recalling determinants (3.1.7) of square matrices and their elementary properties we have the following

Definition. The following apply to any matrix $\mathbf{A}$:

(1) *Gauss–Jordan Rank*: the number of nonzero rows in the GJSF of $\mathbf{A}$
(2) *Row Rank*: the maximum number of LI rows of $\mathbf{A}$
(3) *Column Rank*: the maximum number of LI columns of $\mathbf{A}$
(4) *Range Rank*: dim(Range $\mathbf{A}$)
(5) *Determinant Rank*: the order of the largest submatrix of $\mathbf{A}$ having nonzero determinant

In order to show that the different definitions lead to the same integer let us first establish

The five definitions of rank are equivalent for a GJSF matrix. (3.11)

PROOF. The definitions certainly coincide for zero matrices. Suppose $\mathbf{S}$ is $m \times n$ and has Gauss–Jordan rank $r>0$. Then $\mathbf{S}$ has r nonzero rows and its row rank is r or less. But $\mathbf{S}$ has exactly r LI rows; otherwise $\mathbf{c}^T\mathbf{I}_r = \mathbf{0}^T$ from elements in pivot columns which is impossible by (2.21). Next the column rank is also r because by construction every nonpivot column is a LC of preceding pivot columns and these latter are r in number. The columns of $\mathbf{S}$ span Range$\mathbf{S}$ and range rank equals r. Finally, by construction $\mathbf{I}_r$ is an rth order submatrix with nonzero determinant and every $(r+1)$st order submatrix must have at least one zero row. ■

We need one more intermediate result.

Elementary row operations do not change any of the five kinds of rank. (3.12)

PROOF. Suppose $\mathbf{A}$ and $\mathbf{B}$ are row-equivalent. By (3.9) they have the same Gauss–Jordan rank. By (3.7), $\mathbf{EA}=\mathbf{B}$ and then (2.2.10) the rows of $\mathbf{A}$ are LCs of the rows of $\mathbf{B}$, and conversely. Thus $\mathbf{A}$ and $\mathbf{B}$ have the same row rank. By (2.21), Null$\mathbf{A}$=Null$\mathbf{B}$ so $\mathbf{A}$ and $\mathbf{B}$ have the same nullity and with (2.28) we find that (although their column spaces need not coincide) their column ranks, and also their range ranks coincide. Finally, it is an exercise in using elementary properties of determinants to show that determinant ranks coincide for $\mathbf{A}$ and $\mathbf{B}$ (3.1.9). ■

The end result is the

THEOREM. The five definitions of rank coincide for any matrix. (3.13)

PROOF. The preceding result guarantees in each case

$$\operatorname{rank}\mathbf{A} = \operatorname{rank}\mathbf{S}$$

where $\mathbf{S}$ is the GJSF. By (3.11) the five integers coincide for $\mathbf{S}$. ■

It is customary to use the terms *full row rank* and *full column rank* to mean that the rows are LI, and the columns are LI, respectively. Then *full rank* for an $m \times n$ matrix means that rank equals $\min\{m,n\}$.

PROBLEMS FOR SOLUTION

3.1.1 Zero Rows and Columns in GJSF. (1) Show that there are ten essentially different configurations of GJSFs for matrices $\mathbf{A} \neq \mathbf{O}$. (2) Relate $m-r$, the number of zero rows, and entry to stop. (3) Characterize zero columns in GJSFs.

3.1.2 Rows in GJSF. Replace 3.3(2) by an equivalent condition in terms of rows.

3.1.3 Numerical Exercises. Find the GJSFs of the following, and also of their transposes. (1) [3]. (2) [2,0,1]. (3) [1,1,0,1]. (4) $\mathbf{ab}^T$ where $\mathbf{a}=[1,a_2,\ldots,a_n]^T$ and $\mathbf{b}=[1,b_2,\ldots,b_n]^T$.

3.1.4 Six matrices. Repeat for the following.

$$(1)\begin{bmatrix}1 & 1 & 1 & 7\\ 1 & -1 & 3 & 11\\ 2 & -1 & 1 & 4\end{bmatrix} \quad (2)\begin{bmatrix}1 & 0 & 1 & 1\\ 1 & 1 & 0 & 1\\ 0 & 0 & 1 & 0\end{bmatrix}$$

$$(3)\begin{bmatrix}1 & 1 & 0 & 1 & 0\\ 1 & 1 & 0 & 1 & 0\\ 0 & 0 & 1 & 0 & 1\end{bmatrix} \quad (4)\begin{bmatrix}1 & 0 & -2 & 1\\ 2 & 1 & 4 & -3\\ 5 & 2 & 6 & -5\end{bmatrix}$$

$$(5)\begin{bmatrix}1 & 0 & 1 & 1\\ 2 & 1 & 3 & -4\\ 1 & 1 & 2 & -5\end{bmatrix} \quad (6)\begin{bmatrix}0 & 0 & 2-i & 1-i\\ 0 & 0 & 0 & 1\\ -i & 0 & 1+i & 0\end{bmatrix}$$

3.1.5 Row Equivalence. Establish or resolve: (1) Row equivalence is an equivalence relation for $m \times n$ matrices. (2) True or false? **A** is row-equivalent to **B** iff the set of all LCs of rows of **A** coincides with the set of all LCs of rows of **B**. (3) Repeat for row-equivalence and LCs of columns.

3.1.6 Proof of Uniqueness Theorem. Characterize the matrix **D** appearing in the proof of (3.8).

3.1.7 Determinants. Recall the definition and elementary properties of determinants.

3.1.8 Formulas. Repeat for: (1) Cramer's rule. (2) Order 2 formulas for $\mathbf{A}^{-1}$.

3.1.9 Proof of (3.12). Show: (1) None of the three types of elementary row operations decrease determinant rank. (2) Repeat for increase.

3.1.10 Properties of Rank. Prove the following where sums and products are defined.

(1) $\operatorname{rank}(\mathbf{A}+\mathbf{B}) \leqq \operatorname{rank}\mathbf{A} + \operatorname{rank}\mathbf{B}$.
(2) If **C** is nonsingular, $\operatorname{rank}\mathbf{CA} = \operatorname{rank}\mathbf{A}$ and $\operatorname{rank}\mathbf{BC} = \operatorname{rank}\mathbf{B}$.
(3) $\operatorname{rank}\mathbf{AB} \leqq \min\{\operatorname{rank}\mathbf{A}, \operatorname{rank}\mathbf{B}\}$.
(4) $\operatorname{nullity}\mathbf{AB} \leqq \operatorname{nullity}\mathbf{A} + \operatorname{nullity}\mathbf{B}$.

3.1.11 Range Spaces and Null Spaces. Prove or disprove where **A** is similar to **B**: (1) $\operatorname{Range}\mathbf{A} = \operatorname{Range}\mathbf{B}$. (2) $\operatorname{Null}\mathbf{A} = \operatorname{Null}\mathbf{B}$.

3.2 LINEAR EQUATIONS

Let us obtain the existence theory of solutions of m equations in n *unknowns*, or *variables*, $x_1, x_2, \ldots, x_n$ where we permit $m<n$, $m=n$, or $m>n$, for positive integers m and n.

In *scalar form*, the *system* of *linear equations* is

$$\begin{aligned} a_{11}x_1 + a_{12}x_2 + \cdots + a_{1n}x_n &= b_1 \\ a_{21}x_1 + a_{22}x_2 + \cdots + a_{2n}x_n &= b_2 \\ &\vdots \\ a_{m1}x_1 + a_{m2}x_2 + \cdots + a_{mn}x_n &= b_m \end{aligned}$$

and in *vector form* it is

$$\mathbf{Ax} = \mathbf{b} \tag{3.14}$$

Since the scalar form and vector form both specify that the same equations are to be satisfied we usually do not need to distinguish between them; instead we cite (3.14), or write $\mathbf{Ax}=\mathbf{b}$, to refer to the m equations in n unknowns that are to hold. The $m \times n$ matrix $\mathbf{A}=[a_{ij}]$ is called the *coefficient matrix*, or *matrix*, of the system. The elements a_{ij} are also called the *coefficients* of the variables in the equations. Any column vector $\mathbf{x} \in \mathcal{V}^n$ satisfying (3.14) is called a *solution vector*, or *solution*, and $\mathbf{b} \in \mathcal{V}^m$ is the *RHS vector*, or *RHS*. The *order* of (3.14) is the order of $\mathbf{A}$, namely $m \times n$ and if $m=n$, the system is *square*. The *augmented matrix* of (3.14) is the $m \times (n+1)$ partitioned matrix $[\mathbf{A}, \mathbf{b}]$. If the system has a solution, it is said to be *consistent* and otherwise it is *inconsistent*. The set of all solutions of the system is called its *solution set*; thus the solution set is $\varnothing$ iff the system is inconsistent. Two systems are *equivalent* if their solution sets coincide.

Most often we work with the vector form but let us specify three types of *elementary operations* on the scalar form.

Type I. Interchange of order of appearance of two equations.
Type II. Replacement of an equation by a nonzero multiple of that same equation.
Type III. Replacement of an equation by the sum of that same equation and a multiple of a different equation.

It is easy to verify that elementary operations do not affect solution sets; specifically, any finite succession of elementary operations on the scalar equations for the system produces an equivalent system. There is a one-to-one correspondence between elementary operations on the scalar equations for $\mathbf{Ax}=\mathbf{b}$ and elementary row operations on $[\mathbf{A}, \mathbf{b}]$, and

Any finite succession of elementary row operations on $[\mathbf{A},\mathbf{b}]$ produces the augmented matrix of a system equivalent to $\mathbf{Ax}=\mathbf{b}$. (3.15)

The definitive result on equivalent systems is the following where we must allow for systems in n unknowns that have different numbers of equations.

THEOREM. Two consistent systems are equivalent iff the two GJSFs of augmented matrices have the same submatrix of nonzero rows. (3.16)

PROOF. Let the systems be $\mathbf{C}'\mathbf{x}=\mathbf{d}'$ of order $k\times n$ and $\mathbf{Ax}=\mathbf{b}$ of order $m\times n$ where $m\geqq k$. If $m>k$ we add $k-m$ equations $0=0$ to obtain an $m\times n$ system $\mathbf{Cx}=\mathbf{d}$ that is equivalent to $\mathbf{C}'\mathbf{x}=\mathbf{d}'$. Moreover, the GJSFs of $[\mathbf{C}',\mathbf{d}']$ and $[\mathbf{C},\mathbf{d}]$ have the same submatrix of nonzero rows. If $m=k$ we simply rename $\mathbf{C}'=\mathbf{C}$ and $\mathbf{d}'=\mathbf{d}$. Necessity follows via Null $\mathbf{A}$ = Null $\mathbf{C}$ whereupon, using a special basis, GJSF $\mathbf{A}$ = GJSF $\mathbf{C}$. Conversely, if the specified nonzero submatrices coincide then so do the GJSFs of $[\mathbf{A},\mathbf{b}]$ and $[\mathbf{C},\mathbf{d}]$ and consequently, so do the solution sets. ■

One of the most familiar results in the theory of solutions is the

THEOREM. The system $\mathbf{Ax}=\mathbf{b}$ is consistent iff rank $[\mathbf{A},\mathbf{b}]=\operatorname{rank}\mathbf{A}$. (3.17)

PROOF. The theorem holds for $\mathbf{A}=\mathbf{O}$ because consistency occurs iff $\mathbf{b}=\mathbf{0}$. When $\mathbf{A}\neq\mathbf{O}$ the succession of pivot choices in (3.2) for $[\mathbf{A},\mathbf{b}]$ always coincides with that for $\mathbf{A}$ until the final column is reached. Thus rank $[\mathbf{A},\mathbf{b}]=\operatorname{rank}\mathbf{A}$ iff the final pivot occurs prior to column $n+1$ and this happens iff the final equation corresponding to the GJSF is not "$0=1$". ■

Anticipating a more general concept (introduced in (4.42)) we call variables x_j that correspond to pivot columns in the GJSF of $\mathbf{A}$ *basic variables* and any remaining ones are called *nonbasic variables*. It is convenient to use this terminology in the next proof.

There are two possibilities for $\mathbf{Ax}=\mathbf{b}$ when $\operatorname{rank}[\mathbf{A},\mathbf{b}]=\operatorname{rank}\mathbf{A}=r$.

(1) If $r=n$ there is a unique solution.
(2) If $r<n$ there are an infinite number of solutions. (3.18)

PROOF. Let us suppose we have replaced $[\mathbf{A},\mathbf{b}]$ by its GJSF. All variables are basic in (1) and the unique solution comes from equating $\mathbf{x}$ to the new final column. In (2) we have a solution each time we assign any set of values we wish to the nonbasic variables. ■

When $\mathbf{b}=\mathbf{0}$ the system (3.14) is a *homogeneous system* and otherwise it is *nonhomogeneous*. Then

There are two possibilities for the homogeneous system $\mathbf{Ax}=\mathbf{0}$ where $\operatorname{rank}\mathbf{A}=r$.

(1) If $r=n$ then $\mathbf{x}=\mathbf{0}$ is the unique solution.
(2) If $r<n$ there are an infinite number of nonzero solutions. (3.19)

PROOF. The GJSF of $[\mathbf{A},\mathbf{0}]$ is $[\mathbf{S},\mathbf{0}]$ where $\mathbf{S}$ is the GJSF of $\mathbf{A}$. By (3.17), $\mathbf{Ax}=\mathbf{0}$ is always consistent and the result follows from (3.18). ■

Although homogeneous systems are special, they are of considerable interest because, given any $\mathbf{b}\in\mathcal{V}^m$, there are fundamental connections between solution sets for $\mathbf{Ax}=\mathbf{b}$ and those for $\mathbf{Ax}=\mathbf{0}$, namely

Let $\mathbf{u}_0$ be any solution to $\mathbf{Ax}=\mathbf{b}$. Then $\mathbf{w}$ is also a solution iff

$$\mathbf{w}=\mathbf{u}_0+\mathbf{v}$$

for some $\mathbf{v}$ such that $\mathbf{Av}=\mathbf{0}$. (3.20)

PROOF. Sufficiency is clear because

$$\mathbf{A}(\mathbf{u}_0+\mathbf{v})=\mathbf{Au}_0+\mathbf{Av}$$

Conversely, given both $\mathbf{w}$ and $\mathbf{u}_0$ we have $\mathbf{v}=\mathbf{w}-\mathbf{u}_0$. ■

We call $\mathbf{u}_0$ a *particular solution* of $\mathbf{Ax}=\mathbf{b}$ and the preceding establishes the expression $\mathbf{w}=\mathbf{u}_0+\mathbf{v}$ as the *general solution* since it represents every solution as $\mathbf{v}$ ranges over Null $\mathbf{A}$. The latter also makes it clear how nullity $\mathbf{A}$ "measures" the number of solutions to $\mathbf{Ax}=\mathbf{b}$ when there are an infinite number. (In (9.31) and (10.67) it is appropriate to recall (3.20).)

PROBLEMS FOR SOLUTION

3.2.1 Elementary Properties. Establish for the $m\times n$ system $\mathbf{Ax}=\mathbf{b}$ where $r=\operatorname{rank}\mathbf{A}$:

(1) If the system is consistent then each of the following is necessary and sufficient for a unique solution: (i) $\mathbf{A}$ has full column rank; (ii) $\mathbf{Ax}=\mathbf{0}$ has only the *trivial solution* $\mathbf{x}=\mathbf{0}$.
(2) The system is always consistent if $\mathbf{b}=\mathbf{0}$.
(3) If $r=m$ then the system is consistent for every $\mathbf{b}\in\mathcal{V}^m$.
(4) If there are fewer equations than unknowns in a homogeneous system then there exists a nonzero solution.
(5) If $m=n$ there exists a unique solution iff $\mathbf{A}$ is nonsingular.
(6) If $m=n$ in a homogeneous system then there are nonzero solutions iff $\mathbf{A}$ is singular.
(7) Null $\mathbf{A}$ = Null $\mathbf{B}$ iff the GJSFs of $\mathbf{A}$ and $\mathbf{B}$ have the same submatrices of nonzero rows.

3.2.2 LCs of Equations. An equation in the system $\mathbf{Ax}=\mathbf{b}$ is a LC of other equations in the system if its corresponding row in $[\mathbf{A},\mathbf{b}]$ is a LC of the corresponding other rows. A subset of equations is LI, or LD, if the corresponding set of rows is LI, or LD. Establish:

(1) Any one equation that is a LC of others may be deleted and the solution set will not change.
(2) If the system is consistent then any subset of $r = \operatorname{rank} \mathbf{A}$ LI equations has the same solution set as the system.
(3) At most $n+1$ equations in the system can be LI.
(4) The solution set is empty iff there are $n+1$ LI equations.
(5) The set of m equations is LI if in each equation there appears, with nonzero coefficient, a variable that appears in no preceding equation.
(6) The set is also LI if in each equation there appears, with nonzero coefficient, a variable that appears in no other equation.

3.2.3 One-Sided Inverses. Let us work in terms of matrices rather than linear transformations. If $\mathbf{A}$ is $m \times n$ then any $\mathbf{F}$ satisfying $\mathbf{FA} = \mathbf{I}_n$ is called a *left inverse* of $\mathbf{A}$ and any $\mathbf{G}$ satisfying $\mathbf{AG} = \mathbf{I}_m$ is a *right inverse* of $\mathbf{A}$. Then $\mathbf{A}^{-1}$ is called a *two-sided inverse*. Establish:

(1) $\mathbf{A}$ has both left and right inverses iff it is nonsingular.
(2) $\mathbf{A}$ of order n has a one-sided inverse iff it is nonsingular.

3.2.4 Left Inverses. Establish:

(1) $\mathbf{A}$ has a left inverse iff it has full column rank.
(2) If nonsquare $\mathbf{A}$ has one left inverse then it has an infinite number of such inverses.
(3) A simple example of a nonsquare system $\mathbf{Ax} = \mathbf{b}$ where $\mathbf{A}$ has a left inverse but the system has no solution.
(4) If $\mathbf{Ax} = \mathbf{b}$ is consistent and $\mathbf{A}$ has a left inverse then there is a unique solution.
(5) A procedure for finding a left inverse of $\mathbf{A}$.

3.2.5 Right Inverses. Formulate and establish counterparts to the preceding (1)–(5).

3.3 STANDARD PROCEDURES

In this section we present a set of nine practical procedures for accomplishing the most basic tasks in linear algebra. All procedures are based on GJSFs and all are readily implementable on computers.

The standard basis of $\mathcal{V}^n$ is introduced in (2.6) as the basis $\{\mathbf{e}_1, \mathbf{e}_2, \ldots, \mathbf{e}_n\}$ with matrix $\mathbf{I}_n$. The *standard matrix representation*, or briefly *standard matrix*, for a linear transformation is defined to be the matrix relative to the standard bases in the domain and range. We especially need the following in working with derivatives.

FINDING STANDARD MATRICES. To find the standard matrix of the linear transformation $T: \mathcal{V}^n \to \mathcal{V}^m$, find

$$[T\mathbf{e}_1, T\mathbf{e}_2, \ldots, T\mathbf{e}_n]$$

where $T\mathbf{e}_j \in \mathcal{V}^m$ is the column vector of coordinates with respect to the standard basis in $\mathcal{V}^m$ for $j=1,2,\ldots,n$. (3.21)

PROOF. Established in proof of (2.26). ■

A test for resolving LI for a set of n vectors appearing as the columns of **A** is

The columns of **A** are LI iff the GJSF of **A** has n nonzero rows.

A better test, because it works directly on the columns of **A** and produces a basis for Range **A**, is the following.

RESOLVING LI. If **A** is $m \times n$, find the GJSF of $\mathbf{A}^T$ and use: The columns of **A** are LI iff the GJSF of $\mathbf{A}^T$ has n nonzero rows. (3.22)

PROOF. Column rank, row rank, and Gauss–Jordan rank coincide by (3.13). ■

From the preceding we see that the nonzero columns of the transpose of the GJSF of $\mathbf{A}^T$ form a basis for Range **A**. Since they are uniquely determined as an ordered set of vectors in $\mathcal{V}^m$ we choose the following (recall (3.10)) to cover all cases.

Definition. The *standard basis* of $\mathcal{V}^m$ has matrix $\mathbf{I}_m$ and the *standard basis* of Range $[\mathbf{0}]$ is $\{\mathbf{0}\}$. The *standard basis* of the proper subspace Range $\mathbf{A} \subset \mathcal{V}^m$ is the basis with matrix whose transpose coincides with its own GJSF.

If GJSF $\mathbf{A}^T \neq \mathbf{A}^T$, the matrix of the standard basis of Range **A** is $\mathbf{M}_{11}^T$ where $\mathbf{M}_{11}$ is the submatrix of nonzero rows of **M** the GJSF of $\mathbf{A}^T$. For the example **A** introduced above (3.2),

$$\mathbf{M} = \begin{bmatrix} 1 & 0 & -1 \\ 0 & 1 & 1 \\ 0 & 0 & 0 \end{bmatrix}, \qquad \text{Range}\,\mathbf{A} = \text{Range}\begin{bmatrix} 1 & 0 \\ 0 & 1 \\ -1 & 1 \end{bmatrix}$$

where the final matrix is $\mathbf{M}_{11}^T$. All cases for nonempty sets of vectors from $\mathcal{V}^m$ appearing as columns of any **A** are covered by the following procedure for

FINDING THE STANDARD BASIS. Find **M** the GJSF of $\mathbf{A}^T$. If $\mathbf{M} = \mathbf{O}$ then the standard basis of Range **A** is $\{\mathbf{0}\}$ and otherwise the standard basis is the ordered set of nonzero columns of $\mathbf{M}_{11}^T$ where $\mathbf{M}_{11}$ is the submatrix of nonzero rows of $\mathbf{M}$. (3.23)

PROOF. The GJSF of $\mathbf{O}$ is $\mathbf{O}$. If $\mathbf{A}\neq\mathbf{O}$ then $\mathbf{M}_{11}=\mathbf{I}_m$ iff Range $\mathbf{A}=\mathcal{V}^m$ so via the unique GJSF we find the unique standard basis for any subspace. ■

Notice that the preceding procedure applies to any subspace for which we have a set of vectors (not necessarily forming a basis) that spans the subspace.

In Chapters 5 and 7 it is convenient to use the next procedure for

FINDING THE STANDARD REPRESENTATION OF A SUBSPACE. Find $\mathbf{Z}$ the $m\times r$ matrix of the standard basis of the proper subspace $S\subset\mathcal{V}^m$ and use

$$S=\{\mathbf{y}:\mathbf{y}=\mathbf{Za} \qquad \text{for some } \mathbf{a}\in\mathcal{V}^r\}$$

as the *standard representation* of S. (3.24)

PROOF. This is a matter of definition. ■

We can represent S in the same way using the matrix of any basis. The effect in any case is to be able to work in the lower-dimensional space $\mathcal{V}^r$ and at times this is a nontrivial advantage.

In Chapter 7 we need to find a basis for the null space of a matrix. There are shortcut methods (3.3.4) for handwork, but let us here start with the equation in (3.7) written as

$$\mathbf{M}=\mathbf{DA}^T \tag{3.25}$$

where $\mathbf{M}$ is the GJSF of $\mathbf{A}^T$. We can readily find both $\mathbf{M}$ and $\mathbf{D}$ by applying (3.2) to the $n\times(m+n)$ LHS of

$$[\mathbf{A}^T,\mathbf{I}_n]\overset{R}{\rightarrow}[\mathbf{M},\mathbf{D}] \tag{3.26}$$

Whenever the GJSF of $\mathbf{A}^T$ occupies the first m columns of the RHS we can stop in (3.2) and use the final n columns as $\mathbf{D}$; if we want $\mathbf{D}$ to be a uniquely determined standard factor, we can continue until the RHS of (3.26) is the GJSF of the entire LHS. For the matrix $\mathbf{A}$ displayed below (3.1), we can stop after two pivots

$$\begin{bmatrix} 1 & 2 & 1 & 1 & 0 & 0 \\ 0 & 1 & 1 & 0 & 1 & 0 \\ 1 & 3 & 2 & 0 & 0 & 1 \end{bmatrix} \overset{R}{\rightarrow} \begin{bmatrix} 1 & 0 & -1 & 1 & -2 & 0 \\ 0 & 1 & 1 & 0 & 1 & 0 \\ 0 & 0 & 0 & -1 & -1 & 1 \end{bmatrix}$$

and have (3.25) in the form

$$\begin{bmatrix} 1 & 0 & -1 \\ 0 & 1 & 1 \\ 0 & 0 & 0 \end{bmatrix} = \begin{bmatrix} 1 & -2 & 0 \\ 0 & 1 & 0 \\ -1 & -1 & 1 \end{bmatrix} \begin{bmatrix} 1 & 2 & 1 \\ 0 & 1 & 1 \\ 1 & 3 & 2 \end{bmatrix}$$

Whenever $r = \text{rank}\,\mathbf{A}$ satisfies $0 < r < n$, we partition $\mathbf{D}$ as

$$\mathbf{D} = \begin{bmatrix} \mathbf{D}_{11} \\ \mathbf{D}_{21} \end{bmatrix}$$

where $\mathbf{D}_{11}$ is $r \times n$ and $\mathbf{D}_{21}$ is $(n-r) \times n$. Then we can use the following procedure for

FINDING A BASIS FOR NULL A. If $\mathbf{A}$ is $m \times n$ and $\text{rank}\,\mathbf{A}$ is $r = 0$ or n, then the bases are $\{\mathbf{e}_1, \mathbf{e}_2, \ldots, \mathbf{e}_n\}$ or $\{\mathbf{0}\}$, respectively. If $0 < r < n$, find any $\mathbf{D}$ satisfying (3.26), and then the columns of $\mathbf{D}_{21}^T$ form a basis. (3.27)

PROOF. Suppose $\mathbf{E}$ is the nonsingular matrix from (3.7) for the two row-equivalent $n \times (m+n)$ matrices in (3.26). Then $\mathbf{EI}_n = \mathbf{D}$ shows that $\mathbf{E} = \mathbf{D}$ and consequently (3.25) holds. Now $\mathbf{x} \in \text{Null}\,\mathbf{A}$ iff

$$\mathbf{0}^T = \mathbf{x}^T \mathbf{A}^T$$

and (3.25) shows that each of the final $n-r$ rows of $\mathbf{D}$ can be used as such an $\mathbf{x}^T$. Since these rows are rows of a nonsingular matrix they are also LI. ■

For the preceding

$$\mathbf{D}_{21}^T = \begin{bmatrix} -1 \\ -1 \\ 1 \end{bmatrix}$$

and the vector in the standard basis is $[1, 1, -1]^T$.

The process of determining whether or not the system $\mathbf{Ax} = \mathbf{b}$ has solutions is called *resolving its consistency*. For small m and n this can be done by hand as follows. We find the GJSF of $[\mathbf{A}, \mathbf{b}]$ and then use (3.17): The system is consistent iff the final column is not a pivot column. Actually this procedure works as well by hand for symbols $b_1, b_2, \ldots, b_m$ as for numbers. For the same numerical $\mathbf{A}$, starting with $[\mathbf{A}, \mathbf{b}]$ in

$$\begin{bmatrix} 1 & 0 & 1 & b_1 \\ 2 & 1 & 3 & b_2 \\ 1 & 1 & 2 & b_3 \end{bmatrix} \overset{R}{\to} \begin{bmatrix} 1 & 0 & 1 & b_1 \\ 0 & 1 & 1 & b_2 - 2b_1 \\ 0 & 1 & 1 & b_3 - b_1 \end{bmatrix} \overset{R}{\to} \begin{bmatrix} 1 & 0 & 1 & b_1 \\ 0 & 1 & 1 & b_2 - 2b_1 \\ 0 & 0 & 0 & b_3 - b_2 + b_1 \end{bmatrix}$$

we find: $\mathbf{Ax}=\mathbf{b}$ is consistent iff $b_3-b_2+b_1=0$. For use on computers, or for handwork when we wish to avoid symbol manipulation in the same way we avoid it when we work on the matrix $[\mathbf{A},\mathbf{b}]$ in place of the m scalar equations, we proceed almost the same as for (3.27). We work with $\mathbf{A}$ in place of $\mathbf{A}^T$ and use

$$\mathbf{EA}=\mathbf{S} \tag{3.28}$$

where $\mathbf{S}$ is the GJSF of $\mathbf{A}$ and $\mathbf{E}$ is any matrix satisfying

$$[\mathbf{A},\mathbf{I}_m]\overset{R}{\rightarrow}[\mathbf{S},\mathbf{E}] \tag{3.29}$$

Whenever $r=\operatorname{rank}\mathbf{A}$ satisfies $0<r<m$ we partition $\mathbf{E}$ as

$$\mathbf{E}=\begin{bmatrix}\mathbf{E}_{11}\\ \mathbf{E}_{21}\end{bmatrix}$$

where $\mathbf{E}_{11}$ is $r\times m$ and $\mathbf{E}_{21}$ is $(m-r)\times m$ and we can use the following for

RESOLVING CONSISTENCY OF $\mathbf{Ax}=\mathbf{b}$. If $\mathbf{A}$ is $m\times n$ and $\operatorname{rank}\mathbf{A}$ is $r=0$ or m, then the system is consistent for $\mathbf{b}=\mathbf{0}$ or any $\mathbf{b}\in\mathcal{V}^m$, respectively. If $0<r<m$, find $\mathbf{E}$ in (3.29) and use: The system is consistent iff

$$\mathbf{b}\in\operatorname{Null}\mathbf{E}_{21} \tag{3.30}$$

PROOF. Any matrix $\mathbf{E}$ in (3.29) satisfies (3.28) as shown in the proof of (3.27). The present result follows from (3.17) which can be stated as: the system $\mathbf{Ax}=\mathbf{b}$ is consistent iff the final $m-r$ rows of $[\mathbf{S},\mathbf{Eb}]$ are zero rows. ■

For the preceding

$$\begin{bmatrix}1&0&1&1&0&0\\2&1&3&0&1&0\\1&1&2&0&0&1\end{bmatrix}\overset{R}{\rightarrow}\begin{bmatrix}1&0&1&1&0&0\\0&1&1&-2&1&0\\0&0&0&1&-1&1\end{bmatrix}$$

and $\mathbf{E}_{21}=[1\quad -1\quad 1]$.

Among all systems that are equivalent to $\mathbf{Ax}=\mathbf{b}$, the ones that have particularly advantageous forms both for resolving consistency and for obtaining actual solutions are obtained by

FINDING A GJSF SYSTEM. Find $\mathbf{S}$ and $\mathbf{E}$ by pivoting in (3.2) through at least the first n columns on the LHS of (3.29). Then form

$$\mathbf{Sx}=\mathbf{Eb}$$

which is a *GJSF system for* $\mathbf{Ax}=\mathbf{b}$. (3.31)

In theory, that is without regard to effects of rounding off numbers or deep concerns for economical computing, a good method for solving $\mathbf{Ax}=\mathbf{b}$ is to solve a GJSF system; a refinement for $m=n$ appears in (5.13). When $\mathbf{A}$ is nonsingular the process of finding $\mathbf{E}$ in (3.31) is the familiar one for finding $\mathbf{A}^{-1}$ by using elementary row operations to move $\mathbf{I}_n$ to the left in $[\mathbf{A},\mathbf{I}_n]$.

There is another procedure that goes with the elementary theory of solutions of $\mathbf{Ax}=\mathbf{b}$, namely, the following which applies to the case where m equations are actually required and where there are an infinite number of solutions.

REPRESENTING ALL SOLUTIONS OF $\mathbf{Ax}=\mathbf{b}$. Suppose $\mathbf{A}$ is $m\times n$ and has full row rank $m<n$. Then if $\mathbf{u}_0$ is any particular solution of $\mathbf{Ax}=\mathbf{b}$, all solutions are given by

$$\{\mathbf{x} : \mathbf{x} = \mathbf{u}_0 + \mathbf{Vt} \qquad \text{for some } \mathbf{t}\in\mathcal{V}^{n-m}\}$$

where $\mathbf{V}$ is the matrix of any basis for Null $\mathbf{A}$. (3.32)

PROOF. The procedure implements (3.20). A particular solution $\mathbf{u}_0$ can be found by inspection of $[\mathbf{S},\mathbf{Eb}]$, or by multiplication $\mathbf{Ub}$ where $\mathbf{U}$ is any right inverse of $\mathbf{A}$ found by inspection of $[\mathbf{S},\mathbf{E}]$. ■

No such $\mathbf{U}$ exists when rank $\mathbf{A}<m$. But any solution that can be found, say from $[\mathbf{S},\mathbf{Eb}]$, can always be used in place of $\mathbf{u}_0$ in (3.32). For $\mathbf{A}$ above (3.2) and $\mathbf{b}$ such that $b_3 = b_2 - b_1$, all solutions are given by

$$\mathbf{x} = \begin{bmatrix} b_1 + t \\ b_2 - 2b_1 + t \\ - t \end{bmatrix}$$

where $t\in\mathcal{R}$ is arbitrary.

Let us next introduce some notation for the special case where $\mathbf{I}_r$ appears as the upper LH corner, and where both zero rows and nonpivot columns appear, in the GJSF of $\mathbf{A}$. Here we can partition $\mathbf{A}$ by inserting a cut immediately following column r, and write

$$\mathbf{A}=[\mathbf{A}_{11},\mathbf{A}_{12}]$$

and we can partition the GJSF $\mathbf{S}$ two ways—first with a cut following row r, and second with cuts immediately following row r and column r—and equate

$$\begin{bmatrix} \mathbf{S}_{11} \\ \mathbf{S}_{21} \end{bmatrix} = \begin{bmatrix} \mathbf{I}_r & \mathbf{N} \\ \mathbf{O}^1 & \mathbf{O}^2 \end{bmatrix}$$

where $\mathbf{O}^1 = \mathbf{O}_{m-r,r}$ and $\mathbf{O}^2 = \mathbf{O}_{m-r,n-r}$. The surprising result is that $\mathbf{A}$ can readily be factored into two factors of full rank as follows.

Suppose $\mathbf{A}$ is $m \times n$, $r = \operatorname{rank} \mathbf{A}$ satisfies

$$0 < r < \min\{m, n\}$$

and $\mathbf{I}_r$ stands in the upper LH corner of $\mathbf{S}$ the GJSF of $\mathbf{A}$. Then

$$\mathbf{A} = \mathbf{A}_{11}\mathbf{S}_{11} \tag{3.33}$$

PROOF. The partitionings just displayed are possible and if $\mathbf{EA} = \mathbf{S}$ from (3.7), then also

$$\mathbf{EA}_{11} = \begin{bmatrix} \mathbf{I}_r \\ \mathbf{O}^1 \end{bmatrix}, \qquad \mathbf{EA}_{12} = \begin{bmatrix} \mathbf{N} \\ \mathbf{O}^2 \end{bmatrix}$$

Again multiplying blocks, $\mathbf{EA}_{12} = \mathbf{EA}_{11}\mathbf{N}$, and we conclude $\mathbf{A}_{12} = \mathbf{A}_{11}\mathbf{N}$ since $\mathbf{E}$ is nonsingular. By agreement $\mathbf{S}_{11} = [\mathbf{I}_r, \mathbf{N}]$, and one more multiplication

$$\mathbf{A}_{11}\mathbf{S}_{11} = [\mathbf{A}_{11}, \mathbf{A}_{11}\mathbf{N}]$$

establishes the factorization. ■

In order to show that it is just as simple to factor any matrix we make use of a single type of *elementary column operation*

Type I. Interchange of two columns

The result of any single such operation on an $m \times n$ matrix $\mathbf{A}$ is duplicated by multiplying $\mathbf{A}$ on the right by the corresponding *elementary column matrix* of order n, namely one of

Type I. $\mathbf{F}_{is}$ results from interchanging columns i and s in $\mathbf{I}_n$.

Now $\mathbf{F}_{is}^{-1} = \mathbf{F}_{is}$ so $\mathbf{F}_{is}$ is nonsingular. A *permutation matrix* is a matrix obtained from an identity matrix by performing a finite succession of elementary column operations of Type I. Since we can always use (3.3) to identify the r pivot columns in a GJSF matrix, and since we can successively interchange columns until pivot column p_j occupies column j for $1, 2, \ldots, r$ (3.3.3),

Given any GJSF matrix $\mathbf{S}$ of rank r there exists a permutation matrix $\mathbf{F}$ such that $\mathbf{T} = \mathbf{SF}$ is a GJSF matrix with pivot columns $1, 2, \ldots, r$. (3.34)

Furthermore, columns of $\mathbf{A}$ that are not pivot columns in the reduction procedure (3.2) are merely "skipped over" when they are encountered during selection of successive pivots and consequently

If $\mathbf{S}$ is the GJSF of $\mathbf{A}$ then $\mathbf{T}=\mathbf{SF}$ is the GJSF of $\mathbf{B}=\mathbf{AF}$. (3.35)

We need permutation matrices for the proof but do not need to use them in the statement of the general procedure we need in Example 7.2 for

FACTORING A MATRIX. There is no interest in factoring $\mathbf{A}=\mathbf{O}$. If $\mathbf{A}\neq\mathbf{O}$, find its GJSF $\mathbf{S}$ and express $\mathbf{A}$ as the product of two matrices of full rank

$$\mathbf{A}=\mathbf{B}_{11}\mathbf{S}_{11}$$

where $\mathbf{B}_{11}$ is formed by deleting nonpivot columns, if any, from $\mathbf{A}$ and $\mathbf{S}_{11}$ by deleting zero rows, if any, from $\mathbf{S}$. (3.36)

PROOF. Let us find $\mathbf{F}$ in (3.34) and work with $\mathbf{B}=\mathbf{AF}$ whose GJSF according to (3.35) is $\mathbf{T}=\mathbf{SF}$. First, we show that the result holds for $\mathbf{B}\neq\mathbf{O}$ as

$$\mathbf{B}=\mathbf{B}_{11}\mathbf{T}_{11}$$

where $\mathbf{T}_{11}$ is formed by deleting zero rows from $\mathbf{T}$. If the common rank is $r=m$ then we duplicate the proof of (3.33), without partitioning the rows of $\mathbf{T}$, to show that we can use $\mathbf{T}_{11}=\mathbf{T}$. If $r=n$ we use $\mathbf{B}_{11}=\mathbf{B}$ and $\mathbf{T}_{11}=\mathbf{I}_n$. If $0<r<\min\{m,n\}$ we use (3.33). Second, the result holds for $\mathbf{A}$ because in all three cases

$$\mathbf{BF}=\mathbf{B}_{11}\mathbf{T}_{11}\mathbf{F}$$

and $\mathbf{BF}=\mathbf{A}$ while $\mathbf{T}_{11}\mathbf{F}=\mathbf{S}_{11}$. ■

As an example where $\mathbf{I}_r$ appears in the upper LH corner of $\mathbf{S}$, the matrix we have been using is

$$\begin{bmatrix} 1 & 0 & 1 \\ 2 & 1 & 3 \\ 1 & 1 & 2 \end{bmatrix} = \begin{bmatrix} 1 & 0 \\ 2 & 1 \\ 1 & 1 \end{bmatrix} \begin{bmatrix} 1 & 0 & 1 \\ 0 & 1 & 1 \end{bmatrix} \tag{3.37}$$

As a case where $\mathbf{I}_r$ does not so appear, we have the following where the second column of the LHS is not a pivot column.

$$\begin{bmatrix} 1 & 0 & 1 & 0 & 0 & 1 & 0 \\ 1 & 0 & 1 & 0 & 0 & 1 & 0 \\ 1 & 0 & 2 & 0 & 1 & 2 & 1 \end{bmatrix} = \begin{bmatrix} 1 & 1 \\ 1 & 1 \\ 1 & 2 \end{bmatrix} \begin{bmatrix} 1 & 0 & 0 & 0 & -1 & 0 & -1 \\ 0 & 0 & 1 & 0 & 1 & 1 & 1 \end{bmatrix}$$

(3.38)

Notice that procedure (3.36) can be described as factoring a matrix using its columns.

PROBLEMS FOR SOLUTION

3.3.1 Numerical Exercises. (1) Find the standard matrix for $T:\mathcal{R}^4 \to \mathcal{R}^3$ given: the images of $[1,0,0,0]^T$, $[1,1,0,0]^T$, $[1,1,1,0]^T$, and $[1,1,1,1]^T$ are $[1,0,1]^T$, $[1,1,2]^T$, $[3,1,4]^T$, and $[4,2,6]^T$, respectively. (2) Resolve LI of the columns of the matrices in 3.1.4, and of their transposes, and when LD find a LI set that spans the same subspace. (3) Find the standard basis for each subspace in (2). (4) Find a basis for each null space. (5) Resolve consistency for $\mathbf{b}$ in terms of $b_1, b_2, \ldots$ for each of the matrices in (2). (6) Find a GJSF system for each system. (7) Represent all solutions for each system. (8) Factor each matrix in (2) using (3.36). (9) Factor the LHS of (3.37) using its rows.

3.3.2 Permutation Matrices. Establish: (1) A permutation matrix is a square matrix having elements 0 or 1 such that there is exactly one entry 1 in each row and each column. (2) Repeat for: a matrix obtained by interchanging rows or columns of $\mathbf{I}$. (3) If $\mathbf{F}$ is a permutation matrix then $\mathbf{F}^{-1} = \mathbf{F}^T$. (4) A permutation matrix $\mathbf{F}$ can always be written in the partitioned form (2.13) using some permutation of the columns of $\mathbf{I}$. (5) Given $\mathbf{F}$ in the preceding form, say

$$[\mathbf{e}_2, \mathbf{e}_3, \mathbf{e}_5, \mathbf{e}_4, \mathbf{e}_1, \mathbf{e}_7, \mathbf{e}_6]$$

it is straightforward to write $\mathbf{F}^{-1}$ in the same form.

3.3.3 Procedure. Establish a standard procedure that produces a matrix $\mathbf{F}$ for (3.34).

3.3.4 Null Spaces. The following alternative to (3.27) is often convenient for handwork because it involves working with $\mathbf{A}$ alone rather than the larger $[\mathbf{A}^T, \mathbf{I}_n]$ of (3.26). Establish: (1) Suppose $\mathbf{I}_r$ appears in the upper LH corner of $\mathbf{S}$ the GJSF of $\mathbf{A}$. Then in the notation for (3.33), the columns of

$$\begin{bmatrix} -\mathbf{N} \\ \mathbf{I}_{n-r} \end{bmatrix}$$

form a basis for Null $\mathbf{A}$. (2) The procedure for any $\mathbf{A} \neq \mathbf{O}$.

3.3.5 Linear Transformations on Subspaces. Establish a procedure for finding the standard matrix of the restriction of a linear transformation to a proper subspace $S \subset \mathcal{V}^n$ of dimension p.

3.4 UNITARY SPACE

Here we introduce an inner product and several norms for use in the following chapters.

It is necessary that $\mathcal{R}$ be treated as a special case of $\mathcal{C}$ as follows. First, two fields are *isomorphic* if there is a one-to-one correspondence between elements that preserves the operations of addition and multiplication. Second, a *subfield* of a field is a subset that is itself a field under the given operations. Third, $\mathcal{R}$ is isomorphic to a subfield of $\mathcal{C}$, namely the elements on the real line. Then we can say

All definitions and results in the present section hold for real or complex scalars provided $\mathcal{R}$ is identified as a subfield of $\mathcal{C}$. (3.39)

Less formally, we follow custom and say "z is real" to mean $z=\bar{z}$, "in the real case" for the corresponding specialization, and so on. But, to achieve (3.39) we fix $\mathcal{F}=\mathcal{C}$ in (2.2) and work in spaces of column vectors $\mathcal{C}^n$.

In our usual notation where the same letter is used for column vectors and their components, the *Hermitian transpose* of $\mathbf{u}\in\mathcal{C}^n$ is the $1\times n$ matrix of complex conjugates

$$\mathbf{u}^H=[\bar{u}_1,\bar{u}_2,\ldots,\bar{u}_n]$$

and $(\mathbf{u}^H)^H=\mathbf{u}$. Consequently, $\mathbf{u}^H=\mathbf{u}^T$ iff all u_i are real. The function with values

$$\langle\mathbf{u},\mathbf{v}\rangle=\mathbf{u}^H\mathbf{v} \tag{3.40}$$

is called an *inner product*. In the real case it coincides with the familiar *scalar product* or *dot product* of "position" vectors. For example, in the plane, we have

$$\mathbf{x}^T\mathbf{y}=x_1y_1+x_2y_2$$

More generally we have

The inner product (3.40) satisfies the following for all $\mathbf{u},\mathbf{v},\mathbf{w}\in\mathcal{C}^n$ and all $a,b\in\mathcal{C}$.

(1) Positive definiteness. $\langle\mathbf{u},\mathbf{u}\rangle$ is real, nonnegative, and $\langle\mathbf{u},\mathbf{u}\rangle=0$ iff $\mathbf{u}=\mathbf{0}$.
(2) Hermitian symmetry. $\langle\mathbf{u},\mathbf{v}\rangle=\overline{\langle\mathbf{v},\mathbf{u}\rangle}$.
(3) Linearity in second argument.

$$\langle\mathbf{u},a\mathbf{v}+b\mathbf{w}\rangle=a\langle\mathbf{u},\mathbf{v}\rangle+b\langle\mathbf{u},\mathbf{w}\rangle \tag{3.41}$$

PROOF. The respective properties hold because: a sum of absolute values is zero iff each summand is zero; $\mathbf{u}^H\mathbf{v}$ and $\mathbf{v}^H\mathbf{u}$ are complex conjugates of one another; and, matrix multiplication is distributive with respect to addition. ■

A vector space having complex numbers as scalars and a function that satisfies the conditions in (3.41) is called a *unitary space*. Consequently $\mathcal{C}^n$ is a unitary space, and furthermore, we need consider no other that has nonreal scalars (3.4.3). Hereafter we take the inner product (3.40) as given and use the term

$$\text{unitary space } \mathcal{C}^n \tag{3.42}$$

interchangeably with "vector space $\mathcal{C}^n$."

Given unitary space $\mathcal{C}^n$ the function with values

$$\|\mathbf{v}\| = \langle \mathbf{v}, \mathbf{v} \rangle^{\frac{1}{2}} \tag{3.43}$$

is called a *norm* on account of the general

Definition. A *norm* is any function $\|\cdot\| : \mathcal{C}^n \to \mathfrak{R}$ that satisfies the following conditions for all $\mathbf{u}, \mathbf{v} \in \mathcal{C}^n$ and all $a \in \mathcal{C}$.

(1) Positive Definiteness. $\|\mathbf{u}\| \geqq 0$ and $\|\mathbf{u}\| = 0$ iff $\mathbf{u} = \mathbf{0}$.
(2) Homogeneity. $\|a\mathbf{u}\| = |a| \cdot \|\mathbf{u}\|$.
(3) Triangle Inequality. $\|\mathbf{u} + \mathbf{v}\| \leqq \|\mathbf{u}\| + \|\mathbf{v}\|$. (3.44)

We use different norms and so let us prove the

THEOREM. The norm (3.43) qualifies as a norm according to (3.44) and any norm has the following properties for all $\mathbf{u}, \mathbf{v}, \mathbf{v}_1, \ldots, \mathbf{v}_k \in \mathcal{C}^n$ and $a_1, \ldots, a_k \in \mathcal{C}$.

(1) Cauchy–Schwarz inequality. $|\langle \mathbf{u}, \mathbf{v} \rangle| \leqq \|\mathbf{u}\| \cdot \|\mathbf{v}\|$.
(2) $\|a_1\mathbf{v}_1 + \cdots + a_k\mathbf{v}_k\| \leqq |a_1| \cdot \|\mathbf{v}_1\| + \cdots + |a_k| \cdot \|\mathbf{v}_k\|$. (3.45)

PROOF. Let us first show that (3.43) satisfies (3.44). Positive-definiteness results from (3.41) and homogeneity follows from $\langle a\mathbf{u}, a\mathbf{u} \rangle = \bar{a}a\langle \mathbf{u}, \mathbf{u} \rangle$. In order to establish the triangle inequality let us first prove the Cauchy–Schwarz inequality for (3.43). The inequality holds for $\langle \mathbf{u}, \mathbf{v} \rangle = 0$ so we suppose $\langle \mathbf{u}, \mathbf{v} \rangle > 0$ and consider the following which is nonnegative for $t \in \mathcal{C}$.

$$\|\mathbf{u} + t\mathbf{v}\|^2 = \|\mathbf{u}\|^2 + t\langle \mathbf{u}, \mathbf{v} \rangle + \bar{t}\langle \overline{\mathbf{u}, \mathbf{v}} \rangle + \bar{t}t\|\mathbf{v}\|^2$$

In particular, for $t = -\|\mathbf{u}\|^2 \langle \mathbf{u}, \mathbf{v} \rangle^{-1}$ we obtain

$$0 \leqq -\|\mathbf{u}\|^2 + \|\mathbf{u}\|^4 \|\mathbf{v}\|^2 |\langle \mathbf{u}, \mathbf{v} \rangle|^{-2}$$

which implies the Cauchy–Schwarz inequality for (3.43). Consequently

$$\langle \mathbf{u}+\mathbf{v},\mathbf{u}+\mathbf{v}\rangle = \|\mathbf{u}\|^2+\|\mathbf{v}\|^2+2\,\mathrm{Re}\langle \mathbf{u},\mathbf{v}\rangle$$

serves to establish the triangle inequality because

$$\mathrm{Re}\langle \mathbf{u},\mathbf{v}\rangle \leqq |\langle \mathbf{u},\mathbf{v}\rangle| \leqq \|\mathbf{u}\|\cdot\|\mathbf{v}\|$$

Thus (3.43) qualifies as a norm. The present inequality (1) holds in general because we can use (3.44) to expand $\|\mathbf{u}+t\mathbf{v}\|$ and duplicate the preceding argument for $t\in\mathcal{C}$. The present (2) holds for $k=1$ by homogeneity. Induction proceeds as follows. Suppose (2) holds for k and $\mathbf{v}$ is the vector on the LHS. Then we use the triangle inequality to write

$$\|\mathbf{v}+a_{k+1}\mathbf{v}_{k+1}\| \leqq \|\mathbf{v}\|+\|a_{k+1}\mathbf{v}_{k+1}\|$$

and again use homogeneity for the final term. ■

Whenever we have a norm we define the associated *distance* from $\mathbf{u}$ to $\mathbf{v}$ to be

$$\|\mathbf{u}-\mathbf{v}\|$$

and the *length* of $\mathbf{v}$ is $\|\mathbf{v}\|$, the distance from $\mathbf{v}$ to $\mathbf{0}$. A vector of unit length is said to be *normalized*. Replacing $\mathbf{v}\neq\mathbf{0}$ by $\|\mathbf{v}\|^{-1}\mathbf{v}$, often written $\mathbf{v}/\|\mathbf{v}\|$, is called *normalizing* $\mathbf{v}$.

On the strength of the Cauchy–Schwarz inequality in (3.45), the *angle* θ between $\mathbf{u}$ and $\mathbf{v}$ is defined using

$$\cos\theta = \frac{\langle \mathbf{u},\mathbf{v}\rangle}{\|\mathbf{u}\|\cdot\|\mathbf{v}\|}, \qquad 0\leqq\theta\leqq\pi \tag{3.46}$$

Generalizing the inclusion of a right angle between two "position" vectors, $\mathbf{u}$ and $\mathbf{v}$ are *orthogonal* if $\langle \mathbf{u},\mathbf{v}\rangle=0$. A set of vectors, and in particular a basis, is said to be *orthogonal* if any two distinct vectors in the set are orthogonal, that is, if the vectors are *mutually orthogonal*. If the vectors are furthermore normalized the set is called *orthonormal*; the primary example is the standard basis.

It is readily verified that an orthogonal set of nonzero vectors is LI. Consequently, any such set of n vectors forms a basis that can be normalized to form an orthonormal basis for $\mathcal{C}^n$. What is not so clear is that every subspace of $\mathcal{C}^n$ has an orthonormal basis, namely, one that can efficiently be found using the

GRAM–SCHMIDT PROCEDURE. Start with any LI set $\{\mathbf{a}_1,\mathbf{a}_2,\ldots,\mathbf{a}_k\}\subset\mathcal{C}^n$.

Step 1. Find the intermediate $\mathbf{u}_1 = \mathbf{a}_1$ and normalize as $\mathbf{b}_1 = \mathbf{u}_1/\|\mathbf{u}_1\|$.
Step s for $s = 2,3,\ldots,k$. Find

$$\mathbf{u}_s = \mathbf{a}_s - \langle \mathbf{b}_1, \mathbf{a}_s \rangle \mathbf{b}_1 - \cdots - \langle \mathbf{b}_{s-1}, \mathbf{a}_s \rangle \mathbf{b}_{s-1}$$

and normalize as $\mathbf{b}_s = \mathbf{u}_s/\|\mathbf{u}_s\|$. (3.47)

Not only is the new set $\{\mathbf{b}_1, \mathbf{b}_2, \ldots, \mathbf{b}_k\}$ an orthonormal set but it spans the same subspace as $\{\mathbf{a}_1, \mathbf{a}_2, \ldots, \mathbf{a}_k\}$. Expressed in terms of partitioned matrices,

The columns of $[\mathbf{b}_1, \mathbf{b}_2, \ldots, \mathbf{b}_k]$ form an orthonormal basis for Range $[\mathbf{a}_1, \mathbf{a}_2, \ldots, \mathbf{a}_k]$. (3.48)

PROOF. By construction for $s = 2,3,\ldots,k$, $\langle \mathbf{b}_j, \mathbf{u}_s \rangle = 0$ for $j = 1,2,\ldots,s-1$. Furthermore, $\mathbf{u}_1 \neq \mathbf{0}$ and $\mathbf{u}_s \neq \mathbf{0}$ for $s = 2,3,\ldots,k$ lest $\{\mathbf{a}_1, \ldots, \mathbf{a}_s\}$ be LD. The procedure (3.47) clearly terminates at completion of Step k and, at the end, any LC of the **a**'s is a LC of the **b**'s, and vice versa. ■
See 3.4.6 for an alternative manual procedure.

The Gram–Schmidt procedure is significant on practical grounds because calculations involving inner products are notably brief when orthonormal bases are used. It is also important on theoretical grounds because it guarantees that every subspace has an orthonormal basis and this in turn permits us to restrict our attention to the one unitary space $\mathcal{C}^n$ (3.4.3).

If $S \subset \mathcal{C}^n$ is any subspace its *orthogonal complement* is

$$S^{\perp} = \{\mathbf{v} : \langle \mathbf{v}, \mathbf{s} \rangle = 0 \qquad \text{for every } \mathbf{s} \in S\} \tag{3.49}$$

It is easy to show: $S^{\perp}$ is a subspace, $(S^{\perp})^{\perp} = S$, and so on, through

$$S \oplus S^{\perp} = \mathcal{C}^n \tag{3.50}$$

which has useful consequences; see 3.4.9.

The following concepts are used in $\mathcal{L}(\mathcal{C}^n)$ and associated $\mathcal{M}(n)$. If $T \in \mathcal{L}(\mathcal{C}^n)$ we consider first $\langle \mathbf{x}, \mathbf{y} \rangle$ and second $\langle T\mathbf{x}, \mathbf{y} \rangle$ to imagine

$$\langle \mathbf{x}, \mathbf{y} \rangle \mapsto \langle T\mathbf{x}, \mathbf{y} \rangle$$

as well as $\mathbf{x} \mapsto T\mathbf{x}$. If there is a second linear transformation T^A satisfying for all $\mathbf{x}, \mathbf{y} \in \mathcal{C}^n$

$$\langle T\mathbf{x}, \mathbf{y} \rangle = \langle \mathbf{x}, T^A \mathbf{y} \rangle \tag{3.51}$$

then T^A is called the *adjoint* of T. If $T^A = T$ then T is called *self-adjoint*. While adjoint transformations preserve inner products in the specified sense, *unitary transformations* $U \in \mathcal{L}(\mathcal{C}^n)$ satisfy

$$\langle \mathbf{x}, \mathbf{y} \rangle = \langle U\mathbf{x}, U\mathbf{y} \rangle \tag{3.52}$$

for all $\mathbf{x}, \mathbf{y} \in \mathcal{C}^n$. Consequently, unitary transformations U are the members of $\mathcal{L}(\mathcal{C}^n)$ that leave the norm (3.43) and the associated lengths invariant: $\|\mathbf{x}\| = \|U\mathbf{x}\|$ for all $\mathbf{x} \in \mathcal{C}^n$. We study adjoint and unitary transformations and their associated (square) matrices in Chapter 5.

We need to introduce the concept of a matrix norm for defining derivatives in Chapter 6 and working with matrix exponentials in Chapter 9. It is not effective to use (ordinary or vector) norms (3.44) on $m \times n$ matrices treated as mn-component vectors. The reason is that, while the idea of a norm is to assign a single real number to measure magnitude, it must also provide for consistency or compatibility with other operations. For vectors we have only scalar multiples and sums, but for matrices we have products as well and for most purposes we need (4) in the

Definition. A *matrix norm* is a real-valued function

$$\mathbf{A} \mapsto \|\mathbf{A}\|$$

defined for any $m \times n$ matrix of elements of $\mathcal{C}$ such that for all $a \in \mathcal{C}$, and for sums and products that are defined

(1) $\|\mathbf{A}\| \geqq 0$ and $\|\mathbf{A}\| = 0$ iff $\mathbf{A} = \mathbf{O}$.
(2) $\|a\mathbf{A}\| = |a| \cdot \|\mathbf{A}\|$.
(3) $\|\mathbf{A} + \mathbf{B}\| \leqq \|\mathbf{A}\| + \|\mathbf{B}\|$.
(4) $\|\mathbf{BC}\| \leqq \|\mathbf{B}\| \cdot \|\mathbf{C}\|$. (3.53)

Given a matrix norm we define the corresponding norm of $T \in \mathcal{L}(\mathcal{C}^n, \mathcal{C}^m)$ to be $\|T\| = \|\mathbf{T}\|$ where $\mathbf{T}$ is its standard matrix. Property (4) is reminiscent of the relationship in $\mathcal{C}$

$$|z_1 z_2| = |z_1| \cdot |z_2|$$

but equality would be too restrictive for matrices.

A matrix norm is said to be *compatible* with a (vector) norm if

$$\|\mathbf{Ax}\| \leqq \|\mathbf{A}\| \cdot \|\mathbf{x}\| \tag{3.54}$$

holds for all $m \times n$ $\mathbf{A}$ and $\mathbf{x} \in \mathcal{C}^n$. Notice that the symbol $\|\cdot\|$ represents two different norms, one for vectors and another for matrices. This is typical in working with norms but there should be no difficulty because the different norms can always be identified from context. It can be verified that not all norms are compatible, but there is a straightforward method of deriving a compatible matrix norm for any given norm. We must have

$$\frac{\|\mathbf{Au}\|}{\|\mathbf{u}\|} \leqq \|\mathbf{A}\|$$

for all $\mathbf{u} \neq \mathbf{0}$. This states that $\|\mathbf{A}\|$ is an upper bound so we define the *natural norm* associated with any norm (3.44) as

$$\|\mathbf{A}\| = \sup\{\|\mathbf{Ax}\| : \|\mathbf{x}\| = 1\} \tag{3.55}$$

The geometric interpretation is that $\|\mathbf{A}\|$ represents the maximum relative "stretching" of nonzero vectors under

$$\mathbf{x} \mapsto \mathbf{Ax}$$

In Chapter 6 we develop the properties of matrix norms we need to use.

EXAMPLE 3.1. Particular Norms. We mainly use two norms in this book: the familiar (3.43) called the *standard norm*

$$\|\mathbf{v}\| = (\mathbf{v}^H \mathbf{v})^{\frac{1}{2}}$$

and the associated *standard natural norm* (3.55) for matrices. However, there are norms that are easier to compute and, according to what we find in (6.25) and (6.30), we are free to use any convenient norm in working with limits. Verifications for the following important cases are provided in 6.1.9.

(1) The most common norm (3.43) is at times denoted by $\|\cdot\|_2$ because

$$\|\mathbf{v}\|_2 = \left(|v_1|^2 + |v_2|^2 + \cdots + |v_n|^2\right)^{\frac{1}{2}}$$

exhibits the special role of the integer $p=2$. This norm directly generalizes the familiar Euclidean formula for distance in $\mathcal{R}^n$.

(2) The case $p=1$

$$\|\mathbf{v}\|_1 = |v_1| + |v_2| + \cdots + |v_n|$$

is called the *absolute norm.*

(3) The limiting case $p=\infty$

$$\|\mathbf{v}\|_\infty = \max\{|v_i| : i = 1, 2, \ldots, n\}$$

is called the *maximum norm.* If we apply this norm to $\mathbf{A}$, treated as a vector of dimension mn it fails to be compatible in the sense of (3.54).

(4) The standard natural (matrix) norm associated with (1) is also called the *spectral norm.*
(5) The natural norm associated with (2) is

$$\|\mathbf{A}\|_1 = \max\left\{ \sum_{i=1}^{m} |a_{ij}| : j=1,2,\ldots,n \right\}$$

(6) The natural norm associated with (3) is

$$\|\mathbf{A}\|_\infty = \max\left\{ \sum_{j=1}^{n} |a_{ij}| : i=1,2,\ldots,m \right\}$$

PROBLEMS FOR SOLUTION

3.4.1 Hermitian Linearity. (1) Evaluate $\langle a\mathbf{u}+b\mathbf{v},\mathbf{w}\rangle$. (2) Which of the following determine linear transformations for $\mathbf{x}\in\mathcal{C}^n$: $\langle\mathbf{x},\mathbf{x}\rangle, \langle\mathbf{a},\mathbf{x}\rangle, \langle\mathbf{x},\mathbf{b}\rangle, \mathbf{Ax}, \mathbf{x}^H\mathbf{B}$? (3) Repeat for $\mathcal{R}^n$.

3.4.2 Other Inner Products. Show that the following expressions satisfy the conditions in (3.41). (They are used for differential equations, difference equations, and optimization, respectively.) (1) $\langle\mathbf{f},\mathbf{g}\rangle = \int_{-1}^{1} f(t)g(t)\,dt$ on the vector space $\mathcal{V}$ of all continuous real-valued functions on $[-1,1]$. (2) $\langle\mathbf{f},\mathbf{g}\rangle = \sum_{i=1}^{k} f(i)g(i)$ on the vector space of all real-valued functions on $\{1,2,\ldots,k\}$. (3) $\langle\mathbf{u},\mathbf{v}\rangle = \mathbf{u}^H\mathbf{A}\mathbf{v}$ where $\mathbf{A}^H=\mathbf{A}$, and $\mathbf{z}^H\mathbf{A}\mathbf{z}\geqq 0$ for all $\mathbf{z}\in\mathcal{C}^n$ with $\mathbf{z}^H\mathbf{A}\mathbf{z}=0$ iff $\mathbf{z}=\mathbf{0}$.

3.4.3 Intrinsic Inner Product. Establish the following which, together with (2.11), shows that we need consider only one inner product and one unitary space $\mathcal{C}^n$ for each n.

Relative to an orthonormal basis B for a subspace $S\subset\mathcal{C}^n$, any inner product that satisfies the properties in (3.41) assumes the concrete form

$$\langle\mathbf{u},\mathbf{v}\rangle = \mathbf{x}^H\mathbf{y}$$

where $(\mathbf{u})_B=\mathbf{x}$ and $(\mathbf{v})_B=\mathbf{y}$.

3.4.4 Distance Function. A *distance function*, or *metric*, is a function $d:\mathcal{V}\times\mathcal{V}\to\mathcal{R}$ such that for all $\mathbf{a},\mathbf{b},\mathbf{c}\in\mathcal{V}$: (i) $d(\mathbf{a},\mathbf{b})=0$ iff $\mathbf{a}=\mathbf{b}$, (ii) $d(\mathbf{a},\mathbf{b})=d(\mathbf{b},\mathbf{a})$, (iii) triangle property $d(\mathbf{a},\mathbf{c})\leqq d(\mathbf{a},\mathbf{b})+d(\mathbf{b},\mathbf{c})$. Show that $d(\mathbf{a},\mathbf{b})=\|\mathbf{a}-\mathbf{b}\|$ qualifies as a distance function where $\|\cdot\|$ is any norm.

3.4.5 Orthogonality. Establish: (1) The only vector orthogonal to itself is $\mathbf{0}$. (2) $\langle \mathbf{u}, \mathbf{v} \rangle = 0$ iff $\langle \mathbf{v}, \mathbf{u} \rangle = 0$. (3) An orthogonal set is LI iff it does not contain $\mathbf{0}$. (4) The members of an orthogonal set $\{\mathbf{u}_1, \mathbf{u}_2, \ldots, \mathbf{u}_k\}$ can be multiplied by any set of nonzero scalars $\{a_1, a_2, \ldots, a_k\}$ and the resulting set $\{a_1\mathbf{u}_1, a_2\mathbf{u}_2, \ldots, a_k\mathbf{u}_k\}$ is orthogonal. (5) An orthogonal basis can always be replaced by an orthonormal basis. (6) If $\mathbf{u}$ is orthogonal to every member of a basis for S then (i) $\mathbf{u}$ is orthogonal to every $\mathbf{s} \in S$, and (ii) $\mathbf{u} \notin S$. (7) If $\{\mathbf{b}_1, \mathbf{b}_2, \ldots, \mathbf{b}_k\}$ is an orthonormal set then it can be augmented by $n-k$ vectors to form an orthonormal basis for $\mathcal{C}^n$. (8) If $u_1, u_2, \ldots, u_n$ are the coordinates of $\mathbf{u}$ with respect to an orthonormal basis then

$$|u_i| \leqq \|\mathbf{u}\| \leqq |u_1| + |u_2| + \cdots + |u_n|$$

for $i = 1, 2, \ldots, n$.

3.4.6 Alternative Gram–Schmidt Procedure. (1) Substantially modify (3.47) to simplify hand calculations. (2) Compare procedures for the columns of $\mathbf{A}$: $[1,1,1]^T, [0,2,4]^T, [2,2,-1]^T$.

3.4.7 Standard Orthonormal Basis. Define, and establish a procedure for finding, such a basis for any subspace $S \subset \mathcal{C}^n$.

3.4.8 Gramians. Establish: (1) If $U = \{\mathbf{u}_1, \mathbf{u}_2, \ldots, \mathbf{u}_k\}$ is LD in $\mathcal{C}^n$ then the determinant of the *Gramian*

$$\mathbf{G} = [\langle \mathbf{u}_i, \mathbf{u}_j \rangle] \qquad \text{which is order } k$$

must vanish. (2) If $\det \mathbf{G} = 0$ then U is LD. (3) U is LI iff $\det \mathbf{G} \neq 0$.

3.4.9 Orthogonal Complements. Establish for subspaces S in $\mathcal{C}^n$: (1) $S \cap S^\perp = \{\mathbf{0}\}$. (2) $S^\perp$ is a subspace. (3) $(S^\perp)^\perp = S$. (4) $\dim S^\perp = n - \dim S$. (5) Result (3.50). (6) Each $\mathbf{p} \in \mathcal{C}^n$ has a unique representation $\mathbf{p} = \mathbf{q}_0 + (\mathbf{p} - \mathbf{q}_0)$ where $\mathbf{q}_0 \in S$ and $\mathbf{p} - \mathbf{q}_0 \in S^\perp$. (7) Each $\mathbf{p} \in \mathcal{C}^n$ has the *Fourier expansion*

$$\mathbf{p} = \mathbf{q}_0 + \sum_{j=1}^{k} \langle \mathbf{b}_j, \mathbf{p} \rangle \mathbf{b}_j$$

where $\{\mathbf{b}_1, \mathbf{b}_2, \ldots, \mathbf{b}_k\}$ is an orthonormal basis for S. (8) Bessel's inequality. $\sum_{j=1}^{k} |\langle \mathbf{b}_j, \mathbf{p} \rangle|^2 \leqq \|\mathbf{p}\|^2$.

4
RECTANGULAR MATRICES IN EUCLIDEAN SPACE

We pass freely back and forth between two geometric settings for $m \times n$ matrices $\mathbf{A}$ with real components

$$\text{The columns of } \mathbf{A} \text{ as } n \text{ vectors in } \mathcal{R}^m$$

$$\text{The columns of } \mathbf{A}^T \text{ as } m \text{ vectors in } \mathcal{R}^n$$

to develop mathematics for linear programming and related processes of optimization. Widespread activity in these areas dates from the late 1940s although the historical role of Kantorovich (1939) in problem formulation is noteworthy. Theoretical foundations go back much further. Considerable credit must go to Gale (1960) where the results of this chapter were first brought together.

4.1 GEOMETRIC INTERPRETATIONS

Let us generalize lines and planes from analytic geometry and also introduce the sets shown in Table 4.1.

All matrices in this chapter are *real matrices* in the sense that their elements are real numbers. A vector space having real numbers as scalars and an inner product satisfying (3.41) is called a *Euclidean space*. Treating $\mathcal{R}$ as a special case of $\mathcal{C}$ in the manner prescribed for (3.39) we find that $\mathcal{R}^n$ together with (3.40) is a Euclidean space, and furthermore, we need consider no other (3.4.3). Hereafter we take the inner product (3.40) as given and use the term

$$\text{Euclidean space } \mathcal{R}^n \tag{4.1}$$

TABLE 4.1 CONDITIONS FOR y TO BELONG TO A CONVEX SET

Relation	Convex Set	Reference
$\mathbf{y}=w_1\mathbf{p}+w_2\mathbf{q}$	Line segment	Section 4.1
$\mathbf{y}=\mathbf{p}+t\mathbf{q},\ t\in\mathcal{R}$	Line	(4.7)
$\mathbf{y}=\mathbf{p}+t\mathbf{q},\ t\geqq 0$	Halfline	Section 4.1
$\langle\mathbf{a},\mathbf{y}\rangle=b$	Hyperplane	(4.8)
$\langle\mathbf{a},\mathbf{y}\rangle\leqq b$	Halfspace	(4.12)
$\mathbf{y}\geqq\mathbf{0}$	Nonnegative orthant	Section 4.1
$\mathbf{A}^T\mathbf{y}=\mathbf{c}$	k-Plane	(4.10)
$\mathbf{A}^T\mathbf{y}=\mathbf{c},\ \mathbf{y}\geqq\mathbf{0}$	Solution set	(4.47)
$\mathbf{A}^T\mathbf{y}=\mathbf{0}$	Subspace	Section 4.1
Special LCs	k-Simplex	(4.17)
$\mathbf{y}=\mathbf{A}\mathbf{x},\ \mathbf{x}\geqq\mathbf{0},\ \langle\mathbf{e},\mathbf{x}\rangle=1$	Convex polyhedron	(4.20)
	Polytope	(4.61)
$\mathbf{A}^T\mathbf{y}\leqq\mathbf{c}$	Polytope	(4.21)
	Convex polyhedron if bounded	(4.63)
$\mathbf{A}^T\mathbf{y}\leqq\mathbf{c},\ \mathbf{y}\geqq\mathbf{0}$	Solution set	(4.45)
$\mathbf{y}=\mathbf{A}\mathbf{x},\ \mathbf{x}\geqq\mathbf{0}$	Polyhedral convex cone	(4.24)
	Polytope	(4.60)
$\mathbf{A}^T\mathbf{y}=\mathbf{0},\ \mathbf{y}\geqq\mathbf{0}$	Polyhedral convex cone	(4.52)
	Polytope	(4.60)
$\mathbf{A}^T\mathbf{y}\leqq\mathbf{0}$	Polyhedral convex cone	(4.54)
	Polytope	(4.60)

interchangeably with "vector space $\mathcal{R}^n$." Now the inner product

$$\langle\mathbf{u},\mathbf{v}\rangle=\mathbf{u}^T\mathbf{v} \tag{4.2}$$

has real values and remains positive definite. But in place of the remaining properties in (3.41), it is symmetric, and linear in both arguments. Unless some other norm is specified, we continue to use (3.43) in which case the distance $\|\mathbf{u}-\mathbf{v}\|$ is the same as used in elementary analytic geometry. Whenever $\langle\mathbf{u},\mathbf{v}\rangle=0$ we continue to call $\mathbf{u}$ and $\mathbf{v}$ *orthogonal*, but because $\langle\mathbf{u},\mathbf{v}\rangle$ is real and $\mathcal{R}$ is ordered, we can fully exploit the concept of angle. In fact there are four familiar possibilities in (3.46).

The angle θ between $\mathbf{u},\mathbf{v}\in\mathcal{R}^n$ is called
(1) acute if $\langle\mathbf{u},\mathbf{v}\rangle\geqq 0$
(2) strictly acute if $\langle\mathbf{u},\mathbf{v}\rangle>0$
(3) obtuse if $\langle\mathbf{u},\mathbf{v}\rangle\leqq 0$
(4) strictly obtuse if $\langle\mathbf{u},\mathbf{v}\rangle<0$. (4.3)

This is because $\theta \in [0,\pi/2]$, $[0,\pi/2)$, $[\pi/2,\pi]$, $(\pi/2,\pi]$, respectively.

Let us recall some geometric interpretations for $n=2$ where $\mathfrak{R}^2$ coincides with the set of all "position" vectors in the plane as introduced in Example 2.1.

Figure 4.1(a) illustrates the familiar process of "completing the parallelogram" with broken lines to indicate $\mathbf{x}+\mathbf{y}$; also illustrated is

$$\mathbf{x}-\mathbf{y}=(-1)(\mathbf{y}-\mathbf{x})$$

which is a particular case of the relationship between $\mathbf{x}$ and the scalar multiple $k\mathbf{x}$. Two additional examples, for $k=-1/2$ and $k=2$, are given in

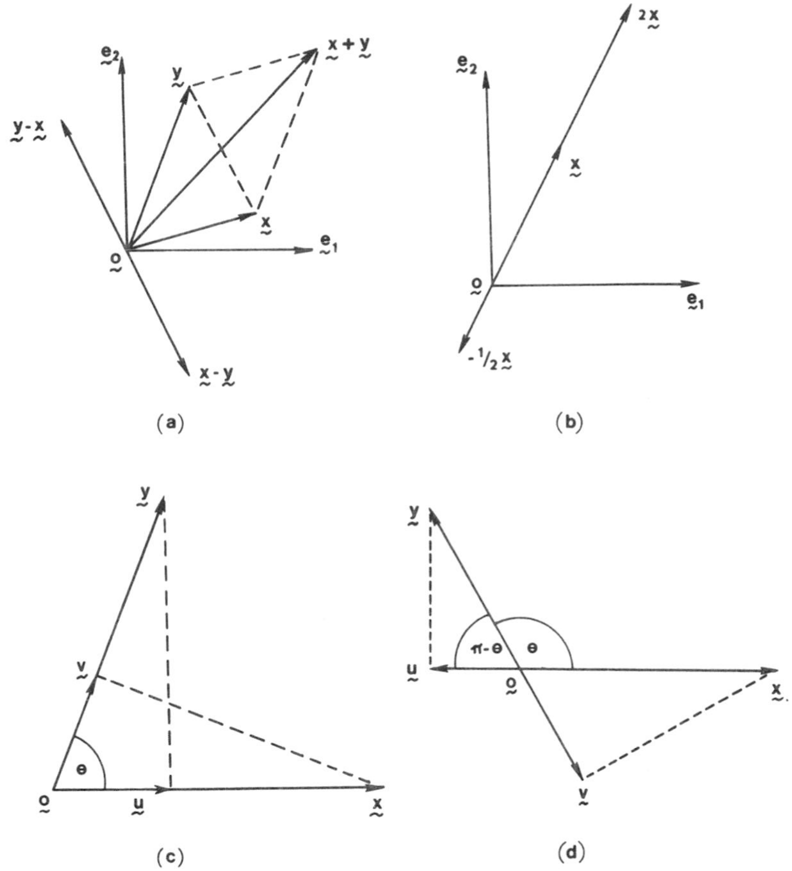

Figure 4.1. Geometric relations in $\mathfrak{R}^2$.

(b). All vectors $k\mathbf{x}$ are on the line from $\mathbf{0}$ through $\mathbf{x}$ if $k>0$ and on its extension through $-\mathbf{x}$ if $k<0$. In (c) the angle θ between $\mathbf{x}$ and $\mathbf{y}$ is an acute angle. The vector $\mathbf{u}$ is the familiar projection of $\mathbf{y}$ on $\mathbf{x}$ and $\mathbf{v}$ is the projection of $\mathbf{x}$ on $\mathbf{y}$. Vectors $\mathbf{x}$ and $\mathbf{y}$ in (d) form an obtuse angle θ. There $\mathbf{u}$ is the projection of $\mathbf{y}$ on $\mathbf{x}$ as before but it is oriented in direction exactly opposite to $\mathbf{x}$. Similarly, $\mathbf{v}$ is a negative multiple of $\mathbf{y}$.

Figure 4.2 illustrates projections on subspaces in $\mathcal{R}^2$ and indicates the pattern for any $\mathcal{R}^n$. The only proper subspaces in $\mathcal{R}^2$ are lines through the origin, S_1 represents such a subspace, and $\mathbf{p}\notin S_1$. Now S_1 has the orthonormal basis $\{\mathbf{c}\}$, and the orthogonal complement

$$S_1^{\perp}=\{\mathbf{x}:\mathbf{x}=t\mathbf{d},\ t\in\mathcal{R}\}$$

is also a line as shown. Vectors $\mathbf{c}$, $\mathbf{d}$, and the familiar $\mathbf{e}_1$ and $\mathbf{e}_2$, lie on the *unit circle*. Then $\mathbf{q}_0$ is the projection of $\mathbf{p}$ on S_1 while $\mathbf{p}-\mathbf{q}_0$ is its projection on $S_1^{\perp}$. The representation

$$\mathbf{p}=\mathbf{q}_0+(\mathbf{p}-\mathbf{q}_0)$$

is portrayed as the sum of the two vectors: $\mathbf{q}_0=\langle\mathbf{p},\mathbf{c}\rangle\mathbf{c}$ and $\mathbf{p}-\mathbf{q}_0=\langle\mathbf{p},\mathbf{d}\rangle\mathbf{d}$. We also see that S_1 has the alternate expression

$$S_1=\{\mathbf{x}:\langle\mathbf{x},\mathbf{d}\rangle=0\}$$

which exhibits $\mathbf{d}$ as the *unit normal*.

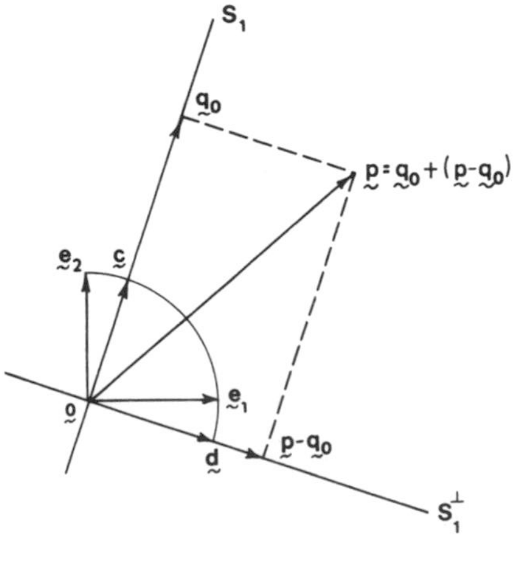

Figure 4.2. Projections in $\mathcal{R}^2$.

There are three distinct types of proper subspaces of $\mathcal{R}^n$.

1. subspaces of dimension 1 consisting of all multiples of some nonzero vector (to be called lines through the origin),
2. subspaces of dimension r where $2 \leqq r \leqq n-2$,
3. subspaces of dimension $n-1$ (to be called hyperplanes through the origin).

Before investigating their individual properties let us first note that any subspace S can be represented as $S = \text{Range}\,\mathbf{A}$ where $\mathbf{A}$ is an appropriate matrix: its columns must span S and we can say that $\mathbf{A}$ is $n \times p$ where $p \geqq \dim S$. It is convenient to have $p = \dim S$ so $\mathbf{A}$ is the matrix of a basis for S, but a fundamental result expressed in terms of any $\mathbf{A}$ is the

THEOREM. If $\mathbf{A}$ has m rows then

(1) $(\text{Range}\,\mathbf{A})^\perp = \text{Null}\,\mathbf{A}^T$.
(2) $\text{Range}\,\mathbf{A} \oplus \text{Null}\,\mathbf{A}^T = \mathcal{R}^m$.
(3) $\text{Null}\,\mathbf{A} = (\text{Range}\,\mathbf{A}^T)^\perp$. (4.4)

PROOF. If $\mathbf{x}$ belongs to the LHS of (1) then in particular $\mathbf{x}$ is orthogonal to every column of $\mathbf{A}$

$$\langle \mathbf{a}_j, \mathbf{x} \rangle = 0 \qquad \text{for all } j$$

that is, $\mathbf{A}^T\mathbf{x} = \mathbf{0}$. Conversely, if $\mathbf{x} \in \text{Null}\,\mathbf{A}^T$ then $\mathbf{x}$ is orthogonal to every column of $\mathbf{A}$ and to all LCs of these columns. Therefore (1) holds and (2) follows from (3.50). Statement (3) results from using transposes in (1). ■
Another result that is used in working with sets of vectors in $\mathcal{R}^n$ is (4.1.6)

If $\mathbf{A}$ has full column rank then $\mathbf{A}^T\mathbf{A}$ is nonsingular. (4.5)

PROOF. If $\mathbf{A}$ is $m \times n$ then $\text{Null}\,\mathbf{A} = \{\mathbf{0}\}$ because $\text{Range}\,\mathbf{A}^T$ has dimension n. But if $\mathbf{x} \in \text{Null}\,\mathbf{A}^T\mathbf{A}$ then also

$$\mathbf{x}^T\mathbf{A}^T\mathbf{A}\mathbf{x} = \langle \mathbf{A}\mathbf{x}, \mathbf{A}\mathbf{x} \rangle = 0$$

whereby $\mathbf{A}\mathbf{x} = \mathbf{0}$ by positive definiteness and we conclude rank $(\mathbf{A}^T\mathbf{A})^T = n$. ■

The *directed line* from $\mathbf{p}$ through $\mathbf{q} \neq \mathbf{p}$ is associated with the vector $\mathbf{q} - \mathbf{p}$ as a generalization of the situation illustrated in Figure 4.1(a) for $\mathcal{R}^2$. In particular, if $\mathbf{p} = \mathbf{0}$ then the vector $\mathbf{q}$ itself is associated with the directed line, or *ray*, from the origin $\mathbf{0}$ through $\mathbf{q}$. Most often we work with vectors as "points" but when we work with them as "rays" we use the following terminology. When $\mathbf{q} \neq \mathbf{0}$ we use *the direction* $\mathbf{q}$, or the *direction of* $\mathbf{q}$, to refer to the set

$$\{\mathbf{x} : \mathbf{x} = t\mathbf{q}, t \geqq 0\}$$

We also use the terminology *halfline through* $\mathbf{q}$ for this set.

In elementary analytic geometry it is shown that the coordinates (x,y) of points on the line through two distinct points, (x_1,y_1) and (x_2,y_2), satisfy

$$x=(1-t)x_1+tx_2$$

$$y=(1-t)y_1+ty_2$$

where $t\in\mathfrak{R}$. This extends to $\mathfrak{R}^n$ as the

Definition. Two-Point Form of a line. If $\mathbf{p}_1,\mathbf{p}_2\in\mathfrak{R}^n$ are distinct then this pair determines the *line*

$$W^1=\{\mathbf{x}:\mathbf{x}=w_1\mathbf{p}_1+w_2\mathbf{p}_2,\ w_1+w_2=1\} \tag{4.6}$$

where the letter W is used to suggest "weighted." Generalizing the situation in $\mathfrak{R}^2$, we call the subset of W^1 corresponding to $w_1,w_2\in[0,1]$ the *line segment* between $\mathbf{p}_1$ and $\mathbf{p}_2$ and $\mathbf{p}_1,\mathbf{p}_2$ are its *endpoints*. If we write $w_2=t$ and $w_1=1-t$ we obtain an instance of the

Definition. Parametric Form of a Line. If $\mathbf{p}_1,\mathbf{q}_1\in\mathfrak{R}^n$ and $\mathbf{q}_1\neq\mathbf{0}$ then this pair determines the *line*

$$P^1=\{\mathbf{x}:\mathbf{x}=\mathbf{p}_1+t\mathbf{q}_1,\ t\in\mathfrak{R}\}$$

passing through $\mathbf{p}_1$ and $\mathbf{p}_1+\mathbf{q}_1$. (4.7)

The following also represents a line in $\mathfrak{R}^2$

$$H^1=\{\mathbf{x}:\langle\mathbf{a},\mathbf{x}\rangle=b\}$$

In order to extend to $\mathfrak{R}^n$ let us introduce the

Definition. To each pair, $\mathbf{a}\neq\mathbf{0}$ and $b\in\mathfrak{R}$, there corresponds

$$H^{n-1}=\{\mathbf{x}:\langle\mathbf{a},\mathbf{x}\rangle=b\}$$

called the *hyperplane* determined by that pair. (4.8)

The vector $\mathbf{a}$ is called *orthogonal*, or *normal*, *to the hyperplane*. This means that if $\mathbf{x},\mathbf{y}\in H^{n-1}$ then $\langle\mathbf{a},\mathbf{x}-\mathbf{y}\rangle=0$ so that unless $b=0$, $\mathbf{a}$ is not orthogonal to vectors in H^{n-1} but is instead orthogonal to the difference of any two such vectors. The vectors $\pm\|\mathbf{a}\|^{-1}\mathbf{a}$ are called the *unit normals* to the hyperplane and the distance to the origin equals $|b|/\|\mathbf{a}\|$. In general we have the

Definition. Hyperplane Form of a Line. If $\mathbf{A}$ is $(n-1)\times n$, rank $\mathbf{A}=n-1$, and $\mathbf{p}_1 \in \mathfrak{R}^n$, then this pair determines the *line*

$$H^1=\{\mathbf{x}:\mathbf{Ax}=\mathbf{Ap}_1\}$$

passing through $\mathbf{p}_1$ and all $\mathbf{p}_1+\mathbf{q}$ for $\mathbf{q}\in \text{Null}\,\mathbf{A}$. (4.9)

We use the letter H because of the use of the hyperplanes corresponding to the rows of $\mathbf{A}$ and corresponding elements of $\mathbf{Ap}_1$. If $\mathbf{p}_1$ and $\mathbf{p}_2$ are distinct points in H^1 then

$$\mathbf{A}(\mathbf{p}_2-\mathbf{p}_1)=\mathbf{b}-\mathbf{b}=\mathbf{0}$$

shows that $\mathbf{p}_2-\mathbf{p}_1$ forms a basis for Null $\mathbf{A}$.

Guided by lines for $k=1$ and hyperplanes for $k=n-1$, we use the

Definition. For $k=0$ and $k=n$ there are the planes $H^0=\{\mathbf{0}\}$ and $H^n=\mathfrak{R}^n$. If $1 \leqq k \leqq n-1$, $\mathbf{A}$ is $(n-k)\times n$, $\mathbf{b}\in\mathfrak{R}^{n-k}$, and rank $\mathbf{A}=n-k$, then this pair determines the *k-plane*

$$H^k=\{\mathbf{x}:\mathbf{Ax}=\mathbf{b}\} \tag{4.10}$$

Other names in use are *flats* and *affine subspaces of dimension k*. Let us include part of (4.10) as the first one of

THREE FUNDAMENTAL FORMS OF k-PLANES. In working with k-planes in $\mathfrak{R}^n$ we lose no generality by selecting one of the following where $1 \leqq k \leqq n-1$.

(1) Hyperplane form. If $\mathbf{A}$ is $(n-k)\times n$, $\mathbf{b}\in R^{n-k}$, and rank $\mathbf{A}=n-k$, then this pair determines

$$H^k=\{\mathbf{x}:\mathbf{Ax}=\mathbf{b}\}$$

(2) Parametric form. If $\mathbf{Q}$ is $n\times k$, $\mathbf{p}\in\mathfrak{R}^n$, and rank $\mathbf{Q}=k$, then this pair determines

$$P^k=\{\mathbf{x}:\mathbf{x}=\mathbf{p}+\mathbf{Qt},\ \mathbf{t}\in\mathfrak{R}^k\}$$

(3) $(k+1)-$Point form. If $\mathbf{P}=[\mathbf{p}_1,\mathbf{p}_2,\ldots,\mathbf{p}_{k+1}]$ where all $\mathbf{p}_i\in\mathfrak{R}^n$, and if

$$\text{rank}\left[\mathbf{p}_2-\mathbf{p}_1,\mathbf{p}_3-\mathbf{p}_1,\ldots,\mathbf{p}_{k+1}-\mathbf{p}_1\right]=k$$

then the $k+1$ points $\mathbf{p}_i$ determine

$$W^k=\{\mathbf{x}:\mathbf{x}=\mathbf{P}\mathbf{w},\ \mathbf{w}\in\mathfrak{R}^{k+1},\ \langle\mathbf{w},\mathbf{e}\rangle=1\} \tag{4.11}$$

See 4.1.9 for the proof.

If M^k denotes a k-plane given in any one of the three forms, there is the unique *associated subspace*

$$S^k=\operatorname{Null}\mathbf{A}=\operatorname{Range}\mathbf{Q}$$

From the parametric form, if $\mathbf{x},\mathbf{y}\in P^k$ then $\mathbf{x}-\mathbf{y}\in S^k$. We call $k=\dim S^k$ the *dimension* of the k-plane even though M^k may not be a subspace of $\mathfrak{R}^n$. We can use the parametric form and write

$$M^k=\mathbf{p}+S^k$$

and refer to M^k as the result of *translating* S^k *by* $\mathbf{p}$. Then it is appropriate to refer to M^k and S^k as being *parallel* and to define M^j and M^k to be *parallel* for $j\neq k$ if the associated subspace for one is a subset of the associated subspace for the other. Two k-planes having points in common are *intersecting* and two nonparallel, nonintersecting planes are called *skew*.

The representation of M^k as a translation of a subspace S^k explains the name "affine subspace": An affine transformation $M:\mathfrak{R}^n\to\mathfrak{R}^n$ has images of the form $M\mathbf{x}=S\mathbf{x}+\mathbf{p}$ where $S:\mathfrak{R}^n\to\mathfrak{R}^n$ is linear and $\mathbf{p}\in\mathfrak{R}^n$. This terminology, and the name "flat," is based on the characteristic property: A k-plane contains all points on the line determined by any two of its points (4.1.21).

Let us next generalize the manner in which any line divides the plane into two parts. We start with the

Definition. If $\mathbf{a}\neq\mathbf{0}$ and $b\in\mathfrak{R}$ then this pair determines the *closed halfspaces* $\{\mathbf{x}:\langle\mathbf{a},\mathbf{x}\rangle\leqq b\}$ and $\{\mathbf{x}:\langle\mathbf{a},\mathbf{x}\rangle\geqq b\}$, and the *open halfspaces* $\{\mathbf{x}:\langle\mathbf{a},\mathbf{x}\rangle<b\}$ and $\{\mathbf{x}:\langle\mathbf{a},\mathbf{x}\rangle>b\}$ in $\mathfrak{R}^n$. (4.12)

The situation is reminiscent of the manner in which any $r\in\mathfrak{R}$ determines four intervals. Figure 4.3(a) provides an illustration for $\mathbf{a}=\mathbf{p}-\mathbf{q}_0\in\mathfrak{R}^2$ which is orthogonal to every vector in

$$H=\{\mathbf{x}:\langle\mathbf{x},\mathbf{p}-\mathbf{q}_0\rangle=0\} \tag{4.13}$$

It is geometrically evident that every vector in $\mathfrak{R}^2$ that makes an obtuse angle with $\mathbf{p}-\mathbf{q}_0$ is on the opposite side of H from $\mathbf{p}-\mathbf{q}_0$. These vectors are denoted by

$$H^- = \{\mathbf{x} : \langle \mathbf{x}, \mathbf{p} - \mathbf{q}_0 \rangle \leqq 0\}$$

which is a closed halfspace in $\mathcal{R}^2$. Notice $H \subset H^-$. The complement of H^- is the open halfspace

$$H^+ = \{\mathbf{x} : \langle \mathbf{x}, \mathbf{p} - \mathbf{q}_0 \rangle > 0\}$$

that contains all vectors making strictly acute angles with $\mathbf{p} - \mathbf{q}_0$. Corresponding halfspaces for any nonzero vector **a** are suggested by (b). Again M^+ consists of all vectors on the same side of the hyperplane M as **a** and $\mathcal{R}^n = M^+ \cup M^-$. We also have the

ELEMENTARY THEOREM OF THE SEPARATING HYPERPLANE. If $S \subset \mathcal{R}^n$ is a subspace and $\mathbf{p} \notin S$ then there exists a hyperplane H that determines a closed halfspace H^- containing S and an open halfspace H^+ containing **p**. (4.14)

PROOF. Given **p** we find $\mathbf{q}_0$ such that $S \subset H$ where H is the hyperplane (4.13). Defining H^- and H^+ as shown above we verify

$$\langle \mathbf{p}, \mathbf{p} - \mathbf{q}_0 \rangle > 0$$

by expanding $\langle \mathbf{p} - \mathbf{q}_0, \mathbf{p} - \mathbf{q}_0 \rangle$. ■

The hyperplane H is called a *separating hyperplane* since **p** is on one side and S is on the other.

One of the most distinctive properties of $\mathcal{R}^n$ is that it possesses an extremely useful order relation. Using our customary notation for components

$$\mathbf{x} \geqq \mathbf{y} \text{ means } x_i \geqq y_i \text{ for all } i$$

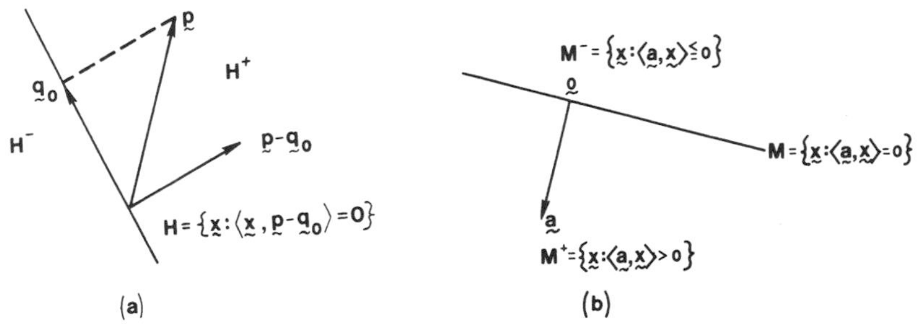

Figure 4.3. Halfspaces in $\mathcal{R}^2$.

which generalizes the ordering of $\mathcal{R}^2$ where "greater than" means "to the right and above." It is at times convenient to use this ordering for one-row matrices $\mathbf{x}^T, \mathbf{y}^T, \ldots$ but we do not use it for other types of matrices. We continue to read "$\mathbf{x}$ is greater than or equal to $\mathbf{y}$," we write $\mathbf{y} \leqq \mathbf{x}$ to mean the same as $\mathbf{x} \geqq \mathbf{y}$, and $\mathbf{x} > \mathbf{y}$ means the same as $\mathbf{y} < \mathbf{x} : x_i > y_i$ for $i = 1, 2, \ldots, n$. But now we find it convenient to use a new symbol

$$\mathbf{x} \geq \mathbf{y} \text{ means } \mathbf{x} \geqq \mathbf{y} \text{ and } \mathbf{x} \neq \mathbf{y}$$

Generally, we are interested in comparisons with the zero vector as follows:

$\mathbf{x}$ is *positive*, $\mathbf{x} > \mathbf{0}$, means all $x_i > 0$
$\mathbf{x}$ is *nonnegative*, $\mathbf{x} \geqq \mathbf{0}$, means all $x_i \geqq 0$
$\mathbf{x}$ is *semipositive*, $\mathbf{x} \geq \mathbf{0}$, means $\mathbf{x} \geqq \mathbf{0}$ and $\mathbf{x} \neq \mathbf{0}$

The terms *negative*, *nonpositive*, and *seminegative* have the corresponding definitions where $>$ is replaced by $<$. The collection of positive vectors

$$\{\mathbf{x} : \mathbf{x} > \mathbf{0}\}$$

is called the *positive orthant*, $\{\mathbf{x} : \mathbf{x} \geqq \mathbf{0}\}$ is the *nonnegative orthant*, and so on, in generalization of the familiar quadrants in $\mathcal{R}^2$ and octants in $\mathcal{R}^3$. Finally, let us consider an arbitrary $S \subset \mathcal{R}^n$. We call S: *bounded from below* if there exists $\mathbf{a} \in \mathcal{R}^n$ such that $\mathbf{a} \leqq \mathbf{s}$ for all $\mathbf{s} \in S$; *bounded from above* if there exists $\mathbf{b} \in \mathcal{R}^n$ such that $\mathbf{s} \leqq \mathbf{b}$ for all $\mathbf{s} \in S$; *bounded* if it is bounded from below and from above; and, *unbounded* if it is not bounded.

The simple concept of a convex set is one of the most important in optimization theory: Given any two of its points, the line segment joining them lies entirely within the set. Using the two-point form we have the

Definition. A set $K \subset \mathcal{R}^n$ is *convex* if $\mathbf{p}, \mathbf{q} \in K$ and $t \in [0, 1]$ imply $t\mathbf{p} + (1 - t)\mathbf{q} \in K$. (4.15)

Examples of convex sets are $\mathcal{R}^n, \varnothing, \{\mathbf{p}\}$ for any $\mathbf{p} \in \mathcal{R}^n$, line segments, subspaces, halfspaces, and k-planes. Two nonconvex sets are shown in Figure 4.4.

A *convex combination* of $X = \{\mathbf{x}_1, \mathbf{x}_2, \ldots, \mathbf{x}_r\} \subset \mathcal{R}^n$ is a LC

$$\mathbf{x} = t_1\mathbf{x}_1 + t_2\mathbf{x}_2 + \cdots + t_r\mathbf{x}_r$$

where $t_i \in [0, 1]$ for all i and $t_1 + t_2 + \cdots + t_r = 1$. If $r = 2$ and $\mathbf{x}_1 \neq \mathbf{x}_2$ the convex combinations are the points on the line segment between $\mathbf{x}_1$ and $\mathbf{x}_2$.

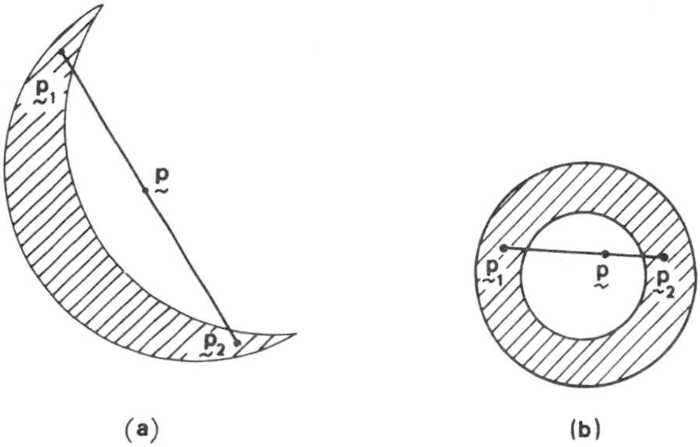

Figure 4.4. Two nonconvex sets.

In general, the *convex hull* of any subset $X \subset \mathfrak{R}^n$ is the union of X and the set of all convex combinations of points in X; the convex hull of the set in Figure 4.4(b) includes all points inside the larger circle. We need to establish that

The convex hull of any set is convex. (4.16)

PROOF. If $\mathbf{x}$ and $\mathbf{y}$ are in the convex hull of X, then using zero coefficients as necessary, we can write $\mathbf{x}$ as displayed previously and $\mathbf{y}$ as

$$\mathbf{y} = u_1\mathbf{x}_1 + u_2\mathbf{x}_2 + \cdots + u_r\mathbf{x}_r$$

Then clearly $\mathbf{z} = c\mathbf{x} + (1-c)\mathbf{y}$ also belongs to the convex hull of X for $c \in [0,1]$. ■

Distinguished examples of convex sets in $\mathfrak{R}^3$ are line segments, triangular areas, and tetrahedral volumes. Each of these geometric figures is a special case of the following

Definition. Let $\mathbf{p}_1, \mathbf{p}_2, \ldots, \mathbf{p}_{k+1} \in \mathfrak{R}^n$ where $k \leqq n$. If

$$\text{rank}\left[\mathbf{p}_2 - \mathbf{p}_1, \mathbf{p}_3 - \mathbf{p}_1, \ldots, \mathbf{p}_{k+1} - \mathbf{p}_1\right] = k$$

then the set X^k of all convex combinations of these $k+1$ points is called the *k-simplex* with *vertexes* $\mathbf{p}_1, \mathbf{p}_2, \ldots, \mathbf{p}_{k+1}$. (4.17)

Simplexes provide motivation for concepts introduced for general convex sets and they are used as special subsets in working with such sets. Simplexes are convex sets by (4.16). The set of vertexes also appears in

(4.11): For $k=1$ the points are distinct and X^1 is a line segment, for $k=2$ they are noncollinear and X^2 is a triangle together with its interior, in general they lie on no plane of dimension $k-1$, and X^k is said to be of *dimension* k because its vertexes determine a k-plane. Each point $\mathbf{x}\in X^k$ can be written uniquely as a convex combination

$$\mathbf{x}=t_1\mathbf{x}_1+t_2\mathbf{x}_2+\cdots+t_{k+1}\mathbf{x}_{k+1}$$

where $t_1, t_2, \ldots, t_{k+1}$ are called the *barycentric* (from center of mass) *coordinates* of $\mathbf{x}$. Use of the name vertex also comes from elementary geometric considerations: Each pair of $\mathbf{p}_i$ determines a line segment in X^k and no $\mathbf{p}_i$ lies on a line segment in X^k for which it is not an endpoint.

Let us use the preceding idea to define a vertex for any convex set. A point $\mathbf{p}\in K$, where K is convex, is a *vertex*, or *extreme point*, of K if

$$\mathbf{p}=t\mathbf{p}_1+(1-t)\mathbf{p}_2$$

for $\mathbf{p}_1, \mathbf{p}_2\in K$, and $t\in[0,1]$ implies $\mathbf{p}=\mathbf{p}_1$ or $\mathbf{p}=\mathbf{p}_2$. A line is an example of a convex set with no vertexes, a line segment has two, a rectangular region has four, and a circular disc has an infinite number.

The union of two convex sets is not necessarily convex. But it is readily verified that

The intersection of any number of convex sets is convex. (4.18)

A criterion of fundamental importance is

A set K is convex iff every convex combination of points in K belongs to K. (4.19)

PROOF. If K is convex then trivially every convex combination of $r=1$ point in K belongs to K. The inductive hypothesis is: Every convex combination of $r \geqq 2$, or fewer, points of K belong to K. Consider now the convex combination

$$\mathbf{p}=t_1\mathbf{p}_1+t_2\mathbf{p}_2+\cdots+t_r\mathbf{p}_r+t_{r+1}\mathbf{p}_{r+1}$$

where all $\mathbf{p}_i\in K$. If $t_{r+1}=1$ then $\mathbf{p}=\mathbf{p}_{r+1}\in K$. Otherwise, $t_{r+1}<1$ and $t=1-t_{r+1}=t_1+t_2+\cdots+t_r>0$. This means

$$\frac{t_1\mathbf{p}_1+t_2\mathbf{p}_2+\cdots+t_r\mathbf{p}_r}{t_1+t_2+\cdots+t_r}=\mathbf{q}$$

belongs to K because it is a convex combination of r or fewer points of K.

Finally,

$$\mathbf{p} = t\mathbf{q} + (1-t)\mathbf{p}_{r+1} \in K$$

because it is a convex combination of $2 \leqq r$ points of K. Conversely, the condition is sufficient because if every convex combination of points in K belongs to K then in particular it is true for any two points. ■

Sets in $\mathfrak{R}^n$ that are not necessarily simplexes, but which are generated by finding all convex combinations of a finite set of points, are specified as follows.

Definition. Let $\mathbf{A}^T = [\mathbf{a}_1, \mathbf{a}_2, \ldots, \mathbf{a}_m]$ be the matrix of any m points in $\mathfrak{R}^n$. Then the convex hull of these points

$$K = \{\mathbf{x} : \mathbf{x} = \mathbf{A}^T\mathbf{y},\ \mathbf{y} \geqq \mathbf{0},\ \langle \mathbf{e}, \mathbf{y} \rangle = 1\}$$

is called a *convex polyhedron*. (4.20)

It is common for authors to omit the modifier "convex" because K is necessarily convex and, presumably, because these are the polyhedra (figures formed by plane faces) that they mainly use. Where the preceding is generated by m points, the following comes from m halfspaces (or hyperplanes).

Definition. Suppose we have $\mathbf{A}^T = [\mathbf{a}_1, \mathbf{a}_2, \ldots, \mathbf{a}_m]$, where all $\mathbf{a}_i$ are nonzero vectors in $\mathfrak{R}^n$, and $\mathbf{b} \in \mathfrak{R}^m$. Then the intersection of the m associated closed halfspaces $\{\mathbf{x} : \langle \mathbf{a}_i, \mathbf{x} \rangle \leqq b_i\}$

$$K = \{\mathbf{x} : \mathbf{A}\mathbf{x} \leqq \mathbf{b}\}$$

is called a *polytope*. (4.21)

Polytopes are necessarily convex due to (4.18); however, they may be empty, bounded, or unbounded.

We call $\mathbf{A}\mathbf{x} \leqq \mathbf{b}$ a *system* of *linear inequalities*. If $\mathbf{b} \neq \mathbf{0}$ it is *nonhomogeneous*, and otherwise it is *homogeneous*. Thus, if $\mathbf{b} = \mathbf{0}$ in (4.21), K is the solution set of a homogeneous system of linear inequalities.

Another type of subset of $\mathfrak{R}^n$ that figures prominently in the present chapter is given by the

Definition. A subset $C \subset \mathfrak{R}^n$ is a *cone* if $\mathbf{c} \in C$ and $k \geqq 0$ imply $k\mathbf{c} \in C$. (4.22)

Subspaces, closed halfspaces where the bounding hyperplane passes through the origin, the nonnegative orthant, and a directed ray from $\mathbf{0}$ through $\mathbf{q}\neq\mathbf{0}$, are examples of cones. The *dimension* of a cone is the maximal number of LI vectors that can be assembled from points in the cone. The union of two rays, as illustrated in Figure 4.5(a) is a two-dimensional cone, and so is the shaded area in (b).

A cone may or may not be convex, that is, it may or may not be a *convex cone*. In Figure 4.5 the cones in (a) and (b) are not convex, but (c) illustrates a convex cone. Three important operations are defined for convex cones. The *sum* of two convex cones C_1 and C_2 is defined by

$$C_1+C_2=\{\mathbf{x}:\mathbf{x}=\mathbf{x}_1+\mathbf{x}_2;\ \mathbf{x}_1\in C_1,\ \mathbf{x}_2\in C_2\}$$

and is a convex cone. The double shaded area in (d) illustrates such a sum. Notice also that the cone C in (c) is the sum of two cones

$$\{\mathbf{x}:\mathbf{x}=a\mathbf{p},\ a\geqq 0\}+\{\mathbf{x}:\mathbf{x}=b\mathbf{q},\ b\geqq 0\}$$

namely, the two halflines, through $\mathbf{p}$ and through $\mathbf{q}$. Clearly, the *intersection* of two convex cones $C_1\cap C_2$ is a convex cone; an illustration of a nonempty intersection is given in (e). The third operation is illustrated in (f) and, unlike the preceding two, it is something new.

Definition. If C is a convex cone then

$$C^*=\left\{\mathbf{x}:\langle\mathbf{x},\mathbf{c}\rangle\leqq 0\quad\text{for all }\mathbf{c}\in C\right\}$$

is its *polar cone*. (4.23)

The cone C^* consists of all vectors $\mathbf{x}$ that make only obtuse angles with vectors $\mathbf{c}\in C$.

Now $C=(C^*)^*$, which holds in Figure 4.5(f), may fail for an arbitrary convex cone and this is one motivation for the

Definition. Let the columns of

$$\mathbf{A}^T=\left[\mathbf{a}_1,\mathbf{a}_2,\ldots,\mathbf{a}_m\right]$$

be any m points in $\mathfrak{R}^n$. Then the set

$$C=\left\{\mathbf{x}:\mathbf{x}=\mathbf{A}^T\mathbf{y},\ \mathbf{y}\geqq\mathbf{0}\right\}$$

is called a *polyhedral convex cone*. (4.24)

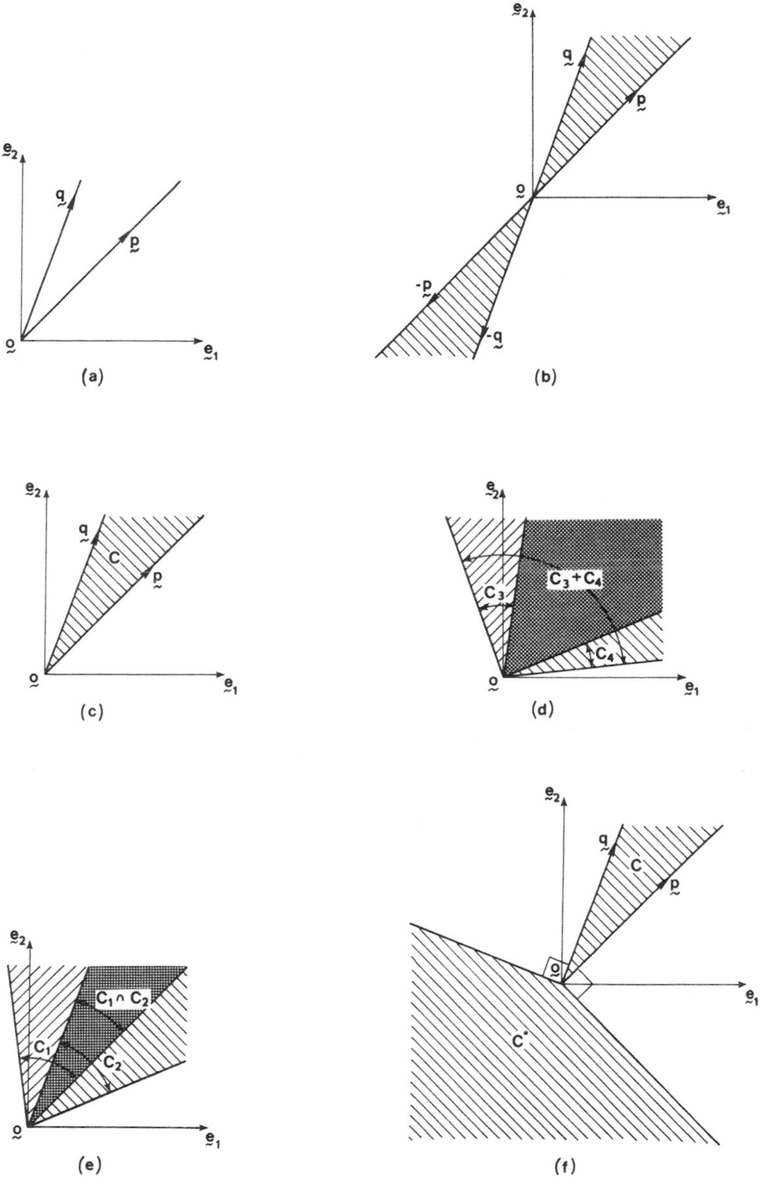

Figure 4.5. Cones in $\mathfrak{R}^2$.

We say that C is *generated* by the m points $\mathbf{a}_1,\ldots,\mathbf{a}_m$. Vectors $\mathbf{x}$ in (4.24) are vectors that can be expressed as LCs of columns of $\mathbf{A}^T$ using nonnegative coefficients; they are also RHSs for which nonnegative solutions $\mathbf{y}$ exist for $\mathbf{A}^T\mathbf{y}=\mathbf{x}$. Geometrically, let us notice that the convex polyhedron (4.20) can be viewed as a "cross section" of (4.24).

A vector $\mathbf{x}$ in a convex cone C is an *extreme vector* if it satisfies two conditions:

1. $\mathbf{x}\neq\mathbf{0}$.
2. $\mathbf{x}=\mathbf{x}_1+\mathbf{x}_2$ is impossible for all LI $\{\mathbf{x}_1,\mathbf{x}_2\}\subset C$.

The vectors $\mathbf{p}$ and $\mathbf{q}$ are extreme vectors of C in Figure 4.5(c). A subspace is an instance of a polyhedral convex cone that has no extreme vectors and neither has a halfspace in $\mathfrak{R}^n$ for $n>2$.

A cone is called *pointed* if it contains no subspace except $\{\mathbf{0}\}$, that is, if it contains no line. For example, in Figure 4.5 only the cone in (b) (which is not convex) is not pointed. In order to obtain a criterion we later use, let us call any set of vectors appearing as the columns of $\mathbf{A}^T$ *semipositively independent* if $\mathbf{A}^T\mathbf{y}=\mathbf{0}$ has no semipositive solution $\mathbf{y}$. Then

If the columns of $\mathbf{A}^T$ are semipositively independent then the associated polyhedral convex cone C is (1) pointed, and (2) the sum of its extreme vectors. (4.25)

PROOF.* Both $\mathbf{a}$ and $-\mathbf{a}$ cannot belong to such a cone unless $\mathbf{a}=\mathbf{0}$ so C is pointed. We can remove, one at a time, any $\mathbf{a}_i$ expressible as an LC, with nonnegative coefficients, of the remaining vectors. At the end we obtain a possibly renumbered set $\{\mathbf{a}_1,\mathbf{a}_2,\ldots,\mathbf{a}_p\}$ from which no vector can be removed, and which defines the same polyhedral convex cone. Let us show that each of these vectors is extreme. If one, say $\mathbf{a}_1$, were not we would have

$$\mathbf{a}_1=u_1\mathbf{a}_1+u_2\mathbf{a}_2+\cdots+u_p\mathbf{a}_p$$

where all $u_i\geqq 0$. If $0\leqq u_1<1$ then $\mathbf{a}_1$ would be an LC with nonnegative coefficients $u_i(1-u_1)^{-1}$ of the remaining $\mathbf{a}$'s. Similarly $1\leqq u_1$ would be impossible because then the full set of $\mathbf{a}$'s would not be semipositively independent. ■

In the proof of the theorem of Farkas we need to project into a polyhedral convex cone. Specifically, we have the

Problem. Given $\mathbf{t},\mathbf{u},\mathbf{v}\in\mathfrak{R}^n$ satisfying $\langle\mathbf{t},\mathbf{u}\rangle\leqq 0$ and $\langle\mathbf{u},\mathbf{v}\rangle>0$, find $\mathbf{t}'\neq\mathbf{0}$ such that

(1) $\langle\mathbf{t}',\mathbf{u}\rangle=0$, and
(2) $\mathbf{t}'$ belongs to the polyhedral convex cone determined by $\mathbf{t}$ and $\mathbf{v}$.

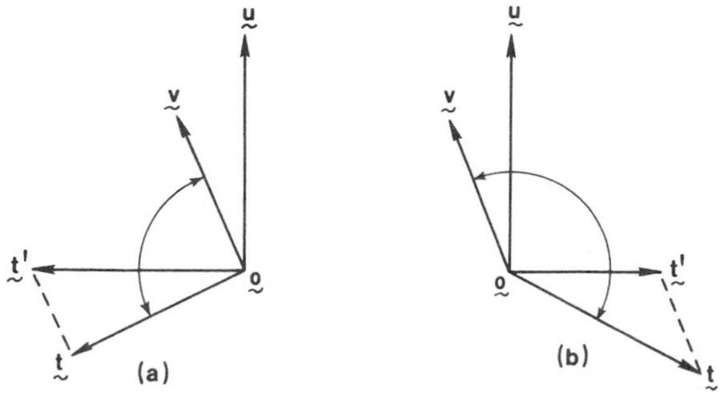

Figure 4.6. Projecting into a polyhedral convex cone.

Thus **t** and **u** form an obtuse angle while that between **u** and **v** is strictly acute. The required vector **t**′ must be orthogonal to **u** according to (1), and (2) requires that **t**′ must lie between **t** and **v** on the side where **t** and **v** form an angle not greater than π. There are two cases as illustrated for $\mathfrak{R}^2$ in Figure 4.6. The condition $\langle \mathbf{u},\mathbf{v}\rangle > 0$ guarantees $\mathbf{v} \neq \mathbf{0}$. Consequently, both coefficients in the LC must be positive and we can use $\mathbf{t}' = \mathbf{t} + k\mathbf{v}$ for some $k > 0$. Then requirement (1) implies $\langle \mathbf{t},\mathbf{u}\rangle + k\langle \mathbf{v},\mathbf{u}\rangle = 0$ from which we solve for k and obtain the solution

If $\langle \mathbf{t},\mathbf{u}\rangle \leqq 0$ and $\langle \mathbf{u},\mathbf{v}\rangle > 0$, then

$$\mathbf{t}' = \mathbf{t} - \frac{\langle \mathbf{t},\mathbf{u}\rangle}{\langle \mathbf{u},\mathbf{v}\rangle}\mathbf{v}$$

is (1) orthogonal to **u**, and (2) a member of the polyhedral convex cone determined by **t** and **v**. (4.26)

PROBLEMS FOR SOLUTION

4.1.1 Projections in $\mathfrak{R}^2$. Find an expression for the projection of **y** on **x** and determine its magnitude and direction.

4.1.2 Gram–Schmidt in $\mathfrak{R}^2$. Express Step 2 of (3.47) in geometric terms.

4.1.3 Elementary Properties. Establish in $\mathfrak{R}^n$:

(1) Parallelogram relation

$$\|\mathbf{x}+\mathbf{y}\|^2 + \|\mathbf{x}-\mathbf{y}\|^2 = 2(\|\mathbf{x}\|^2 + \|\mathbf{y}\|^2)$$

(2) $\|\mathbf{x}-\mathbf{y}\|^2 = \|\mathbf{x}\|^2 + \|\mathbf{y}\|^2 - 2\langle \mathbf{x},\mathbf{y}\rangle$.

(3) Pythagorean theorem. If $\langle \mathbf{x},\mathbf{y}\rangle = 0$ then

$$\|\mathbf{x}+\mathbf{y}\|^2 = \|\mathbf{x}\|^2 + \|\mathbf{y}\|^2$$

4.1.4 Standard bases. Solve where $\mathbf{A}$ is any nonzero matrix of order $m \times n$: (1) Use submatrices of the GJSFs of $[\mathbf{A},\mathbf{I}_m]$ and $[\mathbf{A}^T,\mathbf{I}_n]$ (Section 3.3) to find the standard bases for Range $\mathbf{A}$ and Null $\mathbf{A}$. (2) Continue by finding the standard bases for Range $\mathbf{A}^T$ and Null $\mathbf{A}^T$. (3) Find a matrix $\mathbf{B}$ such that the columns of $\mathbf{A}$ span Null $\mathbf{B}$, that is, Null $\mathbf{B}$ = Range $\mathbf{A}$.

4.1.5 Matrices of Full Rank. Prove or disprove: (1) If $\mathbf{B}$ has full row rank then $\mathbf{B}^T\mathbf{B}$ is nonsingular. (2) If $\mathbf{B}$ has full rank then $\mathbf{B}^T\mathbf{B}$ is nonsingular. (3) If $\mathbf{B}$ has full rank then at least one of $\mathbf{B}^T\mathbf{B}$ and $\mathbf{B}\mathbf{B}^T$ is nonsingular.

4.1.6 Orthogonal Projections. Establish: (1) The definition of an *orthogonal projection P into a subspace S* using the decomposition (3.50) or 4.4(2) whereby any $\mathbf{x} \in \mathfrak{R}^n$ can be written uniquely as $\mathbf{x}=\mathbf{y}+\mathbf{z}$. (2) P is linear. (3) If $\mathbf{B}$ is the matrix of a basis for S then P has matrix

$$\mathbf{P}=\mathbf{B}(\mathbf{B}^T\mathbf{B})^{-1}\mathbf{B}^T$$

(4) If $\mathbf{p} \notin S$ then $\mathbf{q}_0 = \mathbf{P}\mathbf{p}$ is the unique point in S that is closest to $\mathbf{p}$ in the sense of minimum norm. (5) Three additional methods for finding $\mathbf{q}_0$ given $\mathbf{p}$ and a basis for S and apply to $\mathbf{p}=[1,0,0,1,0]^T$ and the basis formed by $[1,1,0,0,1]^T$ and $[1,2,1,0,1]^T$. (6) A real matrix $\mathbf{P}$ of order n is the matrix of an orthogonal projection iff $\mathbf{P}$ is *idempotent*, $\mathbf{P}^2=\mathbf{P}$, and *symmetric*, $\mathbf{P}^T=\mathbf{P}$. (7) A more general projection can be defined. (8) Relate $\mathbf{P}$ to $\mathbf{A}^L$ in Example 7.1.

4.1.7 Lines in $\mathfrak{R}^n$. Establish: (1) Definitions for *identical*, *intersecting*, *parallel*, and *skew* lines. (2) Criteria for the four possibilities for two lines. (3) Points $\mathbf{p}_1$, $\mathbf{p}_2$, $\mathbf{p}_3$ are *collinear*, that is, they lie on a single line, iff $\{\mathbf{p}_2-\mathbf{p}_1,\mathbf{p}_3-\mathbf{p}_1\}$ is LD. (4) Criteria for collinearity using (4.7). (5) A systematic procedure for finding at least three collinear points, provided such exist, from among the columns of $\mathbf{P}=[\mathbf{p}_1,\mathbf{p}_2,\ldots,\mathbf{p}_k]$ and apply to

$$\begin{bmatrix} 0 & 1 & 0 & 2 & -1 \\ 1 & 2 & 1 & 3 & 0 \\ 1 & 3 & 2 & 3 & 1 \end{bmatrix}$$

(6) The three forms (4.6), (4.7), (4.9) are equivalent.

4.1.8 Hyperplanes in $\mathfrak{R}^n$. Establish: (1) The distance from (4.8) to $\mathbf{0}$ is $|b|/\|\mathbf{a}\|$. (2) The interrelations for the three forms for hyperplanes in (4.11) whereby given any one form we can write the other two. (3) The

Gram–Schmidt process can be used to find unit normals. (4) Criteria for two hyperplanes, (4.8) and $\{\mathbf{x}:\langle\mathbf{c},\mathbf{x}\rangle=d\}$ to be parallel. (5) Repeat for coincident. (6) Find at least two separating hyperplanes for $\mathbf{p}=[1,2,3,4]^T$ and Null $\mathbf{A}$ where

$$\mathbf{A}=\begin{bmatrix}1&0&0&1\\0&1&0&2\\0&0&1&3\end{bmatrix}$$

4.1.9 k-Planes. Establish:

(1) If $\mathbf{C}$ is $m\times n$, $\mathbf{d}\in\mathcal{R}^m$, rank $\mathbf{C}=$rank $[\mathbf{C},\mathbf{d}]=n-k$, and $1\leqq k\leqq n-1$ then the pair $\mathbf{C}$ and $\mathbf{d}$ determines the k-plane

$$U^k=\{\mathbf{x}:\mathbf{C}\mathbf{x}=\mathbf{d}\}$$

(2) If $\mathbf{D}$ is $n\times m$, $\mathbf{p}\in\mathcal{R}^n$, and rank $\mathbf{D}=k$ where $1\leqq k\leqq n-1$ then the pair $\mathbf{D}$ and $\mathbf{p}$ determines the k-plane

$$V^k=\{\mathbf{x}:\mathbf{x}=\mathbf{p}+\mathbf{D}\mathbf{t},\ \mathbf{t}\in\mathcal{R}^m\}$$

(3) Consider $\mathbf{P}=[\mathbf{p}_1,\mathbf{p}_2,\ldots,\mathbf{p}_{k+1}]$ where all $\mathbf{p}_i\in\mathcal{R}^n$ and $1\leqq k\leqq n-1$. If the associated matrix

$$\mathbf{Z}=[\mathbf{p}_2-\mathbf{p}_1,\mathbf{p}_3-\mathbf{p}_1,\ldots,\mathbf{p}_{k+1}-\mathbf{p}_1]$$

has full column rank k, then the $k+1$ points $\mathbf{p}_i$ determine the k-plane

$$W^k=\{\mathbf{x}:\mathbf{x}=\mathbf{P}\mathbf{w},\ \mathbf{w}\in\mathcal{R}^{k+1},\ \langle\mathbf{w},\mathbf{e}\rangle=1\}$$

(4) rank$[\mathbf{p}_1,\mathbf{p}_2,\ldots,\mathbf{p}_{k+1}]=k+1$ in (3) iff $\mathbf{0}\notin W^k$.
(5) rank$[\mathbf{p}_1,\mathbf{p}_2,\ldots,\mathbf{p}_{k+1}]=k$ iff $\mathbf{0}\in W^k$.
(6) Result (4.11).
(7) Procedures for passing from any one form in (4.11) to any other.
(8) The three forms in (4.11) are equivalent.
(9) The points $\{\mathbf{p}_1,\mathbf{p}_2,\ldots,\mathbf{p}_k\}$ lie on no j-plane for $j<k-1$ iff

$$\{\mathbf{p}_2-\mathbf{p}_1,\mathbf{p}_3-\mathbf{p}_1,\ldots,\mathbf{p}_k-\mathbf{p}_1\}$$

is LI.

(10) If $\mathbf{p}_1,\mathbf{p}_2,\ldots,\mathbf{p}_{k+1}$ determine a k-plane then they lie on no j-plane for $j<k$.
(11) If $X=\{\mathbf{x}_1,\mathbf{x}_2,\ldots,\mathbf{x}_k\}$ is LI in $\mathcal{R}^n$ then this set determines a k-plane W^{k-1} and $\mathbf{0}\notin W^{k-1}$.

(12) If X is as given in the preceding, and $\mathbf{x}=c_1\mathbf{x}_1+c_2\mathbf{x}_2+\cdots+c_k\mathbf{x}_k$ where $c_1+c_2+\cdots+c_k\neq 1$, then

$$\{\mathbf{x}_1-\mathbf{x},\mathbf{x}_2-\mathbf{x},\ldots,\mathbf{x}_k-\mathbf{x}\}$$

is LI.

4.1.10 Order Relations in $\mathcal{R}^n$. Establish or resolve:

(1) The following hold where R denotes any one of $\geqq, >, \geq$.
  (i) $(\mathbf{x}_1 R\mathbf{y}_1$ and $\mathbf{x}_2 R\mathbf{y}_2)$ implies $(\mathbf{x}_1+\mathbf{x}_2)R(\mathbf{y}_1+\mathbf{y}_2)$.
  (ii) $(a>0$ and $\mathbf{x}R\mathbf{y})$ implies $a\mathbf{x}Ra\mathbf{y}$.
  (iii) $(a<0$ and $(\mathbf{x}R\mathbf{y})$ implies $a\mathbf{y}Ra\mathbf{x}$.
(2) If $\mathbf{z}>\mathbf{0}$ then the following hold.
  (i) $\mathbf{x}\geqq\mathbf{y}$ implies $(\langle\mathbf{z},\mathbf{x}\rangle\geqq\langle\mathbf{z},\mathbf{y}\rangle)$.
  (ii) $\mathbf{x}>\mathbf{y}$ implies $(\langle\mathbf{z},\mathbf{x}\rangle>\langle\mathbf{z},\mathbf{y}\rangle)$.
  (iii) $\mathbf{x}\geq\mathbf{y}$ implies $(\langle\mathbf{z},\mathbf{x}\rangle\geq\langle\mathbf{z},\mathbf{y}\rangle)$.
(3) Prove or disprove the preceding for $\mathbf{z}\geq\mathbf{0}$.
(4) Repeat for $\mathbf{z}\geqq\mathbf{0}$.
(5) Enter the correct response "yes" or "no" for $\langle\mathbf{z},\mathbf{x}\rangle R\langle\mathbf{z},\mathbf{y}\rangle$ in a 3×3 table with row labels $\mathbf{z}\geqq\mathbf{0}$, $\mathbf{z}>\mathbf{0}$, $\mathbf{z}\geq\mathbf{0}$ and column labels $\mathbf{x}\geqq\mathbf{y}$, $\mathbf{x}>\mathbf{y}$, $\mathbf{x}\geq\mathbf{y}$.
(6) Repeat for $>$ in $\mathcal{R}$.
(7) Each of the following is equivalent to $\mathbf{x}>\mathbf{0}$.
  (i) $0<\min\{x_1,x_2,\ldots,x_n\}$.
  (ii) $k\mathbf{x}>\mathbf{e}$ for some $k>0$.
  (iii) $\langle\mathbf{x},\mathbf{y}\rangle=0$ and $\mathbf{y}\geqq\mathbf{0}$ imply $\mathbf{y}=\mathbf{0}$.
(8) Repeat for $\mathbf{x}\geq\mathbf{0}$.
  (i) $\mathbf{x}\geqq\mathbf{0}$ and $0<\max\{x_1,x_2,\ldots,x_n\}$.
  (ii) $\mathbf{x}\geqq\mathbf{0}$ and $\|k\mathbf{x}\|=1$ for some $k\in\mathcal{R}$.
  (iii) $\mathbf{x}\geqq\mathbf{0}$ and $\langle c\mathbf{x},\mathbf{e}\rangle=1$ for some $c\in\mathcal{R}$.
(9) A set $S\subset\mathcal{R}^n$ is bounded iff it contains no ray.

4.1.11 Convex Sets. Establish in $\mathcal{R}^n$: (1) The following are convex sets: $\{\mathbf{p}\}$, subspaces, halfspaces, k-planes. (2) A convex combination of positive, nonnegative, semipositive vectors is, respectively, a positive, nonnegative, semipositive vector. (3) Let $\mathbf{X}=[\mathbf{x}_1,\mathbf{x}_2,\ldots,\mathbf{x}_r]$ where all $\mathbf{x}_i\in\mathcal{R}^n$. Then $\mathbf{x}$ is a convex combination of the columns of $k\mathbf{X}$ for some $k>0$ if $\mathbf{x}=\mathbf{X}\mathbf{t}$ for some semipositive $\mathbf{t}\in\mathcal{R}^r$. (4) A convex set having more than one point must have an infinite number of points. (5) The convex hull of an arbitrary $A\subset\mathcal{R}^n$ is the intersection of all convex sets containing A and is therefore the smallest convex set containing A.

4.1.12 Simplexes. Establish: (1) The medians of a triangle in $\mathcal{R}^2$ intersect at its *barycenter*, that is, at the point whose barycentric coordinates are

$t_1 = t_2 = t_3 = 3^{-1}$. (2) Each $\mathbf{x} \in X^k$ has a unique set of barycentric coordinates. (3) *Standard n-simplex*: $\mathbf{0}, \mathbf{e}_1, \mathbf{e}_2, \ldots, \mathbf{e}_n$ are the vertexes of an n-simplex. (4) Barycentric coordinates of $\mathbf{x}$ in the preceding are the components of the vector ______? (5) A definition of *face* and show: The face of X^k opposite vertex $\mathbf{p}_i$ is a $(k-1)$-simplex. (6) Number of r-dimensional simplex faces of X^k. (7) Use of simplexes to define dimension for any convex set.

4.1.13 Vertexes. Establish: (1) An alternative definition is: a point $\mathbf{p} \in K$, where K is convex, is a vertex of K if it is impossible to find two distinct points $\mathbf{p}_1, \mathbf{p}_2 \in K$ and $t \in (0,1)$ such that $\mathbf{p} = t\mathbf{p}_1 + (1-t)\mathbf{p}_2$. (2) The vertexes of a simplex are vertexes in the sense defined for any convex set. (3) A circular disc in $\mathfrak{R}^2$ has an infinite number of vertexes. (4) A k-plane with $k \geqq 1$ has no vertexes.

4.1.14 Operations on Convex Sets. Establish: (1) The set of products aK, where $a \in \mathfrak{R}$ and K is convex, is a convex set. (2) If K_1 and K_2 are convex, then $K_1 + K_2$ is a convex set.

4.1.15 Carathéodory's Theorem. Establish: (1) The subject theorem: any convex combination of points of $A \subset \mathfrak{R}^n$ can be written as a convex combination of $n+1$ or fewer points of A. (2) If $\mathbf{x} \in X^n$ has all barycentric coordinates positive then $\mathbf{x}$ cannot be written as a convex combination of fewer than $n+1$ vertexes. (3) The geometric interpretation of the preceding in $\mathfrak{R}^2$.

4.1.16 Convex Polyhedra. Establish or resolve.

(1) Convex polyhedra have a finite nonzero number of vertexes.
(2) An n-simplex is a convex polyhedron having exactly $n+1$ vertexes.
(3) Can a convex polyhedron be a subspace of $\mathfrak{R}^n$?
(4) A convex polyhedron is bounded; that is, it is necessarily bounded from below and from above.
(5) The convex polyhedron determined by the columns of $\mathbf{A}^T$, where $\mathbf{A}$ is $m \times n$, has at most m vertexes and they appear among the columns of $\mathbf{A}^T$.
(6) Suppose $\mathbf{A}^T$ is a $2 \times m$, that is, we have m specified points in $\mathfrak{R}^2$. Give a systematic graphical procedure for finding the vertexes of the associated convex polyhedron.
(7) Give an algebraic procedure for finding all vertexes in the preceding.
(8) A convex polyhedron is the convex hull of the set of its extreme points.

4.1.17 Polytopes. Establish: (1) A polytope is a convex set. (2) Examples of polytopes that are (i) empty, (ii) bounded, (iii) unbounded.

4.1.18 Cones. Answer: (1) Are the following cones? (i) Subspace. (ii) Positive orthant. (iii) Simplex. (iv) Unbounded polytope. (v) A bounded set. (vi) $\{\mathbf{x}:\mathbf{Ax}\leqq\mathbf{0}\}$. (vii) $\{\mathbf{x}:\mathbf{Ax}\leqq\mathbf{b}\}$. (2) How can we be sure that the dimension of a cone is a well-defined quantity? (3) Exhibit, that is, classify into types, all cones of the following dimensions in $\mathcal{R}^n$: (i) zero, (ii) one, (iii) two. (4) Can you extend to all dimensions?

4.1.19 Convex Cones. Establish or find: (1) A set $C\subset\mathcal{R}^n$ is a convex cone iff C is closed under addition and under multiplication by nonnegative scalars. (2) The convex hull of the cone in Figure 4.5(b). (3) The following operations on convex cones produce convex cones: (i) sums, (ii) intersections, and (iii) polars. (4) Find an example of a convex cone C for which $(C^*)^*\not\subset C$. (5) Find the polar cones of the cones: (i) Halfline $\mathbf{p}$ where $\mathbf{p}\neq\mathbf{0}$. (ii) Proper subspace S^k. (iii) Closed halfspace where the bounding hyperplane passes through the origin. (iv) Nonnegative orthant.

4.1.20 Polyhedral Convex Cones. Establish or find: (1) Examples of convex cones that are not polyhedral convex cones. (2) The set C in (4.24) is the sum of the m halflines through the $\mathbf{a}_{i\cdot}$. (3) A subspace is an example of a polyhedral convex cone. (4) A subspace has no extreme vector. (5) Yes or no? A halfspace is a polyhedral convex cone. (6) A closed halfspace in $\mathcal{R}^n$ has extreme vectors for $n=2$ but not for $n>2$. (7) Examples of polyhedral convex cones that are not pointed. (8) Conditions such that a polyhedral convex cone will be a subspace.

4.1.21 Closure Conditions. Characterize the following in terms of what they must contain, for example, all LCs of their points, certain types of multiples, and so on, as appropriate. (1) k-plane through $\mathbf{0}$. (2) k-plane. (3) Convex hull. (4) Cone. (5) Polyhedral convex cone.

4.2 NONNEGATIVE SOLUTIONS

When, for example, components of solutions $\mathbf{x}$ to $\mathbf{Ax}=\mathbf{b}$ are to correspond to percentages, then $\mathbf{x}$ must be nonnegative; in fact, $\mathbf{x}$ must be semipositive and must satisfy the additional constraint $\langle\mathbf{x},\mathbf{e}\rangle=100$. In this section we present the existence theory of nonnegative, semipositive, and positive solutions to linear equations and linear inequalities. As can be seen, for example in Gale (1960), this theory is useful for models involving economic considerations and, as shown, for example in Mangasarian (1969), it is of fundamental importance for optimization.

All results in this section are presented in the following format.

THEOREM OF THE ALTERNATIVE. One and only one of Problems I and II has a solution.

Both problems call for the same input data, and for some data it is Problem I that has a solution, while for others it is Problem II. It is efficient to use a fixed procedure for

PROVING A THEOREM OF THE ALTERNATIVE.

Step 1. Show that Problems I and II cannot simultaneously have solutions.
Step 2. Show that if one problem has no solution then the other must have a solution. (4.27)

PROOF. These steps show that if Problem I has a solution then II has none, if I has none then II has one, and we can interchange the roles of I and II. ■
Moreover, we are able to use a fixed procedure for Step 1 in each case we consider. Each problem is based on an ordered pair

(Equations or Inequalities, Constraints)

where the principal objects are matrix-vector products and individual vectors. The typical arrangement is

Problem I	(Matrix-vector, vector)	(4.28)
Problem II	(Matrix-vector, vector)	

Then Step 1 consists of an indirect proof where we suppose each of Problems I and II has a solution and: combine upper LH and lower RH in (4.28) using multiplication to produce a scalar quantity or inequality statement (i), and combine lower LH and upper RH to obtain a contradictory statement (ii). A detailed illustration of this procedure appears in the next proof.

The following theorem of the alternative appears in Gale (1960).

If $\mathbf{A}$ is $m \times n$ and $\mathbf{b} \in \mathfrak{R}^m$, then one and only one of the following has a solution.

I. Find $\mathbf{x} \in \mathfrak{R}^n$ such that $\mathbf{Ax} = \mathbf{b}$.
II. Find $\mathbf{y} \in \mathfrak{R}^m$ such that $\mathbf{A}^T\mathbf{y} = \mathbf{0}$ and $\langle \mathbf{b}, \mathbf{y} \rangle = 1$. (4.29)

PROOF. Step 1 of (4.27) proceeds as follows. Suppose there are solutions $\mathbf{x}$ and $\mathbf{y}$. Combining $\mathbf{Ax} = \mathbf{b}$ and $\langle \mathbf{b}, \mathbf{y} \rangle = 1$ yields $\mathbf{x}^T\mathbf{A}^T\mathbf{y} = 1$ as (i) from (4.28). Then (ii) is the contradictory $\mathbf{x}^T\mathbf{A}^T\mathbf{y} = 0$ and Step 1 is completed. To start Step 2 we note that if Problem I has no solution then necessarily $\mathbf{b} \neq \mathbf{0}$. From (4.4) since $\mathbf{b} \notin \operatorname{Range} \mathbf{A}$

$$\mathbf{b} = \mathbf{u} + \mathbf{v}$$

where $\mathbf{u} \in \text{Range}\,\mathbf{A}$ and $\mathbf{v} \in (\text{Range}\,\mathbf{A})^{\perp}$. Certainly $\mathbf{v} \neq \mathbf{0}$ so also

$$\langle \mathbf{b}, \mathbf{v} \rangle = \langle \mathbf{u}, \mathbf{v} \rangle + \langle \mathbf{v}, \mathbf{v} \rangle = \langle \mathbf{v}, \mathbf{v} \rangle$$

is nonzero. Finally, $\mathbf{y} = \langle \mathbf{v}, \mathbf{v} \rangle^{-1}\mathbf{v}$ is a solution to II because $\langle \mathbf{y}, \mathbf{b} \rangle = 1$ by construction, and $\mathbf{A}^T\mathbf{y} = \mathbf{0}$ since $\mathbf{y} \in \text{Null}\,\mathbf{A}^T$. ■

This theorem applies to any matrix and vector where the number of rows in the former equals the number of components in the latter. There is the

Geometric Interpretation. One and only one of the following alternatives holds in $\mathfrak{R}^m$.

I. The vector $\mathbf{b}$ belongs to the subspace spanned by the columns of $\mathbf{A}$.
II. There exists a vector $\mathbf{y}$ that is orthogonal to every column of $\mathbf{A}$ and makes a strictly acute angle with $\mathbf{b}$.

If we use $\mathbf{A}^T$ in place of $\mathbf{A}$ and replace $\mathbf{b} \in \mathfrak{R}^m$ by $\mathbf{c} \in \mathfrak{R}^n$, the setting changes to $\mathfrak{R}^n$ and we obtain what we call the *standard form* of

GALE'S THEOREM FOR EQUATIONS. If $\mathbf{A}$ is $m \times n$ and $\mathbf{c} \in \mathfrak{R}^n$ then one and only one of the following has a solution.

I. Find $\mathbf{y} \in \mathfrak{R}^m$ such that $\mathbf{A}^T\mathbf{y} = \mathbf{c}$.
II. Find $\mathbf{x} \in \mathfrak{R}^n$ such that $\mathbf{A}\mathbf{x} = \mathbf{0}$ and $\langle \mathbf{c}, \mathbf{x} \rangle = 1$. (4.30)

The proof of (4.29) serves as the proof of (4.30) because the theorems are equivalent. Our choice of nomenclature is that in *standard theorems of the alternative*

1. Problem I deals with finding $\mathbf{y} \in \mathfrak{R}^m$.
2. If either problem calls for solving a system of linear equations then Problem I is such a problem.

We consider "dual" theorems (exemplified by (4.29)) at the end of this section.

Several theorems deserve to be called the "foundation" because, once proved, all remaining theorems readily follow as corollaries. We have selected the following result of Farkas (1902) as the one to be proved first. Notice that it deals with (I) nonnegative solutions to linear equations, and (II) constrained solutions to homogeneous linear inequalities.

FARKAS' THEOREM. If $\mathbf{A}$ is $m \times n$ and $\mathbf{c} \in \mathfrak{R}^n$, then one and only one of the following has a solution.

Farkas I. Find $\mathbf{y} \in \mathfrak{R}^m$ such that $\mathbf{A}^T\mathbf{y} = \mathbf{c}$ and $\mathbf{y} \geqq \mathbf{0}$.
Farkas II. Find $\mathbf{x} \in \mathfrak{R}^n$ such that $\mathbf{A}\mathbf{x} \leqq \mathbf{0}$ and $\langle \mathbf{c}, \mathbf{x} \rangle > 0$. (4.31)

PROOF. Step 1 of (4.27) is immediate. For Step 2 we first establish: If I has no solution, even when $\mathbf{y} \geqq \mathbf{0}$ is waived, then II has a solution. This is true because (4.30) guarantees the existence of $\mathbf{x}$ satisfying $\mathbf{Ax}=\mathbf{0}$ and $\langle \mathbf{c},\mathbf{x}\rangle = 1$. To complete Step 2 we need to prove

> If Farkas I has no solution, but there does exist $\mathbf{y}<\mathbf{0}$ satisfying $\mathbf{A}^T\mathbf{y}=\mathbf{c}$, then Farkas II has a solution.

If $\mathbf{A}^T$ has $m=1$ column, and say $ay=c$ with $y<0$, then, as is readily verified, $x=-y/c$ is a solution to II. The inductive hypothesis is

> Let Farkas' theorem hold for all matrices $\mathbf{A}^T$ of $m-1$ columns or less.

Then Farkas I based on the first $m-1$ columns of the $n\times m$ $\mathbf{A}^T$ can have no nonnegative solution in $\mathcal{R}^{m-1}$ for if it did we could use zero as the mth component to obtain a nonnegative solution in $\mathcal{R}^m$ contrary to assumption. Therefore there is a solution $\mathbf{u}$ for II for the case of the first $m-1$ columns of $\mathbf{A}^T$. This means $\langle \mathbf{a}_j,\mathbf{u}\rangle \leqq 0$ for $j=1,2,\ldots,m-1$ and $\langle \mathbf{c},\mathbf{u}\rangle>0$. If $\langle \mathbf{a}_m,\mathbf{u}\rangle \leqq 0$ we have a solution for II based on m columns; consequently, the only unresolved possibility is

$$\langle \mathbf{a}_m,\mathbf{u}\rangle>0$$

and we need to find a solution for this case. Let us use (4.26) to construct m vectors orthogonal to $\mathbf{u}$

$$\mathbf{a}'_1,\mathbf{a}'_2,\ldots,\mathbf{a}'_{m-1},\mathbf{c}'$$

where $\mathbf{a}'_j=\mathbf{a}_j+k_j\mathbf{a}_m$ for $j=1,2,\ldots,m-1$, and $\mathbf{c}'=\mathbf{c}+k\mathbf{a}_m$. We find for $j=1,2,\ldots,m-1$

$$k_j=-\frac{\langle \mathbf{a}_j,\mathbf{u}\rangle}{\langle \mathbf{a}_m,\mathbf{u}\rangle},\qquad \text{and}\qquad k=-\frac{\langle \mathbf{c},\mathbf{u}\rangle}{\langle \mathbf{a}_m,\mathbf{u}\rangle}$$

so all $k_j\geqq 0$ and $k<0$. Now we assert there can be no solution to Farkas I formulated for matrix

$$[\mathbf{a}'_1,\mathbf{a}'_2,\ldots,\mathbf{a}'_{m-1}]$$

and vector $\mathbf{c}'$. For if $\mathbf{z}\geqq\mathbf{0}$ were a solution, we would have

$$z_1\mathbf{a}'_1+z_2\mathbf{a}'_2+\cdots+z_{m-1}\mathbf{a}'_{m-1}=\mathbf{c}'$$

which we could expand and reassemble as

$$z_1\mathbf{a}_1+\cdots+z_{m-1}\mathbf{a}_{m-1}+\left(-k+\sum_{j=1}^{m-1}k_jz_j\right)\mathbf{a}_m=\mathbf{c}$$

to exhibit a nonnegative solution to Farkas I for all m columns of $\mathbf{A}^T$. By the inductive hypothesis there exists a solution $\mathbf{v} \in \mathfrak{R}^n$ to the most recently specified $(m-1)$-column Problem II. That is, for $j=1,2,\dots,m-1$

$$\langle \mathbf{a}_j', \mathbf{v} \rangle \leqq 0, \qquad \text{and} \qquad \langle \mathbf{c}', \mathbf{v} \rangle > 0$$

If we again use (4.26), this time to construct

$$\mathbf{w} = \mathbf{v} - \frac{\langle \mathbf{v}, \mathbf{a}_m \rangle}{\langle \mathbf{u}, \mathbf{a}_m \rangle} \mathbf{u}$$

orthogonal to $\mathbf{a}_m$, we find for $j=1,2,\dots,m-1$

$$\langle \mathbf{a}_j, \mathbf{w} \rangle = \langle \mathbf{a}_j', \mathbf{v} \rangle$$

while $\langle \mathbf{c}, \mathbf{w} \rangle = \langle \mathbf{c}', \mathbf{v} \rangle$ so $\mathbf{w}$ is a solution to Farkas II for matrix $\mathbf{A}$ and vector $\mathbf{c}$. ■

As might be expected for so fundamental a result, Farkas' Theorem can be expressed in a number of different forms. One example is the following where the senses of the inequalities in II are reversed.

EQUIVALENT FORM OF FARKAS' THEOREM. If $\mathbf{C}$ is $m \times n$ and $\mathbf{d} \in \mathfrak{R}^n$, then one and only one of the following has a solution

I. Find $\mathbf{y} \in \mathfrak{R}^m$ such that $\mathbf{C}^T\mathbf{y} = \mathbf{d}$ and $\mathbf{y} \geqq \mathbf{0}$.
II. Find $\mathbf{x} \in \mathfrak{R}^n$ such that $\mathbf{Cx} \geqq \mathbf{0}$ and $\langle \mathbf{d}, \mathbf{x} \rangle < 0$. (4.32)

PROOF. The original theorem applies to $\mathbf{C} = -\mathbf{A}$ and $\mathbf{d} = -\mathbf{c}$. ■

There are consequently also a number of geometric interpretations of Farkas' theorem.

First Geometric Interpretation. One and only one of the following alternatives holds.

I. The vector $\mathbf{c}$ belongs to the polyhedral convex cone C determined by the columns of $\mathbf{A}^T$.
II. There exists a vector $\mathbf{x}$ that makes an obtuse angle with every column of $\mathbf{A}^T$ and a strictly acute angle with $\mathbf{c}$.

The vector $\mathbf{c}$ either belongs to the cone C or not, but what is particularly significant in II is that when $\mathbf{c} \notin C$ then the solution vector $\mathbf{x}$ determines a separating hyperplane H. All vectors in C make obtuse angles with $\mathbf{x}$ while $\mathbf{c}$ makes a strictly acute angle with $\mathbf{x}$. Thus we also have a

Second Geometric Interpretation. If $C \subset \mathfrak{R}^n$ is a polyhedral convex cone and $\mathbf{c} \notin C$ then there exists a hyperplane H that determines a closed halfspace H^- containing C, and an open halfspace H^+ containing $\mathbf{c}$. (4.33)

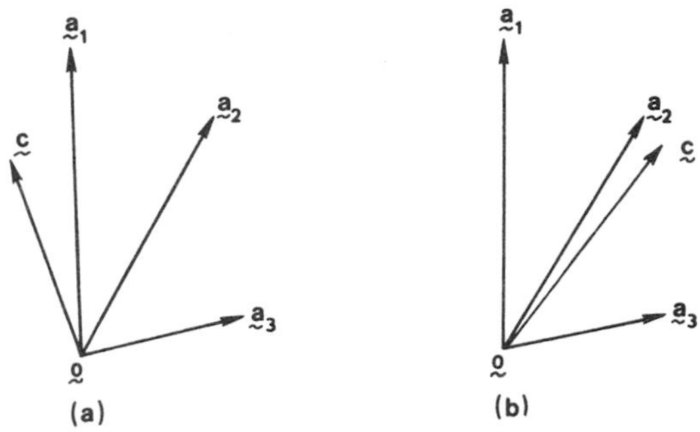

Figure 4.7. Two illustrations of Farkas I.

For this reason Farkas' theorem is often called a "theorem of the separating hyperplane" (4.2.3).

Figure 4.7 illustrates two cases in $\mathcal{R}^2$ for $m=3$. In (a) there is no solution to Farkas I while in (b) there is a solution because $\mathbf{c}$ belongs to the polyhedral convex cone determined by $\mathbf{A}^T=[\mathbf{a}_1,\mathbf{a}_2,\mathbf{a}_3]$. Now Farkas I has a solution iff $\mathbf{c}$ does not lie outside the smaller region determined by $\mathbf{a}_1$ and $\mathbf{a}_3$. The corresponding cases of Farkas II are illustrated in Figure 4.8; we know that the theorem asserts there will be a solution in (a) but none in (b). Arcs are drawn to indicate where solution vectors must lie. Solid arcs identify possible regions for vectors making acute angles with $\mathbf{c}$ while broken arcs identify regions for vectors making obtuse angles with every column of $\mathbf{A}^T$. In Figure 4.8(a) the two regions intersect so II has a solution but there is no such common region in (b).

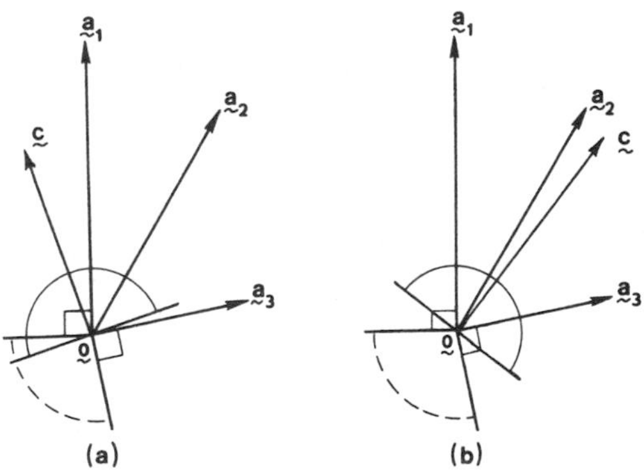

Figure 4.8. Two illustrations of Farkas II.

EXAMPLE 4.1. A Candy Store Problem. This example illustrates Farkas' theorem in a particularly simple context. A candy store stocks three standard mixtures of candies of types A, B, C, and D where percentages by weight are as shown in Table 4.2 for Mixtures 1, 2, 3. For example, an order for two pounds of candy with equal proportions of the four types can be filled by simply measuring out two pounds of Mixture 2. Some orders cannot be filled because of the rule that orders must be filled by simply weighing portions of the different mixtures. For example, an order for one pound of Type A only cannot be filled.

What are the conditions that an order for respective percentages c_1, c_2, c_3, c_4 summing to 100, can be filled? If we associate the order with the vector $\mathbf{c} \in \mathfrak{R}^4$ then the answer according to (2.15) is: the order can be filled iff $\mathbf{c}$ can be written as an LC with nonnegative coefficients of the columns of

$$\mathbf{A}^T = \begin{bmatrix} 10 & 25 & 60 \\ 20 & 25 & 20 \\ 30 & 25 & 20 \\ 40 & 25 & 0 \end{bmatrix}$$

An equivalent answer is: the order can be filled iff the associated Farkas I has a solution.

In order to see how to resolve a particular case of Farkas I, and also to illustrate the ways in which Problems I and II are interrelated, let us first consider the effect of Gauss–Jordan pivoting as follows

$$[\mathbf{A}^T, \mathbf{I}_4] \overset{R}{\rightarrow} \begin{bmatrix} 1 & 0 & 0 & 0 & -\frac{1}{10} & \frac{1}{10} & 0 \\ 0 & 1 & 0 & -\frac{1}{50} & \frac{4}{25} & -\frac{1}{10} & 0 \\ 0 & 0 & 1 & \frac{1}{40} & -\frac{1}{20} & \frac{1}{40} & 0 \\ 0 & 0 & 0 & \frac{1}{2} & 0 & -\frac{3}{2} & 1 \end{bmatrix}$$

TABLE 4.2 PERCENTAGES OF CANDIES BY WEIGHT

Candy Type	Mixture 1	2	3
A	10	25	60
B	20	25	20
C	30	25	20
D	40	25	0

where we have stopped after pivoting in the third column, by which time $\mathbf{M}$ the GJSF of $\mathbf{A}^T$ occupies the first three columns. As in (3.26), the final four columns can be used as $\mathbf{D}$ where $\mathbf{DA}^T=\mathbf{M}$. Therefore $[\mathbf{A}^T,\mathbf{c}]\overset{R}{\rightarrow}[\mathbf{M},\mathbf{Dc}]$ and this final column will be the vector

$$\begin{bmatrix} & - & \frac{1}{10}c_2 & + & \frac{1}{10}c_3 \\ -\frac{1}{50}c_1 & + & \frac{4}{25}c_2 & - & \frac{1}{10}c_3 \\ \frac{1}{40}c_1 & - & \frac{1}{20}c_2 & + & \frac{1}{40}c_3 \\ \frac{1}{2}c_1 & - & \frac{3}{2}c_3 & + & c_4 \end{bmatrix} = \begin{bmatrix} \alpha \\ \beta \\ \gamma \\ \delta \end{bmatrix}$$

We see that there will be solutions of any kind iff the fourth component is zero so that $\operatorname{rank}\mathbf{A}^T=\operatorname{rank}[\mathbf{A}^T,\mathbf{c}]$ and

The order corresponding to $\mathbf{c}$ can be filled from Mixtures 1,2,3 iff

$$\alpha,\beta,\gamma\geqq 0 \qquad \text{and} \qquad \delta=0$$

Given any particular $\mathbf{c}$ it is easy to check these four conditions.

We can also consider Farkas II formulated as follows so we can use GJSFs.

Find $\mathbf{x}\in\mathfrak{R}^4$ such that for some nonnegative $\mathbf{b}\in\mathfrak{R}^3$, $\mathbf{Ax}=-\mathbf{b}$ and $\langle\mathbf{c},\mathbf{x}\rangle=1$.

Setting up the four equations and performing Gauss–Jordan pivoting in the first three columns of the 4×5 matrix we of course find: There is a solution $\mathbf{x}$ iff the order corresponding to $\mathbf{c}$ cannot be filled.

Let us next use Farkas' theorem to establish criteria for semipositive solutions to homogeneous equations. Notice that the following result of Gordan (1873) is formulated in terms of a matrix alone.

GORDAN'S THEOREM. If $\mathbf{A}$ is $m\times n$, then one and only one of the following has a solution

Gordan I. Find $\mathbf{y}\in\mathfrak{R}^m$ such that $\mathbf{A}^T\mathbf{y}=\mathbf{0}$ and $\mathbf{y}\geq\mathbf{0}$, that is, $\mathbf{y}$ is semipositive.

Gordan II. Find $\mathbf{x}\in\mathfrak{R}^n$ such that $\mathbf{Ax}<\mathbf{0}$. (4.34)

PROOF. Step 1 of (4.27) is immediate. Suppose next that Gordan I has no solution. Then Problem I in (4.32) has no nonnegative solution for $\mathbf{C}=[\mathbf{A},\mathbf{e}]$ and $\mathbf{d}=[0,0,\ldots,0,1]^T$ lest Gordan I have a semipositive solution.

This means that the associated II in (4.32) must have a solution

$$\mathbf{z}=\begin{bmatrix}\mathbf{u}\\ t\end{bmatrix}\in\mathcal{R}^{n+1}$$

The vector $\mathbf{z}$ satisfies $\mathbf{Cz}\geqq\mathbf{0}$ which can be written

$$\mathbf{Au}+t\mathbf{e}\geqq\mathbf{0}$$

where $t<0$ since $\langle\mathbf{d},\mathbf{z}\rangle<0$. Consequently $\mathbf{x}=-\mathbf{u}$ is a solution to Gordan II. ■

A problem containing only homogeneous statements can have the senses of all inequality statements freely reversed. Thus Gordan II has a solution $\mathbf{x}$ iff $\mathbf{z}=-\mathbf{x}$ is a solution to $\mathbf{Az}>\mathbf{0}$. If we use this form and express it in geometric terms for the subspace determined by the columns of $\mathbf{A}^T$, Gordan's problems become

I. Null $\mathbf{A}^T$ contains a semipositive vector?
II. Range $\mathbf{A}$ contains a positive vector?

This takes on a convenient form if we use (4.4) and then cite the problems in reverse order, for any subspace, as a

GEOMETRIC VERSION OF GORDAN'S THEOREM. If $S\subset\mathcal{R}^n$ is any subspace, then one and only one of the following holds.

I. There is a positive vector in S.
II. There is a semipositive vector in $S^\perp$. (4.35)

Let us use this form to prove a result of Stiemke (1915).

STIEMKE'S THEOREM. If $\mathbf{A}$ is $m\times n$, then one and only one of the following has a solution.

Stiemke I. Find $\mathbf{y}\in\mathcal{R}^m$ such that $\mathbf{A}^T\mathbf{y}=\mathbf{0}$ and $\mathbf{y}>\mathbf{0}$.
Stiemke II. Find $\mathbf{x}\in\mathcal{R}^n$ such that $\mathbf{Ax}\leq\mathbf{0}$. (4.36)

PROOF. Step 1 of (4.27) is immediate. An equivalent way of stating "Steimke I has no solution" is "Null $\mathbf{A}^T$ has no positive vector." But if this holds, then from (4.35) and (4.4) Range $\mathbf{A}$ must contain a semipositive vector, say $\mathbf{b}=\mathbf{Au}$. This means $\mathbf{x}=-\mathbf{u}$ is a solution to Stiemke II. ■

As shown in the first column of Table 4.3 we now have theorems of the alternative covering existence of solutions to homogeneous linear equations. For nonhomogeneous equations, Farkas' theorem covers semipositive solutions as well as nonnegative because when $\mathbf{c}\neq\mathbf{0}, \mathbf{A}^T\mathbf{y}=\mathbf{c}$ is impossible for $\mathbf{y}=\mathbf{0}$. The final case in Table 4.3 is covered by the

NONHOMOGENEOUS STIEMKE'S THEOREM. If $\mathbf{A}$ is $m\times n$ and $\mathbf{c}\in\mathcal{R}^n$ is nonzero, then one and only one of the following has a solution.

I. Find $\mathbf{y}\in\mathcal{R}^m$ such that $\mathbf{A}^T\mathbf{y}=\mathbf{c}$ and $\mathbf{y}>\mathbf{0}$.

II. Find $\mathbf{x} \in \mathcal{R}^n$ such that

$$\begin{bmatrix} \mathbf{Ax} \\ -\langle \mathbf{c}, \mathbf{x} \rangle \end{bmatrix} \leq \mathbf{0} \tag{4.37}$$

PROOF. Step 1 of (4.27) is immediate. If I has no solution, then Stiemke I also has none for $\mathbf{B}$ where $\mathbf{B}^T = [\mathbf{A}^T, -\mathbf{c}]$ and Stiemke II for this $\mathbf{B}$ coincides with the present II. ■

Table 4.4 cites results on existence of solutions to linear inequalities. Let us consider the four theorems that are cited. The first two are from Gale (1960).

GALE'S THEOREM FOR LINEAR INEQUALITIES. If $\mathbf{A}$ is $m \times n$ and $\mathbf{b} \in \mathcal{R}^m$, then one and only one of the following has a solution.

I. Find $\mathbf{y} \in \mathcal{R}^m$ such that $\mathbf{A}^T\mathbf{y} = \mathbf{0}$, $\langle \mathbf{b}, \mathbf{y} \rangle = -1$, $\mathbf{y} \geqq \mathbf{0}$.
II. Find $\mathbf{x} \in \mathcal{R}^n$ such that $\mathbf{Ax} \leqq \mathbf{b}$. (4.38)

PROOF. Step 1 of (4.27) is immediate. We can write I as Farkas I with matrix $\mathbf{B}^T$ where $\mathbf{B} = [\mathbf{A}, \mathbf{b}]$ and vector $[\mathbf{0}^T, -1]^T$ so if there is no solution, then there exists $\mathbf{z} \in \mathcal{R}^{n+1}$ satisfying $[\mathbf{A}, \mathbf{b}]\mathbf{z} \leqq \mathbf{0}$ and $z_{n+1} < 0$. This means

$$\mathbf{A}[z_1, \ldots, z_n]^T + z_{n+1}\mathbf{b} \leqq \mathbf{0}$$

so $\mathbf{x} = [-z_1/z_{n+1}, \ldots, -z_n/z_{n+1}]^T$ is a solution to the present II. ■
Notice that we could as well use $\mathbf{y} \geq \mathbf{0}$ in I but, as a matter of preference, we use $\mathbf{y} \geqq \mathbf{0}$; similar cases follow.

TABLE 4.3 EXISTENCE OF SOLUTIONS TO LINEAR EQUATIONS $\mathbf{A}^T\mathbf{y} = \mathbf{c}$

	Homogeneous $\mathbf{c} = \mathbf{0}$	Nonhomogeneous $\mathbf{c} \neq \mathbf{0}$
Unrestricted $\mathbf{y}$	At least $\mathbf{y} = \mathbf{0}$	Gale (4.30)
Nonzero $\mathbf{y} \neq \mathbf{0}$	(3.18)	Gale (4.30)
Nonnegative $\mathbf{y} \geqq \mathbf{0}$	At least $\mathbf{y} = \mathbf{0}$	Farkas (4.31)
Semipositive $\mathbf{y} \geq \mathbf{0}$	Gordan (4.34)	Farkas (4.31)
Positive $\mathbf{y} > \mathbf{0}$	Stiemke (4.36)	Nonhomogeneous Stiemke (4.37)

TABLE 4.4 EXISTENCE OF SOLUTIONS TO LINEAR INEQUALITIES $\mathbf{A}^T\mathbf{y} \leqq \mathbf{c}$

	Homogeneous $\mathbf{c}=\mathbf{0}$	Nonhomogeneous $\mathbf{c}\neq\mathbf{0}$
Unrestricted $\mathbf{y}$	At least $\mathbf{y}=\mathbf{0}$	Gale (4.38)
Nonzero $\mathbf{y}\neq\mathbf{0}$	4.3.10	4.3.11
Nonnegative $\mathbf{y}\geqq\mathbf{0}$	At least $\mathbf{y}=\mathbf{0}$	Gale (4.39)
Semipositive $\mathbf{y}\geq\mathbf{0}$	von Neumann (4.40)	4.3.12
Positive $\mathbf{y}>\mathbf{0}$	Tucker (4.41)	4.3.13

GALE'S THEOREM FOR NONNEGATIVE SOLUTIONS. If $\mathbf{A}$ is $m\times n$ and $\mathbf{c}\in\mathfrak{R}^n$ then one and only one of the following has a solution.

I. Find $\mathbf{y}\in\mathfrak{R}^m$ such that $\mathbf{A}^T\mathbf{y}\leqq\mathbf{c}$ and $\mathbf{y}\geqq\mathbf{0}$.
II. Find $\mathbf{x}\in\mathfrak{R}^n$ such that $\mathbf{A}\mathbf{x}\geqq\mathbf{0}$, $\langle\mathbf{c},\mathbf{x}\rangle<0$, $\mathbf{x}\geqq 0$. (4.39)

PROOF. Step 1 of (4.27) is immediate. If I has no solution, then neither does Farkas I for $[\mathbf{A}^T,\mathbf{I}_n]$ and $\mathbf{c}$. Consequently, for some $\mathbf{u}\in\mathfrak{R}^n$, $\mathbf{A}\mathbf{u}\leqq\mathbf{0}$, $\mathbf{I}_n\mathbf{u}\leqq\mathbf{0}$, and $\langle\mathbf{c},\mathbf{u}\rangle>0$ so that $\mathbf{x}=-\mathbf{u}$ is a solution to the present II. ■
In (4.46) we again use this method of changing inequalities to equations. The next result appears in von Neumann and Morgenstern (1944).

VON NEUMANN'S THEOREM FOR SEMIPOSITIVE SOLUTIONS. If $\mathbf{A}$ is $m\times n$ then one and only one of the following has a solution.

I. Find $\mathbf{y}\in\mathfrak{R}^m$ such that $\mathbf{A}^T\mathbf{y}\leqq\mathbf{0}$ and $\mathbf{y}\geq\mathbf{0}$.
II. Find $\mathbf{x}\in\mathfrak{R}^n$ such that $\mathbf{A}\mathbf{x}>\mathbf{0}$ and $\mathbf{x}\geqq\mathbf{0}$. (4.40)

PROOF. Step 1 of (4.27) is immediate. If I has no solution then neither does Problem I in (4.39) for matrix $\mathbf{B}^T$, where $\mathbf{B}=[\mathbf{A},-\mathbf{e}]$, and vector $[\mathbf{0}^T,-1]^T$. Consequently, for some $\mathbf{u}\in\mathfrak{R}^{n+1}$, we can construct a solution

$$\mathbf{x}=\left[\frac{u_1}{u_{n+1}},\ldots,\frac{u_n}{u_{n+1}}\right]^T$$

for the present Problem II. ■
The final theorem cited in Table 4.4 follows from results in Tucker (1956) (4.2.15).

TABLE 4.5 THEOREMS OF THE ALTERNATIVE

		Standard Forms			"Dual" Forms
Gale	I	$\mathbf{A}^T\mathbf{y}=\mathbf{c}$	(4.30)	I	$\mathbf{Ax}=\mathbf{b}$
	II	$\mathbf{Ax}=\mathbf{0}$, $\langle \mathbf{c},\mathbf{x}\rangle=1$		II	$\mathbf{A}^T\mathbf{y}=\mathbf{0}$, $\langle \mathbf{b},\mathbf{y}\rangle=1$
Farkas	I	$\mathbf{A}^T\mathbf{y}=\mathbf{c}$, $\mathbf{y}\geqq\mathbf{0}$	(4.31)	I	$\mathbf{Ax}=\mathbf{b}$, $\mathbf{x}\geqq\mathbf{0}$
	II	$\mathbf{Ax}\leqq\mathbf{0}$, $\langle \mathbf{c},\mathbf{x}\rangle>0$		II	$\mathbf{A}^T\mathbf{y}\leqq\mathbf{0}$, $\langle \mathbf{b},\mathbf{y}\rangle>0$
Gordan	I	$\mathbf{A}^T\mathbf{y}=\mathbf{0}$, $\mathbf{y}\geq\mathbf{0}$	(4.34)	I	$\mathbf{Ax}=\mathbf{0}$, $\mathbf{x}\geq\mathbf{0}$
	II	$\mathbf{Ax}<\mathbf{0}$		II	$\mathbf{A}^T\mathbf{y}<\mathbf{0}$
Stiemke	I	$\mathbf{A}^T\mathbf{y}=\mathbf{0}$, $\mathbf{y}>\mathbf{0}$	(4.36)	I	$\mathbf{Ax}=\mathbf{0}$, $\mathbf{x}>\mathbf{0}$
	II	$\mathbf{Ax}\leq\mathbf{0}$		II	$\mathbf{A}^T\mathbf{y}\leq\mathbf{0}$
Stiemke	I	$\mathbf{A}^T\mathbf{y}=\mathbf{c}$, $\mathbf{y}>\mathbf{0}$	(4.37)	I	$\mathbf{Ax}=\mathbf{b}$, $\mathbf{x}>\mathbf{0}$
$\mathbf{c}\neq\mathbf{0}$	II	$\begin{bmatrix}\mathbf{Ax}\\ -\langle \mathbf{c},\mathbf{x}\rangle\end{bmatrix}\leq\mathbf{0}$		II	$\begin{bmatrix}\mathbf{A}^T\mathbf{y}\\ -\langle \mathbf{b},\mathbf{y}\rangle\end{bmatrix}\leq\mathbf{0}$
Gale	I	$\mathbf{A}^T\mathbf{y}=\mathbf{0}$, $\langle \mathbf{b},\mathbf{y}\rangle=-1$, $\mathbf{y}\geqq\mathbf{0}$	(4.38)	I	$\mathbf{Ax}=\mathbf{0}$, $\langle \mathbf{c},\mathbf{x}\rangle=-1$, $\mathbf{x}\geqq\mathbf{0}$
	II	$\mathbf{Ax}\leqq\mathbf{b}$		II	$\mathbf{A}^T\mathbf{y}\leqq\mathbf{c}$
Gale	I	$\mathbf{A}^T\mathbf{y}\leqq\mathbf{c}$, $\mathbf{y}\geqq\mathbf{0}$	(4.39)	I	$\mathbf{Ax}\leqq\mathbf{b}$, $\mathbf{x}\geqq\mathbf{0}$
	II	$\mathbf{Ax}\geqq\mathbf{0}$, $\langle \mathbf{c},\mathbf{x}\rangle<0$, $\mathbf{x}\geqq\mathbf{0}$		II	$\mathbf{A}^T\mathbf{y}\geqq\mathbf{0}$, $\langle \mathbf{b},\mathbf{y}\rangle<0$, $\mathbf{y}\geqq\mathbf{0}$
von Neumann	I	$\mathbf{A}^T\mathbf{y}\leqq\mathbf{0}$, $\mathbf{y}\geq\mathbf{0}$	(4.40)	I	$\mathbf{Ax}\leqq\mathbf{0}$, $\mathbf{x}\geq\mathbf{0}$
	II	$\mathbf{Ax}>\mathbf{0}$, $\mathbf{x}\geqq\mathbf{0}$		II	$\mathbf{A}^T\mathbf{y}>\mathbf{0}$, $\mathbf{y}\geqq\mathbf{0}$
Tucker	I	$\mathbf{A}^T\mathbf{y}\leqq\mathbf{0}$, $\mathbf{y}>\mathbf{0}$	(4.41)	I	$\mathbf{Ax}\leqq\mathbf{0}$, $\mathbf{x}>\mathbf{0}$
	II	$\mathbf{Ax}\geq\mathbf{0}$, $\mathbf{x}\geqq\mathbf{0}$		II	$\mathbf{A}^T\mathbf{y}\geq\mathbf{0}$, $\mathbf{y}\geqq\mathbf{0}$

TUCKER'S THEOREM FOR POSITIVE SOLUTIONS. If $\mathbf{A}$ is $m\times n$ then one and only one of the following has a solution.

I. Find $\mathbf{y}\in\mathfrak{R}^m$ such that $\mathbf{A}^T\mathbf{y}\leqq\mathbf{0}$ and $\mathbf{y}>\mathbf{0}$.
II. Find $\mathbf{x}\in\mathfrak{R}^n$ such that $\mathbf{Ax}\geq\mathbf{0}$ and $\mathbf{x}\geqq\mathbf{0}$. (4.41)

Table 4.5 lists both standard and "dual" forms of all theorems of the alternative presented previously. The latter result from interchanging $\mathbf{A}^T$ and $\mathbf{A}$, $\mathbf{y}$ and $\mathbf{x}$, $\mathbf{c}$ and $\mathbf{b}$, and they hold by virtue of their equivalence to standard forms. All "dual" forms are furthermore in agreement with those from formal treatments of duality as given in Goldman and Tucker (1956) and Stoer and Witzgall (1970).

PROBLEMS FOR SOLUTION

4.2.1 Fundamental Properties. Resolve: (1) Equivalent format? Problem I has no solution iff II has one. (2) Method of Proof? Step A. If Problem I has a solution then II has none. Step B. If Problem II has no solution then

I must have one. (3) Repeat for Step A′. If Problem II has a solution then I has none. Step B′. If Problem II has no solution then I has one.

4.2.2 Gale's Theorem (4.29). Establish: (1) Whether or not any of the following can replace $\langle \mathbf{b},\mathbf{y}\rangle=1$ in Gale II. (i) $\langle \mathbf{b},\mathbf{y}\rangle=-1$; (ii) $\mathbf{y}\neq\mathbf{0}$; (iii) $\langle \mathbf{b},\mathbf{y}\rangle<0$. (2) The following is an equivalent statement. The system $\mathbf{Ax}=\mathbf{b}$ is consistent iff every solution of $\mathbf{A}^T\mathbf{y}=\mathbf{0}$ satisfies $\langle \mathbf{b},\mathbf{y}\rangle=0$. (3) Repeat for: The system $\mathbf{Ax}=\mathbf{b}$ is inconsistent iff $\mathbf{A}^T\mathbf{y}=\mathbf{0}$ has a solution satisfying $\langle \mathbf{b},\mathbf{y}\rangle\neq 0$. (4) The subject theorem can be expressed as a result holding for any matrix wherein the first column of $\mathbf{A}$ plays the role of $\mathbf{b}$. (5) Which problem has a solution if

$$\mathbf{A}=\begin{bmatrix}1 & 2 & -2\\ 0 & -1 & 2\\ 1 & -2 & 6\end{bmatrix},\qquad \mathbf{b}=\begin{bmatrix}1\\0\\1\end{bmatrix}$$

(6) Necessary and sufficient conditions for $\mathbf{b}$ such that Gale I has a solution for $\mathbf{A}$ in (5). (7) Repeat for Gale II.

4.2.3 Farkas' Theorem (4.31). Establish: (1) Whether or not $\langle \mathbf{c},\mathbf{x}\rangle=1$ can replace $\langle \mathbf{c},\mathbf{x}\rangle>0$ in Farkas II. (2) The following is an equivalent statement. Every $\mathbf{y}$ satisfying $\mathbf{A}^T\mathbf{y}\geqq\mathbf{0}$ also satisfies $\langle \mathbf{b},\mathbf{y}\rangle\geqq 0$ iff $\mathbf{b}=\mathbf{Ax}$ for some $\mathbf{x}\geqq\mathbf{0}$. (3) Repeat for: $\langle \mathbf{c},\mathbf{x}\rangle\geqq 0$ for all $\mathbf{x}$ satisfying $\mathbf{Ax}\geqq\mathbf{0}$ iff $\mathbf{c}=\mathbf{A}^T\mathbf{y}$ for some $\mathbf{y}\geqq\mathbf{0}$. (4) An equivalent statement in terms of the first column and the submatrix composed of the remaining columns of $\mathbf{A}$. (5) The region for $\mathbf{c}$ such that Farkas II will have a solution for $\mathbf{a}_1$ and $\mathbf{a}_2$ shown in Figure 4.9(a). (6) Which Farkas problem has a solution if $[\mathbf{A}^T,\mathbf{c}]$ is given by

$$\begin{bmatrix}1 & 0 & 0 & 1 & 2\\ 0 & 1 & 0 & 0 & 1\\ 0 & 0 & 1 & 0 & 1\end{bmatrix}$$

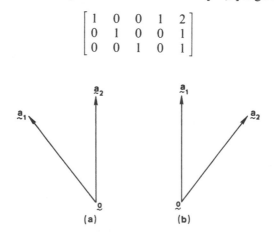

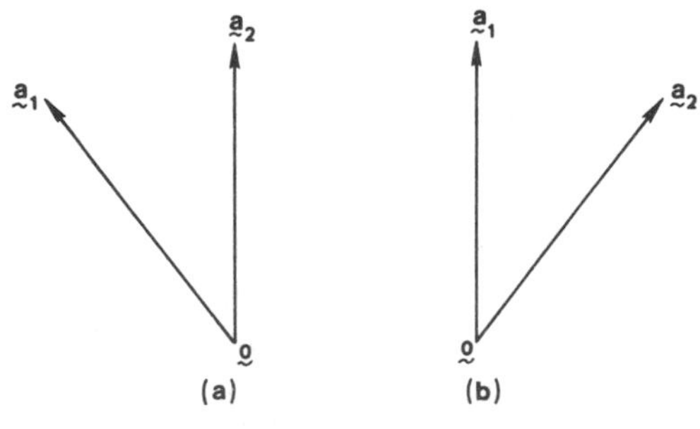

Figure 4.9. Problems in $\Re^2$.

(7) Repeat for

$$\begin{bmatrix} 1 & 0 & -1 & 1 & 1 \\ -1 & 1 & 2 & -1 & 1 \\ -1 & 0 & 2 & -1 & 0 \end{bmatrix}$$

(8) Necessary and sufficient conditions for $\mathbf{b}$ such that Farkas I has a solution for $\mathbf{A}$ in 4.2.2(5). (9) Repeat for Farkas II. (10) Specific consequences of specifying $\operatorname{rank}\mathbf{A} = n < m$. (11) Repeat for $\operatorname{rank}\mathbf{A} = m < n$.

4.2.4 Example 4.1. Which of the following correspond to orders that can be filled?

1. $\mathbf{c} = [22, 24, 26, 28]^T$
2. $\mathbf{c} = [40, 10, 20, 30]^T$
3. $\mathbf{c} = [40, 20, 24, 16]^T$

4.2.5 Gordan's Theorem (4.34). Establish: (1) Which problem has a solution if $\mathbf{A}$ is (i) nonsingular; (ii) rank m; (iii) rank n. (2) Where $\mathbf{a}_3$ must lie in order that Gordan II have a solution in Figure 4.9(b). (3) Which problem has a solution for the matrix in 4.2.2(5), and find a solution. (4) $\mathbf{Ax} > \mathbf{0}$ has a solution $\mathbf{x}$ iff $\mathbf{Az} > \mathbf{e}$ has a solution $\mathbf{z}$. (5) An equivalent form that bears the same relation to (4.34) that 4.2.3(2) does to (4.31). (6) Repeat for 4.2.3(3).

4.2.6 Stiemke's Theorem (4.36). Establish: (1) A geometric version that is a counterpart to (4.35). (2) The counterpart to 4.2.5(5). (3) Repeat for 4.2.5(6).

4.2.7 Nonhomogeneous Stiemke's Theorem (4.37). Establish (1) The counterpart to 4.2.5(5). (2) Repeat for 4.2.5(6). (3) Neither of the following can replace the two conditions in Problem II.

(i) $\mathbf{Ax} \leq \mathbf{0}$ and $\langle \mathbf{c}, \mathbf{x} \rangle > 0$
(ii) $\mathbf{Ax} \leq \mathbf{0}$ and $\langle \mathbf{c}, \mathbf{x} \rangle \geqq 0$

(4) There are three distinct possibilities for conditions on the pair $\mathbf{Ax}$ and $\langle \mathbf{c}, \mathbf{x} \rangle$ when Problem II has a solution. These can be used to replace II by a set of three problems.

4.2.8 Table 4.3. Accomplish the following. (1) Resolve existence for the five categories of solutions for the homogeneous systems with matrices given in 3.1.4(1)–(5). (2) Repeat for nonhomogeneous systems with RHSs (i) $[1, 2, 3]^T$; (ii) $[1, 1, 2]^T$; (iii) $[1, 0, 0]^T$. (3) Show how the subject table can be used to resolve existence of restricted solutions as follows for homogeneous systems. (i) Nonpositive; (ii) seminegative; (iii) negative. (4) Repeat for nonhomogeneous systems. (5) Use (3.32) to write all solutions in (1) and (2).

4.2.9 Nonzero Solutions. Show that each of the following is necessary and sufficient for $\mathbf{A}^T\mathbf{y}=\mathbf{0}$ to have a nonzero solution in $\mathcal{R}^m$. (1) There is a solution $\mathbf{y}$ such that $\langle\mathbf{u},\mathbf{y}\rangle\neq 0$ for at least one $\mathbf{u}\in\mathcal{R}^m$. (ii) There is a solution $\mathbf{y}$ such that $\langle\mathbf{v},\mathbf{y}\rangle=1$ for at least one $\mathbf{v}\in\mathcal{R}^m$. (iii) There is a solution $\mathbf{y}$ such that $\langle\mathbf{w},\mathbf{y}\rangle=-1$ for at least one $\mathbf{w}\in\mathcal{R}^m$. (iv) There is a solution $\mathbf{y}$ such that $\|\mathbf{y}\|=1$.

4.2.10 Equivalent Systems of Equations. Establish the equivalence of $\mathbf{Ax}=\mathbf{b}$ and (1) The $2m$ inequalities in $\mathbf{Ax}\leqq\mathbf{b}$ and $\mathbf{Ax}\geqq\mathbf{b}$. (2) The $m+1$ inequalities in $\mathbf{Ax}\leqq\mathbf{b}$ and $\langle\mathbf{e},\mathbf{Ax}\rangle\geqq\langle\mathbf{e},\mathbf{b}\rangle$.

4.2.11 Gale's Theorem for Inequalities (4.38). Establish: (1) Geometric interpretations using convex polyhedra, polytopes, and cones as appropriate in addition to subspaces and k-planes. (2) Conditions under which it is predictable which problem has a solution. (3) Where appropriate vector must lie in Figure 4.9(a) or (b) in order that Problem II have a solution. (4) Solutions for numerical exercises in 4.2.8(2).

4.2.12 Gale's Theorem for Nonnegative Solutions (4.39). Repeat the preceding, noting that input data consist of $\mathbf{A}$ and $\mathbf{c}$.

4.2.13 Von Neumann's Theorem for Semipositive Solutions (4.40). Repeat, noting that input data consists of $\mathbf{A}$ alone.

4.2.14 Tucker's Theorem for Positive Solutions (4.41). Repeat for this case.

4.2.15 Proof of (4.41). Supply a proof.

4.2.16 Typical Uses. Describe typical uses of theorems of the alternative as can be deduced from the present section.

4.3 BASIC SOLUTIONS

Here we present results that are most closely associated with linear programming. They are used for Example 7.4.

Both algebraic and geometric properties of solution sets for linear equations and inequalities are expressible in terms of the

Definition. A solution $\mathbf{x}$ of $\mathbf{Ax}=\mathbf{b}$ is a *basic solution* if the columns of $\mathbf{A}$ corresponding to nonzero components are LI. (4.42)

There are no solutions of any kind unless $\mathbf{Ax}=\mathbf{b}$ is consistent. If there is a unique solution it must be basic because, according to 3.2.1, $\mathbf{A}$ has full column rank and every collection of its columns must then be LI. This means that (4.42) provides nothing new unless there are an infinite number of solutions. But in every case

The number of basic solutions is finite. (4.43)

PROOF. The number cannot exceed the number of ways of selecting subsets of the set of columns of **A**. This is true because if two basic solutions **y** and **z** had identically located nonzero components, say, their first r, then from

$$y_1\mathbf{a}_1 + y_2\mathbf{a}_2 + \cdots + y_r\mathbf{a}_r = \mathbf{b}$$

we could subtract

$$z_1\mathbf{a}_1 + z_2\mathbf{a}_2 + \cdots + z_r\mathbf{a}_r = \mathbf{b}$$

to show that $\{\mathbf{a}_1, \mathbf{a}_2, \ldots, \mathbf{a}_r\}$ would be LD. ■

The relationship between basic solutions and subsets of columns is characterized in the following where we shift our emphasis to zero coordinates.

Suppose **A** is $m \times n$ and $\operatorname{rank}\mathbf{A} = r$. If **x** is a basic solution of $\mathbf{Ax} = \mathbf{b}$ then **A** has at least one $m \times r$ submatrix **R** such that
(1) $\operatorname{rank}\mathbf{R} = r$, and
(2) The components of **x** corresponding to columns not appearing in **R** are zero.
Conversely, to each $m \times r$ submatrix **R** of **A** having rank r there corresponds a unique basic solution. (4.44)

PROOF. If there are k nonzero components in the basic solution **x**, then $k \leqq r$ and we can adjoin $r - k$ columns of **A** to those corresponding to the nonzero components so as to find **R** satisfying (1) and (2). The converse follows from (3.18). ■

Given an $m \times r$ submatrix **R** of rank r, the associated basic solution is called the *basic solution of* $\mathbf{Ax} = \mathbf{b}$ *corresponding to* **R**. The name derives from the fact that the columns of **R** form a basis for Range **A**. The components of **x** associated with the columns of **R** are called *basic variables* and the remainder are the *nonbasic variables*; this generalizes the usage introduced for (3.18). (We do not define basic variables relative to a nonbasic solution.) If at least one basic variable has value zero, then the basic solution is called a *degenerate basic solution*; otherwise it is *nondegenerate*. (Degeneracy is not defined for nonbasic solutions.) There is possible ambiguity as to which are the nonbasic variables when we are given a basic solution with **R** unspecified.

Let us suppose $\operatorname{rank}\mathbf{A}=m<n$ and consider some examples. If $n=2$, then there can be only $m=1$ equation. Given $\mathbf{A}=[1,0]$ and $b=3$, the system is $x_1=3$ and there is a unique basic solution $\mathbf{x}=[3,0]^T$ and it is nondegenerate. The point $[3,1]^T$ represents a nonbasic solution. For the system $3x_1+x_2=9$ there are two basic nondegenerate solutions, $[0,9]^T$ and $[3,0]^T$; but if we change to $b=0$ all basic solutions are degenerate.

Basic solutions have applications to the linear inequality

Problem. If $\mathbf{A}$ is $m\times n$ and $\mathbf{b}\in\mathcal{R}^m$, find $\mathbf{x}\in\mathcal{R}^n$ such that $\mathbf{Ax}\leqq\mathbf{b}$ and $\mathbf{x}\geqq\mathbf{0}$. (4.45)

This is an important problem because it concerns a subset of $\mathcal{R}^n$ that often appears in optimization problems. As formulated it is Gale's Problem I of (4.39) and there is a solution iff the associated Problem II has no solution. When $\mathbf{b}=\mathbf{0}$ Problem (4.45) is concerned with the intersection of a convex cone and the nonnegative orthant. For $\mathbf{b}\neq\mathbf{0}$ the cone is replaced by a polytope.

We can introduce *slack variables* $z_i\geqq 0$ (as in the proof of (4.39)) to replace $\mathbf{Ax}\leqq\mathbf{b}$ and $\mathbf{x}\geqq\mathbf{0}$ by

$$\mathbf{Ax}+\mathbf{I}_m\mathbf{z}=\mathbf{b},\qquad \mathbf{x}\geqq\mathbf{0},\ \mathbf{z}\geqq\mathbf{0} \tag{4.46}$$

In practice we can find a solution

$$\begin{bmatrix}\mathbf{x}\\ \mathbf{z}\end{bmatrix}\in\mathcal{R}^{m+n}$$

or show that none exists, by using the simplex method of Example 7.4. Sometimes in simple cases we can easily replace (4.46) by a new problem whose RHS is $\mathbf{b}'\geqq\mathbf{0}$. Then starting with $\mathbf{0}$ we can find basic solutions to the original (4.45) as illustrated in the next example.

EXAMPLE 4.2. Finding Basic Solutions by Gauss–Jordan Pivoting. The shaded area in Figure 4.10(a) corresponds to

$$\begin{aligned} 3x_1+\ x_2&\leqq 9\\ x_1+2x_2&\leqq 8\\ 2x_1-\ x_2&\leqq 6\\ x_1\quad\ &\leqq 3\\ x_1,x_2&\geqq 0 \end{aligned}$$

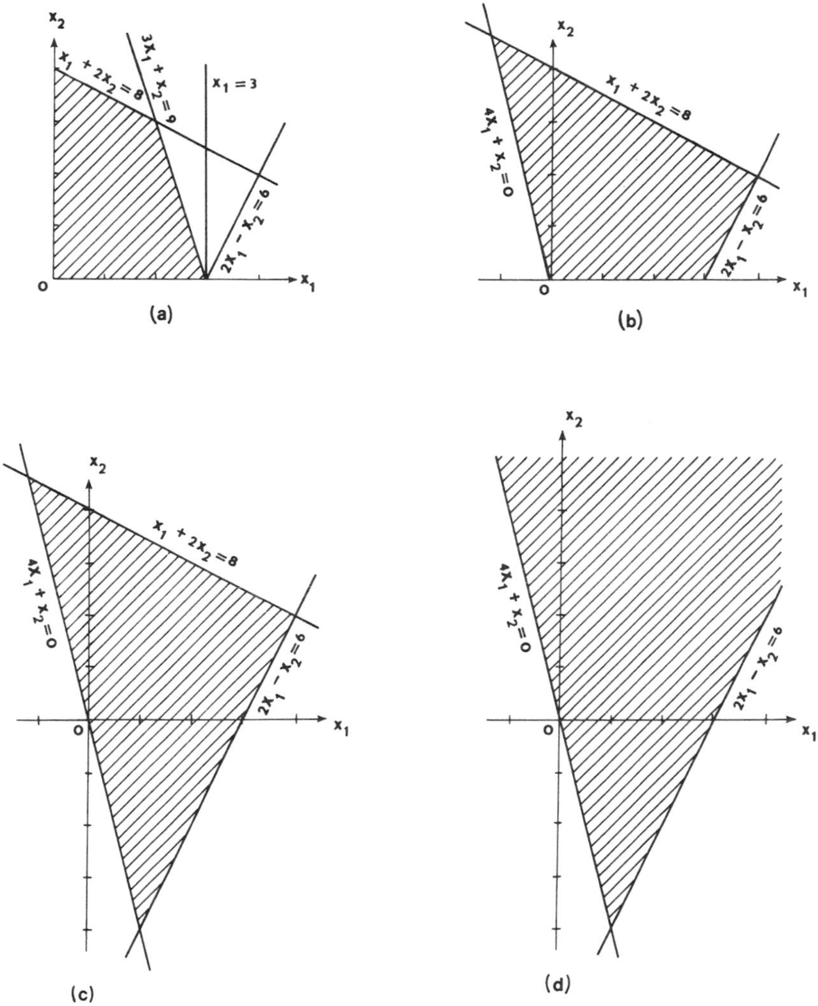

Figure 4.10. Problem settings in $\Re^2$.

If we introduce slack variables $z_i \geqq 0$ we obtain

(1)
$$
\begin{aligned}
3x_1 + x_2 + z_1 \qquad\qquad\quad &= 9\\
x_1 + 2x_2 \quad + z_2 \qquad\quad &= 8\\
2x_1 - x_2 \qquad\quad + z_3 \quad &= 6\\
x_1 \qquad\qquad\qquad\quad + z_4 &= 3
\end{aligned}
$$

$$x_1, x_2 \geqq 0;\ z_1, z_2, z_3, z_4 \geqq 0$$

Notice that we have adjoined $\mathbf{I}_4$ to the matrix of the original LHSs so that now we surely have a coefficient matrix of rank 4. Table 4.6 shows the correspondence between zero values of the six variables in (1) and hyperplanes in $\mathfrak{R}^2$. The shaded area in Figure 4.10(a) has four vertexes formed by intersections of hyperplanes and we see from (1) that they are the following points $[x_1, x_2, z_1, z_2, z_3, z_4]^T$

$$
\begin{gathered}
[0,0,9,8,6,3]^T\\
[3,0,0,5,0,0]^T\\
[2,3,0,0,5,1]^T\\
[0,4,5,0,10,3]^T
\end{gathered}
$$

Each point is nonnegative and each is readily shown to correspond to a basic solution of (1) by showing that four LI columns remain after deleting two columns from the coefficient matrix corresponding to nonbasic variables. Notice that there is ambiguity in the case of the second point as to which are the basic variables; this is the algebraic symptom of a degener-

TABLE 4.6 HYPERPLANES IN FIGURE 4.10(a)

Value of Variable	Associated Hyperplane
$x_1 = 0$	x_2-axis
$x_2 = 0$	x_1-axis
$z_1 = 0$	$3x_1 + x_2 = 9$
$z_2 = 0$	$x_1 + 2x_2 = 8$
$z_3 = 0$	$2x_1 - x_2 = 6$
$z_4 = 0$	$x_1 = 3$

ate basic solution and the geometric interpretation, as shown in (a), is that more than two hyperplanes intersect at the associated point.

Let us remove the two redundant equations and their associated variables z_3 and z_4. Then the conditions defining the shaded area in (a) are

(2)
$$\begin{aligned} 3x_1 + x_2 + z_1 \quad &= 9 \\ x_1 + 2x_2 \quad + z_2 &= 8 \\ x_1, x_2 \geqq 0;\ z_1, z_2 &\geqq 0 \end{aligned}$$

This problem of finding $[x_1, x_2, z_1, z_2]^T$ is conveniently represented by annotating the augmented matrix as follows in the format of a *tableau*

(3)
$$\begin{array}{c} \begin{matrix} \mathbf{a}_1 & \mathbf{a}_2 & \mathbf{e}_1 & \mathbf{e}_2 & \mathbf{b} \end{matrix} \\ \begin{bmatrix} 3 & 1 & 1 & 0 & 9 \\ 1 & 2 & 0 & 1 & 8 \end{bmatrix} \begin{matrix} \mathbf{e}_1 \\ \mathbf{e}_2 \end{matrix} \end{array}$$

where in this case it is understood that all variables are to be nonnegative. Column labels identify the vectors occupying the columns. What is new is the use of row labels to identify the basis being used for Range$[\mathbf{a}_1, \mathbf{a}_2]$. Let us write

$$\begin{aligned} \mathbf{a}_1 &= 3\mathbf{e}_1 + \mathbf{e}_2 \\ \mathbf{a}_2 &= \mathbf{e}_1 + 2\mathbf{e}_2 \\ \mathbf{e}_1 &= \mathbf{e}_1 \\ \mathbf{e}_2 &= \mathbf{e}_2 \\ \mathbf{b} &= 9\mathbf{e}_1 + 8\mathbf{e}_2 \end{aligned}$$

From (3) we immediately read off the basic solution $[0,0,9,8]^T$ corresponding to the associated basis $\{\mathbf{e}_1, \mathbf{e}_2\}$.

It is readily seen that the effect of a Gauss–Jordan pivot at location (i,j), $j \neq 5$, in (3) is to bring the vector labeling column j into the basis replacing the previous vector labeling row i. Column labels remain unchanged during Gauss–Jordan pivoting and only their coordinates change as basis vectors, and hence row labels, change. For example, if we pivot on the

(1, 1) position the new tableau is

$$\begin{array}{ccccc} \mathbf{a}_1 & \mathbf{a}_2 & \mathbf{e}_1 & \mathbf{e}_2 & \mathbf{b} \end{array}$$
$$\begin{bmatrix} 1 & \frac{1}{3} & \frac{1}{3} & 0 & 3 \\ 0 & \frac{5}{3} & -\frac{1}{3} & 1 & 5 \end{bmatrix} \begin{array}{c} \mathbf{a}_1 \\ \mathbf{e}_2 \end{array}$$

This yields the nonnegative basic solution $[3,0,0,5]^T$. If we pivot at (2, 1) in (3) we obtain

$$\begin{array}{ccccc} \mathbf{a}_1 & \mathbf{a}_2 & \mathbf{e}_1 & \mathbf{e}_2 & \mathbf{b} \end{array}$$
$$\begin{bmatrix} 0 & -5 & 1 & -3 & -15 \\ 1 & 2 & 0 & 1 & 8 \end{bmatrix} \begin{array}{c} \mathbf{e}_1 \\ \mathbf{a}_1 \end{array}$$

Here the basic solution $[8,0,-15,0]^T$ is not a solution to (2) because it is not nonnegative. Its geometric interpretation is that it represents the intersection $[8,0]^T$ of the second and fourth hyperplanes in Table 4.6, two hyperplanes that do not intersect at a vertex for the region of interest (4.3.5).

Let us note that in this simple example

$$\mathbf{A} = \begin{bmatrix} 3 & 1 \\ 1 & 2 \end{bmatrix}$$

is nonsingular and so (2) has solutions

$$\mathbf{x} = \mathbf{A}^{-1}\mathbf{b} - \mathbf{A}^{-1}\mathbf{z}$$

where both $\mathbf{z}$ and the RHS are nonnegative. This approach is not available in exactly this form for rectangular $\mathbf{A}$, and, even for large nonsingular $\mathbf{A}$, it is not efficient. The simplex method of solving linear programming problems introduced in Example 7.4 is an effective method of solving (4.46) in order to solve (4.45).

Let us name a new standard problem to cover (4.46). In place of the coefficient matrix $[\mathbf{A}, \mathbf{I}_m]$ let us simply write $\mathbf{A}$ and let us also revert to use of $\mathbf{x}$ alone in the

Problem. If $\mathbf{A}$ is $m \times n$ and $\mathbf{b} \in \mathfrak{R}^m$, find $\mathbf{x} \in \mathfrak{R}^n$ such that $\mathbf{Ax} = \mathbf{b}$ and $\mathbf{x} \geqq \mathbf{0}$. (4.47)

which is the "dual" Farkas Problem I in Table 4.5. When $\mathbf{b} = \mathbf{0}$ it deals with the intersection of a subspace and the nonnegative orthant. When $\mathbf{b} \neq \mathbf{0}$ the subspace is replaced by a k-plane where $k = n - \operatorname{rank} \mathbf{A}$.

We should note that it is conventional in optimization theory to use the term *feasible point* for solution points of Problems (4.45) and (4.47). Such terminology is used throughout Chapter 7 for points lying in constraint sets of which the above sets of solution points provide important examples. Solution points in Chapter 7 are feasible points at which "objective functions" achieve extrema. We have no objective functions in the present chapter and we continue to use the term solution point, or solution vector, for Problems (4.45) and (4.47). This should cause no confusion because we use this terminology only in relation to clearly identified problems.

It may be surprising that inequality Problem (4.45) can be replaced by (4.47); after all, we learn how to solve linear equations long before meeting linear inequalities. What may be even more surprising is that Problem (4.47) qualifies as a standard form for any problem of finding a vector satisfying a finite number of linear inequalities, with both $\leqq$ and $\geqq$ permitted, and linear equations, with signs for individual x_i constrained or not. By this last we mean that some or all of the variables x_i may be *free variables* in that neither $x_i \geqq 0$ nor $x_i \leqq 0$ is specified. In general we have cases of the

Problem. If $\mathbf{B}, \mathbf{C}, \mathbf{D}$ are $p_1 \times q, p_2 \times q, p_3 \times q$, respectively, and $\mathbf{f} \in \mathfrak{R}^{p_1}$, $\mathbf{g} \in \mathfrak{R}^{p_2}$, $\mathbf{h} \in \mathfrak{R}^{p_3}$ find $\mathbf{z} \in \mathfrak{R}^q$ such that

$$\mathbf{Bz} \leqq \mathbf{f}, \ \mathbf{Cz} = \mathbf{g}, \ \mathbf{Dz} \geqq \mathbf{h} \tag{4.48}$$

and the result described above is

Any Problem (4.48) can be replaced by an instance of (4.47). (4.49)

Let us specify the transformations that can be used in practice and upon which proof can be based.

First, we have already seen in (4.46) how slack variables are used to convert inequalities $\leqq$ to equations. Second, we can adapt the same idea to change inequalities $\geqq$ to equations. Given an inequality $\langle \mathbf{c}, \mathbf{x} \rangle \geqq d$ we add $-z_i$ to the LHS to obtain the equivalent

$$\langle \mathbf{c}, \mathbf{x} \rangle - z_i = d, \qquad z_i \geqq 0$$

and call z_i a *surplus variable*. (Equivalently, we can multiply both sides of the inequality by -1 to change it to $\leqq$, and then add a slack variable z_i to obtain $\langle -\mathbf{c}, \mathbf{x} \rangle + z_i = -d$.) Third, if a variable x_i is a free variable, that is, if it is unrestricted in sign, there are two procedures that can be used.

PROCEDURE (1). The free variable x_i can be eliminated by Gauss–Jordan pivoting on one of its nonzero coefficients in the augmented matrix for the final set of equations. If there exists a solution for the new problem (having one less variable and one less equation) then we find the associated value for x_i from the original pivot row.

PROCEDURE (2). The free variable x_i can be replaced through $x_i = u_i - v_1$ where $u_i \geqq 0$ and $v_1 \geqq 0$ are specified. Linearity is preserved in all constraining equations and there is a net increase of one variable in the new problem. In fact, if there are k free variables, then, as readily verified, they can be replaced by $k+1$ nonnegative variables u_i and v_1 where

$$x_i = u_i - v_1, \qquad u_i \geqq 0$$

for $i = 1, 2, \ldots, k$, and $v_1 \geqq 0$.

Fourth, we can guarantee that the column rank of the coefficient matrix for the new problem coincides with that for the old when we transform with new variables to replace free variables.

EXAMPLE 4.3. Transforming Problems into Instances of (4.47). The shaded area in Figure 4.10(b) is defined by

$$\begin{aligned} 2x_1 - \ x_2 &\leqq 6 \\ x_1 + 2x_2 &\leqq 8 \\ 4x_1 + \ x_2 &\geqq 0 \\ x_2 &\geqq 0 \end{aligned}$$

where x_1 is a free variable. An equivalent set of equations is

$$\begin{aligned} 2x_1 - \ x_2 + z_1 \qquad\qquad &= 6 \\ x_1 + 2x_2 \qquad + z_2 \qquad &= 8 \\ 4x_1 + \ x_2 \qquad\qquad - z_3 &= 0 \\ x_2 \geqq 0;\ z_1, z_2, z_3 &\geqq 0 \end{aligned}$$

where z_3 is a surplus variable. If we follow Procedure (1) and use the first equation to eliminate x_1 we obtain

$$\begin{aligned} 5x_2 - z_1 + 2z_2 &= 10 \\ 3x_2 - 2z_1 - z_3 &= -12 \\ x_2 \geqq 0;\ z_1, z_2, z_3 &\geqq 0 \end{aligned}$$

which provides an instance of (4.47).

Both x_1 and x_2 are free variables in the following specifications for the shaded area in (c).

$$\begin{aligned} 2x_1 - x_2 + z_1 \qquad &= 6 \\ x_1 + 2x_2 \quad + z_2 \quad &= 8 \\ 4x_1 + x_2 \qquad - z_3 &= 0 \\ z_1, z_2, z_3 &\geqq 0 \end{aligned}$$

If we follow Procedure (2) and introduce $u_1 - v_1 = x_1$, $u_2 - v_1 = x_2$ we obtain

$$\begin{aligned} 2u_1 - u_2 - v_1 + z_1 \qquad &= 6 \\ u_1 + 2u_2 - 3v_1 \quad + z_2 \quad &= 8 \\ 4u_1 + u_2 - 5v_1 \qquad - z_3 &= 0 \\ u_1, u_2 \geqq 0;\ v_1 \geqq 0;\ z_1, z_2, z_3 &\geqq 0 \end{aligned}$$

which is another instance of (4.47).

The principal feature of the shaded area in (d) insofar as we are presently concerned, is that it is an unbounded region. It is a polytope, it is not a convex polyhedron, and it is not a cone. We could follow (2) to replace the free variables x_1 and x_2 by nonnegative u_1, u_2 and v_1 but, as treated in 4.3.4(5), it is preferable to follow (1) and eliminate both x_1 and x_2.

Whenever a system of linear equations has a solution it must have a basic solution; this is shown in (3.18). A deeper result, and one that is important in the simplex method of solving linear programming problems, is that if there is a nonnegative solution then there is one that is also a basic solution. This is the

THEOREM. If there exists a solution to (4.47) then there exists a basic solution. (4.50)

PROOF. Let us use strong induction on n. The theorem holds for $n=1$ and we suppose it holds for all $k<n$. We may limit our attention to solutions $\mathbf{x}>\mathbf{0}$ since with even one zero component we can use the inductive hypothesis to guarantee that there is a basic solution. If the columns of $\mathbf{A}$ are LI then $\mathbf{x}>\mathbf{0}$ is a basic solution and so we suppose them to be LD. This means $\mathbf{Au}=\mathbf{0}$ for some $\mathbf{u}\neq\mathbf{0}$ and we lose no generality by supposing not only $u_1>0$ but also

$$\frac{u_1}{x_1}=\max\left\{\frac{u_1}{x_1},\frac{u_2}{x_2},\ldots,\frac{u_n}{x_n}\right\}$$

If we use $\mathbf{Ax}=\mathbf{b}$ and $-(x_1/u_1)\mathbf{Au}=\mathbf{0}$ to form $\mathbf{A}\{\mathbf{x}-(x_1/u_1)\mathbf{u}\}=\mathbf{b}$ then the first component of the vector $\{\cdot\}$ is zero and all of them are nonnegative. Thus we have once again reduced the argument to a case covered by the inductive hypothesis. ■

The main result in this section is the

THEOREM. A vector $\mathbf{x}$ is a vertex of the convex set

$$K=\{\mathbf{x}:\mathbf{Ax}=\mathbf{b},\ \mathbf{x}\geqq 0\}$$

iff $\mathbf{x}$ is a basic solution to (4.47). (4.51)

PROOF. We confine our attention to $\mathbf{b}\neq\mathbf{0}$ because the contrary case has $\mathbf{x}=\mathbf{0}$ as the only vertex and is readily handled. (See 4.3.7 for the preceding and for other elementary details.) If $\mathbf{x}$ is a basic solution and rank $\mathbf{A}=r$ then we lose no generality by assuming that the first r columns of $\mathbf{A}$ are LI and the first r components of $\mathbf{x}$ are positive. If $\mathbf{x}$ is not a vertex of K then

$$\mathbf{x}=t\mathbf{y}+(1-t)\mathbf{z}$$

where $t\in(0,1)$ and $\mathbf{y},\mathbf{z}\in K$ are distinct. But then the final $n-r$ components of $\mathbf{y}$ and $\mathbf{z}$ are zero because those of $\mathbf{x}$ have this property and, as further consequences of $\mathbf{y},\mathbf{z}\in K$, we can use the subtraction technique employed in the proof of (4.43) to show $\mathbf{y}=\mathbf{z}$. We can also use an indirect proof for the converse. If $\mathbf{x}$ is a vertex of K we lose no generality by assuming

$$x_1\mathbf{a}_1+x_2\mathbf{a}_2+\cdots+x_k\mathbf{a}_k=\mathbf{b}$$

where each of $x_1,x_2,\ldots,x_k$ is positive. Suppose $\{\mathbf{a}_1,\mathbf{a}_2,\ldots,\mathbf{a}_k\}$ is LD. Then there exist k scalars c_i, not all zero, such that

$$c_1\mathbf{a}_1+c_2\mathbf{a}_2+\cdots+c_k\mathbf{a}_k=\mathbf{b}$$

Since $x_1>0$ we can choose $\varepsilon_1>0$ as follows depending on the sign of c_1; if $c_1>0$ we choose ε_1 to satisfy $x_1>\varepsilon_1 c_1$ and then of course also $x_1>-\varepsilon_1 c_1$; if $c_1=0$ then for specificity $\varepsilon_1=1$ satisfies $x_1>\pm\varepsilon_1 c_1$; if $c_1<0$ we choose ε_1 to satisfy $x_1>-\varepsilon_1 c_1$ and then of course also $x_1>\varepsilon_1 c_1$. If we repeat the preceding for $x_2,\ldots,x_k$ and define $\varepsilon=\min\{\varepsilon_1,\varepsilon_2,\ldots,\varepsilon_k\}$ we are led to a contradiction of the hypothesis that x is a vertex: $\mathbf{x}=(1/2)(\mathbf{x}+\varepsilon\mathbf{c})+(1/2)(\mathbf{x}-\varepsilon\mathbf{c})$ where $\mathbf{c}=[c_1,c_2,\ldots,c_k,0,\ldots,0]^T$. ■

PROBLEMS FOR SOLUTION

4.3.1 Fundamental Properties. Establish: (1) Definition (4.42) does not define basic variables. (2) An alternative for (4.42) that uses a definition of basic variables. (3) If $\mathbf{b}=\mathbf{0}$ then $\mathbf{x}=\mathbf{0}$ is the only basic solution. (4) The analysis of all basic solutions in the text for $\mathbf{A}$ $m\times n$, $\operatorname{rank}\mathbf{A}=m<n$, can readily be extended to $n=3$. (5) Any system that is row-equivalent to $\mathbf{Ax}=\mathbf{b}$ has the same set of basic solutions. (6) If $\operatorname{rank}\mathbf{A}=\operatorname{rank}[\mathbf{A},\mathbf{b}]<n$ then we can replace $\mathbf{Ax}=\mathbf{b}$ by an equivalent system having (i) no contradictory equations, (ii) no redundant equations, and (iii) fewer equations than unknowns. (7) The preceding can be used to find a necessary and sufficient condition that the consistent system $\mathbf{Ax}=\mathbf{b}$ have $C(n,r)$ nondegenerate basic solutions where $r=\operatorname{rank}\mathbf{A}$.

4.3.2. Numerical Exercises. Use Gauss–Jordan pivoting in the manner of Example 4.2 to find all basic solutions for the systems of equations in x_1 and x_2 representing the shaded area in the following parts of Figure 4.10. (1) Figure (a) as nearly completed in Example 4.2. (2) Figure (b) where note $x_1<0$ is possible. (3) Figure (c). (4) Figure (d).

4.3.3 Two Interpretations. Select one of the preceding parts (1), (2), or (3) and interpret each pivot operation two ways: as changing variables from nonbasic to basic, and as bringing a new vector into the basis.

4.3.4 Problem (4.48). Establish: (1) If there are k free variables they can be replaced by $k+1$ nonnegative variables, u_i and v_1, as stated in Procedure (2). (2) The column rank of the coefficient matrix is not changed in replacing free variables by new nonnegative ones. (3) Procedure (1) can be used to replace the free variables x_1 and x_2 in the representation of the shaded area in Figure 4.10(c). (4) The shaded area in Figure 4.10(d) is a polytope but is not a convex polyhedron or a cone. (5) Procedure (1) can be used to eliminate both free variables from the representation of the shaded area in Figure 4.10(d). (6) Find in this way a formula for $\mathbf{x}$ in terms of $\mathbf{z}$.

4.3.5 Nonnegative Gauss–Jordan Pivoting. Often it is desirable to use special criteria for selecting successive pivot elements; an earlier example is the maximum modulus strategy noted above (3.9) but the primary example is provided by the simplex method of Example 7.4. The present procedure applies to Problems (4.45) and (4.47) given $\mathbf{b}\geqq\mathbf{0}$. For brevity we formulate the procedure for choice of initial pivot in the latter. (For (4.45) we consider $[\mathbf{A},\mathbf{I}_m,\mathbf{b}]$, and for successive pivots for either problem we replace a_{ij} and b_i by $a_{ij}^{(k)}$ and $b_i^{(k)}$ at the kth stage.) There are two rules. (i) Do not pivot in the final column. (ii) It is permissible to select a_{ij} as pivot provided $a_{ij}>0$ and

$$\frac{b_i}{a_{ij}}=\min\left\{\frac{b_k}{a_{kj}}:a_{kj}>0,\ 1\leqq k\leqq m\right\}$$

(1) Show that under such a choice of pivot the final column remains nonnegative. (2) Use the preceding to find all nonnegative basic solutions, if any, in 4.3.2.

4.3.6 Theorem (4.50). Provide: (1) A simple example where (4.47) has no basic solution. (2) Geometric interpretation in $\mathcal{R}^2$ using sketches of possible cases.

4.3.7 Theorem (4.51). Establish (1) The set of solutions to Problem (4.47) is a convex set. (2) If $K\neq\varnothing$ then it has at least one vertex. (3) A simple example in $\mathcal{R}^2$ where K has a unique vertex. (4) The general result in $\mathcal{R}^n$ concerning a unique vertex for K. (5) An illustration of (4.51) for $\mathbf{A}=[\mathbf{a}_1,\mathbf{a}_2]$ where $\mathbf{a}_1=\mathbf{a}_2=\mathbf{b}=[1,1]^T$. (6) Proof for the case $\mathbf{b}=\mathbf{0}$. (7) We lose no generality by assuming that the first r columns of $\mathbf{A}$ are LI and $x_1,x_2,\ldots,x_r$ are positive as stated in the proof. (8) Whether or not the initial r components of $\mathbf{y}$ and $\mathbf{z}$ must be nonzero in the proof. (9) We lose no generality by assuming that $x_1,x_2,\ldots,x_k$ are positive in the proof.

4.3.8 Additional Exercises. Find the GJSF of $[\mathbf{A},\mathbf{b}]$ as given and then find all solutions of Problem (4.47)

$$(1)\ \begin{bmatrix}1&0&1&-3&4\\1&1&0&4&3\\0&0&1&-4&3\end{bmatrix}\qquad(2)\ \begin{bmatrix}1&0&0&1&0&1\\0&1&0&3&1&2\\0&0&1&-4&2&3\end{bmatrix}$$

4.3.9 Another Method. Use nonnegative Gauss–Jordan pivoting (4.3.5) to find some solutions for the preceding two problems.

4.3.10 Nonzero Solutions to $\mathbf{A}^T\mathbf{y}\leqq\mathbf{0}$. Establish for Table 4.4: (1) If rank $\mathbf{A}^T<m$ then nonzero solutions exist. (2) If rank $\mathbf{A}^T=m$ then Stiemke's theorem (4.36) can be used to resolve existence of nonzero solutions. (3) A simple example for the preceding where a solution exists and another where there is none. (4) If $\mathbf{A}$ is square then there exists a nonzero solution to $\mathbf{A}^T\mathbf{y}\leqq\mathbf{0}$. (5) A geometric interpretation for the preceding in (i) $\mathcal{R}^2$, and (ii) $\mathcal{R}^3$. (6) Result (4.62) can be used to state existence criteria in $\mathcal{R}^m$.

4.3.11 Nonzero Solutions to $\mathbf{A}^T\mathbf{y}\leqq\mathbf{c}$. Establish for $\mathbf{c}\neq\mathbf{0}$ in Table 4.4: (1) If $\mathbf{c}$ has components of both signs then Gale's theorem (4.38) can be used to resolve existence of nonzero solutions. (2) Repeat for the case $\mathbf{c}\leq\mathbf{0}$. (3) Details for a resolution procedure as follows for $\mathbf{c}\geq\mathbf{0}$. If $\mathbf{A}^T\mathbf{y}=\mathbf{c}$ is consistent then there is a nonzero solution. If not then the system $\mathbf{A}^T\mathbf{y}+\mathbf{I}_n\mathbf{z}=\mathbf{c}$ has solution $\mathbf{z}=\mathbf{c}\geq 0$, $\mathbf{y}=\mathbf{0}$, on which we can use nonnegative Gauss–Jordan pivoting (4.3.5) to determine whether or not a column of $\mathbf{A}^T$ can be brought into the basis (See Example 7.4 for use of the simplex method.) (4) Gale's (4.30) and (4.38) can be used to handle some of the possibilities given $\mathbf{c}\geq \mathbf{0}$, and inconsistent $\mathbf{A}^T\mathbf{y}=\mathbf{c}$. (5) What can be determined given only that $\mathbf{A}$ is square. (6) A geometric interpretation of the problem of resolving existence of the subject solutions.

4.3.12 Semipositive Solutions to $\mathbf{A}^T\mathbf{y}\leqq \mathbf{c}$. Establish for Table 4.4: (1) Existence of solutions for the three cases (i) $\mathbf{c}$ has components of both signs. (ii) $\mathbf{c}\leq \mathbf{0}$. (iii) $\mathbf{c}\geq \mathbf{0}$. (2) Repeat of preceding (6).

4.3.13 Positive Solutions to $\mathbf{A}^T\mathbf{y}\leqq \mathbf{c}$. Establish for Table 4.4: (1) Relevance of existence of a semipositive solution. (2) Existence of positive solutions given that at least one semipositive solution exists. (3) Repeat of 4.3.11(5).

4.3.14 Extensions of Table 4.4. Supply: (1) Additional rows with entries covering (i) Nonpositive $\mathbf{y}\leqq \mathbf{0}$. (ii) Seminegative $\mathbf{y}\leq \mathbf{0}$. (iii) Negative $\mathbf{y}<\mathbf{0}$. (2) A counterpart table, extended as in the preceding, for solutions to $\mathbf{A}^T\mathbf{y}\leq \mathbf{c}$. (3) Repeat for $\mathbf{A}^T\mathbf{y}<\mathbf{c}$.

4.4 LINEAR INEQUALITIES

The practical key to finding solutions, or showing that none exist, is (4.49) because the simplex method can be used for (4.47) as described in Example 7.4. The theoretical key is that sets of solutions have both "point forms" and "halfspace forms"; five such pairs are exhibited in Table 4.1 (and related results are in Section 7.6).

Let us first show that when the RHS in Problem (4.47) is $\mathbf{0}$ the solution set is a polyhedral convex cone.

The set $U=\{\mathbf{y}:\mathbf{A}^T\mathbf{y}=\mathbf{0},\ \mathbf{y}\geqq \mathbf{0}\}$ is a polyhedral convex cone. (4.52)

PROOF. We work with the related set

$$S=\left\{\mathbf{z}:\mathbf{A}^T\mathbf{z}=\mathbf{0},\ \langle \mathbf{z},\mathbf{e}\rangle=1,\ \mathbf{z}\geqq \mathbf{0}\right\}$$

that is, the solution set for Problem (4.47) with matrix $\mathbf{B}^T$, where $\mathbf{B}=[\mathbf{A},\mathbf{e}]$, and vector $\mathbf{d}=[\mathbf{0}^T,1]^T$. If $\mathbf{y}\in U$ is nonzero then it is a positive multiple of some $\mathbf{z}\in S$. Therefore we use strong induction on m with hypothesis: Every $\mathbf{z}\in S$ consisting of LCs of fewer than m vectors is an LC, with nonnegative coefficients, of basic solutions. Verification for $m=1,2$ is immediate, and, just as in the proof of (4.50) we can limit our attention to

$\mathbf{z} > \mathbf{0}$ and the case where the columns of $\mathbf{B}^T$ are LD for a general value of m. Then by (4.50) there exists a basic solution $\mathbf{u}$ and $u_i = 0$ for at least one i because LD implies that there is at least one nonbasic variable. Also we can be sure that $u_i > z_i$ for some i and we may, in fact, suppose, as for (4.50)

$$\frac{u_1}{z_1} = \max\left\{\frac{u_1}{z_1}, \frac{u_2}{z_2}, \ldots, \frac{u_m}{z_m}\right\}$$

so that $(u_1/z_1) > 1$. We may then define

$$\mathbf{v} = \frac{z_1}{u_1 - z_1}\left(\frac{u_1}{z_1}\mathbf{z} - \mathbf{u}\right)$$

and verify $\mathbf{v} \in S$ with $v_1 = 0$. By the inductive hypothesis, $\mathbf{v}$ is an LC, with nonnegative coefficients, of basic solutions. This means that if we solve the above displayed equation, then

$$\mathbf{z} = \left(\frac{z_1}{u_1}\right)\mathbf{u} + \left(\frac{u_1 - z_1}{u_1}\right)\mathbf{v}$$

is expressed as the sum of a positive multiple of the basic solution $\mathbf{u}$ and a positive multiple of $\mathbf{v}$. ■

Another way of expressing this result is

The intersection of a subspace and the nonnegative orthant is a polyhedral convex cone. (4.53)

We are now able to establish the following from Minkowski (1896) and (1910).

MINKOWSKI'S THEOREM. If $\mathbf{A}$ is any matrix, then

$$K = \{\mathbf{y} : \mathbf{A}^T\mathbf{y} \leqq \mathbf{0}\}$$

is a polyhedral convex cone. (4.54)

PROOF. The following is a polyhedral convex cone by (4.53).

$$(\text{Range } \mathbf{A}^T) \cap \{\mathbf{x} : \mathbf{x} \geqq \mathbf{0}\}$$

Suppose it is generated by $\mathbf{x}_1, \mathbf{x}_2, \ldots, \mathbf{x}_r$ where, since all $\mathbf{x}_i \in \text{Range } \mathbf{A}^T$, we may assign

$$-\mathbf{x}_i = \mathbf{A}^T\mathbf{u}_i$$

for $i=1,2,\ldots,r$. Next, let C_1 denote the polyhedral convex cone generated by $\mathbf{u}_1,\mathbf{u}_2,\ldots,\mathbf{u}_r$. Now $\operatorname{Null}\mathbf{A}^T$ is a subspace and consequently also a polyhedral convex cone; let us identify it by C_2 and suppose it is generated by $\mathbf{v}_1,\mathbf{v}_2,\ldots,\mathbf{v}_s$. To complete the proof we show $K=C_1+C_2$. If $\mathbf{y}\in K$ then we may write

$$-\mathbf{A}^T\mathbf{y}=b_1\mathbf{x}_1+b_2\mathbf{x}_2+\cdots+b_r\mathbf{x}_r$$

Using (2.15) and (2.16), this last can be rewritten in our usual notation as

$$\mathbf{A}^T\mathbf{y}=\mathbf{A}^T\mathbf{U}\mathbf{b}$$

Consequently $\mathbf{y}-\mathbf{U}\mathbf{b}\in\operatorname{Null}\mathbf{A}^T$ has the representation

$$\mathbf{y}-\mathbf{U}\mathbf{b}=\mathbf{V}\mathbf{c}$$

for some $\mathbf{c}\geqq\mathbf{0}$ which shows $\mathbf{y}\in C_1+C_2$. ■

Minkowski's theorem establishes the more difficult half of

A polytope $K=\{\mathbf{y}:\mathbf{A}^T\mathbf{y}\leqq\mathbf{c}\}$ is a polyhedral convex cone iff $c=\mathbf{0}$. (4.55)

PROOF. It remains to prove "only if." Actually, if K is any kind of cone, $\mathbf{0}\in K$ and necessarily $\mathbf{c}\geqq\mathbf{0}$. But if $\mathbf{y}\in K$ and $t>0$ then $t\mathbf{y}\in K$ and this requires $\mathbf{c}=\mathbf{0}$. ■

Here is another

ALTERNATIVE FORM OF FARKAS' THEOREM. If C is a polyhedral convex cone, then $C^{**}=C$. (4.56)

PROOF. If $C=\{\mathbf{c}:\mathbf{c}=\mathbf{A}^T\mathbf{y},\ \mathbf{y}\geqq\mathbf{0}\}$ then the polar (4.23) is

$$C^*=\{\mathbf{x}:\langle\mathbf{y},\mathbf{A}\mathbf{x}\rangle\leqq 0\text{ for all }\mathbf{y}\geqq\mathbf{0}\}$$

Consequently the polar of the polar is

$$C^{**}=\{\mathbf{c}:\mathbf{A}\mathbf{x}\leqq\mathbf{0}\text{ implies }\langle\mathbf{c},\mathbf{x}\rangle\leqq 0\}$$

Now let us refer to the Farkas problems (4.31). If $\mathbf{c}\in C$ then Farkas I has a solution, II has none, and $\mathbf{c}\in C^{**}$. Conversely, if $\mathbf{c}\in C^{**}$ then II has none, I has one, and $\mathbf{c}\in C$. ■

One of the principal ways in which this result is used for polyhedral convex cones is in the form of the following

THEOREM. Each of $K=\{\mathbf{y}:\mathbf{A}^T\mathbf{y}\leqq\mathbf{0}\}$ and

$$C=\{\mathbf{y}:\mathbf{y}=\mathbf{A}\mathbf{x},\ \mathbf{x}\geqq\mathbf{0}\}$$

is the polar of the other. (4.57)

PROOF. First of all, by (4.54) each of K and C is a polyhedral convex cone. By definition, $\mathbf{y} \in C^*$ iff

$$\langle \mathbf{y}, \mathbf{A}\mathbf{x} \rangle = \langle \mathbf{A}^T\mathbf{y}, \mathbf{x} \rangle \leqq 0$$

for all $\mathbf{x} \geqq \mathbf{0}$, and this happens iff $\mathbf{A}^T\mathbf{y} \leqq \mathbf{0}$. Thus $C^* = K$ and then $C = C^{**} = K^*$ from (4.56). ■

A notable consequence of this theorem is

The polar of a polyhedral convex cone is also a polyhedral convex cone. (4.58)

Additional properties that also follow readily from the results we have established are contained in the so-called

MODULARITY THEOREM. If C_1 and C_2 are polyhedral convex cones in $\mathcal{R}^m$ then so also are $C_1 + C_2$ and $C_1 \cap C_2$, and the following relations hold:

(1) $(C_1 + C_2)^* = C_1^* \cap C_2^*$.

(2) $(C_1 \cap C_2)^* = C_1^* + C_2^*$. (4.59)

Most of the results in this chapter commencing with Section 4.2 can be readily established using polyhedral convex cones (4.4.1).

Minkowski's theorem (4.54) states that every polytope having $\mathbf{0}$ as the defining RHS vector is a polyhedral convex cone. The converse is from Weyl (1935).

WEYL'S THEOREM. Every polyhedral convex cone is a polytope. (4.60)

PROOF. Let C and $K = C^*$ be as given in (4.57). Then by (4.54) C^* has the alternative representation

$$\{\mathbf{y} : \mathbf{y} = \mathbf{B}\mathbf{v},\ \mathbf{v} \geqq \mathbf{0}\}$$

and $C^{**} = C = \{\mathbf{y} : \mathbf{B}^T\mathbf{y} \leqq \mathbf{0}\}$ by (4.57). ■

See Uzawa (1958) for a formula expressing the columns of $\mathbf{B}$ in terms of those of $\mathbf{A}^T$. It can be shown (see Stoer and Witzgall (1970), for example) that Weyl's theorem implies those of Farkas and Minkowski; in our approach the latter two are used to prove the former.

Weyl's theorem and consequence (4.55) of Minkowski's theorem establish representational relationships between polytopes and polyhedral convex cones. In order to complete the program of establishing pairs of forms we need to relate convex polyhedra and polytopes. That every convex hull of a finite set of points is representable as the intersection of a finite number of halfspaces is assured by the

THEOREM. Every convex polyhedron is a polytope. (4.61)

PROOF. First let us establish a one-to-one correspondence between nonempty convex polyhedra $K \subset \mathfrak{R}^n$ and certain polyhedral convex cones $C \subset \mathfrak{R}^{n+1}$. Given K generated by the columns of $\mathbf{A}^T$ there corresponds C generated by the columns of $[\mathbf{A}, \mathbf{e}]^T$. Notice that C is contained in the halfspace $S = \{\mathbf{z} : z_{n+1} \geqq 0\}$ and it intersects the hyperplane $H = \{\mathbf{z} : z_{n+1} = 1\}$. Conversely, given any polyhedral convex cone $D \subset S$ such that $D \cap H \neq \varnothing$ there corresponds the convex polyhedron

$$\{\mathbf{x} : [\mathbf{x}^T, 1]^T \in D \cap H\} \subset \mathfrak{R}^n$$

Let us now use this correspondence. If K is given then C^* is a polyhedral convex cone by (4.58); with (4.57) and (4.60) we can write

$$C^* = \{\mathbf{d} : \mathbf{d} = \mathbf{B}\mathbf{v},\ \mathbf{v} \geqq \mathbf{0}\}$$

Then we can intersect $C = C^{**} = \{\mathbf{d} : \mathbf{B}^T\mathbf{d} \leqq \mathbf{0}\}$ with H to obtain a representation of K as the intersection of a finite number of halfspaces. ■

As illustrated by Figure 4.10(d) a polytope can be unbounded in which case it clearly cannot be a convex polyhedron. In order to prove the converse to (4.61) for bounded polytopes we need some criteria for unboundedness.

The following statements hold.

(1) A convex set S is unbounded iff it contains a halfline.
(2) A polytope $K = \{\mathbf{y} : \mathbf{A}^T\mathbf{y} \leqq \mathbf{c}\}$ is unbounded iff the polyhedral convex cone

$$C = \{\mathbf{y} : \mathbf{A}^T\mathbf{y} \leqq \mathbf{0}\}$$

contains a nonzero vector. (4.62)

PROOF. The nontrivial part of (1) can be established as: if S contains no halfline then S is bounded. Since S is convex it contains the line segment between any two of its distinct points, say $\mathbf{p}$ and $\mathbf{q}$. If no halfline determined by $\mathbf{p}$ and $\mathbf{q}$ belongs to S, there exists $\mathbf{r} \notin S$ on such a halfline and $\|\mathbf{p}\|, \|\mathbf{q}\| < \|\mathbf{r}\|$. If K is unbounded in (2), it contains a halfline, $\mathbf{c}$ cannot have components of both signs, and $C \neq \{\mathbf{0}\}$. Conversely, if $\mathbf{y}_2 \in C$ is nonzero and $\mathbf{y}_1 \in K$ then

$$\mathbf{y}_1 + k\mathbf{y}_2 \in K$$

for all $k > 0$ and K is unbounded. ■

Using (4.62) together with an appropriate one-to-one correspondence between polytopes in $\mathfrak{R}^m$ and polyhedral convex cones in $\mathfrak{R}^{m+1}$ we can adapt the procedure used to prove (4.61) to establish

A polytope is a convex polyhedron iff the polytope is bounded. (4.63)

The definitive result along this line is (4.4.2)

A polytope is the sum of a convex polyhedron and a polyhedral convex cone. (4.64)

Let us turn to some additional matters connected with the theory of solutions of homogeneous linear inequalities. The solution set

$$C = \left\{ \mathbf{y} : \mathbf{A}^T \mathbf{y} \leqq \mathbf{0} \right\} \tag{4.65}$$

is a polyhedral convex cone by Minkowski's theorem (4.54) and it is natural to ask if we can establish a counterpart to (4.51) whereby all $\mathbf{y} \in C$ can be shown to be expressible as LCs, with nonnegative coefficients, of extreme vectors of C. The answer is "no" because any subspace S can be expressed in the form (4.65) and (4.1.20) S has no extreme vectors. Thus, Minkowski's theorem (4.54) does guarantee that C can be expressed as the set of all LCs, with nonnegative coefficients, of a finite set of points but, according to the subspace counterexample, these latter need not be extreme vectors.

Polyhedral convex cones possessing extreme vectors appear in the form of the pointed cones in (4.25). As we next show, (4.65) also possesses such vectors under certain restrictions. Since we are focusing on solutions, it is convenient to adopt some special terminology. We call an extreme vector of (4.65) an *extreme solution* of $\mathbf{A}^T\mathbf{y} \leqq \mathbf{0}$. If $\mathbf{y} \in C$, then any one of the n individual inequalities in which equality holds is called a *binding inequality* for $\mathbf{y}$, in contrast to *slack inequalities* in which equality does not hold. That is, if $\mathbf{A} = [\mathbf{a}_1, \mathbf{a}_2, \ldots, \mathbf{a}_n]$ then the individual inequalities are $\langle \mathbf{a}_i, \mathbf{y} \rangle \leqq 0$ and the ith inequality is binding if either of the following equivalent conditions hold:

1. The vectors $\mathbf{a}_i$ and $\mathbf{y}$ are orthogonal, and
2. The vector $\mathbf{y}$ lies on the hyperplane (4.8) through the origin determined by $\mathbf{a}_i$.

Whereas basic solutions of $\mathbf{A}^T\mathbf{y} = \mathbf{0}$ are LCs, with nonzero coefficients, of LI columns of $\mathbf{A}^T$, extreme solutions of $\mathbf{A}^T\mathbf{y} \leqq \mathbf{0}$ are characterized by their satisfying exactly $m-1$ LI equations. In other words, extreme solutions of homogeneous inequalities $\mathbf{A}^T\mathbf{y} \leqq \mathbf{0}$ lie on $m-1$ independent hyperplanes according to

A point in (4.65) is extreme iff the coefficient matrix of its binding inequalities has rank $m-1$. (4.66)

PROOF. If the rank is $m-1$ as stated and $\mathbf{p}=\mathbf{q}+\mathbf{r}$ is not extreme, then for the coefficient matrix $\mathbf{B}^T$ of the binding inequalities

$$\mathbf{B}^T\mathbf{p}=\mathbf{B}^T\mathbf{q}+\mathbf{B}^T\mathbf{r}$$

which implies $\mathbf{q},\mathbf{r}\in\operatorname{Null}\mathbf{B}^T$. This latter space has dimension one and so $\{\mathbf{q},\mathbf{r}\}$ is LD contradicting the assumption that $\mathbf{p}$ is an extreme solution. Conversely, suppose $\mathbf{p}\in C$ and let us rearrange the rows of $\mathbf{A}^T$ as necessary so that there are two submatrices $\mathbf{B}^T$ and $\mathbf{D}^T$ satisfying $\mathbf{B}^T\mathbf{p}=\mathbf{0}$ and $\mathbf{D}^T\mathbf{p}<\mathbf{0}$. If $\mathbf{p}\neq\mathbf{0}$ then rank $\mathbf{B}^T<m$. Suppose further rank $\mathbf{B}^T<m-1$ in which case for some $\mathbf{q}\in\operatorname{Null}\mathbf{B}^T$, $\{\mathbf{p},\mathbf{q}\}$ is LI. But then we can find $\varepsilon\neq0$ with ε sufficiently small so that

$$\{(\mathbf{p}+\varepsilon\mathbf{q})/2,\ (\mathbf{p}-\varepsilon\mathbf{q})/2\}$$

is LI, $\mathbf{D}\{(\mathbf{p}+\varepsilon\mathbf{q})/2\}<0$, and we see that $\mathbf{p}$ cannot be extreme. This means rank $\mathbf{B}^T=m-1$. ■

(The proof of the converse is reminiscent of that for (4.51).)

Let us call (4.65) a *nondegenerate polyhedral convex cone* if the columns of $\mathbf{A}^T$ are LI, that is, if rank $\mathbf{A}^T=m$. Then

A nondegenerate polyhedral convex cone is generated by its extreme vectors. (4.67)

PROOF. Let C be generated by $\mathbf{b}_1,\mathbf{b}_2,\ldots,\mathbf{b}_r$ and suppose these vectors are semipositively dependent

$$u_1\mathbf{b}_1+u_2\mathbf{b}_2+\cdots+u_r\mathbf{b}_r=\mathbf{0}$$

where all $u_i\geqq 0$ and at least one, say u_1, is positive. If we solve for u_1 and form

$$-\mathbf{A}^T\mathbf{b}_1=\frac{1}{u_1}\left(u_2\mathbf{A}^T\mathbf{b}_2+\cdots+u_r\mathbf{A}^T\mathbf{b}_r\right)$$

then the RHS is nonpositive while $\mathbf{b}\in C$ requires the LHS to be nonnegative. We conclude that both sides would have to be zero, which is impossible. Consequently, the $\mathbf{b}_i$ cannot be semipositively dependent and the result follows from (4.25). ■

PROBLEMS FOR SOLUTION

4.4.1 Fundamental Proofs. Establish: (1) (4.57). (2) (4.59). (3) The following using (4.56) through (4.59). (i) Gordan's theorem (4.34); (ii) Stiemke's theorem (4.36); (iii) von Neumann's theorem (4.40). (4) (4.63).

4.4.2 Proof of (4.64). Supply the proof.

4.4.3 Nondegenerate Polyhedral Convex Cones. Suppose we call a set of vectors $U=\{\mathbf{u}_1,\mathbf{u}_2,\ldots,\mathbf{u}_p\}\subset\mathfrak{R}^m$ *nondegenerate* if rank $\mathbf{U}=m$. (1) Give the geometrical interpretation in (i) $\mathfrak{R}^2$; (ii) $\mathfrak{R}^3$; (iii) $\mathfrak{R}^m$. (2) Show that (4.65) is nondegenerate iff every set of generators for C^* is nondegenerate. (3) If (4.65) is not nondegenerate, does (4.67) necessarily fail?

5

SQUARE MATRICES IN UNITARY SPACE

Square matrices are frequently encountered in virtually all fields where quantitative methods are used. They go with

changes of coordinates in any one vector space
systems of n linear equations in n unknowns
representations of second derivatives
systems of differential and difference equations

and so on. Here we work with many different kinds of square matrices.

5.1 SQUARE MATRICES

Table 5.1 displays the special kinds of square matrices that are most important in this book. Let us take the shortcut of supposing that the reader will refer to this table as necessary to find definitions and locations of properties as we proceed.

We work in unitary space $\mathcal{C}^n$ because—except for Section 5.3 where it is merely efficient to work there in place of $\mathcal{R}^n$—we need to use zeros of polynomials. Most often $\mathbf{A}$ is a *complex matrix* of order n in the sense that its elements are complex numbers. Results in this chapter may or may not hold for a particular real matrix. The *complex conjugate* of $\mathbf{A}=[a_{ij}]$ is

$$\overline{\mathbf{A}}=[\,\overline{a_{ij}}\,]$$

$\operatorname{Re}\mathbf{A}=[\operatorname{Re}a_{ij}]$ is the *real part* of $\mathbf{A}$, and $\operatorname{Im}\mathbf{A}=[\operatorname{Im}a_{ij}]$ is the *imaginary part* of $\mathbf{A}$, where all matrices are of order n. Whenever $\mathbf{A}\neq\overline{\mathbf{A}}$ we follow custom and call $\mathbf{A}$ a *nonreal matrix*. The matrix $\mathbf{A}^H=(\overline{\mathbf{A}})^T$ is the *Hermitian transpose* of $\mathbf{A}$. If $\mathbf{A}^H=\mathbf{A}$ then $\mathbf{A}$ is a *Hermitian matrix*; in the real case $\mathbf{A}$ is called a *real symmetric matrix*.

TABLE 5.1 SPECIAL $\mathbf{A}=[a_{ij}]$ OF ORDER n

Type	Definition or Distinction	Reference
Block diagonal	A special partitioned matrix	5.1.12
Block triangular	A special partitioned matrix	5.1.12
Companion	Has prescribed form so characteristic polynomial can be written down at once	5.1.6
Defective	Possesses fewer than n LI eigenvectors; equivalently, it cannot be diagonalized	(5.7)
Diagonalizable	$\mathbf{P}^{-1}\mathbf{AP}=\mathbf{diag}(\lambda_1,\lambda_2,\ldots,\lambda_n)$ for some $\mathbf{P}$; equivalently, it posseses a full set of n LI eigenvectors	(5.5)
Gaussian pivot	Effects the operation if multiplies on left	5.2.7
Hermitian	$\mathbf{A}^H=\mathbf{A}$; specializes to real symmetric matrix	5.1.7
Jordan block	Possesses one LI eigenvector $\mathbf{x}=c\mathbf{e}_1$ and has prescribed form	5.4.2
JCF	Diagonal or "nearly diagonal" canonical form for any complex $\mathbf{A}$ of order n	(5.38)
Lower-triangular	$a_{ij}=0$ for $i<j$	5.1.11
Nilpotent	$\mathbf{A}^{p-1}\neq\mathbf{O}$ and $\mathbf{A}^p=\mathbf{O}$ for $p\geqq 1$; provides key to JCF	5.4.1
Normal	$\mathbf{AA}^H=\mathbf{A}^H\mathbf{A}$; can be diagonalized by a unitary transformation	5.1.10
Orthogonal	$\mathbf{A}^T=\mathbf{A}^{-1}$; real case of unitary matrix	5.1.8
Symmetric	$\mathbf{A}=\mathbf{A}^T$; real case of Hermitian matrix	5.1.7
Triangular	Lower-triangular or upper-triangular matrix	5.1.11
Unitary	$\mathbf{A}^H=\mathbf{A}^{-1}$; specializes to real orthogonal matrix	5.1.8
Upper-triangular	$a_{ij}=0$ for $i>j$	5.1.11

Let us introduce some scalars and vectors that are used for many purposes in working with square matrices. If

$$\mathbf{Ax}=\lambda\mathbf{x} \qquad \text{where } \mathbf{x}\neq\mathbf{0}$$

then $\mathbf{x}\in\mathcal{C}^n$ is an *eigenvector* and $\lambda\in\mathcal{C}$ is an *eigenvalue* of $\mathbf{A}$. For $A\in\mathcal{L}(\mathcal{C}^n)$ we simply replace $\mathbf{A}$ by A. Other modifiers in use for "eigen" are *characteristic*, *latent*, *proper*, and *secular*. When λ is real there is the following geometric interpretation. If $\lambda>0$ and $\lambda\neq 1$, then for eigenvectors $\mathbf{x}$

$$\mathbf{x}\mapsto\mathbf{Ax}$$

effects a change in magnitude, $\|\mathbf{x}\|$ to $\lambda\|\mathbf{x}\|$, which is either a contraction or an expansion, but there is no change in direction. If $\lambda<0$ there is a change to the direction opposite to $\mathbf{x}$.

Clearly $\mathbf{x}$ is an eigenvector and λ is an eigenvalue of $\mathbf{A}$ iff $\mathbf{x}$ is a nontrivial solution of

$$(\mathbf{A}-\lambda\mathbf{I})\mathbf{x}=\mathbf{0} \tag{5.1}$$

which we recognize as a square homogeneous system of linear equations in unknowns $x_1,x_2,\ldots,x_n$. If λ is an eigenvalue we can find $\mathbf{x}$ by solving (5.1). There can be no unique solution for if $\mathbf{x}$ satisfies (5.1) so also does $k\mathbf{x}$ for every $k\in\mathcal{C}$, that is,

Eigenvectors are determined only to within scalar multiples.

If $\mathbf{x}$ is an eigenvector we can find λ by dividing the first nonzero component of $\mathbf{Ax}$ by the corresponding component of $\mathbf{x}$.

The subspace of $\mathcal{C}^n$ spanned by all eigenvectors corresponding to a given eigenvalue λ is called the *eigenspace* for λ. It is the subspace

$$\mathrm{Null}(\mathbf{A}-\lambda\mathbf{I}) \tag{5.2}$$

and its dimension d is called the *geometric multiplicity* of λ. Every nonzero vector in (5.2) is an eigenvector of $\mathbf{A}$ corresponding to λ. There can be nonzero vectors in (5.2) iff there can be nontrivial solutions in (5.1) iff

$$\det(\mathbf{A}-\lambda\mathbf{I})=0 \tag{5.3}$$

because of (3.13) and (3.19). The LHS of (5.3) is a polynomial of degree n in λ called the *characteristic polynomial* of $\mathbf{A}$ and (5.3) itself is called the *characteristic equation* of $\mathbf{A}$. In this way we arrive at the

THEOREM. The following hold for pairs $\mathbf{A}\in\mathfrak{M}(n)$ and $A\in\mathcal{L}(\mathcal{C}^n)$.

(1) There are n eigenvalues of $\mathbf{A}$ and they coincide with the roots of (5.3).

(2) The characteristic equation, eigenvalues, eigenvectors, and determinant of $\mathbf{A}$ do not depend on the choice of basis for $\mathcal{C}^n$. (5.4)

PROOF. Any root of (5.3) must be an eigenvalue of $\mathbf{A}$ and conversely any eigenvalue of $\mathbf{A}$ must be a root of (5.3) so (1) follows from the fundamental theorem of algebra (1.8). By (2.26) and (2.27) we can prove (2) by showing that the same objects correspond to $\mathbf{A}$ and to any similar $\mathbf{Q}^{-1}\mathbf{AQ}$. The characteristic polynomial of $\mathbf{Q}^{-1}\mathbf{AQ}$ equals

$$\mathbf{Q}^{-1}(\mathbf{AQ}-\lambda\mathbf{IQ})=\mathbf{Q}^{-1}(\mathbf{A}-\lambda\mathbf{I})\mathbf{Q}$$

and so the characteristic equations and eigenvalues, and hence also the eigenvectors, coincide. Finally

$$\det \mathbf{Q}^{-1}\mathbf{A}\mathbf{Q} = (\det \mathbf{Q}^{-1})(\det \mathbf{A})(\det \mathbf{Q})$$

so determinants coincide by commutativity of multiplication in $\mathcal{C}$. ■

If $\lambda_1, \lambda_2, \ldots, \lambda_n$ are the eigenvalues of $\mathbf{A}$ then the characteristic polynomial is

$$(-1)^n(\lambda-\lambda_1)(\lambda-\lambda_2)\cdots(\lambda-\lambda_n)$$

and the multiplicity of a zero μ is called the *algebraic multiplicity*, briefly *multiplicity*, of the eigenvalue μ. The set of eigenvalues

$$\{\lambda_1, \lambda_2, \ldots, \lambda_n\}$$

is called the *spectrum* of both $\mathbf{A}$ and any associated linear transformation A.

The preceding theorem depends on the fundamental theorem of algebra in an essential way. For example, the real matrix

$$\mathbf{M} = \begin{bmatrix} 0 & 1 \\ -1 & 0 \end{bmatrix}$$

has characteristic polynomial $\lambda^2 + 1$ and so has no eigenvalues in $\mathcal{R}$. This shows why we must work in $\mathcal{C}^n$ with $\mathcal{R}$ identified as a subfield of $\mathcal{C}$ whenever we work with eigenvalues. The matrix $\mathbf{M}$ has eigenvalues $\lambda_1 = i$, $\lambda_2 = -i$ in $\mathcal{C}$ and associated eigenvectors are

$$\mathbf{x}_1 = \begin{bmatrix} 1 \\ i \end{bmatrix}, \qquad \mathbf{x}_2 = \begin{bmatrix} 1 \\ -i \end{bmatrix}$$

where from all possible scalar multiples we have chosen vectors whose first nonzero components are unity. While $\mathbf{M}$ has two LI eigenvectors

$$\mathbf{J} = \begin{bmatrix} 1 & 1 \\ 0 & 1 \end{bmatrix}$$

has but one since any eigenvector is a multiple of $\mathbf{e}_1$. This observation is significant because of the

THEOREM. A matrix of order n is similar to a diagonal matrix iff it has n LI eigenvectors. (5.5)

PROOF. If $\mathbf{A}$ is the matrix and $\mathbf{P} = [\mathbf{p}_1, \mathbf{p}_2, \ldots, \mathbf{p}_n]$ is the matrix of n LI eigenvectors then $\mathbf{AP}$ equals

$$[\lambda_1\mathbf{p}_1, \lambda_2\mathbf{p}_2, \ldots, \lambda_n\mathbf{p}_n] = \mathbf{P}\,\mathbf{diag}(\lambda_1, \lambda_2, \ldots, \lambda_n)$$

where $\lambda_1, \lambda_2, \ldots, \lambda_n$ are the (not necessarily distinct) eigenvalues of **A**. Of course **P** is nonsingular and

$$\mathbf{P}^{-1}\mathbf{A}\mathbf{P} = \mathbf{diag}(\lambda_1, \lambda_2, \ldots, \lambda_n) = \mathbf{D} \tag{5.6}$$

Conversely, if $\mathbf{D}'$ equals

$$\mathbf{diag}(d_1, d_2, \ldots, d_n) = \mathbf{Q}^{-1}\mathbf{A}\mathbf{Q}$$

then also $\mathbf{Q}\mathbf{D}' = \mathbf{A}\mathbf{Q}$ which in our usual notation means $d_i\mathbf{q}_i = \mathbf{A}\mathbf{q}_i$ for $i = 1, 2, \ldots, n$. Thus the d_i are eigenvalues, $\mathbf{q}_i$ are eigenvectors, and they are LI because **Q** is nonsingular. ■

A matrix **A** of order n is called *diagonalizable* if it has n LI eigenvectors. Whenever (5.6) holds we say that **A** has been *diagonalized* by **P**. We call the form implied by (5.6)

$$\mathbf{A} = \mathbf{P}\,\mathbf{diag}(\lambda_1, \lambda_2, \ldots, \lambda_n)\mathbf{P}^{-1}$$

a *diagonalized form* of **A**; note that the RHS depends on both the order in which the eigenvalues are prescribed and the scalar multiples selected as respective eigenvectors. From the theorem we see that if **A** can be diagonalized then the diagonal matrix must be the diagonal matrix of eigenvalues of **A** and the matrix **P** must be a matrix of n LI eigenvectors of **A**. If **A** has fewer than n LI eigenvectors it is a *defective matrix*. Thus

A square matrix is diagonalizable iff it is not defective. (5.7)

is a restatement of (5.5).

EXAMPLE 5.1. Finding the kth Power of a Matrix. The main reason we are interested in eigenvalues and eigenvectors in this book is that they make it possible to find general formulas for the nth order matrix $\mathbf{A}^k$ where k is any positive integer. Such formulas are required for the direct algebraic methods of solving linear differential equations in Chapter 9 and linear difference equations in Chapter 10.

Let us suppose **A** is diagonalizable as in the first part of the proof of (5.5). Then

$$\mathbf{A} = \mathbf{P}\mathbf{D}\mathbf{P}^{-1}$$

is a diagonalized form of **A** and $\mathbf{A}^2$ equals

(1) $$\mathbf{P}\mathbf{D}\mathbf{P}^{-1}\mathbf{P}\mathbf{D}\mathbf{P}^{-1} = \mathbf{P}\mathbf{D}^2\mathbf{P}^{-1}$$

The higher powers clearly "telescope" in the same way. Moreover, $\mathbf{D}^k$ can be written at once so

(2) $$\mathbf{A}^k = \mathbf{P}\,\mathbf{diag}(\lambda_1^k, \lambda_2^k, \ldots, \lambda_n^k)\mathbf{P}^{-1}$$

for $k = 1, 2, \ldots$.

When **A** is defective there are less than n LI eigenvectors which means that the nonsingular matrix **P** in (1) cannot be formed from eigenvectors alone. By the end of this chapter we find a "nearly diagonal" matrix **J** whose kth powers can be written almost as easily as those for **D** to produce the formula

(3) $$\mathbf{A}^k = \mathbf{T}\mathbf{J}^k\mathbf{T}^{-1}$$

of which (2) is a special case.

Once we have established (3), or its special case (2), there are a number of important consequences over and above the facility for solving differential and difference equations. For example, we can prove a special case of the general result which is often described as "a square matrix satisfies its own characteristic equation."

SPECIAL CAYLEY–HAMILTON THEOREM. If **A** is diagonalizable and has characteristic equation

$$\lambda^n + p_{n-1}\lambda^{n-1} + \cdots + p_1\lambda + p_0 = 0$$

then also

(4) $$\mathbf{A}^n + p_{n-1}\mathbf{A}^{n-1} + \cdots + p_1\mathbf{A} + p_0\mathbf{I} = \mathbf{O}$$

PROOF. Let us write $p(\lambda)$ for the characteristic polynomial and $p(\mathbf{A})$ for the matrix in question. Then using (2) for **D** and its special case

$$\mathbf{I} = \mathbf{P}\mathbf{D}^0\mathbf{P}^{-1}$$

for $k = 0$ we can factor $p(\mathbf{A}) = \mathbf{P}p(\mathbf{D})\mathbf{P}^{-1}$ and rewrite it as

$$\mathbf{P}\,\mathbf{diag}(p(\lambda_1), p(\lambda_2), \ldots, p(\lambda_n))\mathbf{P}^{-1}$$

This last diagonal matrix is **O** since each $p(\lambda_i) = 0$. ■

In consequence of the Cayley–Hamilton theorem, any power $\mathbf{A}^k$ for $k \geqq n$, and any polynomial $q(\mathbf{A})$ of degree k, can be expressed in terms of powers $\mathbf{A}^{n-1}$ or less. (This is what is important for differential equations in Chapter 9.) All we do is use $p(\mathbf{A}) = \mathbf{O}$ to write $\mathbf{A}^k$ as a LC of powers of degrees less than n. For example

$$\mathbf{A} = \begin{bmatrix} 1 & 2 \\ 2 & 3 \end{bmatrix}$$

has characteristic polynomial $\lambda^2-4\lambda-1$ so

$$\mathbf{A}^2=4\mathbf{A}+\mathbf{I}$$

$$\mathbf{A}^3=17\mathbf{A}+4\mathbf{I}$$

and so on for any power of **A**.

Using the first displayed expression we find

$$\mathbf{A}(\mathbf{A}-4\mathbf{I})=\mathbf{I}$$

which shows $\mathbf{A}^{-1}=\mathbf{A}-4\mathbf{I}$, that is

$$\mathbf{A}^{-1}=\begin{bmatrix} -3 & 2 \\ 2 & -1 \end{bmatrix}$$

which illustrates how the Cayley–Hamilton theorem can be used to find inverse matrices.

Unitary transformations are special kinds of similarity transformations that are particularly convenient for numerical work such as finding the eigenvalues of an arbitrary square matrix **A**. This is true because their associated matrices **U** are defined by

$$\mathbf{U}^{-1}=\mathbf{U}^H$$

so $\mathbf{U}^{-1}$ is always readily available for finding matrices

$$\mathbf{U}^{-1}\mathbf{A}\mathbf{U}$$

that are similar to **A**. Similar matrices have the same eigenvalues by (5.4) and so the idea is to use the convenient unitary transformations to find a "simpler" matrix that has the same eigenvalues as **A**. By (5.5) we know that "simpler" cannot always mean the ultimate, namely diagonal, but the following shows that at least theoretically we can do almost as well.

SCHUR'S THEOREM. Any square complex matrix **A** can be reduced by a unitary transformation to an upper-triangular matrix that has the eigenvalues of **A** arranged on its principal diagonal in any prescribed order. (5.8)

PROOF. Since similar matrices have the same eigenvalues by (5.4) and since (5.1.11) upper-triangular matrices have their eigenvalues as elements on their main diagonals, it suffices to show that we can reduce **A** to upper-triangular form by a unitary transformation. There will be at least one eigenvector of **A** corresponding to the first prescribed eigenvalue λ_1. If it is $\mathbf{x}_1$ we may suppose

$$\mathbf{A}\mathbf{x}_1=\lambda_1\mathbf{x}_1 \qquad \text{and} \qquad \|\mathbf{x}_1\|=1$$

Then we can use the Gram–Schmidt procedure (3.47) to construct a unitary matrix

$$\mathbf{Q}=[\mathbf{x}_1,\mathbf{y}_2,\ldots,\mathbf{y}_n]=[\mathbf{x}_1,\mathbf{Y}]$$

for which $\mathbf{Q}^H\mathbf{Q}=\mathbf{I}$ implies $\mathbf{Y}^H\mathbf{x}_1=\mathbf{0}$ and

$$\mathbf{Q}^H\mathbf{A}\mathbf{Q}=\begin{bmatrix}\lambda_1 & \mathbf{F}\\ \mathbf{0} & \mathbf{G}\end{bmatrix}$$

If $n=2$ the above shows that the theorem holds. We make the inductive hypothesis that it holds for all matrices of order $n-1$. Since $\mathbf{G}$ is order $n-1$, there exists a unitary matrix $\mathbf{P}$ such that $\mathbf{P}^H\mathbf{G}\mathbf{P}$ is upper-triangular. Then

$$\mathbf{M}=\begin{bmatrix}1 & \mathbf{0}^T\\ \mathbf{0} & \mathbf{P}\end{bmatrix}$$

is readily verified as unitary, as is $\mathbf{QM}$, and

$$\begin{bmatrix}1 & \mathbf{0}^T\\ \mathbf{0} & \mathbf{P}^H\end{bmatrix}\begin{bmatrix}\lambda_1 & \mathbf{F}\\ \mathbf{0} & \mathbf{G}\end{bmatrix}\begin{bmatrix}1 & \mathbf{0}^T\\ \mathbf{0} & \mathbf{P}\end{bmatrix}=\begin{bmatrix}\lambda_1 & \mathbf{FP}\\ \mathbf{0} & \mathbf{P}^H\mathbf{G}\mathbf{P}\end{bmatrix}$$

shows $(\mathbf{QM})^H\mathbf{A}(\mathbf{QM})$ to be upper-triangular. ■

We call the procedure in Schur's theorem *upper-triangularization* of $\mathbf{A}$ by a unitary transformation. Whenever

$$\mathbf{Q}^{-1}\mathbf{A}\mathbf{Q}=\mathbf{U}$$

is upper-triangular, where $\mathbf{Q}$ is any nonsingular matrix, we say that $\mathbf{A}$ has been *upper-triangularized* by a similarity transformation. In both cases we refer to

$$\mathbf{A}=\mathbf{QUQ}^{-1}$$

as an *upper-triangularized form* of $\mathbf{A}$.

Whereas (5.7) is the definitive result on diagonalization by any matrix, the corresponding result for the convenient unitary transformations is

A square matrix is diagonalizable by a unitary transformation iff it is normal. (5.9)

PROOF. If $\mathbf{A}$ is normal then the upper-triangular $\mathbf{T}=\mathbf{Q}^H\mathbf{A}\mathbf{Q}$ from (5.8) is also normal (5.1.10). Let us consider the (1,1) elements of $\mathbf{T}\mathbf{T}^H=\mathbf{T}^H\mathbf{T}$, namely

$$|t_{11}|^2+|t_{12}|^2+\cdots+|t_{1n}|^2=|t_{11}|^2$$

We conclude $t_{12}=t_{13}=\cdots=t_{1n}=0$. The (2,2) elements are

$$|t_{22}|^2+|t_{23}|^2+\cdots+|t_{2n}|^2=|t_{12}|^2+|t_{22}|^2=|t_{22}|^2$$

which implies $t_{23}=\cdots=t_{2n}=0$. Continuing, we conclude that $\mathbf{T}$ is a diagonal matrix. Conversely, if $\mathbf{U}^H\mathbf{A}\mathbf{U}=\mathbf{D}$ is the stated diagonalization, then

$$\mathbf{A}=\mathbf{U}\mathbf{D}\mathbf{U}^H \qquad \text{and} \qquad \mathbf{A}\mathbf{A}^H=\mathbf{U}\mathbf{D}\bar{\mathbf{D}}\mathbf{U}^H$$

But $\mathbf{A}^H\mathbf{A}$ coincides with the latter because $\mathbf{D}\bar{\mathbf{D}}=\bar{\mathbf{D}}\mathbf{D}$ when $\mathbf{D}$ is a diagonal matrix. ■

PROBLEMS FOR SOLUTION

5.1.1 Eigenvalues and Eigenvectors. Establish for complex matrices of order n:

(1) Eigenvalues of $\mathbf{A}^H$ are complex conjugates of those of $\mathbf{A}$. Eigenvectors of $\mathbf{A}^H$ and $\mathbf{A}$ need not coincide.
(2) $\mathbf{A}$ and $\mathbf{B}$ can have the same eigenvalues and yet not be similar.
(3) There is at least one eigenvector corresponding to each eigenvalue.
(4) The set of all eigenvectors of $\mathbf{A}$ corresponding to the eigenvalue λ spans the subspace (5.2).
(5) $\mathbf{A}$ cannot have more than n LI eigenvectors.
(6) A scalar λ is an eigenvalue of $\mathbf{A}$ iff $\mathbf{A}-\lambda\mathbf{I}$ is singular.
(7) Zero is an eigenvalue of algebraic multiplicity n of $\mathbf{A}=\mathbf{O}$.
(8) Zero is an eigenvalue of $\mathbf{A}$ iff $\mathbf{A}$ is singular.
(9) $\mathbf{A}$ may be singular and yet have a nonsingular matrix of eigenvectors.
(10) The same eigenvector cannot correspond to each of two distinct eigenvalues.
(11) If $\mathbf{A}$ is nonsingular and λ is an eigenvalue of $\mathbf{AB}$ then λ is also an eigenvalue of $\mathbf{BA}$.
(12) If $\mathbf{A}\mathbf{x}=\lambda\mathbf{x}$ for $\mathbf{x}\neq\mathbf{0}$ then $\mathbf{A}^k\mathbf{x}=\lambda^k\mathbf{x}$ where k is any positive integer.
(13) If $\mathbf{A}$ has spectrum $\{\lambda_1,\lambda_2,\ldots,\lambda_n\}$ then
  (i) $c\mathbf{A}$ has spectrum $\{c\lambda_1,c\lambda_2,\ldots,c\lambda_n\}$ where $c\neq 0$.
  (ii) $\mathbf{A}+k\mathbf{I}$ has spectrum $\{\lambda_1+k,\lambda_2+k,\ldots,\lambda_n+k\}$
  (iii) Whenever $\mathbf{A}$ is nonsingular, $\mathbf{A}^{-1}$ has spectrum

$$\{\lambda_1^{-1},\lambda_2^{-1},\ldots,\lambda_n^{-1}\}$$

  (iv) $\det\mathbf{A}=\lambda_1\lambda_2\cdots\lambda_n$
  (v) $\operatorname{trace}\mathbf{A}=\lambda_1+\lambda_2+\cdots+\lambda_n$ where by definition for $\mathbf{A}=[a_{ij}]$ of order n $\operatorname{trace}\mathbf{A}=a_{11}+a_{22}+\cdots+a_{nn}$.

(14) If **A** has n distinct eigenvalues then it has n LI eigenvectors.
(15) The geometric multiplicity d_i of eigenvalue λ_i satisfies $1 \leqq d_i \leqq k_i$ where k_i is the algebraic multiplicity.

5.1.2 Finding All Eigenvalues. Establish a manual procedure for $n=2$ or 3.

5.1.3 Finding All Eigenvectors. Establish the following standard procedure.

Given an eigenvalue λ of **A**, use (3.27) to find a basis for Null$(\mathbf{A}-\lambda\mathbf{I})$.

5.1.4 Numerical Exercises. Find the characteristic equation, eigenvalues, and a complete set of standard eigenvectors where, in this problem, standard means first nonzero component unity

$$(1)\ \begin{bmatrix} 4 & -2 \\ 3 & -1 \end{bmatrix} \qquad (2)\ \begin{bmatrix} 0 & 1 \\ 1 & 0 \end{bmatrix} \qquad (3)\ \begin{bmatrix} 3 & -2 \\ 4 & 3 \end{bmatrix}$$

$$(4)\ \begin{bmatrix} 0 & 1 \\ -2 & 3 \end{bmatrix} \qquad (5)\ \begin{bmatrix} 1 & i \\ i & 1 \end{bmatrix} \qquad (6)\ \begin{bmatrix} i & 1 \\ 0 & i \end{bmatrix}$$

5.1.5 More Exercises. Repeat for the following.

$$(1)\ \begin{bmatrix} 1 & 1 & 0 \\ -1 & 1 & 0 \\ 0 & 0 & 1 \end{bmatrix} \qquad (2)\ \begin{bmatrix} 1 & 1 & 0 \\ 0 & 1 & 0 \\ 1 & 0 & 1 \end{bmatrix}$$

$$(3)\ \begin{bmatrix} 1 & 0 & 1 \\ 0 & 1 & 0 \\ 1 & 0 & 1 \end{bmatrix} \qquad (4)\ \begin{bmatrix} 0 & 0 & 0 \\ 0 & 1 & 1 \\ 1 & 0 & 1 \end{bmatrix}$$

5.1.6 Companion Matrix. The matrix of order n

$$\mathbf{C}=\begin{bmatrix} 0 & 1 & 0 & \cdots & 0 & 0 \\ 0 & 0 & 1 & \cdots & 0 & 0 \\ \vdots & \vdots & \vdots & & \vdots & \vdots \\ 0 & 0 & 0 & \cdots & 0 & 1 \\ -a_0 & -a_1 & -a_2 & \cdots & -a_{n-2} & -a_{n-1} \end{bmatrix}$$

is called the *companion matrix* of the polynomial

$$p(\lambda)=\lambda^n+a_{n-1}\lambda^{n-1}+\cdots+a_1\lambda+a_0$$

Show that the characteristic equation of **C** is $p(\lambda)=0$.

5.1.7 Hermitian and Symmetric Matrices. A matrix $\mathbf{A}$ is *Hermitian* if $\mathbf{A}^H=\mathbf{A}$ and *symmetric* if $\mathbf{A}^T=\mathbf{A}$; consequently, such matrices are necessarily square. Any property established for a Hermitian matrix holds for a real symmetric matrix. Since the same cannot be said for nonreal matrices it is prudent to replace "Hermitian" by "real symmetric" in stating results inferred for real matrices in this manner. Establish the following where $\mathbf{A}$ and $\mathbf{B}$ are Hermitian of order n:

(1) For any $\mathbf{M}$ of order $m\times n$ the matrices $\mathbf{MM}^H$ and $\mathbf{M}^H\mathbf{M}$ are both Hermitian.
(2) $\mathbf{AB}$ is Hermitian iff $\mathbf{AB}=\mathbf{BA}$.
(3) If $\mathbf{A}$ is nonsingular then $\mathbf{A}^{-1}$ is Hermitian.
(4) All main diagonal elements of $\mathbf{A}$ are real.
(5) The complex conjugate $\overline{\mathbf{A}}$, $\mathbf{A}^H$, and $\mathbf{A}^p$ where p is a positive integer, are Hermitian.
(6) If $\mathbf{Q}$ is of order n then $\mathbf{Q}^H\mathbf{AQ}$ is Hermitian.
(7) If $\mathbf{x}\in\mathcal{C}^n$ then $\mathbf{x}^H\mathbf{Ax}$ is real.
(8) A matrix $\mathbf{A}$ of order n is Hermitian iff $\langle\mathbf{x},\mathbf{Ay}\rangle=\langle\mathbf{Ax},\mathbf{y}\rangle$ for all $\mathbf{x},\mathbf{y}\in\mathcal{C}^n$.
(9) The eigenvalues of $\mathbf{A}$, and $\det\mathbf{A}$ are real.
(10) Eigenvectors corresponding to distinct eigenvalues of $\mathbf{A}$ are orthogonal.
(11) It is always possible to diagonalize $\mathbf{A}$.
(12) The *spectral decomposition*

$$\mathbf{A}=\lambda_1\mathbf{X}_1+\lambda_2\mathbf{X}_2+\cdots+\lambda_n\mathbf{X}_n$$

where $\lambda_1,\lambda_2,\ldots,\lambda_n$ are the eigenvalues of $\mathbf{A}$, $\mathbf{X}_i=\mathbf{x}_i\mathbf{x}_i^H$ for $i=1,2,\ldots,n$ where $\{\mathbf{x}_1,\mathbf{x}_2,\ldots,\mathbf{x}_n\}$ is an orthonormal set of eigenvectors of $\mathbf{A}$.
(13) The matrices $\mathbf{X}_i$ just above satisfy
  (i) $\mathbf{X}_i^2=\mathbf{X}_i$ for $i=1,2,\ldots,n$.
  (ii) $\mathbf{X}_i\mathbf{X}_j=\mathbf{O}$ for $i\neq j$.
  (iii) $\mathbf{X}_1+\mathbf{X}_2+\cdots+\mathbf{X}_n=\mathbf{I}$.
(14) Not all of (1)–(13) hold if "Hermitian" is replaced by "symmetric" and nonreal matrices are involved.

5.1.8 Unitary and Orthogonal Matrices. A matrix $\mathbf{U}$ is *unitary* if $\mathbf{U}^H=\mathbf{U}^{-1}$ and *orthogonal* if $\mathbf{U}^T=\mathbf{U}^{-1}$; consequently such matrices are necessarily nonsingular. Any property established for a unitary matrix holds for a real orthogonal matrix. Since the same cannot be said for nonreal matrices, it is prudent to replace "unitary" by "real orthogonal" in stating results inferred for real matrices in this manner. Establish the following where $\mathbf{U}$ and $\mathbf{V}$ are unitary of order n.

(1) A complex matrix of order n is unitary iff its columns form an orthonormal basis for $\mathcal{C}^n$.
(2) The complex conjugate $\overline{\mathbf{U}}$, $\mathbf{U}^H$, and $\mathbf{U}^p$ where p is a positive integer, are unitary.
(3) det $\mathbf{U}$ lies on the unit circle.
(4) $\mathbf{UV}$ is unitary.
(5) A permutation matrix is a real orthogonal matrix.
(6) If $\mathbf{A}$ is Hermitian then $\mathbf{U}^{-1}\mathbf{AU}$ is Hermitian.
(7) A matrix $\mathbf{P}$ is unitary iff $\langle \mathbf{x},\mathbf{y}\rangle = \langle \mathbf{Px},\mathbf{Py}\rangle$ for all $\mathbf{x},\mathbf{y}\in\mathcal{C}^n$.
(8) All eigenvalues of $\mathbf{U}$ lie on the unit circle in $\mathcal{C}$.
(9) If $\{\mathbf{a}_1,\mathbf{a}_2,\ldots,\mathbf{a}_p\}$ is an orthonormal set in $\mathcal{C}^n$ there exist $n-p$ vectors $\mathbf{b}_{p+1},\ldots,\mathbf{b}_n$ such that

$$\left[\mathbf{a}_1,\ldots,\mathbf{a}_p,\mathbf{b}_{p+1},\ldots,\mathbf{b}_n\right]$$

is unitary.
(10) $\mathbf{AA}^H=\mathbf{BB}^H$ iff $\mathbf{A}=\mathbf{BU}$ for some unitary $\mathbf{U}$.
(11) It is always possible to diagonalize a unitary matrix.
(12) Not all of (1)–(10) hold if "unitary" is replaced by "orthogonal" and nonreal matrices are involved.

5.1.9 Skew-Hermitian and Skew-Symmetric Matrices. A matrix $\mathbf{A}$ is *skew-Hermitian* if $\mathbf{A}^H=-\mathbf{A}$ and *skew-symmetric* if $\mathbf{A}^T=-\mathbf{A}$. The concepts apply to square matrices, and counterpart remarks to those in 5.1.7 for specializations to real matrices apply here. Establish:

(1) All main diagonal elements of a skew-Hermitian matrix are imaginary.
(2) If $\mathbf{A}$ is skew-Hermitian then $i\mathbf{A}$ is Hermitian.
(3) Any square matrix can be written as the sum of a Hermitian matrix and a skew-Hermitian matrix.
(4) Any square matrix $\mathbf{A}$ can be written as $\mathbf{A}=\mathbf{B}+i\mathbf{C}$ where $\mathbf{B}$ and $\mathbf{C}$ are Hermitian.
(5) All eigenvalues of a skew-Hermitian matrix are imaginary.
(6) A skew-Hermitian matrix has even rank.

5.1.10 Normal Matrices. A matrix $\mathbf{A}$ is *normal* if $\mathbf{AA}^H=\mathbf{A}^H\mathbf{A}$; consequently such matrices are necessarily square. The definition applies to real matrices as $\mathbf{AA}^T=\mathbf{A}^T\mathbf{A}$. Establish:

(1) The following types of matrices are normal: diagonal, Hermitian, skew-Hermitian, unitary, and $\mathbf{A}^H$ where $\mathbf{A}$ is normal.
(2) If $\mathbf{P}$ is unitary then $\mathbf{A}$ is normal iff $\mathbf{P}^H\mathbf{AP}$ is normal.
(3) A matrix is normal iff its real and imaginary parts commute.
(4) $\mathbf{A}$ is normal iff $\|\mathbf{Ax}\|=\|\mathbf{A}^H\mathbf{x}\|$ for all $\mathbf{x}\in\mathcal{C}^n$.

(5) If **A** is normal, then:
   (i) **A** is Hermitian iff all eigenvalues of **A** are real, and
   (ii) **A** is unitary iff all eigenvalues lie on the unit circle in $\mathcal{C}$.
(6) Geometric and algebraic multiplicities coincide for each eigenvalue of a normal matrix.
(7) If **A** and **B** are normal of order n then $\mathbf{A}=\mathbf{PB}$ for some unitary **P** iff **A** and **B** have the same eigenvalues.

5.1.11 Triangular Matrices. A matrix $\mathbf{A}=[a_{ij}]$ is *upper-triangular* if it is square and $a_{ij}=0$ for $i>j$; it is *lower-triangular* if it is square and $a_{ij}=0$ for $i<j$; it is *triangular* if it is upper-triangular or lower-triangular. Establish the following where for brevity we mainly deal with upper-triangular matrices.

(1) The transpose of an upper-triangular matrix is lower-triangular.
(2) A matrix that is both upper- and lower-triangular must be a diagonal matrix.
(3) The set of all upper-triangular matrices is closed under addition, multiplication by scalars, and matrix multiplication which, in general, is not commutative.
(4) The diagonal elements of the product of two upper-triangular matrices equal the products of the corresponding diagonal elements.
(5) An upper-triangular matrix is nonsingular iff no zeros appear on the main diagonal.
(6) The inverse of any nonsingular upper-triangular matrix is upper-triangular, and has main diagonal elements which are the reciprocals of the corresponding elements in the given matrix.
(7) The determinant of a triangular matrix equals the product of its main diagonal elements.
(8) The eigenvalues of a triangular matrix are its main diagonal elements.
(9) Any square matrix can be reduced to upper-triangular form using only elementary row operations of Type III.

5.1.12 Block-Triangular and Block-Diagonal Matrices. Some square matrices **A** can be written in *block upper-triangular form*

$$\mathbf{A}=\begin{bmatrix} \mathbf{A}_{11} & \mathbf{A}_{12} & \cdots & \mathbf{A}_{1t} \\ \mathbf{O}^{21} & \mathbf{A}_{22} & \cdots & \mathbf{A}_{2t} \\ \vdots & \vdots & & \vdots \\ \mathbf{O}^{t1} & \mathbf{O}^{t2} & \cdots & \mathbf{A}_{tt} \end{bmatrix}$$

where $t>1$ (so the representation is nontrivial); where $\mathbf{A}_{ii}$ are square matrices for $i=1,2,\ldots,t$; where $\mathbf{A}_{ij}=\mathbf{O}^{ij}$ are zero matrices of various orders for $i>j$, and; where $\mathbf{A}_{ij}$ are matrices of various orders for $i<j$. *Block lower-triangular form* has the corresponding definition and *block triangular form* means block upper-triangular or block lower-triangular. A form that is both block upper-triangular and block lower-triangular is called *block diagonal* and is denoted by

$$\mathbf{diag}(\mathbf{A}_{11},\mathbf{A}_{22},\ldots,\mathbf{A}_{tt})$$

Establish:

(1) True or False? If $\mathbf{A}$ is in block-triangular form then

$$\det\mathbf{A}=(\det\mathbf{A}_{11})(\det\mathbf{A}_{22})\cdots(\det\mathbf{A}_{tt})$$

(2) Formulas for the following when $\mathbf{A}$ is in block-diagonal form: $\mathbf{A}^H,\mathbf{A}^p$ where p is a positive integer, and $\mathbf{A}^{-1}$ where $\mathbf{A}$ is nonsingular.

5.1.13 An Upper-Triangular Matrix. The following is an exercise in providing for all possible cases for a relatively simple object. (1) Find the eigenvalues and all associated LI eigenvectors, with first nonzero components unity, given a, d, f real and distinct in

$$\mathbf{A}=\begin{bmatrix} a & b & c \\ 0 & d & e \\ 0 & 0 & f \end{bmatrix}$$

(2) Repeat for $a=d=f=\lambda$. (3) Repeat for $a=d=\lambda, f\neq\lambda$. (4) Repeat for $d=f=\lambda$, $a\neq\lambda$. (5) Repeat for $a=f=\lambda$, $d\neq\lambda$.

5.1.14 Tri-Diagonal Matrices. A square matrix $\mathbf{A}=[a_{ij}]$ is a *tri-diagonal* matrix if $a_{ij}=0$ for $|i-j|>1$. Another name is *band matrix* of *width three*. Show that the determinant of the following tri-diagonal matrix of order n

$$\mathbf{T}_n=\begin{bmatrix} b_1 & c_1 & 0 & 0 & 0 & \cdots & 0 & 0 \\ a_2 & b_2 & c_2 & 0 & 0 & \cdots & 0 & 0 \\ 0 & a_3 & b_3 & c_3 & 0 & \cdots & 0 & 0 \\ \vdots & \vdots & \vdots & \vdots & \vdots & & \vdots & \vdots \\ 0 & 0 & 0 & 0 & 0 & \cdots & a_n & b_n \end{bmatrix}$$

satisfies $\det\mathbf{T}_n=b_n\det\mathbf{T}_{n-1}-a_nc_{n-1}\det\mathbf{T}_{n-2}$.

5.1.15 Adjugate Matrices. The *adjugate* of a square matrix $\mathbf{A}=[a_{ij}]$, $\operatorname{adj}\mathbf{A}$, is the transpose of the matrix obtained from $\mathbf{A}$ by replacing a_{ij} by its cofactor c_{ij}. (The name *adjoint* was used in former years.) Establish

$$\mathbf{A}(\operatorname{adj}\mathbf{A})=(\operatorname{adj}\mathbf{A})\mathbf{A}=(\det\mathbf{A})\mathbf{I}$$

5.1.16 Gauss–Jordan Pivot Matrices. Such a matrix is $\mathbf{G}_{ij}$ of order m such that if $\mathbf{X}=[x_{ij}]$ is $m\times n$ and $x_{ij}\neq 0$ then $\mathbf{G}_{ij}\mathbf{X}$ coincides with the matrix that results when (3.1) is applied to $\mathbf{X}$. Find $\mathbf{G}_{ij}$.

5.1.17 Idempotent Matrices. A matrix $\mathbf{A}$ is *idempotent* if $\mathbf{A}^2=\mathbf{A}$; consequently such matrices are necessarily square. Establish for idempotent $\mathbf{A}$ of order n:

(1) If λ is an eigenvalue of $\mathbf{A}$ then necessarily $\lambda=0$ or $\lambda=1$.
(2) If $\mathbf{U}$ is unitary of order n then $\mathbf{U}^H\mathbf{A}\mathbf{U}$ is idempotent.

5.1.18 Outer Product Matrices. If $\mathbf{a},\mathbf{b}\in\mathcal{C}^n$ then the associated *outer product* matrix is $\mathbf{a}\mathbf{b}^H$. Establish:

(1) If $\mathbf{a}^H\mathbf{b}\neq 0$ then $\operatorname{rank}\mathbf{a}\mathbf{b}^H=1$.
(2) $\det(\mathbf{I}+\mathbf{a}\mathbf{b}^H)=1+\mathbf{a}^H\mathbf{b}$.
(3) If $\mathbf{a}^H\mathbf{b}\neq 0$ and $\mathbf{P}=\mathbf{I}_n-\mathbf{a}\mathbf{b}^H(\mathbf{a}^H\mathbf{b})^{-1}$ then (i) $\mathbf{P}\mathbf{a}=\mathbf{0}$; (ii) $\mathbf{P}^H\mathbf{b}=\mathbf{0}$; (iii) $\mathbf{P}$ is idempotent.

5.1.19 Scalar Matrices. If $a\in\mathcal{C}$ then the associated *scalar matrix* is $a\mathbf{I}_n$. Establish:

(1) The scalar matrices of order n form a field
(2) A scalar matrix of order n commutes with any matrix of order n.

5.1.20 Stochastic Matrices. A *stochastic matrix* is a real matrix $\mathbf{A}=[a_{ij}]$ of order n such that all elements satisfy $0\leqq a_{ij}\leqq 1$ and all column sums equal unity. (If all row sums also equal unity then $\mathbf{A}$ is called *doubly stochastic*.) (1) Establish the following instance of a *fixed point theorem*. If $\mathbf{A}$ is a stochastic matrix then there is some nonzero $\mathbf{x}\in\mathcal{R}^n$ that is mapped into itself under the associated linear transformation $\mathbf{x}\mapsto\mathbf{A}\mathbf{x}$. (2) Verify that a permutation matrix is a doubly stochastic matrix. (3) Express the definitions in terms of the vector $\mathbf{e}$.

5.1.21 Unimodular Matrices. A matrix $\mathbf{A}$ is *unimodular* if $|\det\mathbf{A}|=1$; consequently such matrices are necessarily square. Show that if all elements of a nonsingular matrix $\mathbf{A}$ are integers then $\mathbf{A}$ is unimodular iff all elements of $\mathbf{A}^{-1}$ are integers.

5.1.22 Vandermonde Matrices. If $a_1,a_2,\ldots,a_n\in\mathcal{C}$ the associated *Vandermonde matrix* is the following matrix of order n

$$\mathbf{V}=\begin{bmatrix} 1 & 1 & \cdots & 1\\ a_1 & a_2 & \cdots & a_n\\ a_1^2 & a_2^2 & \cdots & a_n^2\\ \vdots & \vdots & & \vdots\\ a_1^{n-1} & a_2^{n-1} & \cdots & a_n^{n-1}\end{bmatrix}$$

although the name is frequently given to $\mathbf{V}^T$. Establish $\det\mathbf{V}=0$ iff no two of the a_i are equal and hence

If no two of the a_i are equal then $\mathbf{Vc}=\mathbf{0}$ has only the trivial solution $\mathbf{c}=\mathbf{0}$ in $\mathcal{C}^n$

5.2 UNIQUE SOLUTIONS

In this section we introduce Gaussian pivoting and obtain some results that are useful in numerical work.

Our title indicates that we are concerned with solutions $\mathbf{x}$ for $\mathbf{Ax}=\mathbf{b}$ where $\mathbf{A}$ is nonsingular. Because we are interested in $[\mathbf{A},\mathbf{b}]$ and other nonsquare matrices let us use the following.

Definition. If $\mathbf{B}=[b_{ij}]$ is an $m\times n$ matrix with elements from $\mathcal{R}$ or $\mathcal{C}$, where $m\leqq n$ and where the submatrix formed by columns $1,2,\ldots,m$ is nonzero, then a *Gaussian pivot operation in* $\mathbf{B}$ is performed as follows.

(1) Select a nonzero element b_{jj} as *Gaussian pivot* and, equivalently call row j the *Gaussian pivot row* and column j the *Gaussian pivot column.*

(2) Answer: $j<m$? If "yes," reduce elements in column j and rows $j+1$, $j+2,\ldots,m$ to zero by adding appropriate multiples of row j to these rows, and stop. If "no," stop. (5.10)

We also say we have *performed a Gaussian pivot on* b_{jj}, or *on position* (j,j). In words, we do not change the Gaussian pivot row and, in contrast to Gauss–Jordan pivoting (3.1), we "sweep" downward only. The resulting matrix has rows $1,2,\ldots,j$ unchanged and rows $j+1, j+2,\ldots,m$ coincide with those that would have resulted from a Gauss–Jordan pivot on b_{jj}. Only elementary row operations of Type III are used.

In the following procedure we plan to perform Gaussian pivoting successively on positions $(1,1),(2,2),\ldots,(m,m)$. Whenever we cannot continue because of a zero, say at (j,j), we plan to add to row j the next appearing row with a nonzero element in column j. If we stop before reaching (m,m) then the submatrix formed by the first m columns has a zero determinant.

GAUSSIAN REDUCTION PROCEDURE. To start for any real or complex $\mathbf{B}\neq\mathbf{O}$ of order $m\times n$ where $m\leqq n$, set $k=1$, designate the submatrix formed by the first m columns as $\mathbf{B}_1$ and go to Step 1(1) in the following list of steps for $k=1,2,\ldots,m$.

Step 1(k). Answer: $\mathbf{B}_k$ has a zero row or zero column? If "yes," assign $d=0$ (as the value of $\det\mathbf{B}_1$), and go to Step 5(k). If "no," go to Step 2(k).

Step 2(k). Answer: element (k,k) zero? If "yes," add to row k of the $m\times n$ matrix the next appearing row with nonzero element in column k, and go to Step 3(k). If "no," go to Step 3(k).
Step 3(k). Designate element (k,k) as kth *Gaussian pivot* d_k, perform the Gaussian pivot operation in the $m\times n$ matrix, and go to Step 4(k).
Step 4(k). Answer: $k<m$? If "yes," designate the submatrix formed by elements (i,j) for $i,j=k+1,k+2,\ldots,m$ as $\mathbf{B}_{k+1}$, and go to Step 1($k+1$). If "no," assign $d=d_1d_2\cdots d_m$ (as the value of $\det\mathbf{B}_1$) and go to Step 5(k).
Step 5(k). Designate the $m\times n$ matrix as $\mathbf{G}$ the *Gaussian Reduced Form* (GRF) of $\mathbf{B}$, and stop.

In summary, input is $\mathbf{B}$ and output consists of d and the GRF $\mathbf{G}$. (5.11)

GRFs are easily found by hand for small m and n but in practice it is advantageous to use a computer program. Procedure (5.11) is readily validated as a well-defined procedure for any nonzero matrix of elements from $\mathcal{C}$: It is always possible to start, it is always certain what to do next, and it always stops after at most m Gaussian pivots, counting the operation at (m,m) which calls for no arithmetic at all. The GRF $\mathbf{G}$ is not uniquely determined in a broad sense comparable to that which holds in (3.8) for the GJSF (5.2.3). In order to complete the justification of (5.11) we need to show

$$\det\mathbf{B}_1=d \tag{5.12}$$

PROOF. Type III operations do not alter determinants so the determinant of the first m columns remains unchanged throughout (5.11). If there are no Gaussian pivots the result holds because $\mathbf{B}_1$ has either a zero row or a zero column, and $d=0$ is assigned. Otherwise,

$$\det\mathbf{B}_1=d_1d_2\cdots d_j\det\mathbf{B}_{j+1}$$

for $j=1,2,\ldots,m-1$ and the final Gaussian pivot is d_j because $\det\mathbf{B}_{j+1}=0$ or else $d_m\neq 0$ and the result holds by assignment. ■

The preceding shows that without regard to effects of rounding off numbers or deep concerns for economical computing, (5.11) provides a reasonable, systematic method for evaluating determinants. Less systematically, we could use any finite succession of Type III operations to reduce to triangular form and then evaluate the determinant by forming the product of principal diagonal elements.

Triangular systems of linear equations are easy to solve because we can use *back-substitution*: We substitute from the equation with one variable to the next which has two, and so on until we finish. Indeed, we have the following procedure for real or complex systems.

SOLVING Ax=b BY GAUSSIAN REDUCTION AND BACK-SUBSTITUTION. Given **A** is nonsingular of order n, find the unique solution $\mathbf{x}_0$ by

(1) Reducing $[\mathbf{A},\mathbf{b}]\overset{R}{\rightarrow}[\mathbf{G},\mathbf{Eb}]$ where **G** is the GRF of **A**, and

(2) Solving $[\mathbf{G},\mathbf{Eb}]\overset{R}{\rightarrow}[\mathbf{I}_n,\mathbf{x}_0]$ by back-substitution. (5.13)

As shown in 5.2.6 it can be said that for large n, solving with GJSFs in (3.31) requires about 150% of the effort in using (5.13); finding and using $\mathbf{A}^{-1}$ requires about 300%.

In some cases the GRF can be factored as follows.

If **A** is nonsingular of order n then its GRF is

$$\mathbf{G}=\mathbf{DU}$$

where $\mathbf{D}=\mathbf{diag}(d_1,d_2,\ldots,d_n)$ and **U** is upper-triangular with all elements unity on its main diagonal. (5.14)

PROOF. Since **G** has no zeros on its main diagonal by (5.12), its diagonal elements can be used for **D**. Then **U** is found by multiplying the ith row of **G** by d^{-1}. ■

Under additional restrictions we have the

LDU FACTORIZATION THEOREM. Suppose **A** is nonsingular of order n and, moreover, no zeros are encountered in positions $(1,1),(2,2),\ldots,(n-1,n-1)$ during the Gaussian reduction procedure (5.11). Then **A** can be factored uniquely

$$\mathbf{A}=\mathbf{LDU}$$

where $\mathbf{L}=\mathbf{E}^{-1}$ is lower-triangular with all elements unity on its main diagonal, where $\mathbf{EA}=\mathbf{G}$ is the GRF of **A**, and where **DU** is given by (5.14). (5.15)

PROOF. If no zeros are encountered as stated then also $d_n\neq 0$. Consider $[\mathbf{A},\mathbf{I}_n]\overset{R}{\rightarrow}[\mathbf{G},\mathbf{E}]$ through (5.11). Multiples of Gaussian pivot rows are only added to lower rows so **E** is lower-triangular with all elements unity on its main diagonal; consequently $\mathbf{E}^{-1}$ has the same two properties. To prove uniqueness, suppose

$$\mathbf{A}=\mathbf{L}'\mathbf{D}'\mathbf{U}'$$

is a second such factorization. From $\mathbf{LDU}=\mathbf{L}'\mathbf{D}'\mathbf{U}'$ we have

$$\mathbf{DU}(\mathbf{U}')^{-1}=\mathbf{L}^{-1}\mathbf{L}'\mathbf{D}'$$

where the LHS is upper-triangular and the RHS is lower-triangular. This can happen only if both sides are diagonal matrices and then, since units occupy all diagonal positions in each of

$$\mathbf{U}, (\mathbf{U}')^{-1}, \mathbf{L}^{-1}, \mathbf{L}',$$

actually $\mathbf{U}(\mathbf{U}')^{-1} = \mathbf{I} = \mathbf{L}^{-1}\mathbf{L}'$. We conclude $\mathbf{U} = \mathbf{U}'$ and $\mathbf{L} = \mathbf{L}'$ since inverses are unique. ■

See 5.2.5 for what can be done for any nonsingular matrix.

PROBLEMS FOR SOLUTION

5.2.1 More on Determinants. Establish:

(1) Elementary matrices of Types I, II, and III have determinants -1, k, and 1, respectively.
(2) $\det \mathbf{EA} = (\det \mathbf{E})(\det \mathbf{A})$ where $\mathbf{E}$ is any elementary matrix.
(3) Elementary row operations of Type III do not alter determinants.
(4) A matrix is nonsingular iff its determinant is nonzero.
(5) If $\mathbf{A}$ and $\mathbf{B}$ are of order n then $\det \mathbf{AB} = (\det \mathbf{A})(\det \mathbf{B})$.
(6) The preceding extends to any finite number of matrices of order n.

5.2.2 Evaluating Determinants. Establish a standard procedure.

5.2.3 Gaussian Reduced Forms. Establish or furnish:

(1) An example where changing the rule for overcoming zeros changes the GRF.
(2) A different procedure for Step $2(k)$ in (5.11).

5.2.4 Numerical Exercises. Apply (5.11) to the matrices in 5.1.4 and 5.1.5 and also determine the ones to which (5.15) applies.

5.2.5 LDU Factorization Theorem (5.15). Establish or furnish:

(1) A procedure for factoring $\mathbf{A}$ when (5.15) applies.
(2) An example of a nonsingular matrix to which (5.15) does not apply.
(3) A procedure for producing a row-equivalent matrix $\mathbf{TA}$ that is $\mathbf{LDU}$ factorable when (5.15) does not apply to $\mathbf{A}$ even though $\mathbf{A}$ is nonsingular.
(4) If $\mathbf{A}$ is Hermitian and (5.15) applies then $\mathbf{D}$ is real and $\mathbf{U} = \mathbf{L}^H$.

5.2.6 Counting Multiplications and Divisions. The total count of multiplications and divisions has traditionally been used as a measure of the amount of time, and hence effort, required to solve $\mathbf{Ax} = \mathbf{b}$ for $\mathbf{x}$. Consider $\mathbf{A}$ nonsingular of order n, where n is a large number, and find approximate total counts for

(1) Evaluating $\mathbf{A}^{-1}\mathbf{b}$ given $\mathbf{A}^{-1}$.
(2) Evaluating $\mathbf{A}^2$.
(3) Solving by (3.2).
(4) Solving by (5.13).
(5) Finding $\mathbf{A}^{-1}$ in (3.29).
(6) Solving by (5) and (1).

5.2.7 Gaussian Pivot Matrices. Such a matrix is $\mathbf{E}$ of order m such that if $\mathbf{B}$ is of order $m \times n$ then $\mathbf{EB}$ is the matrix resulting in (5.10). Establish: (1) The form of $\mathbf{E}$ for position (j,j). (2) $\mathbf{E}$ is lower-triangular and nonsingular.

5.3 DEFINITENESS

Readers who are especially interested in optimization can enter this section directly because it depends very little on the preceding two. The subject of definiteness arises frequently, and procedures that merely restate classical criteria involving determinants are of limited use. In this section we give a thorough treatment culminating in practical procedures based on Gaussian pivoting.

Concepts of definiteness are most naturally introduced for certain functions. A *Hermitian quadratic form*, briefly a *quadratic form* or a *form*, is a function on $\mathcal{C}^n$ to $\mathcal{R}$

$$\mathbf{x} \mapsto \langle \mathbf{x}, \mathbf{A}\mathbf{x} \rangle = \mathbf{x}^H \mathbf{A}\mathbf{x} \tag{5.16}$$

where $\mathbf{A}$ is a Hermitian matrix called the *matrix of the form*. There is a one-to-one correspondence between forms (5.16) and Hermitian matrices $\mathbf{A}$; for example, $5x_1^2 + 3x_1x_2 + x_2x_1 + x_2^2$ can be rewritten $5x_1^2 + 2x_1x_2 + 2x_2x_1 + x_2^2$ so the matrix is

$$\begin{bmatrix} 5 & 2 \\ 2 & 1 \end{bmatrix}$$

and the same can be done for any RHS of (5.16). The *rank* of the form (5.16) is rank $\mathbf{A}$ and the form is called *singular* or *nonsingular* according as $\mathbf{A}$ is singular or nonsingular. We have the

Definition. The following terminology applies both to quadratic forms and to Hermitian matrices $\mathbf{A}$
(1) *Positive definite* if $\langle \mathbf{x}, \mathbf{A}\mathbf{x} \rangle > 0$ for all $\mathbf{x} \neq \mathbf{0}$,
(2) *Positive semidefinite* if $\langle \mathbf{x}, \mathbf{A}\mathbf{x} \rangle \geqq 0$ for all $\mathbf{x}$,
(3) *Positive semidefinite-singular* if (i) positive semidefinite, and (ii) $\langle \mathbf{x}, \mathbf{A}\mathbf{x} \rangle = 0$ for at least one $\mathbf{x} \neq \mathbf{0}$,

(4) *Negative definite* if $-\mathbf{A}$ is positive definite,
(5) *Negative semidefinite* if $-\mathbf{A}$ is positive semidefinite,
(6) *Negative semidefinite-singular* if $-\mathbf{A}$ is positive semidefinite-singular,
(7) *Indefinite* if none of the preceding six holds, and *definite* if at least one of (1) through (6) holds. *Definiteness classes* or *classifications* (1) through (7) are defined by the respective criteria. Determining the classes to which an object belongs is called *resolving*, or *resolution of*, *definiteness*. (5.17)

Classes (4), (5), and (6) can be specified directly in terms of $\langle \mathbf{x}, \mathbf{Ax} \rangle$ by replacing $>$ by $<$ in (1), (2), and (3), respectively, and in practice we stress classes (1), (2), (3), and (7). Use of "singular" is justified in (5.24). Class (1) is a subset of (2), and so is (3), and there are corresponding inclusions in (5).

An example of a positive definite form is provided by $x_1^2 + x_2^2$ on $\mathcal{R}^2$ while $x = [1, -1]^T$ shows $x_1^2 + 2x_1x_2 + x_2^2$ goes with a positive semidefinite-singular form. Finally, $x_1^2 - x_2^2$ is positive for $\mathbf{e}_1$ and negative for $\mathbf{e}_2$; we see that it fits into no one of (1) through (6) and it must therefore be indefinite. This illustrates a convenient criterion for

INDEFINITENESS. A form is indefinite iff there exist vectors $\mathbf{u}$ and $\mathbf{v}$ such that

$$\langle \mathbf{u}, \mathbf{Au} \rangle > 0 \qquad \text{and} \qquad \langle \mathbf{v}, \mathbf{Av} \rangle < 0 \tag{5.18}$$

PROOF. Sufficiency is demonstrated by the argument used for the third form just above. Conversely, let the form be indefinite. If all of its values were nonpositive we have the contradictory conclusion that it must be negative semidefinite; if all were nonnegative it would have to be positive semidefinite. ■

Let us next assemble several necessary conditions that are useful in practice (just as are all necessary conditions) because when any one can be shown to fail for a particular object it rules out the associated property.

POSITIVE DEFINITENESS. Each of the following is necessary for a Hermitian matrix $\mathbf{A} = [a_{pq}]$ of order n to be positive definite.
(1) $a_{pp} > 0$ for $p = 1, 2, \ldots, n$.
(2) $\mathbf{A}$ is nonsingular and moreover $\det \mathbf{A} > 0$.
(3) $\mathbf{A}^{-1}$ is positive definite.
(4) For all $p \neq q$,

$$\det \begin{bmatrix} a_{pp} & a_{pq} \\ \bar{a}_{pq} & a_{qq} \end{bmatrix} > 0$$

(5) $\max\{a_{pp} : p=1,2,\ldots,n\} \geqq \max\{|a_{pq}| : p,q=1,2,\ldots,n\}$.
(6) Each leading principal submatrix of **A** is also positive definite. (5.19)

PROOF. Property (1) follows by considering $\mathbf{x}=c\mathbf{e}_p$ because then

$$\langle \mathbf{x}, \mathbf{A}\mathbf{x}\rangle = |c|^2 a_{pp} > 0$$

since $\mathbf{x} \neq \mathbf{0}$. Properties (2) and (3) follow directly from the fact that $\det \mathbf{A}$ equals the product of the eigenvalues (5.1.1). Selecting

$$\mathbf{x} = a_{pq}\mathbf{e}_p + t\mathbf{e}_q$$

where $t \in \mathfrak{R}$ we can prove (4) as follows. The quadratic form has value

$$a_{pp}|a_{pq}|^2 + 2|a_{pq}|^2 t + a_{qq}t^2$$

and for this to be positive for all t, an elementary necessary condition is that it vanish for no t, and

$$|a_{pq}|^4 - a_{pp}a_{qq}|a_{pq}|^2 < 0$$

which implies (4). If (5) failed and the largest modulus appeared at (p,q), then using (1) we would obtain a contradiction to (4). Suppose that (6) fails for the leading principal submatrix of order k. Then we could conclude that **A** is not positive definite by using the first k components of a vector determined by the submatrix and using zero as components $k+1, k+2, \ldots, n$. ■

POSITIVE SEMIDEFINITENESS. Each of the following is necessary for a Hermitian matrix $\mathbf{A}=[a_{pq}]$ of order n to be positive semidefinite.
(1) $a_{pp} \geqq 0$ for $p=1,2,\ldots,n$.
(2) If $a_{pp}=0$ then all elements in the pth row and the pth column must also be zero. (5.20)

PROOF. Property (1) is proved in the same way as in the preceding theorem. For (2) we suppose $a_{pq} \neq 0$ and again use

$$\mathbf{x} = a_{pq}\mathbf{e}_p + t\mathbf{e}_q$$

for $t \in \mathfrak{R}$ whereupon

$$\mathbf{x}^H\mathbf{A}\mathbf{x} = t\left(2|a_{pq}|^2 + a_{qq}t\right)$$

There are three cases, and in each we can find two intervals for t such that the value of the form is negative for t belonging to the first, and positive for the second. If $a_{qq}=0$ the intervals are $(-\infty, 0)$ and $(0,\infty)$. If $a_{qq}<0$ the intervals are $(0,u)$ and (u,∞) where $u=-2|a_{pq}|^2/a_{qq}$. If $a_{qq}>0$ the intervals are $(u,0)$ and $(0,\infty)$. ■

The following restatement and immediate consequence of the preceding (2) are used to establish our ultimate procedure (5.28).

INDEFINITENESS. The following hold for any Hermitian $\mathbf{A}=[a_{pq}]$.

(1) If $a_{pp}=0$ and $a_{pq}\neq 0$ for some p and q, then $\mathbf{A}$ is indefinite.

(2) If $\mathbf{A}$ belongs to the same definiteness class as

$$\begin{bmatrix} \mathbf{U} & \mathbf{V}^H \\ \mathbf{V} & \mathbf{W} \end{bmatrix}$$

where $\mathbf{W}$ is indefinite, then $\mathbf{A}$ is indefinite. (5.21)

PROOF. It remains to prove the present (2). If $\mathbf{u}$ and $\mathbf{v}$ are vectors from (5.18) for $\mathbf{W}$ then

$$\begin{bmatrix} \mathbf{0} \\ \mathbf{u} \end{bmatrix} \quad \text{and} \quad \begin{bmatrix} \mathbf{0} \\ \mathbf{v} \end{bmatrix}$$

are the vectors that show $\mathbf{A}$ to be indefinite. ■

The key concept in the theory of definiteness classes is an equivalence relation that is remarkably easy to use. If we replace $\mathbf{x}$ in the form (5.16) by $\mathbf{Py}$ where $\mathbf{P}$ is nonsingular of order n and $\mathbf{y}\in\mathcal{C}^n$ we obtain

$$\langle \mathbf{y}, \mathbf{P}^H\mathbf{APy}\rangle = \mathbf{y}^H\mathbf{P}^H\mathbf{APy}$$

which is also a form because $\mathbf{P}^H\mathbf{AP}$ is Hermitian. These two forms, the one with matrix $\mathbf{A}$ and the other with matrix $\mathbf{P}^H\mathbf{AP}$ are called *Hermitian-congruent*, or briefly *congruent*. Whenever there exists a nonsingular matrix $\mathbf{P}$ such that

$$\mathbf{B}=\mathbf{P}^H\mathbf{AP}$$

then the same terminology applies to $\mathbf{A}$ and $\mathbf{B}$, and we say that $\mathbf{B}$ results from $\mathbf{A}$ by a *Hermitian-congruence*, or briefly *congruence, transformation*. Since $\mathbf{I}$ is Hermitian, congruence is reflexive; it is symmetric because $\mathbf{P}^{-1}$ is nonsingular iff $\mathbf{P}$ is nonsingular; and it is easy to verify that it is transitive. Congruence is therefore an equivalence relation and its relevance is revealed in the

THEOREM. If $\mathbf{A}$ and $\mathbf{B}$ are Hermitian-congruent then they belong to the same definiteness class. (5.22)

PROOF. Suppose $\mathbf{A}$ is positive definite and $\mathbf{y}\neq\mathbf{0}$ is arbitrary. Then $\mathbf{x}=\mathbf{Py}$ is nonzero by (2.21) and

$$(\mathbf{Py})^H\mathbf{A}(\mathbf{Py})=\mathbf{y}^H(\mathbf{P}^H\mathbf{AP})\mathbf{y}$$

is positive so **B** is positive definite. The argument is clearly reversible to show: If $\mathbf{B}=\mathbf{P}^H\mathbf{A}\mathbf{P}$ is positive definite then so is **A**. Essentially the same technique establishes the result for the remaining six classes in (5.17). In particular, for indefiniteness we use pairs of vectors guaranteed by (5.18). ■

Before proceeding to show how to find congruence transformations it will be helpful to have the companion result to (5.22) which is the

THEOREM. If **A** is congruent to any

$$\mathbf{D}=\mathbf{diag}(d_1,d_2,\ldots,d_n)$$

then its definiteness is resolved by classifying $d_1,d_2,\ldots,d_n$ as follows.

(1) Positive definite iff all are positive.
(2) Positive semidefinite iff all are nonnegative.
(3) Positive semidefinite-singular iff all are nonnegative and at least one is zero.
(4) Negative definite iff all are negative.
(5) Negative semidefinite iff all are nonpositive.
(6) Negative semidefinite-singular iff all are nonpositive and at least one is zero.
(7) Indefinite iff at least one is positive and at least one is negative. (5.23)

PROOF. It suffices to show that each of (1) through (7) is true for **D**. Let **D** be positive definite. Then if one d_i, say d_1, were nonpositive, choosing $\mathbf{x}=\mathbf{e}_1$ would show

$$\mathbf{x}^H\mathbf{D}\mathbf{x}=d_1$$

Conversely, if all d_i are positive then for $\mathbf{x}\in\mathcal{C}^n$

$$\mathbf{x}^H\mathbf{D}\mathbf{x}=d_1|x_1|^2+d_2|x_2|^2+\cdots+d_n|x_n|^2$$

which is clearly positive for $\mathbf{x}\neq\mathbf{0}$. This completes the proof of (1), and (2) follows in the same way. Sufficiency in (3) is also immediate. For necessity it is clearly impossible for d_1 to be negative, or for all d_i to be positive, when **D** is positive semidefinite-singular. Then (4), (5), and (6) result by interchanging $>$ and $<$ in the preceding arguments. For (7) let **D** be indefinite. Then if no d_i is positive we have a contradiction from (5) and if none is negative we have one from (2). Conversely, if $d_1>0$ and $d_2<0$ then $\mathbf{x}^H\mathbf{D}\mathbf{x}$ is positive for $\mathbf{x}=\mathbf{e}_1$ and negative for $\mathbf{x}=\mathbf{e}_2$ so **D** is indefinite by (5.18). ■

As an immediate consequence we have the following which justifies our terminology.

A Hermitian matrix is positive semidefinite-singular iff it is positive semidefinite and singular. (5.24)

The next result is of substantial theoretical interest.

If all eigenvalues of a Hermitian matrix are known then its definiteness is resolved by applying the criteria in (5.23) to the eigenvalues. (5.25)

PROOF. A Hermitian matrix is a normal matrix so by (5.9) it can be diagonalized by a unitary transformation. The latter is nonsingular and hence is a congruence transformation. ■
We soon find a procedure that is much easier than finding all eigenvalues and then using the preceding.

EXAMPLE 5.2. Resolving Definiteness for $n=1,2,3$. This example is a prelude that also illustrates elementary manipulation of nonreal matrices.

Resolution is trivial for $n=1$ because the quadratic form is

$$a_{11}|x|^2$$

where $\mathbf{A}=[a_{11}]$ Hermitian requires a_{11} to be real. Classifications (1)–(6) in (5.17) depend on the sign of a_{11} and (7) cannot occur.

The general Hermitian matrix of order 2 is

$$\mathbf{A}=\begin{bmatrix} a_{11} & \bar{a}_{21} \\ a_{21} & a_{22} \end{bmatrix}$$

where $a_{11}, a_{22} \in \mathfrak{R}$. There are two cases. First, if $a_{11}=0$ then $\mathbf{A}$ cannot be positive definite by (5.19) and there are two possibilities. If $a_{21} \neq 0$ then $\mathbf{A}$ is indefinite by (5.21); if $a_{21}=0$ we have

$$\mathbf{diag}(0, a_{22})$$

Second, if $a_{11} \neq 0$ we can perform a Gaussian pivot and the equation $\mathbf{EA}=\mathbf{G}$ is

$$(1) \qquad \begin{bmatrix} 1 & 0 \\ -\dfrac{a_{21}}{a_{11}} & 1 \end{bmatrix} \begin{bmatrix} a_{11} & \bar{a}_{21} \\ a_{21} & a_{22} \end{bmatrix} = \begin{bmatrix} a_{11} & \bar{a}_{21} \\ 0 & a_{22} - \dfrac{a_{21}\bar{a}_{21}}{a_{11}} \end{bmatrix}$$

Now it is important to observe that if we take the complex conjugate of the multiplier used on row 1, that is, if we take $-\bar{a}_{21}/a_{11}$, and use it to multiply column 1 and add the result to column 2, we obtain

$$(2) \qquad \mathbf{EAE}^H = \mathbf{diag}\left(a_{11}, a_{22} - \frac{a_{21}\bar{a}_{21}}{a_{11}} \right)$$

Thus **A** is congruent to the diagonal matrix formed by the elements on the main diagonal of **G**. The important point is that having reached the upper-triangular **G** in (1) we need not proceed to carry out (2); instead, we merely inspect the main diagonal of **G** and apply (5.23). In summary,

(3) Resolution for $n=2$. There are two cases.

(i) $a_{11}=0$. If $a_{21}\neq 0$ then **A** is indefinite. If $a_{21}=0$, resolve definiteness of

$$\mathbf{diag}(0, a_{22})$$

(ii) $a_{11}\neq 0$. Perform a Gaussian pivot on a_{11} and resolve definiteness of

$$\mathbf{diag}(a_{11}, b_{22})$$

where b_{22} is the resulting $(2,2)$ element.

Let us also notice $\det \mathbf{A} = a_{11}b_{22}$ which suggests how we can establish classical criteria based on determinants (5.3.4).

The general Hermitian matrix of order 3 is

$$\mathbf{A}=\begin{bmatrix} a_{11} & \bar{a}_{21} & \bar{a}_{31} \\ a_{21} & a_{22} & \bar{a}_{32} \\ a_{31} & a_{32} & a_{33} \end{bmatrix}$$

where the main diagonal elements are real. As in the preceding, if $a_{11}=0$ then there are two cases. First, column 1 may be nonzero in which case **A** is indefinite. Second, column 1 may be zero in which case the definiteness of **A** coincides with that of the 2×2 submatrix obtained by deleting the (zero) row 1 and (zero) column 1. Thus, this second case reduces to that for $n=2$.

If $a_{11}\neq 0$ we can perform a Gaussian pivot and the equation $\mathbf{EA}=\mathbf{G}$ is

$$\text{(4)} \quad \begin{bmatrix} 1 & 0 & 0 \\ -\dfrac{a_{21}}{a_{11}} & 1 & 0 \\ -\dfrac{a_{31}}{a_{11}} & 0 & 1 \end{bmatrix}\mathbf{A}=\begin{bmatrix} a_{11} & \bar{a}_{21} & \bar{a}_{31} \\ 0 & a_{22}-a_{21}\dfrac{\bar{a}_{21}}{a_{11}} & a_{32}-a_{21}\dfrac{\bar{a}_{31}}{a_{11}} \\ 0 & a_{32}-a_{31}\dfrac{\bar{a}_{21}}{a_{11}} & a_{33}-a_{31}\dfrac{\bar{a}_{31}}{a_{11}} \end{bmatrix}$$

Notice that the lower RH 2×2 submatrix of **G**, let us call it **B**, is Hermitian. Notice also that we can use the conjugates of the multipliers used to produce rows 2 and 3 of **G** to reduce row 1 of **G** to $[a_{11}, 0, 0]$. In matrix terms, if we multiply (4) on the right by $\mathbf{E}^H$ we obtain

$$\mathbf{EAE}^H = \begin{bmatrix} a_{11} & \mathbf{0}^T \\ \mathbf{0} & \mathbf{B} \end{bmatrix} \tag{5}$$

where $\mathbf{B}$ is unchanged from (4) because $\mathbf{G}$ has zeros in positions (1,2) and (1,3). From (5) it follows that $\mathbf{x}^H\mathbf{Ax}$ is congruent to

$$a_{11}x_1^2 + \mathbf{y}^H\mathbf{By}$$

where $x_1 \in \mathfrak{R}$ and $\mathbf{y} \in \mathfrak{R}^2$ and so we see the problem for $n=3$ can be solved using the results for $n=2$. Let us proceed.

Consider the (1,1) element in $\mathbf{B}$. If $b_{11}=0$ there are the familiar cases depending on whether column 1 of $\mathbf{B}$ is zero or not. If $b_{11} \neq 0$ we can perform a Gaussian pivot and obtain the counterpart of (5), say

$$(\mathbf{FE})\mathbf{A}(\mathbf{FE})^H = \mathbf{diag}(a_{11}, b_{11}, c_{11}) \tag{6}$$

which is congruent to $\mathbf{A}$ and to which (5.23) applies. In summary,

(7) Resolution for $n=3$. There are two cases.

(i) $a_{11}=0$. If column 1 is nonzero then $\mathbf{A}$ is indefinite. If column 1 is zero, use (3) to resolve the definiteness of (5) for $a_{11}=0$. If $b_{11}=0$ and $\mathbf{B}$ is indefinite then $\mathbf{A}$ is indefinite. If $b_{11} \neq 0$ use (5.23) for the RHS of (6).

(ii) $a_{11} \neq 0$. Perform a Gaussian pivot and resolve the definiteness of (5). If $b_{11}=0$ and $\mathbf{B}$ is indefinite then $\mathbf{A}$ is indefinite. If $b_{11} \neq 0$ use (5.23) for the RHS of (6).

We see that the procedure is the same for both $n=2$ and $n=3$. We perform Gaussian pivots and act as follows if we encounter a zero on the main diagonal: If there is a nonzero element standing below the zero in the same column we stop because the matrix is indefinite; if only zeros stand below we merely skip to the next main diagonal position and continue.

Guided by Example 5.2 let us formalize a variation of (5.11) that will be shown to resolve the definiteness of any Hermitian matrix.

MODIFIED GAUSSIAN REDUCTION PROCEDURE. To start for Hermitian $\mathbf{C} \neq \mathbf{O}$ of order n, set $k=1$, designate $\mathbf{C}_1 = \mathbf{C}$, and go to Step 1(1) in the following list of steps for $k=1,2,\ldots,n$.

Step 1(k). Designate element (k,k) as q_k and answer: $q_k=0$? If "yes," go to Step 2(k). If "no," perform a Gaussian pivot on q_k and go to Step 3(k).
Step 2(k). Answer: first column of $\mathbf{C}_k$ zero? If "yes," go to Step 3(k). If "no," go to Step 4(k).

Step 3(k). Answer: $k<n$? If "yes," designate the submatrix formed by elements (i,j) for $i,j=k+1,k+2,\ldots,n$ as $\mathbf{C}_{k+1}$, and go to Step 1($k+1$). If "no," go to Step 4(n).
Step 4(k). Assign $t=k$ (the terminal value of k), and stop.

In summary, input is $\mathbf{C}$ and output consists of t and the set

$$\{q_1,q_2,\ldots,q_t\}$$

whose nonzero members, if any, are the Gaussian pivots. (5.26)

In order to make use of the preceding modification, we need to prove the major result in Example 5.2, namely the

THEOREM. If $t<n$ in the modified Gaussian reduction procedure (5.26) then $\mathbf{C}$ is indefinite. If $t=n$ then $\mathbf{C}$ is congruent to

$$\mathbf{diag}(q_1,q_2,\ldots,q_n) \tag{5.27}$$

PROOF. Let us use the notation of (5.26) including $q_1=c_{11}$. For $t=1<n$ it is necessary and sufficient that $q_1=0$ while column 1 of $\mathbf{C}$ is nonzero so $\mathbf{C}$ is indefinite by 5.21(1). For $t=2<n$ it is necessary and sufficient that $q_2=0$ while column 1 of $\mathbf{C}_2$ is nonzero; let us show that $\mathbf{C}$ is congruent to

$$\begin{bmatrix} q_1 & \mathbf{0}^T \\ \mathbf{0} & \mathbf{C}_2 \end{bmatrix}$$

for any q_1. If $q_1=0$ the preceding coincides with $\mathbf{C}$. If $q_1\neq 0$ then in the first place

$$\mathbf{E}_1\mathbf{C}=\begin{bmatrix} q_1 & \mathbf{T}_1 \\ \mathbf{0} & \mathbf{C}_2 \end{bmatrix}$$

where $\mathbf{E}_1$ is the lower-triangular Gaussian pivot matrix (5.2.7) for position (1, 1) and $[q_1,\mathbf{T}_1]$ is the first row of $\mathbf{C}$. Consequently,

$$\mathbf{E}_1\mathbf{C}\mathbf{E}_1^H=\begin{bmatrix} q_1 & \mathbf{0}^T \\ \mathbf{0} & \mathbf{C}_2 \end{bmatrix}$$

because row 1 of $\mathbf{E}_1\mathbf{C}$ is the Hermitian transpose of column 1 of $\mathbf{C}$. Finally, the theorem holds for $t=2$ by 5.21(2) because $\mathbf{C}_2$ is then indefinite. The general theorem follows by induction. Suppose first, for any $r\leqq n-1$

$$\mathbf{E}_r\mathbf{E}_{r-1}\cdots\mathbf{E}_1\mathbf{C}=\begin{bmatrix} \mathbf{Q}_r & \mathbf{T}_r \\ \mathbf{O} & \mathbf{C}_{r+1} \end{bmatrix}$$

where all $\mathbf{E}_i$ are lower-triangular Gaussian pivot matrices, where $\mathbf{Q}_r$ is the upper-triangular matrix of order r that coincides with the upper LH submatrix of $\mathbf{C}$ on and above the main diagonal, and where $\mathbf{T}_r$ denotes the submatrix standing above $\mathbf{C}_{r+1}$ when the latter is designated in (5.26). Second, as verified for $r=1$

$$\mathbf{E}_r\mathbf{E}_{r-1}\cdots\mathbf{E}_1\mathbf{C}\mathbf{E}_1^H\mathbf{E}_2^H\cdots\mathbf{E}_r^H=\begin{bmatrix}\mathbf{D}_r & \mathbf{O}^T\\ \mathbf{O} & \mathbf{C}_{r+1}\end{bmatrix}$$

where $\mathbf{D}_r=\mathbf{diag}(q_1,q_2,\ldots,q_r)$ whether or not any $q_i=0$. If $r<n-1$ the conclusion follows from 5.21(2) and if $r=n$ it follows because of the congruence transformation exhibited just above. ■

We can now state our ultimate procedure for

RESOLVING DEFINITENESS. Suppose $\mathbf{A}$ is Hermitian of order n. If $\mathbf{A}=\mathbf{O}$ it is positive semidefinite-singular. If $\mathbf{A}\neq\mathbf{O}$ complete (5.26) for $\mathbf{A}$ and use

(1) If $t<n$ then $\mathbf{A}$ is indefinite.
(2) If $t=n$ use (5.23) to resolve the definiteness of

$$\mathbf{D}_n=\mathbf{diag}(q_1,q_2,\ldots,q_n)$$

which coincides with that of $\mathbf{A}$. (5.28)

PROOF. The case $\mathbf{A}=\mathbf{O}$ is trivial and all others follow from (5.27). ■

In optimization theory in Chapter 7 we need the concept of definiteness on a subspace. This merely denotes the definiteness of the restriction of the quadratic form to a subspace $S\subset\mathcal{C}^n$.

Definition. The following terminology applies both to quadratic forms and to Hermitian matrices. If $S\subset\mathcal{C}^n$ is a subspace there are seven *definiteness classes for S* defined by adjoining $\mathbf{x}\in S$ to the definitions in (5.17).

We use the terminology *resolving definiteness on S* and the following procedure for

RESOLVING DEFINITENESS ON A SUBSPACE. Suppose $S\subset\mathcal{C}^n$ is a proper subspace of dimension r and $\mathbf{A}$ is Hermitian of order n. Use (3.24) to find $\mathbf{Z}$ the $n\times r$ matrix of the standard basis and then use (5.28) to resolve the definiteness of

$$\mathbf{Z}^H\mathbf{A}\mathbf{Z}$$

on $\mathcal{C}^r$ which coincides with that of $\mathbf{A}$ on S. (5.29)

PROOF. There is a one-to-one correspondence

$$\mathbf{x}^H\mathbf{A}\mathbf{x} \leftrightarrow \mathbf{z}^H\mathbf{Z}^H\mathbf{A}\mathbf{Z}\mathbf{z}$$

between forms as well as between vectors. ■

EXAMPLE 5.3. Resolving Definiteness by Gaussian Pivoting. The power of (5.28) results from the convenience of its application as compared with finding eigenvalues to use (5.25) or evaluating determinants to use other classical criteria (5.3.4). For example,

$$\mathbf{A}=\begin{bmatrix} 1 & 0 & 1 & -1 \\ 0 & 2 & 1 & 0 \\ 1 & 1 & 2 & 1 \\ -1 & 0 & 1 & 3 \end{bmatrix}$$

is indefinite because $\mathbf{D}_4=\mathbf{diag}(1,2,\frac{1}{2},-6)$.

If $\mathbf{B}$ denotes the matrix obtained from $\mathbf{A}$ by changing the (4, 4) element to $t\in\mathfrak{R}$, then

$$\mathbf{D}_4=\mathbf{diag}\left(1,2,\frac{1}{2},t-9\right)$$

shows that $\mathbf{B}$ is indefinite for $t<9$, positive semidefinite-singular for $t=9$, and positive definite for $t>9$.

If we reduce the following where $u\in\mathfrak{R}$

$$C=\begin{bmatrix} 1 & 0 & 1 & 1 \\ 0 & 2 & 2 & 0 \\ 1 & 2 & 3 & u \\ 1 & 0 & u & 1 \end{bmatrix}$$

we arrive at the following after two Gaussian pivots.

$$\begin{bmatrix} 1 & 0 & 1 & 1 \\ 0 & 2 & 2 & 0 \\ 0 & 0 & 0 & u-1 \\ 0 & 0 & u-1 & 0 \end{bmatrix}$$

Consequently $\mathbf{C}$ is positive semidefinite-singular for $u=1$ and indefinite for $u\neq 1$.

As an observation of a different kind, it is easy to verify that using Gaussian pivoting is not as convenient as using eigenvalues for finding conditions on v such that $\mathbf{A}$ positive definite implies $\mathbf{A}-v\mathbf{I}$ positive definite.

Although we have determined that the preceding matrix **A** is indefinite (on its domain $\mathcal{C}^4$) it may or may not be indefinite on a particular subspace. For example, if S is the subspace determined by the columns of

$$\begin{bmatrix} 1 & 1 & -1 \\ 0 & 2 & 1 \\ 0 & 2 & 1 \\ 0 & 0 & 0 \end{bmatrix}$$

then **A** is positive definite on S. This is true because the standard basis (3.23) has matrix

$$\mathbf{Z}=\begin{bmatrix} 1 & 0 \\ 0 & 1 \\ 0 & 1 \\ 0 & 0 \end{bmatrix}$$

and $\mathbf{Z}^H\mathbf{A}\mathbf{Z}$ as specified in (5.29) is

$$\begin{bmatrix} 1 & 1 \\ 1 & 6 \end{bmatrix}$$

which is positive definite on $\mathcal{C}^2$.

PROBLEMS FOR SOLUTION

5.3.1 Fundamental Properties. Establish:

(1) **A** is indefinite iff neither **A** nor $-\mathbf{A}$ is positive semidefinite.

(2) "Completing the square" can be used to establish (2) in Example 5.2.

(3) True or false? If there exists $\mathbf{r}\neq\mathbf{0}$ and $\mathbf{r}^T\mathbf{A}\mathbf{r}=0$ where **A** is a real symmetric matrix then **A** is singular.

(4) True or false? If **A** is positive definite and **B** is Hermitian, both of order n, then the eigenvalues of **AB** are real.

(5) If **A** is positive definite and **B** is positive semidefinite then zero is a root of $\det(\mathbf{B}-\lambda\mathbf{A})=0$.

(6) An explicit test to determine whether or not $\mathbf{A}-v\mathbf{I}$ is positive definite given that **A** is positive definite.

(7) **A** is Hermitian iff it is congruent to a real diagonal matrix.

(8) Each of the following is necessary and sufficient for **A** to be positive definite.

   (i) $\mathbf{A}^{-1}$ exists and is positive definite.

   (ii) $\mathbf{A}^p$ is positive definite for all integers p.

   (iii) There exists a unique nonsingular Hermitian **B**, called the *square root* of **A**, such that $\mathbf{B}^2=\mathbf{A}$.

5.3.2 Numerical Exercises. Resolve the definiteness where $\varepsilon, \delta, t \in \mathfrak{R}$

(1) $\begin{bmatrix} 1 & 1+i \\ 1-i & 2 \end{bmatrix}$

(2) $\begin{bmatrix} 1 & 2 & 2 & 0 \\ 2 & 5 & 1 & 1 \\ 2 & 1 & 10 & 4 \\ 0 & 1 & 4 & 12 \end{bmatrix}$

(3) $q = x_1^2 + 2x_2^2 + 2x_3^2 + 3x_4^2 + 2x_1x_3 - 2x_1x_4 + 2x_2x_3 + 2x_3x_4$

(4) $q = 2x_1^2 + 10x_2^2 + 40x_3^2 + 5x_1x_2 + 7x_1x_3 + 3x_2x_1 + 10x_2x_3 + 9x_3x_1 + 30x_2x_3$

(5) $\begin{bmatrix} 1 & 2 & 2 & 0 \\ 2 & 5 & 1 & 1 \\ 2 & 1 & 18 & 1 \\ 0 & 1 & 1 & \delta \end{bmatrix}$

(6) $\begin{bmatrix} 1+\varepsilon & 1+\varepsilon & \varepsilon \\ 1+\varepsilon & 2+\varepsilon & 1+\varepsilon \\ \varepsilon & 1+\varepsilon & 4+\varepsilon \end{bmatrix}$

(7) $\begin{bmatrix} t & -1 & 0 & 0 & \cdots & 0 & 0 \\ -1 & t & -1 & 0 & \cdots & 0 & 0 \\ 0 & -1 & t & -1 & \cdots & 0 & 0 \\ \vdots & \vdots & \vdots & \vdots & & \vdots & \vdots \\ 0 & 0 & 0 & 0 & \cdots & -1 & t \end{bmatrix}$ positive definite?

(8) $\begin{bmatrix} 2 & 0 \\ 0 & 1 \end{bmatrix}$ on Null[4, 0].

(9) $\begin{bmatrix} 1 & 1 & 0 \\ 1 & 1 & 0 \\ 0 & 0 & 3 \end{bmatrix}$ on Null[2, −2, 0].

(10) $\begin{bmatrix} 2 & \lambda & 0 \\ \lambda & 2 & 0 \\ 0 & 0 & 2-2\lambda \end{bmatrix}$ on Range$\begin{bmatrix} 1 & 0 \\ 1 & 0 \\ 0 & 1 \end{bmatrix}$.

(11) $\begin{bmatrix} 0 & \frac{-k}{3} & \frac{-k}{3} \\ \frac{-k}{3} & 0 & \frac{-k}{3} \\ \frac{-k}{3} & \frac{-k}{3} & 0 \end{bmatrix}$ on Null[1, 1, 1].

(12) **diag**(−36, −18, −12) on Range$\left[\frac{1}{6}, \frac{1}{3}, \frac{1}{2}\right]^T$.

5.3.3 Definiteness of Products. Establish the following where $\mathbf{X}$ is $m \times n$, $\mathbf{Y}$ is positive definite of order m, $\mathbf{Z}=\mathbf{X}^H\mathbf{Y}\mathbf{X}$, and $\mathbf{A}$ is of order n.

(1) $\mathbf{Z}$ is positive semidefinite.
(2) If $\mathbf{X}$ has full column rank n then $\mathbf{Z}$ is positive definite.
(3) If $\operatorname{rank}\mathbf{X}<n$ then $\mathbf{Z}$ is positive semidefinite-singular.
(4) If $\operatorname{rank}\mathbf{X}=n$ then $\mathbf{X}^H\mathbf{X}$ is positive definite.
(5) If $\operatorname{rank}\mathbf{X}<n$ then $\mathbf{X}^H\mathbf{X}$ is positive semidefinite-singular.
(6) If $\mathbf{A}$ is nonsingular then $\mathbf{A}^H\mathbf{A}$ and $\mathbf{A}\mathbf{A}^H$ are positive definite.
(7) If $\mathbf{A}$ is singular then $\mathbf{A}^H\mathbf{A}$ and $\mathbf{A}\mathbf{A}^H$ are positive semidefinite-singular.

5.3.4 Classical Criteria. Establish for Hermitian $\mathbf{A}$ of order n:

(1) $\mathbf{A}$ is positive definite iff every leading principal submatrix has a positive determinant.
(2) The preceding fails to be true if we translate to "positive semidefinite" using "nonnegative."
(3) $\mathbf{A}$ is positive semidefinite iff every principal submatrix (not merely the leading ones) has a nonnegative determinant.
(4) $\mathbf{A}$ is negative definite iff the determinants of the leading principal submatrices, proceeding in order down the main diagonal, alternate in sign with the first being negative.

5.3.5 Sylvester's Law of Inertia. Suppose $\mathbf{D}$ is any diagonal matrix that is congruent to a given Hermitian matrix $\mathbf{A}$ and let $\mathbf{D}$ have p positive, q negative, and r zero diagonal elements. (1) Prove the subject law: Any other diagonal matrix that is congruent to $\mathbf{A}$ must also have p positive, q negative, and r zero diagonal elements. (2) Is the law a simple consequence of (5.23)?

5.3.6 Use of the GJSF. Suppose that we find the GJSF of the Hermitian matrix $\mathbf{A}$ of order n with the following results. There are u positive pivots, v negative pivots, and w zero pivots where $w=n-\operatorname{rank}\mathbf{A}$ is the number of zero rows in the GJSF. Prove that $\mathbf{A}$ has u positive, v negative, and w zero eigenvalues.

5.3.7 Simultaneous Diagonalization of Two Quadratic Forms. Establish: (1) If $\mathbf{A}$ is positive definite and $\mathbf{B}$ is Hermitian, both of order n, then there exists a nonsingular $\mathbf{Q}$ such that (i) $\mathbf{Q}^H\mathbf{A}\mathbf{Q}=\mathbf{I}$, and (ii) $\mathbf{Q}^H\mathbf{B}\mathbf{Q}=\mathbf{D}$ where $\mathbf{D}$ is a diagonal matrix. (2) If $\mathbf{A}$ and $\mathbf{B}$ are real symmetric matrices then $\mathbf{Q}$ can also be chosen to be real.

5.4 DIAGONALIZATION

Readers who are especially interested in linear systems can enter this section directly from Section 5.1. We use the results in Sections 9.4 and 10.5.

In the present section we solve the

Matrix Diagonalization Problem. Given a complex matrix **A** of order n, find a nonsingular matrix **T** such that

$$\mathbf{T}^{-1}\mathbf{A}\mathbf{T}$$

is diagonal whenever **A** is diagonalizable, and "nearly diagonal" whenever **A** is defective. (5.30)

Here is how we replace "nearly diagonal" by specific properties. We first decide that of all properties possessed by any diagonal matrix **D**, two are particularly remarkable, namely

1. Its eigenvalues lie on its main diagonal.
2. Its powers $\mathbf{D}^p$ can be found at once for any positive integer p.

Next we recall that (1) holds for triangular matrices so we decide that "nearly diagonal" will in particular require upper triangularity. From (2) we conclude that there should be a minimal number of nonzero elements above the main diagonal, all should be "close" to the diagonal, and if possible all should be unity to simplify multiplications.

The method of solution we use depends on properties of nilpotent matrices (5.4.1). Zero matrices $\mathbf{O}_n$ are trivially nilpotent. Primary examples are triangular matrices with zeros on their main diagonals; however, nontriangular matrices can be nilpotent. Special significance attaches to matrices having the form of

$$\mathbf{N}_r = \begin{bmatrix} 0 & 1 & \cdots & 0 \\ 0 & 0 & \cdots & 0 \\ \vdots & \vdots & & \vdots \\ 0 & 0 & \cdots & 1 \\ 0 & 0 & \cdots & 0 \end{bmatrix} \tag{5.31}$$

and called *nilpotent canonical blocks of order r*. More precisely, let us define the *first superdiagonal* of a matrix of order $r>1$ to consist of positions $(1,2),(2,3),\ldots,(r-1,r)$. Then $\mathbf{N}_r$ is specified as $[0]$ for $r=1$, and $\mathbf{N}_r$ for $r>1$ has entries unity on its first superdiagonal and zeros everywhere else. It is easy to verify that $\mathbf{N}_r$ is nilpotent of index equal to its order r, all of its eigenvalues equal zero, and it is defective. What we mainly require is the substantial

THEOREM. Suppose **Z** is of order n and nilpotent of index p. Then there exists a nonsingular matrix **X** with the following properties.
(1) The first s columns form a basis for Null **Z**.
(2) The first $(s+t)$ columns form a basis for Null $\mathbf{Z}^2$ where $t \leqq s$, and

furthermore

$$\mathbf{Z}\mathbf{x}_{s+i}=\mathbf{x}_i$$

for $i=1,2,\ldots,t$.

(3) The first $(s+t+u)$ columns form a basis for Null $\mathbf{Z}^3$ where $u \leqq t$, and furthermore

$$\mathbf{Z}\mathbf{x}_{s+t+i}=\mathbf{x}_{s+i}$$

for $i=1,2,\ldots,u$.

⋮

(p) The columns of

$$\mathbf{X}=\left[\mathbf{x}_1,\ldots,\mathbf{x}_v,\mathbf{x}_{v+1},\ldots,\mathbf{x}_{v+w},\mathbf{x}_{v+w+1},\ldots,\mathbf{x}_n\right]$$

form a basis for Null $\mathbf{Z}^p$ (where the first $(v+w)$ form a basis for Null $\mathbf{Z}^{p-1}$), where $n-(v+w)\leqq w$, and furthermore

$$\mathbf{Z}\mathbf{x}_{v+w+i}=\mathbf{x}_{v+i}$$

for $i=1,2,\ldots,n-(v+w)$. (5.32)

PROOF. *Stage 1* consists of finding a basis $\{\mathbf{u}_1,\ldots,\mathbf{u}_s\}$ for Null $\mathbf{Z}$. If $p \geqq 2$ we begin *Stage 2* by extending this basis to a basis for Null $\mathbf{Z}^2$

$$\{\mathbf{u}_1,\ldots,\mathbf{u}_s,\mathbf{u}_{s+1},\ldots,\mathbf{u}_{s+t}\}$$

Now $\{\mathbf{Z}\mathbf{u}_{s+1},\ldots,\mathbf{Z}\mathbf{u}_{s+t}\}$ is LI since, if not,

$$c_1\mathbf{Z}\mathbf{u}_{s+1}+\cdots+c_t\mathbf{Z}\mathbf{u}_{s+t}=\mathbf{0}$$

where not all $c_i=0$, would lead to a contradiction as follows. The vector

$$c_1\mathbf{u}_{s+1}+\cdots+c_t\mathbf{u}_{s+t}$$

would be nonzero because these $\mathbf{u}$'s are LI, it by assumption would belong to Null $\mathbf{Z}$, and hence it would also be representable as an LC of $\{\mathbf{u}_1,\ldots,\mathbf{u}_s\}$. This last is impossible because all $(s+t)\mathbf{u}$'s form a basis. Since this subset of t elements of Null $\mathbf{Z}$ is LI we must have $t \leqq s$. Let us rename

$$\mathbf{Z}\mathbf{u}_{s+i}=\mathbf{v}_i \qquad \text{for } i=1,2,\ldots,t$$

and extend this set to form a basis for Null $\mathbf{Z}$

$$\{\mathbf{v}_1,\ldots,\mathbf{v}_t,\mathbf{v}_{t+1},\ldots,\mathbf{v}_s\}$$

If we now adjoin the vectors determined at the beginning of this stage, we obtain

$$\{\mathbf{v}_1,\dots,\mathbf{v}_s,\mathbf{u}_{s+1},\dots,\mathbf{u}_{s+t}\}$$

which forms a basis for Null $\mathbf{Z}^2$. This completes Stage 2, and also the proof if $p=2$.

If $p \geqq 3$ we begin *Stage 3* by extending the basis produced by the preceding stage to a basis for Null $\mathbf{Z}^3$

$$\{\mathbf{v}_1,\dots,\mathbf{v}_s,\mathbf{u}_{s+1},\dots,\mathbf{u}_{s+t},\mathbf{u}_{s+t+1},\dots,\mathbf{u}_{s+t+u}\}$$

Now $\{\mathbf{Z}\mathbf{u}_{s+t+1},\dots,\mathbf{Z}\mathbf{u}_{s+t+u}\}$ is LI and $u \leqq t$ by the same arguments as used in the preceding stage. We rename

$$\mathbf{Z}\mathbf{u}_{s+t+i}=\mathbf{v}_{s+i} \qquad \text{for } i=1,2,\dots,u$$

and by extension form the basis

$$\{\mathbf{v}_1,\dots,\mathbf{v}_s,\mathbf{v}_{s+1},\dots,\mathbf{v}_{s+t}\}$$

for Null $\mathbf{Z}^2$. Once again, we can show that

$$\{\mathbf{Z}\mathbf{v}_{s+1},\dots,\mathbf{Z}\mathbf{v}_{s+t}\}$$

is LI; we rename

$$\mathbf{Z}\mathbf{v}_{s+1}=\mathbf{w}_i \qquad \text{for } i=1,2,\dots,t$$

and form the basis $\{\mathbf{w}_1,\dots,\mathbf{w}_s\}$ for Null $\mathbf{Z}$. In this way we obtain

$$\{\mathbf{w}_1,\dots,\mathbf{w}_s,\mathbf{v}_{s+1},\dots,\mathbf{v}_{s+t},\mathbf{u}_{s+t+1},\dots,\mathbf{u}_{s+t+u}\}$$

which forms a basis for Null $\mathbf{Z}^3$. This completes Stage 3, and also the proof if $p=3$. The proof for any p follows by proceeding by induction for *Stage* p. ■

EXAMPLE 5.4. Some Nilpotent Matrices. As a first illustration consider

$$\mathbf{C}=\begin{bmatrix} 0 & 1 & 1 \\ 0 & 0 & 1 \\ 0 & 0 & 0 \end{bmatrix}$$

where $p=3$. Nullity $\mathbf{C}$ is unity and $\mathbf{x}_1=\mathbf{e}_1$ provides a basis for Null $\mathbf{C}$. Guided by the end result in (5.32), we try to solve $\mathbf{C}\mathbf{x}_2=\mathbf{x}_1$ and we can do so to obtain $\mathbf{x}_2=\mathbf{e}_2$. In the same way, $\mathbf{C}\mathbf{x}_3=\mathbf{x}_2$ yields $\mathbf{x}_3=[0,-1,1]^T$ and we have $\mathbf{X}=[\mathbf{x}_1,\mathbf{x}_2,\mathbf{x}_3]$ as specified.

Sometimes the straightforward approach of the preceding paragraph must be modified as is the case for

$$\mathbf{B}=\begin{bmatrix} 2 & -1 & -1 \\ 2 & -1 & -1 \\ 2 & -1 & -1 \end{bmatrix}$$

Here both index p and nullity s equal 2. The standard basis for Null $\mathbf{B}$ has matrix

$$[\mathbf{x}_1,\mathbf{x}_2]=\begin{bmatrix} 1 & 0 \\ 0 & 1 \\ 2 & -1 \end{bmatrix}$$

and what is different is

$$[\mathbf{B},\mathbf{x}_1,\mathbf{x}_2]\overset{R}{\to}\begin{bmatrix} 2 & -1 & -1 & 1 & 0 \\ 0 & 0 & 0 & -1 & 1 \\ 0 & 0 & 0 & 1 & -1 \end{bmatrix}$$

which shows (recall (3.17)) that neither $\mathbf{Bx}=\mathbf{x}_1$ nor $\mathbf{Bx}=\mathbf{x}_2$ can be solved for $\mathbf{x}$. This means that we cannot use the shortcut that worked for $\mathbf{C}$. There are two alternatives. First, we can find a basis for Null $\mathbf{B}^2$ and go through Stage 2 of the proof for (5.32). Second, we can use the shortcut of solving $\mathbf{Bx}_3=c_1\mathbf{x}_1+c_2\mathbf{x}_2$ for appropriate c_1,c_2. Here any nonzero $c_1=c_2$ suffices and one choice leads to

$$\mathbf{X}=\begin{bmatrix} 1 & 1 & 0 \\ 1 & 0 & -1 \\ 1 & 2 & 0 \end{bmatrix}$$

as specified in (5.32).

A third example where $p=4$ and $s=2$ is provided by

$$\mathbf{A}=\begin{bmatrix} 0 & 2 & 0 & 1 & -2 \\ 0 & 0 & 1 & 0 & 0 \\ 0 & 0 & 0 & 0 & 0 \\ 0 & 1 & 0 & 0 & -1 \\ 0 & 0 & 0 & 0 & 0 \end{bmatrix}$$

for which we find

$$\mathbf{X}=\begin{bmatrix} 1 & 0 & 0 & 0 & 0 \\ 0 & 1 & 0 & 1 & 0 \\ 0 & 0 & 0 & 0 & 1 \\ 0 & 0 & 1 & -2 & 4 \\ 0 & 1 & 0 & 0 & 2 \end{bmatrix}$$

using essentially the shortcuts that worked for **C**. Let us use this example to show why we are interested in the format of (5.32) for **X**. The relations satisfied by the columns of **X** can be displayed as follows where there are $p=4$ rows and $s=2$ columns

$$\begin{array}{ll} \mathbf{Ax}_1=\mathbf{0} & \mathbf{Ax}_2=\mathbf{0} \\ \mathbf{Ax}_3=\mathbf{x}_1 & \\ \mathbf{Ax}_4=\mathbf{x}_3 & \\ \mathbf{Ax}_5=\mathbf{x}_4 & \end{array}$$

Since any matrix $\mathbf{A}=[\mathbf{a}_1,\mathbf{a}_2,\ldots,\mathbf{a}_n]$ satisfies $\mathbf{Ae}_i=\mathbf{a}_i$ for all i, the preceding display suggests forming the nonsingular matrix $\mathbf{P}=[\mathbf{x}_1,\mathbf{x}_3,\mathbf{x}_4,\mathbf{x}_5,\mathbf{x}_2]$ for which

$$\mathbf{AP}=\mathbf{PN}$$

and in this example

$$\mathbf{N}=[\mathbf{0},\mathbf{e}_1,\mathbf{e}_2,\mathbf{e}_3,\mathbf{0}]=\left[\begin{array}{cccc|c} 0 & 1 & 0 & 0 & 0 \\ 0 & 0 & 1 & 0 & 0 \\ 0 & 0 & 0 & 1 & 0 \\ 0 & 0 & 0 & 0 & 0 \\ \hline 0 & 0 & 0 & 0 & 0 \end{array}\right]$$

Notice that **N** is constructed from $s=2$ nilpotent canonical blocks $\mathbf{N}_4$ and $\mathbf{N}_1$, and the larger of these is of order $p=4$. This means that the difference between $\mathbf{N}_5$ and the present **N** is that the latter has both 0's and 1's on the first superdiagonal. Now **N** is nilpotent of the same index $p=4$ as **A** because

$$\mathbf{P}^{-1}\mathbf{AP}=\mathbf{N}$$

shows **A** and **N** to be similar and nilpotency is preserved under similarity transformations. It is easy to show that we can permute the **x**'s to cause $\mathbf{N}_1$ to appear first, and $\mathbf{N}_4$ second, in $\mathbf{P}^{-1}\mathbf{AP}$, and that no third arrangement of blocks $\mathbf{N}_i$ of any size can be made to appear through rearranging the **x**'s.

In order to illustrate the preceding phenomenon of permuting blocks $\mathbf{N}_i$ in $\mathbf{N}=\mathbf{P}^{-1}\mathbf{FP}$ let us suppose that **F** is nilpotent of index $p=3$, its nullity is $s=3$, and its order is 6. Then (5.32) guarantees that there exists **X** whose columns satisfy relations we can display in $p=3$ rows and $s=3$ columns as

$$\begin{array}{lll} \mathbf{Fx}_1=\mathbf{0} & \mathbf{Fx}_2=\mathbf{0} & \mathbf{Fx}_3=\mathbf{0} \\ \mathbf{Fx}_4=\mathbf{x}_1 & \mathbf{Fx}_5=\mathbf{x}_2 & \\ \mathbf{Fx}_6=\mathbf{x}_4 & & \end{array}$$

If we construct $\mathbf{P}=[\mathbf{x}_1,\mathbf{x}_4,\mathbf{x}_6,\mathbf{x}_2,\mathbf{x}_5,\mathbf{x}_3]$ then

$$\mathbf{P}^{-1}\mathbf{FP}=\left[\begin{array}{ccc|cc|c} 0 & 1 & 0 & 0 & 0 & 0 \\ 0 & 0 & 1 & 0 & 0 & 0 \\ 0 & 0 & 0 & 0 & 0 & 0 \\ \hline 0 & 0 & 0 & 0 & 1 & 0 \\ 0 & 0 & 0 & 0 & 0 & 0 \\ \hline 0 & 0 & 0 & 0 & 0 & 0 \end{array}\right]$$

where we say the blocks appear in order $\mathbf{N}_3,\mathbf{N}_2,\mathbf{N}_1$. We verify that $\mathbf{P}=[\mathbf{x}_2,\mathbf{x}_5,\mathbf{x}_1,\mathbf{x}_4,\mathbf{x}_6,\mathbf{x}_3]$ will produce blocks in order $\mathbf{N}_2,\mathbf{N}_3,\mathbf{N}_1$. In general, we can obtain all six orders and these six matrices constitute the only six matrices similar to $\mathbf{F}$ having unity appearing on their first superdiagonals as the only nonzero entries.

In summary, the displayed $\mathbf{P}^{-1}\mathbf{FP}$ is the uniquely determined (up to permutations of the three blocks) canonical form of the specified nilpotent matrix $\mathbf{F}$. This is a special case of the form in the ultimate result (5.38) where eigenvalues appear on the main diagonal, unity may occur on the first superdiagonal, all other entries are zero, and the number of blocks equals the number of LI eigenvectors.

Schur's theorem (5.8) guarantees that any complex matrix $\mathbf{A}$ of order n can be upper-triangularized by a similarity transformation because a unitary transformation is a special kind of similarity transformation. Furthermore, any prescribed order of appearance can be specified for the eigenvalues of $\mathbf{A}$ which occupy the main diagonal positions in the upper-triangular matrix. Let us relabel the n eigenvalues of $\mathbf{A}$ and specify a *standard order of appearance*

$$(\mu_1,\ldots,\mu_1,\mu_2,\ldots,\mu_2,\ldots,\mu_k,\ldots,\mu_k) \tag{5.33}$$

as follows. First, there are k sets as indicated where $\mathbf{A}$ has distinct eigenvalues $\mu_1,\mu_2,\ldots,\mu_k$. Second, the associated algebraic multiplicities m_i satisfy

$$m_1 \geqq m_2 \geqq \cdots \geqq m_k$$

Third, where multiplicities, but not moduli, coincide, say $m_i=m_j$ and $|\mu_i|>|\mu_j|$, then the eigenvalue with larger modulus precedes in (5.33). Fourth, where multiplicities and moduli coincide, say $m_i=m_j$ and $|\mu_i|=|\mu_j|$, then the order of precedence coincides with the order of appearance of the eigenvalues as the circle $\{z:|z|=|\mu_i|\}$ is traversed counterclockwise starting at $(|\mu_i|,0)$ in $\mathcal{C}$. The fourth convention is the most arbitrary but it breaks ties. For example, if the eigenvalues are $1,1,i,i,-1,-1,-i,-i$ then this is actually the standard order of appearance. The next result is implied by Schur's theorem but we need it because, as we show, the transformation need not be unitary in

Any square complex matrix $\mathbf{A}$ can be reduced by a similarity transformation to an upper-triangular matrix with main diagonal (5.33). (5.34)

PROOF. As in the proof of Schur's theorem (5.8), we construct $\mathbf{Q}=[\mathbf{x}_1, \mathbf{Y}]$ where now $\mathbf{Q}$ is merely nonsingular and

$$\mathbf{Q}^{-1}\mathbf{A}\mathbf{Q}=\begin{bmatrix} \mu_1 & \mathbf{F} \\ \mathbf{0} & \mathbf{G} \end{bmatrix}$$

shows that the theorem holds for $n=2$. Under the inductive hypothesis there exists a nonsingular $\mathbf{P}$ such that $\mathbf{P}^{-1}\mathbf{G}\mathbf{P}$ is upper-triangular and it has the same eigenvalues as $\mathbf{G}$, namely the remaining eigenvalues of $\mathbf{A}$. We can specify that the first appearing eigenvalue in $\mathbf{G}$ is the second one in (5.33). Then $\mathbf{M}$ defined with the notation used for (5.8) is nonsingular, as is $\mathbf{QM}$, and it is straightforward to verify that $(\mathbf{QM})^{-1}\mathbf{A}(\mathbf{QM})$ is upper-triangular. ■

EXAMPLE 5.5. Upper-triangularization by Similarity Transformations. Let us illustrate (5.34) for

$$\mathbf{A}=\begin{bmatrix} 2 & 1 & 0 \\ 0 & 2 & 0 \\ 1 & 0 & 1 \end{bmatrix}$$

which has eigenvalues 2, 2, 1 and only two LI eigenvectors. The standard order (5.33) has $\mu_1=2$ and we can use any associated eigenvector as the first column of $\mathbf{Q}$. Second and third columns can be selected in any manner provided $\mathbf{Q}$ is nonsingular. For example, let us use

$$\mathbf{Q}=\begin{bmatrix} 1 & 0 & 0 \\ 0 & 1 & 0 \\ 1 & 0 & 1 \end{bmatrix}, \qquad \mathbf{Q}^{-1}=\begin{bmatrix} 1 & 0 & 0 \\ 0 & 1 & 0 \\ -1 & 0 & 1 \end{bmatrix}$$

so that the result of operating on the first column of $\mathbf{A}$ is

$$\mathbf{Q}^{-1}\mathbf{A}\mathbf{Q}=\begin{bmatrix} 2 & 1 & 0 \\ 0 & 2 & 0 \\ 0 & -1 & 1 \end{bmatrix}$$

We next turn our attention to the submatrix

$$\mathbf{G}=\begin{bmatrix} 2 & 0 \\ -1 & 1 \end{bmatrix}$$

Continuing to use the notation from the proof of (5.34),

$$\mathbf{P}=\begin{bmatrix} -1 & 0 \\ 1 & 1 \end{bmatrix}$$

has an eigenvector corresponding to $\mu_1=2$ as column 1 while column 2 is an eigenvector corresponding to $\mu_2=1$. The final pair of matrices is

$$\mathbf{M}=\begin{bmatrix} 1 & 0 & 0 \\ 0 & -1 & 0 \\ 0 & 1 & 1 \end{bmatrix}, \qquad \mathbf{M}^{-1}=\begin{bmatrix} 1 & 0 & 0 \\ 0 & -1 & 0 \\ 0 & 1 & 1 \end{bmatrix}$$

and the end result is that we have found **QM** and the upper-triangular

$$(\mathbf{QM})^{-1}\mathbf{A}(\mathbf{QM})=\begin{bmatrix} 2 & -1 & 0 \\ 0 & 2 & 0 \\ 0 & 0 & 1 \end{bmatrix}$$

It should be noticed that, while systematic, this approach is computationally formidable because of the many eigenvector determinations and matrix multiplications. (Alternative approaches appear in Example 5.8.)

The next result shows that we can further reduce **A** to a special block diagonal form where k square submatrices $\mathbf{C}_1, \mathbf{C}_2, \ldots, \mathbf{C}_k$ appear on the main diagonal.

Let **B** be the upper-triangular matrix of order n resulting from the similarity transformation on **A** in (5.34). Then there exists a nonsingular matrix **Q** such that

$$\mathbf{Q}^{-1}\mathbf{B}\mathbf{Q}=\mathbf{diag}(\mathbf{C}_1, \mathbf{C}_2, \ldots, \mathbf{C}_k)$$

where $\mathbf{C}_i$ is upper-triangular and contains μ_i in every position of its main diagonal for $i=1,2,\ldots,k$. (5.35)

PROOF. If b_{pq} is to be replaced by zero then $p<q$ and $b_{pp}\neq b_{qq}$. It is readily verified that the following two operations replace b_{pq} by zero. First, replace row p by its sum with c times row q where

$$c=\frac{b_{pq}}{b_{pp}-b_{qq}}$$

Second, replace column q by its sum with $-c$ times row p. The total result is the similarity transformation

$$\mathbf{E}_{pq}(c)\mathbf{B}\mathbf{E}_{pq}^{-1}(c)$$

and the only elements that can be altered besides b_{pq} are those to its right in row p and those above it in column q. We consequently start at the left in the first row above the μ_k block, work all the way to the right, and then start on the next row. In this way there is determined a succession of similarity transformations producing the stated effect. ■

Our last preliminary objective is to show that each $\mathbf{C}_i$ can be transformed into a block diagonal matrix containing the following kinds of blocks. A *Jordan block matrix* or *Jordan block* is a square matrix

$$\mathbf{J}=\begin{bmatrix} \mu & 1 & 0 & \cdots & 0 \\ 0 & \mu & 1 & \cdots & 0 \\ \vdots & \vdots & \vdots & & \vdots \\ 0 & 0 & 0 & \cdots & \mu \end{bmatrix} \tag{5.36}$$

where μ occupies every main diagonal position, unity occupies every position on the first superdiagonal, and all other entries are zero.

For each $\mathbf{C}_i$ in (5.35) there exists a nonsingular matrix $\mathbf{R}_i$ such that

$$\mathbf{R}_i^{-1}\mathbf{C}_i\mathbf{R}_i=\mathbf{diag}(\mathbf{J}_1,\mathbf{J}_2,\ldots,\mathbf{J}_{s(i)})$$

where each block is a Jordan block with $\mu=\mu_i$, where the order of the largest block $\mathbf{J}_1$ equals the index of nilpotency for $\mathbf{C}_i-\mu_i\mathbf{I}$, and where $s(i)$ equals the number of LI eigenvectors of $\mathbf{C}_i$. (5.37)

PROOF. For each $\mathbf{C}_i$ the matrix $\mathbf{C}_i-\mu_i\mathbf{I}$ is nilpotent and by (5.32) there exists a nonsingular $\mathbf{R}_i$ such that

$$\mathbf{R}_i^{-1}(\mathbf{C}_i-\mu_i\mathbf{I})\mathbf{R}_i=\mathbf{N}(i)$$

Then $\mathbf{R}_i^{-1}\mathbf{C}_i\mathbf{R}_i=\mu_i\mathbf{I}+\mathbf{N}(i)$ has the stated form. In particular, $s(i)=\text{nullity}(\mathbf{C}_i-\mu_i\mathbf{I})$. ■

The order of the largest block $\mathbf{J}_1$ is called the *index* of μ_i.

EXAMPLE 5.6. Reducing Upper-Triangular Blocks to Jordan Blocks. This example illustrates (5.37). The end-product of Example 5.5 is already in the format produced in (5.35), that is

$$\mathbf{C}_1=\begin{bmatrix} 2 & -1 \\ 0 & 2 \end{bmatrix}, \qquad \mathbf{C}_2=[1]$$

The associated matrices from (5.32) are

$$\mathbf{R}_1=\begin{bmatrix} 1 & 0 \\ 0 & -1 \end{bmatrix}, \qquad \mathbf{R}_2=[1]$$

and finally

$$\mathbf{R}_1^{-1}\mathbf{C}_1\mathbf{R}_1=\begin{bmatrix}2&1\\0&2\end{bmatrix},\qquad \mathbf{R}_2^{-1}\mathbf{C}_2\mathbf{R}_1=[1]$$

are the Jordan blocks in (5.37).

It is now easy to prove the main result, namely the

JORDAN CANONICAL FORM THEOREM. If **A** is any complex matrix of order n there exists a nonsingular matrix **T** such that

$$\mathbf{T}^{-1}\mathbf{A}\mathbf{T}=\mathbf{diag}(\mathbf{J}_1,\mathbf{J}_2,\ldots,\mathbf{J}_q)$$

is called the Jordan Canonical Form (JCF) of **A** and has the following properties.

(1) The main diagonal is (5.33).
(2) Each **J** is a Jordan block (5.36).
(3) The number of **J**'s containing μ_i equals the number of LI eigenvectors corresponding to μ_i.
(4) The total number q of blocks equals the total number of LI eigenvectors of **A**. (5.38)

PROOF. If **P**, **Q**, and all $\mathbf{R}_i$ are determined by (5.34), (5.35), and (5.37), respectively, and if

$$\mathbf{R}=\mathbf{diag}(\mathbf{R}_1,\mathbf{R}_2,\ldots,\mathbf{R}_k)$$

then $(\mathbf{PQR})^{-1}\mathbf{A}(\mathbf{PQR})$ is the JCF of **A**. Property (1) results from (5.34) while (2)–(4) follow from (5.37). ■

The JCF theorem establishes the existence of a solution to the matrix diagonalization problem (5.30).

EXAMPLE 5.7. Using the Proof of the JCF Theorem to Calculate JCFs. If we change to the notation used in the proof, $\mathbf{T}=\mathbf{PQR}$ is determined from the results of the two preceding examples as

$$\begin{bmatrix}1&0&0\\0&1&0\\1&-1&1\end{bmatrix}=\begin{bmatrix}1&0&0\\0&-1&0\\1&1&1\end{bmatrix}\begin{bmatrix}1&0&0\\0&1&0\\0&0&1\end{bmatrix}\begin{bmatrix}1&0&0\\0&-1&0\\0&0&1\end{bmatrix}$$

where $\mathbf{Q}=\mathbf{I}$ because the reduction (5.35) is not needed for this particular **A**. Then the JCF is

$$\mathbf{T}^{-1}\mathbf{A}\mathbf{T}=\begin{bmatrix}2&1&0\\0&2&0\\0&0&1\end{bmatrix}$$

and we see that it has the properties guaranteed by (5.38). Somewhat more convenient methods for finding JCFs are presented in Example 5.8.

Practical methods for finding the JCF of matrices of orders larger than three or four are developed in works on numerical analysis. Before we present a few ideas that can be used for small matrices, let us explain where difficulties may lie. If **A** has n LI eigenvectors there are no difficulties so far as we are concerned here; the columns of **T** are eigenvectors and the JCF is the diagonal matrix of eigenvalues. Otherwise, when **A** is defective, there are too few eigenvectors to make up the columns of **T** and we must find other vectors for the remaining columns. There are shortcuts provided we know the structure of the Jordan blocks in the JCF, and this brings us to a possible difficulty we wish to explain. If no μ_i appears in more than one block then the familiar type of equation

$$\mathbf{AT}=\mathbf{TJ}$$

can be used at once without difficulty; this is illustrated in Example 5.8. The same can be said if there is no "preliminary ambiguity" about the structure of the different blocks in which a given eigenvalue appears. For example, if $\mu=1$ occurs with multiplicity 3 and there are two associated LI eigenvectors then the only possibility is that two blocks

$$\begin{bmatrix} 1 & 1 \\ 0 & 1 \end{bmatrix} \quad \text{and} \quad [1]$$

appear in the JCF. But if we have multiplicity 4, and two eigenvectors then there is what we above term "preliminary ambiguity" because there are the two possibilities

$$\begin{bmatrix} 1 & 1 & 0 \\ 0 & 1 & 1 \\ 0 & 0 & 1 \end{bmatrix} \quad \text{and} \quad [1], \qquad \begin{bmatrix} 1 & 1 \\ 0 & 1 \end{bmatrix} \quad \text{and} \quad \begin{bmatrix} 1 & 1 \\ 0 & 1 \end{bmatrix}$$

There are at least two ways to resolve the situation for any given matrix. First, all possibilities can be checked out to determine the one that applies. Second, we can use (5.37) in a way that amounts to using the method illustrated in Example 5.7. That is, for the present case under discussion there is one 3×3 and one 1×1 Jordan block iff the index of $\mu=1$ is 3. This would be determined by producing a 4×4 block **C** and finding the index of nilpotency for $\mathbf{C}-1\mathbf{I}$. In summary, what we have termed "preliminary ambiguity" actually refers to whether or not it is immediately obvious as to how the different blocks containing a given eigenvalue are configured.

EXAMPLE 5.8. Shortcuts for Finding JCFs. Let us first note that we generally need to find both the JCF and the nonsingular matrix **T**. We first must find the eigenvalues $\mu_1, \mu_2, \ldots, \mu_k$ and the numbers of the LI eigenvectors associated with each. If the latter total n, then **T** is a matrix of eigenvectors and the JCF is the diagonal matrix of eigenvalues. If there are fewer than n LI eigenvectors, then there are two possibilities.

Case 1 occurs when there are no "preliminary ambiguities" as described just above and necessarily the JCF theorem tells us exactly what we seek. Consider, for example, the matrix **A** treated in Examples 5.5–5.7. From (5.38) we can at once write the JCF and then determine $\mathbf{T}=[\mathbf{t}_1, \mathbf{t}_2, \mathbf{t}_3]$ from

$$\mathbf{AT}=\mathbf{T}\begin{bmatrix} 2 & 1 & 0 \\ 0 & 2 & 0 \\ 0 & 0 & 1 \end{bmatrix}$$

This can be written as three vector equations

$$\mathbf{At}_1 = 2\mathbf{t}_1$$

(1) $$\mathbf{At}_2 = 2\mathbf{t}_2 + \mathbf{t}_1$$

$$\mathbf{At}_3 = \mathbf{t}_3$$

We see that $\mathbf{t}_1$ and $\mathbf{t}_3$ are eigenvectors, say $\mathbf{t}_1=[1,0,1]^T$ and $\mathbf{t}_3=\mathbf{e}_3$. The vector $\mathbf{t}_2=[0,1,-1]^T$ is found by solving the second vector equation, say via the GJSF

$$[\mathbf{A}-2\mathbf{I}, \mathbf{t}_1] \overset{R}{\rightarrow} \begin{bmatrix} 1 & 0 & -1 & 1 \\ 0 & 1 & 0 & 1 \\ 0 & 0 & 0 & 0 \end{bmatrix}$$

This is simpler than following the proof of the JCF theorem as in Example 5.7.

Case 2 occurs when the given matrix is defective and there are "preliminary ambiguities" so we are not immediately certain as to the configuration of the JCF. For example,

$$\mathbf{B}=\begin{bmatrix} 1 & 1 & 1 & 0 \\ 0 & 1 & 1 & 0 \\ 0 & 0 & 1 & 0 \\ 0 & 1 & -1 & 1 \end{bmatrix}$$

has $\mu=1$ with multiplicity 4 and only two LI eigenvectors. Under the method of trying all possibilities we would try to solve $\mathbf{BT}=\mathbf{TJ}$ for **T** under the two possibilities for **J**

$$\begin{bmatrix} 1 & 1 & 0 & 0 \\ 0 & 1 & 1 & 0 \\ 0 & 0 & 1 & 0 \\ 0 & 0 & 0 & 1 \end{bmatrix}, \qquad \begin{bmatrix} 1 & 1 & 0 & 0 \\ 0 & 1 & 0 & 0 \\ 0 & 0 & 1 & 1 \\ 0 & 0 & 0 & 1 \end{bmatrix}$$

and solution is possible only for the first matrix.

The preceding "preliminary ambiguity" can also be resolved via (5.37) as follows. Computing successive powers $(\mathbf{B}-\mathbf{I}), (\mathbf{B}-\mathbf{I})^2, (\mathbf{B}-\mathbf{I})^3$ we find the latter is the first zero matrix, so the index of $\mu=1$ is 3 and therefore the larger Jordan block is 3×3. In more complicated situations we might be forced to rely on trying all possibilities, or on proceeding systematically but using some of the preceding shortcuts.

In general, once we have determined the exact configuration of the JCF we can use $\mathbf{AT}=\mathbf{TJ}$ as a practical relation to determine the columns of **T**. In terms of vector equations we have equations of the form

(2) $$\mathbf{A}\mathbf{t}_i = \mu_i \mathbf{t}_i + \nu_i \mathbf{t}_{i-1}$$

where ν_i is 0 or 1. If $\nu_i = 0$ then $\mathbf{t}_i$ is an eigenvector corresponding to μ_i. If $\nu_i = 1$ then $\mathbf{t}_i$ is called a *generalized eigenvector*; an example is $\mathbf{t}_2$ in (1).

Provided we know the eigenvalues and eigenvectors we can deduce the JCF for small matrices from the properties listed in (5.38) and then find **T** from (2) with less total effort than in following the proof of the JCF theorem.

The JCF can be used to prove the general form of (4) in Example 5.1 above, namely the

CAYLEY–HAMILTON THEOREM. If **A** is any square matrix with characteristic equation

$$\lambda^n + p_{n-1}\lambda^{n-1} + \ldots + p_1\lambda + p_0 = 0$$

then also

$$\mathbf{A}^n + p_{n-1}\mathbf{A}^{n-1} + \cdots + p_1\mathbf{A} + p_0\mathbf{I} = \mathbf{O} \tag{5.39}$$

PROOF. Let $p(\lambda)$ denote the characteristic polynomial as in Example 5.1 (where the proof is given for **A** diagonalizable). When **A** is defective, then in the notation of the JCF theorem the diagonal matrix with entries $p(\lambda_i)$ is the block diagonal matrix

$$\mathbf{diag}\big(p(\mathbf{J}_1), p(\mathbf{J}_2), \ldots, p(\mathbf{J}_q)\big)$$

By elementary properties of Jordan blocks (5.4.2) each of the blocks on the main diagonal has the form

$$\begin{bmatrix} p(\lambda_i) & p'(\lambda_i) & p''(\lambda_i)/2! & \cdots & p^{(q-1)}(\lambda_i)/(q-1)! \\ 0 & p(\lambda_i) & p'(\lambda_i) & \cdots & p^{(q-2)}(\lambda_i)/(q-2)! \\ \vdots & \vdots & \vdots & & \vdots \\ 0 & 0 & 0 & \cdots & p(\lambda_i) \end{bmatrix}$$

where q is the order of the block, and by the elementary properties of polynomials and their derivatives, each is a zero matrix. ■

PROBLEMS FOR SOLUTION

5.4.1 Nilpotent Matrices. A matrix $\mathbf{Z}$ is *nilpotent of index p*, where p is a positive integer, if $\mathbf{Z}^{p-1} \neq \mathbf{O}$ and $\mathbf{Z}^p = \mathbf{O}$; consequently such matrices are necessarily square. Establish:

(1) An example of a nilpotent nontriangular matrix of order 2 or more.
(2) Any triangular matrix with zeros on its main diagonal is nilpotent.
(3) Matrices $\mathbf{N}_k$ in (5.31) have the following properties.
  (i) Nilpotent of index k.
  (ii) Zero is an eigenvalue of multiplicity k.
  (iii) There is only one LI eigenvector.
  (iv) $\mathbf{N}_k^r$ has zero entries except for unity in positions $(1, r+1), (2, r+2), \ldots, (k-r, k)$ which by definition are the positions on the rth *superdiagonal*.
(4) A matrix is nilpotent iff all of its eigenvalues equal zero.
(5) Nilpotency is preserved under similarity transformations; that is, if $\mathbf{Z}$ is nilpotent of index p and $\mathbf{Z}$ is similar to $\mathbf{B}$, then $\mathbf{B}$ is nilpotent of index p.
(6) If $\mathbf{Z}$ is of order n and nilpotent of index p then
  (i) $p \leqq n$.
  (ii) $\mathbf{Z}^{p-1}\mathbf{x} \neq \mathbf{0}$ implies $\{\mathbf{x}, \mathbf{Zx}, \mathbf{Z}^2\mathbf{x}, \ldots, \mathbf{Z}^{p-1}\mathbf{x}\}$ is LI.
  (iii) $\{\mathbf{0}\} \neq \text{Null}\,\mathbf{Z} \subset \text{Null}\,\mathbf{Z}^2 \subset \cdots \subset \text{Null}\,\mathbf{Z}^p = \mathcal{C}^n$.
  (iv) Nullity $\mathbf{Z}$ equals the number of LI eigenvectors of $\mathbf{Z}$.
(7) Any nonzero nilpotent matrix is defective.

5.4.2 Jordan Block Matrices. Establish for blocks $\mathbf{J}$ of order q defined by (5.36):

(1) $\mathbf{J}$ is nilpotent iff $\mu = 0$; however

$$\mathbf{J} - \mu\mathbf{I}_q = \mathbf{N}_q$$

where $\mathbf{N}_q$ is the nilpotent canonical block of (5.31).

(2) The characteristic equation $\det(\mathbf{J}-\lambda\mathbf{I})=0$ is $(\mu-\lambda)^q=0$.

(3) The quantity μ is an eigenvalue of multiplicity q.

(4) There is one and only one LI eigenvector, namely $\mathbf{x}=c\mathbf{e}_1$ where $c\in\mathcal{C}$ is nonzero but otherwise arbitrary.

(5) If m is a positive integer, then

$$\mathbf{J}^m=\begin{bmatrix} \mu^m & C(m,1)\,\mu^{m-1} & C(m,2)\,\mu^{m-2} & \cdots & C(m,q-1)\,\mu^{m-q+1} \\ 0 & \mu^m & C(m,1)\,\mu^{m-1} & \cdots & C(m,q-2)\,\mu^{m-q+2} \\ \vdots & \vdots & \vdots & & \vdots \\ 0 & 0 & 0 & \cdots & \mu^m \end{bmatrix}$$

(6) The matrix equation $\mathbf{T}^{-1}\mathbf{A}\mathbf{T}=\mathbf{J}$, where $\mathbf{T}=[\mathbf{t}_1,\mathbf{t}_2,\ldots,\mathbf{t}_q]$ is equivalent to the set of q vector equations

$$\begin{aligned} \mathbf{A}\mathbf{t}_1 &= \mu\mathbf{t}_1 \\ \mathbf{A}\mathbf{t}_2 &= \mu\mathbf{t}_2+\mathbf{t}_1 \\ \mathbf{A}\mathbf{t}_3 &= \mu\mathbf{t}_3+\mathbf{t}_2 \\ &\vdots \\ \mathbf{A}\mathbf{t}_q &= \mu\mathbf{t}_q+\mathbf{t}_{q-1} \end{aligned}$$

(7) The preceding equations are consistent iff μ is an eigenvalue of $\mathbf{A}$ of multiplicity q and $\mathbf{A}$ has one and only one LI eigenvector.

5.4.3 Possibilities for JCFs. Establish:

(1) All possibilities for the first few values of n for the pair—geometric multiplicity and index—for order n where there is a single eigenvalue of algebraic multiplicity n.

(2) True or false? If the preceding pair is known for a particular n, it determines the configuration of the Jordan block.

(3) The number of essentially different JCF configurations for the first few values of n.

5.4.4 Numerical Exercises. Find JCFs and associated Ts for

(1) Matrices in 5.1.4

(2) Matrices in 5.1.5

(3) $\begin{bmatrix} 1 & 1 & 1 \\ 0 & 1 & 1 \\ 0 & 0 & 1 \end{bmatrix}$ (4) $\begin{bmatrix} 1 & 1 & 1 & 0 & 0 \\ 0 & 1 & 1 & 0 & 0 \\ 0 & 0 & 1 & 0 & 0 \\ 0 & 0 & 0 & 0 & 1 \\ 0 & 0 & 0 & 0 & 0 \end{bmatrix}$

6

DIFFERENTIAL CALCULUS ON $\mathcal{R}^n$

Here we start using limits and associated methods of analysis together with algebraic results from earlier chapters. The reader who is knowledgeable in differential calculus on $\mathcal{R}^n$ for at least $n=2,3$ can proceed rapidly to Section 6.3. Prerequisites are Section 4.1 and (for Section 6.4) Section 4.2.

6.1 CONTINUOUS TRANSFORMATIONS

In this section we show how to work with limits and continuity on $\mathcal{R}^n$ for any n. We use a few results for $n=1$, and several for $\mathcal{R}^n$ as a vector space, to extend what is done in elementary calculus.

Only minor changes are needed in definitions and proofs to extend the theory of sequences from $\mathcal{R}$ to $\mathcal{R}^n$. A *vector sequence* $\{\mathbf{x}_k\}$, that is

$$k \mapsto \mathbf{x}_k \in \mathcal{R}^n$$

has the *limit* $\mathbf{a} \in \mathcal{R}^n$, and we write $\lim\{\mathbf{x}_k\} = \mathbf{a}$, or $\mathbf{x}_k \to \mathbf{a}$ as $k \to \infty$, if

To each $\varepsilon > 0$ there corresponds an integer k' such that $\|\mathbf{x}_k - \mathbf{a}\| < \varepsilon$ whenever $k > k'$. (6.1)

If $\{\mathbf{x}_k\}$ has a limit it is *convergent* and otherwise it is *divergent*. In the notation for components introduced below (2.13)

$$\mathbf{x}_k = [x_{1k}, x_{2k}, \ldots, x_{nk}]^T$$

$$\mathbf{a} = [a_1, a_2, \ldots, a_n]^T$$

so a vector sequence $\{\mathbf{x}_k\}$ has associated *component sequences* $\{x_{ik}\}$ for

$i=1,2,\ldots,n$ which are familiar sequences of scalars. Convergence in $\mathcal{R}^n$ depends on convergence in $\mathcal{R}$ because of the

THEOREM. A vector sequence converges iff all of its component sequences converge. (6.2)

PROOF. If $\varepsilon>0$ and k' is determined from $\lim\{\mathbf{x}_k\}=\mathbf{a}$, then from 3.4.5(8)

$$|x_{ik}-a_i|\leqq \|\mathbf{x}_k-\mathbf{a}\|<\varepsilon$$

for $k>k'$ which shows $\lim\{x_{ik}\}=a_i$. Conversely, if each component sequence converges and $\varepsilon>0$ is specified, then there exist integers k_i' for $i=1,2,\ldots,n$ such that

$$|x_{ik}-a_i|<\frac{\varepsilon}{n} \qquad \text{for } k>k_i'.$$

Consequently, if $k'=\max\{k_1',k_2',\ldots,k_n'\}$ then again from 3.4.5(8)

$$\|\mathbf{x}_k-\mathbf{a}\|<\varepsilon$$

whenever $k>k'$. ■

Most elementary properties of scalar sequences extend to vector sequences; for products we use inner products, and notable exceptions are results involving quotients. In order to establish the properties of limits of vector sequences, and indeed to work with other limiting operations in $\mathcal{R}^n$ for any n, we need to generalize some concepts that are familiar for $n=1,2,3$. If $\mathbf{x}_0\in\mathcal{R}^n$ and $\delta>0$ then the *δ-neighborhood* of $\mathbf{x}_0$, or the *neighborhood* of $\mathbf{x}_0$ of *radius* δ, is the set

$$N_\delta(\mathbf{x}_0)=\{\mathbf{x}:\|\mathbf{x}-\mathbf{x}_0\|<\delta\} \tag{6.3}$$

which, in brief, we refer to as a *neighborhood* of $\mathbf{x}_0$. Another name is *open ball of radius δ centered* at $\mathbf{x}_0$ because it generalizes an open interval in $\mathcal{R}$, an open disc in $\mathcal{R}^2$, and an open spherical ball in $\mathcal{R}^3$.

All limiting operations in $\mathcal{R}^n$ can be defined in terms of neighborhoods. For the one instance introduced above, clearly $\{\mathbf{x}_k\}\to\mathbf{a}$ as $k\to\infty$ iff

To each neighborhood $N_\varepsilon(\mathbf{a})$ there corresponds an integer k' such that

$$\mathbf{x}_k\in N_\varepsilon(\mathbf{a})$$

whenever $k>k'$.

Use of neighborhoods makes precise the concept of "nearby" or "neighboring" points in $\mathcal{R}^n$ where distance is measured by the norm $\|\cdot\|$ (3.4.4).

Let us now introduce the first part of the rather considerable body of terminology we need to work with limiting operations in $\mathcal{R}^n$. A set $G \subset \mathcal{R}^n$ is an *open set* if each of its elements has a neighborhood that lies entirely within the set. Thus every point in an open set has a neighborhood that contains only points belonging to the set. A set $F \subset \mathcal{R}^n$ is a *closed set* if its complement

$$F^c = \mathcal{R}^n - F$$

is an open set. The *interior* of a set A, $\operatorname{int} A$, consists of all points having at least one neighborhood lying within the set. The *exterior*, $\operatorname{ext} A$, consists of all points having at least one neighborhood lying within the complement A^c. All points such that every neighborhood intersects both A and A^c comprise the *boundary* of A, $\operatorname{bdry} A$. We find

$$\operatorname{int} A \subset A \qquad \text{and} \qquad \operatorname{ext} A \subset A^c$$

but points in $\operatorname{bdry} A$ can belong either to A or to A^c. If we adjoin the boundary to A we obtain the *closure* of A

$$\operatorname{clos} A = A \cup \operatorname{bdry} A$$

An equivalent procedure is to define the closure as the union of the set and its *limit points*, which are points $\mathbf{x}$ for which every *deleted neighborhood*

$$N_\delta(\mathbf{x}) - \{\mathbf{x}\}$$

contains at least one point (and hence an infinite number of points) of the set. The idea is that points in $\operatorname{clos} A$, whether they belong to A or not, are arbitrarily close to A as measured by $\|\cdot\|$. If $\mathbf{x} \in A$ is not a limit point then it is an *isolated point*.

The general notion of a connected open set is that it cannot be expressed as the union of two nonempty disjoint open sets. For purposes in this book we define a *connected set* in $\mathcal{R}^n$ to be an open set such that any two points in the set can be joined by a *polygonal path*, which is a sequence of line segments where each segment lies entirely within the set. An open convex set is connected, a finite set is not connected, and while an interval (a,b) is connected in $\mathcal{R}$, it is not connected in $\mathcal{R}^n$ for $n > 1$ because it is not open.

A limit point of the set of images

$$\{\mathbf{x}_k : k = 1, 2, \ldots\}$$

is called a *limit point* of the sequence $\{\mathbf{x}_k\}$ and

Every limit point $\mathbf{x}$ of a sequence $\{\mathbf{x}_k\}$ is the limit of some subsequence of $\{\mathbf{x}_k\}$. (6.4)

PROOF. Using the sequence of neighborhoods corresponding to $\delta = 1, 1/2, 1/3, \ldots$ we can find $k_1 < k_2 < k_3 < \cdots$ such that

$$\|\mathbf{x}_{k_i} - \mathbf{x}\| < \frac{1}{i}$$

for $i = 1, 2, 3, \ldots$. The subsequence determined in this way clearly converges to $\mathbf{x}$. ■

We can take advantage of (6.4) by supposing the sequence $\{\mathbf{x}_k\}$ itself converges to a particular limit point; see (7.22) for an example.

Starting in this chapter, transformations are denoted by lowercase as well as capital letters and we continue to call the transformation

$$f: D \subset \mathcal{R}^n \to \mathcal{R}^m$$

a function if $m = 1$. In general, we write images

$$f(\mathbf{x}) = [f_1(\mathbf{x}), f_2(\mathbf{x}), \ldots, f_m(\mathbf{x})]^T$$

and we call the functions $f_i: D \subset \mathcal{R}^n \to \mathcal{R}$ for $i = 1, 2, \ldots, m$ the *coordinate functions* of the transformation f. Always we require that the domain D be an open set. The entire space $\mathcal{R}^n$ is open but often there are difficulties —as, for example, at $x = 0$ for $f(x) = 1/x$, or less seriously, as for piecewise continuous functions on $\mathcal{R}$—and D is then of necessity a proper subset of $\mathcal{R}^n$. The interior of any set is open so if f is defined on A and $\operatorname{int} A \neq \varnothing$ then we use $D = \operatorname{int} A$ to ensure that every $\mathbf{x} \in D$ has a neighborhood that contains only points of D. If necessary, we can work with limit points of A by intersecting their neighborhoods with D in the manner used to define one-sided limits in $\mathcal{R}$.

If $\mathbf{x}_0 \in D$ where D is open, then $\mathbf{y}_0 \in \mathcal{R}^m$ is the *limit* of the transformation f as $\mathbf{x}$ *approaches* $\mathbf{x}_0$, written $\mathbf{y}_0 = \lim_{\mathbf{x} \to \mathbf{x}_0} f(\mathbf{x})$ or $f(\mathbf{x}) \to \mathbf{y}_0$ as $\mathbf{x} \to \mathbf{x}_0$ if

To each $\varepsilon > 0$ there corresponds $\delta > 0$ such that $f(\mathbf{x}) \in N_\varepsilon(\mathbf{y}_0)$ whenever $\mathbf{x} \in N_\delta(\mathbf{x}_0) - \{\mathbf{x}_0\}$. (6.5)

Note that the value $f(\mathbf{x}_0)$ is irrelevant to the limit, just as in elementary calculus.

Limits of transformations depend on limits of vector sequences because of the

THEOREM. Let $\mathbf{x}_0 \in D$ where D is open and let $\{\mathbf{x}_k\}$ be any sequence in $\mathcal{R}^n$ where $\mathbf{x}_k = \mathbf{x}_0$ holds for no $k = 1, 2, \ldots$. Then

$$\lim_{\mathbf{x} \to \mathbf{x}_0} f(\mathbf{x}) = \mathbf{y}_0 \quad \text{iff} \quad \lim\{f(\mathbf{x}_k)\} = \mathbf{y}_0$$

whenever $\lim\{\mathbf{x}_k\} = \mathbf{x}_0$. (6.6)

PROOF. If $\varepsilon > 0$ and then $\delta > 0$ is determined by (6.5), and k' by (6.1) for $\{\mathbf{x}_k\}$,

$$\|f(\mathbf{x}_k) - \mathbf{y}_0\| < \varepsilon$$

whenever $k > k'$. Conversely, suppose

$$f(\mathbf{x}) \not\to \mathbf{y}_0 \text{ as } \mathbf{x} \to \mathbf{x}_0$$

For some $\varepsilon > 0$, each $\delta = \delta_k = k^{-1}$ determines $\mathbf{x}_k$ such that

$$\|f(\mathbf{x}_k) - \mathbf{y}_0\| \geqq \varepsilon \qquad \text{and} \qquad 0 < \|\mathbf{x}_k - \mathbf{x}_0\| < \frac{1}{k}$$

Thus $\{\mathbf{x}_k\}$ converges to $\mathbf{x}_0, \mathbf{x}_k = \mathbf{x}_0$ for no k, and $f(\mathbf{x}_k) \not\to \mathbf{y}_0$. ■

The preceding theorem enables us to establish the counterpart of (6.2)—a transformation has a limit at $\mathbf{x}_0$ iff each coordinate function has a limit at $\mathbf{x}_0$ (6.1.6). This is important in practice, as is (6.2), because it replaces an m-dimensional process (in the range) by m separate one-dimensional ones.

Another frequently used property—but weaker than (6.6) because, as shown below, it is not iff—replaces the n-dimensional process (in the domain) by a single limit in $\mathfrak{R}$. In words, whenever $f(\mathbf{x}_0) \to \mathbf{y}_0$ as $\mathbf{x} \to \mathbf{x}_0$ in $\mathfrak{R}^n$ then $\mathbf{y}_0$ must also be the limit as

$$t \to 0 \text{ in } \mathfrak{R}$$

whenever this causes $\mathbf{x} \to \mathbf{x}_0$ along a line in D through $\mathbf{x}_0$; that is,

If $\lim_{\mathbf{x} \to \mathbf{x}_0} f(\mathbf{x}) = \mathbf{y}_0$ then for any $\mathbf{h} \neq \mathbf{0}$ such that D contains the line segment between $\mathbf{x}_0$ and $\mathbf{x}_0 + \mathbf{h}$

$$\lim_{t \to 0} f(\mathbf{x}_0 + t\mathbf{h}) = \mathbf{y}_0 \tag{6.7}$$

PROOF. Suppose $\varepsilon > 0$ is specified and δ is determined from (6.5). Then for any $\mathbf{h}$ as specified and any nonzero t such that $|t| \cdot \|\mathbf{h}\| < \delta$

$$0 < \|\mathbf{x}_0 + t\mathbf{h} - \mathbf{x}_0\| < \delta$$

and consequently $f(\mathbf{x}_0 + t\mathbf{h}) \in N_\varepsilon(\mathbf{y}_0)$. ■

The example on $\mathfrak{R}^2$

$$\mathbf{x} \mapsto \frac{x_1^2}{x_1^2 + x_2^2}, \qquad \mathbf{x} \neq \mathbf{0}$$

$$\mapsto 0, \qquad \mathbf{x} = \mathbf{0}$$

has limit 1 as $\mathbf{x} \to \mathbf{0}$ for $\mathbf{h} = \mathbf{e}_1$, and limit 0 for $\mathbf{h} = \mathbf{e}_2$. Consequently, the two-dimensional limit does not exist as $\mathbf{x} \to \mathbf{0}$ and we see that the condition in (6.7), while necessary, is not sufficient for existence of $\lim_{\mathbf{x} \to \mathbf{x}_0} f(\mathbf{x})$. When the limit exists, (6.7) is often invoked so that the simple setting of a line through $\mathbf{x}_0$ can be used as, for example, in (6.35).

The transformation f is *continuous at* $\mathbf{x}_0 \in D$, where D is open, if

$$f(\mathbf{x}_0) = \lim_{\mathbf{x} \to \mathbf{x}_0} f(\mathbf{x})$$

There are two independent conditions—the limit must exist, and its value must equal the value of f at $\mathbf{x}_0$—just as in elementary calculus. The transformation is *continuous on D* if it is continuous at all $\mathbf{x} \in D$. Whenever f is not continuous it is *discontinuous*.

All continuous functions have the following property of *persistence of sign*.

Suppose f is a continuous function at $\mathbf{x}_0$ and $f(\mathbf{x}_0) \neq 0$. Then there is some $N_\delta(\mathbf{x}_0)$ such that $f(\mathbf{x})$ has the same sign as $f(\mathbf{x}_0)$ whenever $\mathbf{x} \in N_\delta(\mathbf{x}_0)$. (6.8)

PROOF. It suffices to specify $\varepsilon = |f(\mathbf{x}_0)|/2$ and determine δ from (6.5). ■

Typical uses occur for (7.9) and (7.23).

Let us return to transformations, that is, to $f(\mathbf{x}) \in \mathcal{R}^m$ where $m \geqq 1$, and consider two conditions that are equivalent to continuity. The first is a straightforward consequence of (6.2) and (6.6): A transformation is continuous at a point iff each of its coordinate functions is continuous at the point (6.1.7). The second is

The transformation f is continuous on its open domain D iff

$$f^{-1}(V) \subset \mathcal{R}^n \text{ is open whenever } V \subset \mathcal{R}^m \text{ is open} \tag{6.9}$$

PROOF. Suppose f is continuous on D, V is open, and $f(\mathbf{x}') \in V$. Since V is open there exists $\varepsilon > 0$ such that $\mathbf{y} \in V$ whenever

$$\|f(\mathbf{x}') - \mathbf{y}\| < \varepsilon$$

By continuity, there exists $\delta > 0$ such that

$$\|f(\mathbf{x}) - f(\mathbf{x}')\| < \varepsilon$$

whenever $\|\mathbf{x} - \mathbf{x}'\| < \delta$. Consequently $\mathbf{x}'$ has a neighborhood entirely contained in $f^{-1}(V)$ because

$$\|\mathbf{x} - \mathbf{x}'\| < \delta \text{ implies } \mathbf{x} \in f^{-1}(V)$$

For the converse, fix $\varepsilon > 0$ and $\mathbf{x}' \in D$, and define

$$V = \{\mathbf{y} : \|\mathbf{y} - f(\mathbf{x}')\| < \varepsilon\}$$

Then V is open, and consequently so also is $f^{-1}(V)$ and there exists $\delta > 0$ such that again

$$\|\mathbf{x} - \mathbf{x}'\| < \delta \text{ implies } \mathbf{x} \in f^{-1}(V)$$

That is, $f(\mathbf{x}) \in V$ so $\|f(\mathbf{x}) - f(\mathbf{x}')\| < \varepsilon$ and f is continuous at $\mathbf{x}'$. ■

The preceding condition deserves close attention. We cannot replace f^{-1} by f; for example, the constant transformation

$$\mathbf{x} \mapsto \mathbf{a} \in \mathfrak{R}^m$$

shows that $f(V)$ is not necessarily open when f is continuous and V is open (6.4.2). Moreover, the function on $\mathfrak{R}$

$$x \mapsto 1 + \frac{1}{x}$$

maps the closed interval $[1, \infty)$ into $(1, 2]$ so it is not necessarily true that a continuous function maps closed sets into closed sets. We can replace open sets in (6.9) by closed sets (6.1.7(5)); again, we cannot replace f^{-1} by f in the result as

$$x \mapsto \arctan x$$

demonstrates.

We find it worthwhile to adopt some standard notation for the large number of objects involved in work with "continuous transformations of continuous transformations." The notation is presented in Figure 6.1 and in the remainder of this chapter we reference

$$\gamma : U \subset \mathfrak{R}^p \to \mathfrak{R}^r$$

with the understanding that both domains, U and V, are open sets. To the extent possible we use $\mathbf{u}, \mathbf{v}, \mathbf{w}$ as notation for general points in $U, V, \mathfrak{R}^r$, respectively. Thus

$$\mathbf{v} = \beta(\mathbf{u}) \qquad \text{and} \qquad \mathbf{w} = \gamma(\mathbf{u}) = \alpha(\beta(\mathbf{u}))$$

The first result using this notation is the

THEOREM. If β is continuous at $\mathbf{u}_0$ and α is continuous at $\mathbf{v}_0 = \beta(\mathbf{u}_0)$, then γ is continuous at $\mathbf{u}_0$. (6.10)

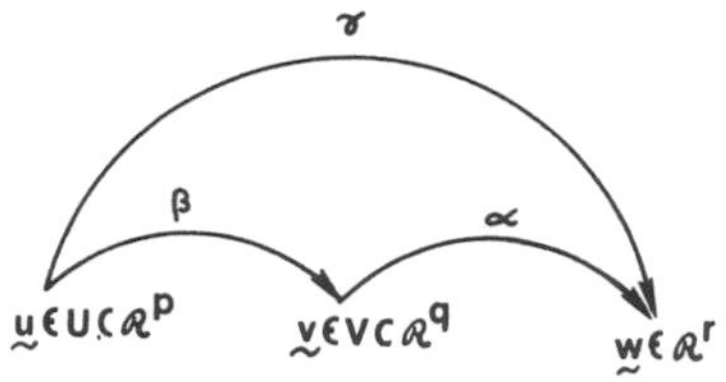

Figure 6.1. The composite transformation $\gamma: U \subset \mathfrak{R}^p \to \mathfrak{R}^r$.

PROOF. Given $\varepsilon > 0$ there exists $\delta_1 > 0$ such that

$$\|\alpha(\mathbf{v}) - \alpha(\mathbf{v}_0)\| < \varepsilon \text{ whenever } \|\mathbf{v} - \mathbf{v}_0\| < \delta_1$$

Given this δ_1 there exists $\delta_2 > 0$ such that

$$\|\beta(\mathbf{u}) - \beta(\mathbf{u}_0)\| < \delta_1 \text{ whenever } \|\mathbf{u} - \mathbf{u}_0\| < \delta_2$$

Consequently $\|\gamma(\mathbf{u}) - \gamma(\mathbf{u}_0)\| < \varepsilon$ whenever $\|\mathbf{u} - \mathbf{u}_0\| < \delta_2$. ■
With this theorem we readily establish (6.1.7)

The transformation f is continuous at any point $\mathbf{x}_0$ in its open domain D iff the associated transformation

$$\mathbf{h} \mapsto f(\mathbf{x}_0 + \mathbf{h})$$

is continuous at the origin. (6.11)

Additional uses of (6.10) appear in the next example.

EXAMPLE 6.1. Particular Continuous Transformations and Functions. Because of 6.1.7, the calculus of continuous transformations covering "sums," "inner products," and so on, follows readily from corresponding results for continuous functions on $\mathfrak{R}^n$ to $\mathfrak{R}$. In particular, we have

(1) Let f and g be transformations that are continuous at $\mathbf{x}_0 \in D$ where D is open. Then the transformations $f+g$ and kf where $k \in \mathfrak{R}$, and the functions corresponding to $\langle f, g \rangle$ and $\|f\|$, are continuous at $\mathbf{x}_0$.

Combining this result, which concerns "continuity in all variables jointly," with (6.10) often serves to establish continuity in particular cases. Many other cases follow from the continuity of elementary functions (polynomials, exponentials, trigonometric functions, and so on) on $\mathfrak{R}$. However, directly from the definition of continuity, a constant transformation $\mathbf{x} \mapsto \mathbf{a} \in \mathfrak{R}^n$ is continuous on all of $\mathfrak{R}^n$ as is $\mathbf{x} \mapsto \mathbf{x}$. Consequently, $\mathbf{x} \mapsto \langle \mathbf{a}, \mathbf{x} \rangle$ is a continuous function, $\mathbf{x} \mapsto \mathbf{A}\mathbf{x}$ is a continuous transformation where $\mathbf{A}$ is $m \times n$, and

(2) Any linear transformation on $\mathcal{R}^n$ to $\mathcal{R}^m$ is continuous.
Using $\langle \mathbf{x}, \mathbf{Ax} \rangle$ together with (6.10) establishes
(3) Any quadratic form on $\mathcal{R}^n$ is continuous on $\mathcal{R}^n$.
If we consider $\mathbf{x}^T\mathbf{I}\mathbf{x}$ and again use (6.10) we find
(4) The function $\mathbf{x} \mapsto \|\mathbf{x}\|$ is continuous on $\mathcal{R}^n$.
By successive applications of (1) in much the same way as for 1.3.1 we can prove that every polynomial in $x_1, x_2, \ldots, x_n$ is continuous on $\mathcal{R}^n$, and so is every corresponding rational function wherever its denominator is nonzero. From this latter we can establish the following.
(5) Suppose $\mathbf{A} = [a_{ij}]$ is nonsingular of order m on D, all

$$a_{ij} : D \subset \mathcal{R}^n \to \mathcal{R}$$

are continuous on D, and $\mathbf{w} \in \mathcal{R}^m$ is a constant vector. Then the solution of

$$\mathbf{Az} = \mathbf{w}$$

is the value of a continuous transformation on $D \subset \mathcal{R}^n$ to $\mathcal{R}^m$.

PROOF. The transformation $\mathbf{x} \mapsto \mathbf{z} = \mathbf{A}^{-1}\mathbf{w}$ is defined for each $\mathbf{x} \in D$. By Cramer's rule, the components z_i are rational functions, their denominators are nonzero, and hence they are continuous as noted previously. ■
In Example 8.3 we use complex variables to prove
(6) The zeros of a polynomial are continuous functions of its coefficients.
Continuity here means continuity on $\mathcal{C}^n$ where a set of n coefficients $a_0, a_1, \ldots, a_{n-1}$ is treated as a point in $\mathcal{C}^n$. From this result the continuity (as functions of n^2 elements) of the eigenvalues, and hence of the trace and the determinant, of a square complex matrix of order n follows directly.

The next several results culminate in (6.16). An *open cover* of any set $A \subset \mathcal{R}^n$ is a collection of open sets

$$\mathcal{G} = \{ G(x) : x \in X \}$$

such that A is a subset of the union of all sets in $\mathcal{G}$; the number of elements in X, which determines the number of sets, may be finite or infinite. We say that A has a *finite subcover* from $\mathcal{G}$ if A is a subset of the union of a finite number of the sets in $\mathcal{G}$. The set A is *compact* if it has the

Heine–Borel Property. Every open cover can be replaced by a finite subcover.

For example, $\varnothing$ is compact as is any finite set, but $\mathcal{R}^n$ is not compact. We first show that the property of compactness in $\mathcal{R}^n$ is associated with a set being closed.

The following hold if $A \subset \mathcal{R}^n$ is compact.
(1) A is closed.
(2) If $B \subset A$ is closed, then B is compact.
(3) If $B \subset A$ is infinite then B has a limit point in A. (6.12)

PROOF. Suppose $\mathbf{p} \in A^c$ is fixed and with each $\mathbf{a} \in A$ we associate the neighborhood $N_\delta(\mathbf{a})$ where

$$3\delta = \|\mathbf{p} - \mathbf{a}\|$$

All such neighborhoods constitute an open cover of A and if A is compact there is a finite subcover corresponding to say

$$\mathbf{a}_1, \mathbf{a}_2, \ldots, \mathbf{a}_m$$

If we now use the associated $\delta_1, \delta_2, \ldots, \delta_m$ as radii of neighborhoods of $\mathbf{p}$, their intersection is a neighborhood of $\mathbf{p}$ that does not intersect A. This shows that A^c is open. Suppose in (2) we obtain an open cover of A by adjoining B^c to $\mathcal{G}$ where the latter is an open cover of B. Then there exists a finite subcover of $A \supset B$ and at most we must remove B^c (if it appears) to obtain a finite subcover of B. If B in (3) has no limit point then each $\mathbf{b} \in B$ has a neighborhood containing no other point of B. Because B is infinite, this exhibits an open cover of B that has no finite subcover of B. ■

If $\mathbf{a}, \mathbf{b} \in \mathcal{R}^n$ and $\mathbf{a} < \mathbf{b}$ then this pair determines an *n-dimensional interval*, briefly *n-cell*, in $\mathcal{R}^n$

$$C = \{\mathbf{x} : a_i \leqq x_i \leqq b_i,\ i = 1, 2, , \ldots, n\}$$

In words, C is the Cartesian product of the n finite closed intervals $[a_i, b_i]$ determined by pairs of corresponding components of $\mathbf{a}$ and $\mathbf{b}$.

Every n-cell C is compact. (6.13)

PROOF. Suppose C is not compact and let $\mathcal{G}$ be an open cover of C for which no finite subcover is possible. Now the real intervals $[a_i, c_i]$ and $[c_i, b_i]$, where $c_i = (a_i + b_i)/2$ is the midpoint of $[a_i, b_i]$ for $i = 1, 2, \ldots, n$, determine 2^n n-cells whose union equals C. At least one of them, say C_1, cannot be covered by any finite subcover of $\mathcal{G}$. Midpoints can also be used to subdivide the n-cell C_1 and determine C_2, and so on, such that for $m = 1, 2, 3, \ldots$ we obtain sets for which no finite subcovers of $\mathcal{G}$ exist

$$C \supset C_1 \supset C_2 \supset \cdots \supset C_m \supset \cdots$$

Furthermore, applying (1.5) to suprema of LH endpoints of associated real intervals we find for each m

$$C \cap C_1 \cap \cdots \cap C_m \neq \varnothing$$

If $\mathbf{x}$ belongs to this intersection then $\mathbf{x} \in G_k$ for some set G_k in the open cover and $N_\delta(\mathbf{x}) \subset G_k$ for some $\delta > 0$. For any $\mathbf{x}, \mathbf{y} \in C_m$ we have

$$\|\mathbf{x}-\mathbf{y}\| \leqq \frac{1}{2^m} \|\mathbf{b}-\mathbf{a}\|$$

and we can choose m sufficiently large so that

$$\frac{1}{2^m} \|\mathbf{b}-\mathbf{a}\| < \delta$$

whereupon $C_m \subset G_k$. This is a contradiction because by definition C_m cannot be covered by any finite subcover of $\mathcal{G}$. ■

Let us next show that the property of boundedness in $\mathcal{R}^n$ is also closely associated with compactness. As introduced for use in (4.62), boundedness of $S \subset \mathcal{R}^n$ means existence of $\mathbf{u}, \mathbf{v} \in \mathcal{R}^n$ such that $\mathbf{u} \leqq \mathbf{s} \leqq \mathbf{v}$ for all $\mathbf{s} \in S$. In the context of the present chapter

Each of the following is necessary and sufficient for $A \subset \mathcal{R}^n$ to be bounded.

(1) Each $\mathbf{a} \in A$ has a neighborhood that contains A.

(2) There exists $M \in \mathcal{R}$ such that $\|\mathbf{a}\| \leqq M$ for all $\mathbf{a} \in A$. (6.14)

The next result gives a sufficient condition.

If every infinite subset of $A \subset \mathcal{R}^n$ has a limit point in A, then A is closed and bounded. (6.15)

PROOF. Let us use an indirect proof for each of the two conclusions. If A is not closed there exists $\mathbf{p} \in A^c$ and

$$L = \left\{ \mathbf{a}_i : \|\mathbf{a}_i - \mathbf{p}\| < \frac{1}{i},\ i = 1, 2, \ldots \right\}$$

clearly must be an infinite subset of A. Moreover, L has $\mathbf{p}$ as its only limit point because if $\mathbf{q} \neq \mathbf{p}$, then (starting with the triangle inequality) for i sufficiently large

$$\|\mathbf{a}_i - \mathbf{q}\| \geqq \|\mathbf{p}-\mathbf{q}\| - \|\mathbf{a}_i - \mathbf{p}\| \geqq \|\mathbf{p}-\mathbf{q}\| - \frac{1}{i} \geqq \frac{1}{2}\|\mathbf{p}-\mathbf{q}\|$$

Consequently $\mathbf{q}$ would not be a limit point of L and we would have the contradiction of L having no limit point in A. It is also impossible for the hypothesis to hold if A is not bounded. The reason is that we could exhibit

$$\{\mathbf{b}_i : \|\mathbf{b}_i\| > i,\ i = 1, 2, \ldots\}$$

which is infinite and has no limit point anywhere in $\mathcal{R}^n$. ■

We now have all we need to prove the fundamental

THEOREM. Each of the following is necessary and sufficient for $A \subset \mathcal{R}^n$ to be compact.
(1) Every infinite subset of A has a limit point in A.
(2) A is closed and bounded. (6.16)

PROOF. If A is compact then (1) holds by 6.12(3). Property (1) implies (2) by (6.15). Finally, (6.13) and 6.12(2) show that (2) implies A is compact. ■

Also see 6.1.8.

With the preceding theorem we can establish several results that are needed for optimization theory in Chapter 7. There we deal with extrema of functions on $\mathcal{R}^n$ to $\mathcal{R}$ and the first result applies to the space where functional values lie.

If $A \subset \mathcal{R}$ is compact then both

$$m = \inf A \qquad \text{and} \qquad M = \sup A$$

belong to A. (6.17)

PROOF. If $M \notin A$ then, since A is bounded above, 1.2.4(1) shows that every neighborhood of M contains a point of A distinct from M. Consequently, M would be a limit point of A not belonging to A which is impossible because A is closed. In the same way, $m \in A$. ■

The following establishes the key relationship between compactness and continuity for any transformation, that is, for any $m \geqq 1$.

If $f: D \subset \mathcal{R}^n \to \mathcal{R}^m$ is continuous on its open domain D and $A \subset D$ is compact, then $f(A)$ is also compact. (6.18)

PROOF. If $\mathcal{G}$ denotes an open cover of $f(A)$, then by (6.9) $f^{-1}(G(x))$ is open for every x. Since the collection of all $f^{-1}(G(x))$ is an open cover of A, there is a finite subcover

$$\{f^{-1}(G(1)), \ldots, f^{-1}(G(m))\}$$

of A. Finally, $f(f^{-1}(G(x))) = G(x)$ and so

$$\{G(1), \ldots, G(m)\}$$

is a finite subcover of $f(A)$. ■

By (6.16), $f(A)$ in the preceding is closed and bounded. If f is a function on $\mathcal{R}^n$ then, just as in elementary calculus, it achieves both its minimum and maximum on A according to the

EXTREME VALUE THEOREM. If $m=1$ in (6.18) then there exist $\mathbf{p},\mathbf{q}\in A$ such that

$$f(\mathbf{p})=\inf f(A), \qquad \text{and} \qquad f(\mathbf{q})=\sup f(A). \tag{6.19}$$

PROOF. The set of images $f(A)$ is again compact and the conclusion follows from (6.17). ■

Just as in elementary calculus a function

$$f: D\subset \mathcal{R}^n \to \mathcal{R}$$

is a *bounded function* if its set of images $f(D)$ is a bounded set; otherwise it is *unbounded*. Thus (6.18) states

The restriction of a continuous function to a compact subset of its domain is a bounded function. (6.20)

The elementary example of a continuous function $x \mapsto x^{-1}$ on $(0,1)$ shows that compactness is necessary. So also does the unbounded function

$$x \mapsto x \text{ on } \mathcal{R}$$

and this is an example of a linear transformation, in fact of

$$\mathbf{x} \mapsto \mathbf{I}\mathbf{x}$$

From the latter we conclude that the definition of boundedness for $m=1$ is too restrictive to be used for any m; we use the

Definition. A transformation $f: D\subset \mathcal{R}^n \to \mathcal{R}^m$, where $m>1$, is a *bounded transformation* if there exists $M\in\mathcal{R}$ such that

$$\|f(\mathbf{x})\| \leqq M\|\mathbf{x}\|$$

for all $\mathbf{x}\in D$. (6.21)

The number M is called a *bound* for the transformation. It is necessary to use the full expressions *bounded transformation* and *bounded function* in order to avoid difficulty.

Any linear transformation $T: \mathcal{R}^n \to \mathcal{R}^m$ is a bounded transformation. (6.22)

PROOF. Consider the orthonormal basis $\{\mathbf{b}_1,\mathbf{b}_2,\ldots,\mathbf{b}_n\}$ for the domain and define

$$M' = \max\{\|T\mathbf{b}_i\|: i = 1,2,\ldots,n\}$$

Then using the Cauchy–Schwarz inequality of (3.45) and $\|\mathbf{b}_i\|=1$ for all i

$$\|T\mathbf{x}\| \leqq M'(\|\mathbf{b}_1\|+\|\mathbf{b}_2\|+\cdots+\|\mathbf{b}_n\|)\,\|\mathbf{x}\|$$

shows $M=nM'$ is a bound for T. ■
If we consider the standard matrix $\mathbf{T}$ we find

The least bound that serves for a linear transformation is the natural norm of its standard matrix. (6.23)

PROOF. Always $T\mathbf{0}=\mathbf{0}$ so only nonzero vectors need be considered in (6.22). If $\mathbf{x}\neq\mathbf{0}$ then the matrix $\mathbf{T}$ satisfies

$$\|\mathbf{Tx}\| \leqq M\|\mathbf{x}\| \quad \text{iff} \quad \left\|\mathbf{T}\frac{\mathbf{x}}{\|\mathbf{x}\|}\right\| \leqq M$$

again because $\mathbf{x}\mapsto\mathbf{Tx}$ is linear. ■

As revealed by their proofs, (6.22) and (6.23) hold for any norm. More than this, let us show that all norms (3.44) define the same theory of limits on $\mathcal{R}^n$ in the sense that $\{\mathbf{x}_k\}$ converges according to (6.1) and $f(\mathbf{x})\to\mathbf{y}_0$ as $\mathbf{x}\to\mathbf{x}_0$ according to (6.5), and so on, when an arbitrary norm $\|\cdot\|^*$ is used iff the same happens when the standard norm $\|\cdot\|$ is used. The basic idea of the proof is that given a neighborhood defined by one norm we can always insert a second neighborhood, defined by any second norm, as a subset. While we can deal with any two norms, it suffices to suppose that one is the standard norm.

If $\|\cdot\|^*$ is any norm on $\mathcal{R}^n$ then there exist $a,b\in\mathcal{R}$, $0<a\leqq b$, such that for every $\mathbf{x}\in\mathcal{R}^n$

$$a\|\mathbf{x}\| \leqq \|\mathbf{x}\|^* \leqq b\|\mathbf{x}\|$$

where $\|\cdot\|$ is the standard norm. (6.24)

PROOF. Clearly we can concentrate on $\mathbf{x}\neq\mathbf{0}$. Using 3.45(2), which applies to any norm,

$$\|\mathbf{x}\|^* \leqq \sum_{i=1}^{n} |x_i|\cdot\|\mathbf{e}_i\|^* \leqq \frac{b}{n}\sum_{i=1}^{n}|x_i| \leqq b\|\mathbf{x}\|$$

where $b=n\cdot\max\{\|\mathbf{e}_1\|^*,\ldots,\|\mathbf{e}_n\|^*\}$ and where the final inequality holds because

$$|x_i| \leqq \|\mathbf{x}\|$$

for each i according to 3.4.5(8). Consequently

$$\|\mathbf{x}-\mathbf{y}\|^* \leqq b\|\mathbf{x}-\mathbf{y}\|$$

for every $\mathbf{x}$ and $\mathbf{y}$ so $\|\cdot\|^*$ is continuous on $\mathcal{R}^n$. By (6.19) $\|\cdot\|^*$ must achieve its minimum a on the closed and bounded set

$$S=\{\mathbf{x}:\|\mathbf{x}\|=1\}$$

Moreover, $a>0$ by 3.44(1). If $k=\|\mathbf{x}\|^{-1}$ then $k\mathbf{x}\in S$,

$$\|k\mathbf{x}\|^*\geqq a$$

and multiplying both sides by $\|\mathbf{x}\|$ produces $\|\mathbf{x}\|^*\geqq a\|\mathbf{x}\|$. ■
If $\mathbf{x}$ and $\mathbf{x}_0$ are any vectors in $\mathcal{R}^n$, then the preceding yields

$$a\|\mathbf{x}-\mathbf{x}_0\|\leqq\|\mathbf{x}-\mathbf{x}_0\|^*\leqq b\|\mathbf{x}-\mathbf{x}_0\|$$

which can be used to establish the equivalence of the limits defined by $\|\cdot\|$ and by $\|\cdot\|^*$. Let us use asterisk superscripts in open* and $N_\delta^*(\mathbf{x}_0)$ to denote use of $\|\cdot\|^*$. That is, a set $G\subset\mathcal{R}^n$ is open* if each $\mathbf{x}_0\in G$ has a neighborhood

$$N_\delta^*(\mathbf{x}_0)=\{\mathbf{x}:\|\mathbf{x}-\mathbf{x}_0\|^*<\delta\}$$

that lies entirely within G.

EQUIVALENCE OF VECTOR NORMS. If $\|\cdot\|^*$ is any norm on $\mathcal{R}^n$ then it defines the same collection of open sets, and hence the same theory of limits, as does the standard norm. (6.25)

PROOF. If a and b are determined by (6.24) we can use the three-member inequality displayed just previously to establish the following. For any $\delta>0$

$$N_{\frac{\delta}{b}}(\mathbf{x}_0)\subset N_\delta^*(\mathbf{x}_0)$$

which shows that open* implies open. Continuing,

$$N_{a\delta}^*(\mathbf{x}_0)\subset N_\delta(\mathbf{x}_0)$$

shows that the converse is also valid. ■

In working with derivatives we need to use matrix norms and it is convenient to establish some equivalent expressions for (3.55). Since every linear transformation is continuous (Example 6.1), the extreme value theorem (6.19) guarantees that the supremum is actually achieved and can be replaced by the maximum over the closed and bounded unit sphere

$$\|\mathbf{A}\|=\max\{\|\mathbf{A}\mathbf{x}\|:\|\mathbf{x}\|=1\} \tag{6.26}$$

Notice that if we use standard norms then there are three different norms denoted by $\|\cdot\|$ just previously: the natural norm on the LHS and the standard norms in $\mathcal{R}^m$ and $\mathcal{R}^n$, respectively, on the RHS. A particularly useful alternative expression is the maximum over the closed unit ball

$$\|\mathbf{A}\| = \max\{\|\mathbf{Ax}\| : \|\mathbf{x}\| \leqq 1\} \tag{6.27}$$

PROOF. If $S = \{\mathbf{x} : \|\mathbf{x}\| = 1\}$ and $B = \{\mathbf{x} : 0 < \|\mathbf{x}\| \leqq 1\}$, then using brief notation

$$\max_Y g(\mathbf{x}) = \max\{g(\mathbf{x}) : \mathbf{x} \in Y\}$$

for functions g and subsets Y of their domains

$$\max_B \frac{\|\mathbf{Ax}\|}{\|\mathbf{x}\|} = \max_S \|\mathbf{Ax}\| \leqq \max_B \|\mathbf{Ax}\| \leqq \max_B \frac{\|\mathbf{Ax}\|}{\|\mathbf{x}\|}$$

The first relation follows from $\|\mathbf{Ax}\|/\|\mathbf{x}\| = \|\mathbf{A}\|\mathbf{x}\|^{-1}\mathbf{x}\|$ which holds for $\mathbf{x} \neq \mathbf{0}$. Then $S \subset B$ implies the second while the third holds because

$$\|\mathbf{Ax}\| \leqq \|\mathbf{x}\|^{-1}\|\mathbf{Ax}\|$$

whenever $\mathbf{x} \in B$. Finally, considering $\mathbf{x} = \mathbf{0}$ in (6.27) cannot change the value of the maximum. ■

We also see that a third convenient expression is

$$\|\mathbf{A}\| = \max\{\|\mathbf{Ax}\| : \mathbf{x} \neq \mathbf{0}\} \tag{6.28}$$

Now it is easy to establish

The natural norm associated with any norm qualifies as a matrix norm. (6.29)

PROOF. Properties (1) and (2) of (3.53) are readily verified using any one of the three equivalent expressions for a natural norm. Since the maximum is achieved in (6.26), say at $\mathbf{x}_0$, we have

$$\|\mathbf{A}+\mathbf{B}\| = \|\mathbf{Ax}_0 + \mathbf{Bx}_0\| \leqq \|\mathbf{Ax}_0\| + \|\mathbf{Bx}_0\| \leqq \|\mathbf{A}\| + \|\mathbf{B}\|$$

which establishes (3). Next, from (6.26)

$$\|\mathbf{BC}\| = \|\mathbf{BCy}_0\|$$

for some $\mathbf{y}_0$ with unit norm, the RHS equals the first member in

$$\|\mathbf{B}(\mathbf{Cy}_0)\| \leqq \|\mathbf{B}\|\|\mathbf{Cy}_0\| \leqq \|\mathbf{B}\|\|\mathbf{C}\|\|\mathbf{y}_0\|$$

where the final member equals $\|\mathbf{B}\|\|\mathbf{C}\|$, and so (4) holds. ■

For derivatives we need to use limits in $\mathcal{L}(\mathcal{R}^n, \mathcal{R}^m)$ and the associated $\mathfrak{M}(m,n)$. We nearly always work in the latter vector space so it is essential to establish the

EQUIVALENCE OF NATURAL NORMS. If $\|\cdot\|^*$ and $\|\cdot\|'$ are any two natural norms then they define the same collection of open sets, and hence the same theory of limits. (6.30)

PROOF. Any two of $\mathcal{L}(\mathcal{R}^n, \mathcal{R}^m)$, $\mathfrak{M}(m,n)$, $\mathcal{R}^{mn}$ are isomorphic by (2.26) and it suffices to work with the latter two where we can use the one-to-one correspondence

$$[\mathbf{a}_1, \mathbf{a}_2, \ldots, \mathbf{a}_n] \leftrightarrow \begin{bmatrix} \mathbf{a}_1 \\ \mathbf{a}_2 \\ \vdots \\ \mathbf{a}_n \end{bmatrix}$$

Given any norm $\|\cdot\|$ on $\mathcal{R}^m$, let $\|\cdot\|'$ denote the natural norm on $\mathfrak{M}(m,n)$, and let $\|\cdot\|''$ denote the norm on $\mathcal{R}^{mn}$ obtained by assigning $\|\mathbf{z}\|'' = \|\mathbf{A}\|'$ where $\mathbf{z} \leftrightarrow \mathbf{A}$. If $\|\cdot\|^*$ is a second natural norm, then values of $\|\mathbf{A}\|'/\|\mathbf{A}\|^*$ for nonzero $\mathbf{A}$ coincide with those of $\|\mathbf{z}\|''/\|\mathbf{z}\|^{**}$ for nonzero $\mathbf{z}$. Consequently, adapting the proof of (6.24) to $\mathcal{R}^{mn}$ we obtain

$$a\|\mathbf{A}\|' \leqq \|\mathbf{A}\|^* \leqq b\|\mathbf{A}\|'$$

and the desired conclusion results as in (6.25). ■

PROBLEMS FOR SOLUTION

6.1.1 Limits of Sequences. Establish in $\mathcal{R}^n$:

(1) If $\{\mathbf{x}_k\}$ converges then its limit is unique.
(2) If $\{\mathbf{x}_k\}$ converges then it is *bounded* in the sense that its set of images is a bounded subset of $\mathcal{R}^n$.
(3) If $\lim\{\mathbf{x}_k\} = \mathbf{a}$ and $\lim\{\mathbf{y}_k\} = \mathbf{b}$ then $\lim\{\mathbf{x}_k + \mathbf{y}_k\} = \mathbf{a} + \mathbf{b}$.
(4) Counterpart for $\{\langle \mathbf{x}_k, \mathbf{y}_k \rangle\}$.
(5) If $\{\mathbf{x}_k\}$ is bounded and $\lim\{\mathbf{y}_k\} = \mathbf{0}$ then $\lim\{\langle \mathbf{x}_k, \mathbf{y}_k \rangle\} = 0$.

6.1.2 Neighborhoods. Establish:

(1) $\mathbf{y} \in N_\delta(\mathbf{p})$ iff $\mathbf{y} = \mathbf{p} + \mathbf{h}$ for some $\mathbf{h} \in N_\delta(\mathbf{0})$.
(2) Any neighborhood $N_\delta(\mathbf{x}_0)$ is open and furthermore connected.
(3) Any neighborhood $N_\delta(\mathbf{x}_0)$ is a convex set.
(4) True or false? Deleted neighborhoods are open sets.

6.1.3 Topology. If we take any set, call it a *space* of *points*, and determine a class of subsets called the *open sets* satisfying

(I) the null set and the entire space are open sets;
(II) the intersection of any finite number of open sets is an open set; and
(III) the union of any number of open sets is an open set

then we say we have defined a *topology* on the space. Once we have done this we can proceed to define associated notions of limit, continuity, compactness, and connectedness which are *topological notions* in the sense that they are defined in terms of open sets. From these notions, we can proceed to derivatives, integrals, and their associated concepts. Show that the definitions in the text in terms of neighborhoods (6.3) define a topology in $\mathcal{R}^n$, and develop some basic properties, by proceeding as follows. Establish:

(1) Properties I, II, and III hold in $\mathcal{R}^n$.
(2) The intersection of an infinite number of open sets may fail to be open.
(3) Properties I, II, and III can be replaced by equivalent properties involving closed sets and obtained by taking complements in I, II, and III.
(4) The union of any infinite number of closed sets may fail to be closed.

6.1.4 Neighborhood Topology. This is the name given to the topology defined by (6.3) and validated in the preceding (1). Establish the following for any $A \subset \mathcal{R}^n$:

(1) $\operatorname{int} A \subset A$.
(2) $\operatorname{ext} A \subset A^c$.
(3) In general, $\operatorname{bdry} A \not\subset A$ and $\operatorname{bdry} A \not\subset A^c$.
(4) Every neighborhood of a limit point of A contains an infinite number of points of A.
(5) $\operatorname{clos} A$ is the union of A and the set of limit points of A.
(6) $\operatorname{int} A$ is open.
(7) $\operatorname{clos} A$ and $\operatorname{bdry} A$ are closed.
(8) A is open iff $A = \operatorname{int} A$.
(9) A is closed iff A contains all of its limit points.
(10) A is closed iff $\operatorname{bdry} A \subset A$.
(11) A and $\operatorname{clos} A$ have the same closure.
(12) A and A^c have the same boundary.
(13) $(\operatorname{clos} A)^c = \operatorname{int} A^c$.
(14) $\operatorname{bdry} A$ consists of limit points and isolated points.
(15) True or false given A open and B closed?
  (i) $A - B$ is open.
  (ii) $B - A$ is neither open nor closed.

6.1.5 Exercises. Find the interior, boundary, and closure for each set and determine whether open, closed, connected, or compact. Work in $\mathfrak{R}^n$ unless a particular n is specified.

(1) The eight intervals defined in Section 1.2 for $\mathfrak{R}$.
(2) $\{\mathbf{x}:\|\mathbf{x}\|<1\}$, the *open unit ball.*
(3) $\{\mathbf{x}:\|\mathbf{x}\|=1\}$, the *unit sphere.*
(4) $\{\mathbf{x}:\|\mathbf{x}\|\leqq 1\}$, the *closed unit ball.*
(5) $\{\mathbf{x}_1,\mathbf{x}_2,\ldots,\mathbf{x}_m\}$.
(6) $\{\mathbf{x}:x_1>0,\ x_2>0\}\subset\mathfrak{R}^2$.
(7) $\{\mathbf{x}:x_1<1,\ x_2<1+x_1,\ x_2>-1-x_1\}\subset\mathfrak{R}^2$.
(8) $\{\mathbf{x}:x_1 \text{ or } x_2 \text{ is irrational}\}\subset\mathfrak{R}^2$.
(9) A k-simplex.
(10) An n-cell.

6.1.6 Limits of Transformations. Establish for $f:D\subset\mathfrak{R}^n\to\mathfrak{R}^m$ where D is open and $\mathbf{x}_0\in D$:

(1) $|f_k(\mathbf{x})-y_{k0}|\leqq\|f(\mathbf{x})-\mathbf{y}_0\|\leqq\sum_{i=1}^{m}|f_i(\mathbf{x})-y_{i0}|^2$ for $k=1,2,\ldots,m$.
(2) $f(\mathbf{x})\to\mathbf{y}_0$ as $\mathbf{x}\to\mathbf{x}_0$ iff for $i=1,2,\ldots,m$, $f_i(\mathbf{x})\to y_{i0}$ as $\mathbf{x}\to\mathbf{x}_0$.
(3) $\lim_{\mathbf{x}\to\mathbf{x}_0}f(\mathbf{x})=\mathbf{0}$ iff $\lim_{\mathbf{x}\to\mathbf{x}_0}\|f(\mathbf{x})\|=0$.
(4) $\lim_{\mathbf{x}\to\mathbf{x}_0}f(\mathbf{x})=\mathbf{y}_0$ implies $\lim_{\mathbf{x}\to\mathbf{x}_0}\|f(\mathbf{x})\|=\|\mathbf{y}_0\|$.
(5) If $\mathbf{y}_0\neq\mathbf{0}$ then the converse in (4) need not hold.
(6) $\lim_{\mathbf{x}\to\mathbf{x}_0}f(\mathbf{x})=\mathbf{y}_0$ iff $\lim_{\mathbf{h}\to\mathbf{0}}f(\mathbf{x}_0+\mathbf{h})=\mathbf{y}_0$.
(7) Various authors find it convenient to use the terminology "*limits as* $\|\mathbf{h}\|\to 0$" and the notation

$$\lim_{\|\mathbf{h}\|\to 0}f(\mathbf{x}_0+\mathbf{h})$$

These limits can be defined so as to be equivalent to the limits in (7).

6.1.7 Continuous Transformations. Establish for f, D, and $\mathbf{x}_0$ as specified in the preceding.

(1) A transformation is continuous at a point iff each of its coordinate functions is continuous at the point.
(2) "If f is continuous at $\mathbf{x}_0$ then it is continuous in each variable separately."
(3) The converse in (2) need not hold.
(4) Property (6.11).
(5) The transformation f is continuous iff $f^{-1}(A)\subset\mathfrak{R}^n$ is closed whenever $A\subset\mathfrak{R}^m$ is closed.
(6) If f is a continuous function and $a\in\mathfrak{R}$ then

$$\{\mathbf{x}:f(\mathbf{x})<a\}\qquad\text{and}\qquad\{\mathbf{x}:f(\mathbf{x})>a\}$$

are open subsets of $\mathcal{R}^n$. Consequently their complements

$$\{\mathbf{x}:f(\mathbf{x})\geqq a\} \qquad \text{and} \qquad \{\mathbf{x}:f(\mathbf{x})\leqq a\}$$

are closed.

6.1.8 Compact Sets. Establish for $A\subset\mathcal{R}^n$:

(1) Bolzano–Weierstrass Theorem. If A is bounded and infinite then A has a limit point.

(2) Finite Intersection Property. A collection of sets has the subject property if every finite subcollection has a nonempty intersection. The set A is compact iff every collection F of closed subsets of A that has the finite intersection property has a nonempty intersection.

(3) If $\{A_k\}$ is a sequence of nonempty compact sets such that

$$A_1\supset A_2\supset\cdots\supset A_k\supset\cdots$$

then there is at least one point that belongs to every set in the sequence.

(4) The *diameter* of any set $X\subset\mathcal{R}^n$ is

$$\operatorname{diam} X=\sup\{\|\mathbf{u}-\mathbf{v}\|:\mathbf{u},\mathbf{v}\in X\}$$

which is finite iff X is bounded. If $\lim\{\operatorname{diam} A_k\}=0$ in (3) then the intersection of all A_k is a single point.

(5) The *distance* from $\mathbf{u}\in\mathcal{R}^n$ to any set $X\subset\mathcal{R}^n$ is

$$d(\mathbf{u},X)=\inf\{\|\mathbf{u}-\mathbf{x}\|:\mathbf{x}\in X\}$$

If A is compact then $d(\mathbf{u},A)=\|\mathbf{u}-\mathbf{a}\|$ for some $\mathbf{a}\in A$.

(6) An Inverse Function Theorem. Suppose $m=1$ and f is one-to-one and continuous on D. If $A\subset D$ is compact then f^{-1} is continuous on $f(A)$.

6.1.9 Verifications for Example 3.1. Establish:

(1) Any natural (matrix) norm is compatible in the sense of (3.54) with its associated (vector) norm.

(2) The functions $\|\cdot\|_1$ and $\|\cdot\|_\infty$ qualify as norms (3.44).

(3) The norm $\|\cdot\|_\infty$ fails to be compatible in the sense of (3.54).

(4) The expressions in (5) and (6) of Example 3.1 correspond to $\|\mathbf{x}\|_1$ and $\|\mathbf{x}\|_\infty$, respectively, and qualify as matrix norms (3.53).

(5) The standard natural norm is

$$\|\mathbf{A}\|=\lambda^{\frac{1}{2}}$$

where λ is the largest eigenvalue of $\mathbf{A}^H\mathbf{A}$.

(6) The standard natural norm satisfies

$$\alpha \leqq \|\mathbf{A}\| \leqq (mn)^{\frac{1}{2}} \alpha$$

where $\alpha = \max\{|a_{ij}| : i = 1,2,\ldots,m; j = 1,2,\ldots,n\}$.

6.1.10 More on Norms. Establish:

(1) The value of the standard natural norm $\|\mathbf{a}^T\|$ equals the value of the standard norm $\|\mathbf{a}\|$ for every $\mathbf{a} \in \mathfrak{R}^n$.

(2) If $\mathbf{A}$ is nonsingular of order n then for all $\mathbf{x} \in \mathfrak{R}^n$

$$\|\mathbf{x}\| \leqq \|\mathbf{A}^{-1}\| \|\mathbf{Ax}\|$$

(3) For any natural norm

$$\|\mathbf{Ax} - \mathbf{Ay}\| \leqq \|\mathbf{A}\| \|\mathbf{x} - \mathbf{y}\|$$

(4) Suppose $\|\cdot\|^*$ and $\|\cdot\|'$ are any two norms on $\mathfrak{R}^n$ and consider $f: D \subset \mathfrak{R}^n \to \mathfrak{R}^n$, where D is open. Then

   (i) f is bounded with respect to $\|\cdot\|^*$ iff it is bounded with respect to $\|\cdot\|'$.

   (ii) f is continuous with respect to $\|\cdot\|^*$ iff it is continuous with respect to $\|\cdot\|'$.

6.1.11 Working in $\mathfrak{R}^n$. At times it is possible to generalize procedures for $\mathfrak{R}^2$ so as to obtain proofs in $\mathfrak{R}^n$ for any n. The following illustrates some techniques.

(1) Proceed as follows to prove from first principles that an open half-space is an open set. Establish:

   (i) We lose no generality by supposing $\|\mathbf{a}\| = 1$ and working with

$$H^+ = \{\mathbf{x} : \langle \mathbf{a}, \mathbf{x} \rangle > b\}$$

   We use 6.1.2(1) and given $\mathbf{p} \in H^+$ we specify two nonzero real numbers: k is defined by

$$\langle \mathbf{a}, \mathbf{p} \rangle = b + k$$

   and $\delta = |k|/2$.

   (ii) If $\mathbf{h} \in N_\delta(\mathbf{0})$ then

$$|\langle \mathbf{a}, \mathbf{h} \rangle| \leqq \|\mathbf{a}\| \cdot \|\mathbf{h}\| < \frac{|k|}{2}$$

(iii) If $y \in N_\delta(\mathbf{p})$ then

$$\langle \mathbf{a}, \mathbf{y} \rangle = \langle \mathbf{a}, \mathbf{p} \rangle + \langle \mathbf{a}, \mathbf{h} \rangle$$

is guaranteed to lie between $b + k/2$ and $b + 3k/2$ so $\mathbf{y} \in H^+$. ∎

(2) Use the continuity of $\langle \mathbf{a}, \cdot \rangle$ to prove that a closed halfspace is a closed set; this is an alternative approach to (1) that relies on results in Section 6.1 rather than on first principles.

(3) Show that if $\mathbf{p} \in A \subset \mathfrak{R}^n$ and $\mathbf{q} \in A^c$ then the line segment between $\mathbf{p}$ and $\mathbf{q}$ intersects bdry A.

6.1.12 Convex Sets. Establish:

(1) An open convex set is connected.
(2) The interior of a convex set K is convex.
(3) The convex hull of an open convex set is the set itself.
(4) The convex hull of an open set is open.
(5) The closure of a convex set is convex.
(6) Extreme points of a convex set are boundary points.
(7) A simplex in $\mathfrak{R}^n$ has interior points iff it is an n-simplex.

6.1.13 Affine Hull. Where the convex hull consists of all convex combinations (of not necessarily a finite number) of points of C, the *affine hull* A consists of all *affine combinations*.

$$(1-t)\mathbf{p} + t\mathbf{q}, \qquad t \in \mathfrak{R}$$

of points $\mathbf{p}, \mathbf{q} \in C$.

(1) Contrast the definition of A with that of the convex hull of C.
(2) Show that A is uniquely determined by C and that it coincides with the intersection of all affine subspaces containing C.

6.1.14 Relative Topology. In general, and in contrast to 6.1.3, a *relative topology* is obtained by working with *relative neighborhoods* of points $\mathbf{p} \in C$ which are defined to be intersections of neighborhoods of $\mathbf{p}$ in $\mathfrak{R}^n$ with the set C itself. This amounts to ignoring the complement $\mathfrak{R}^n - C$. If a set is open in the relative topology, it is called *relatively open*, and similarly for *relatively closed, relatively connected*, and so on, for topological notions in general. Establish:

(1) The interior of the *unit disc*

$$C = \left\{ \mathbf{x} : x_1^2 + x_2^2 < 1 \right\} \subset \mathfrak{R}^2$$

is convex but it has no interior points in $\mathfrak{R}^3$.

(2) C is relatively open in the relative topology of the affine hull of C.

(3) The relative interior (relative to its affine hull) of a nonempty convex set C is nonempty.

6.2 DERIVATIVES

We need to establish standard matrices for derivatives, and other results such as the chain rule (6.64), for use in Chapter 7.

A function $f: D \subset \mathcal{R}^n \to \mathcal{R}$, where D is open, is *differentiable at* $\mathbf{x}_0 \in D$ if there exists a linear transformation $F: \mathcal{R}^n \to \mathcal{R}$ such that

$$\lim_{\mathbf{h}\to\mathbf{0}} \frac{f(\mathbf{x}_0+\mathbf{h})-f(\mathbf{x}_0)-F\mathbf{h}}{\|\mathbf{h}\|}=0 \tag{6.31}$$

If f is differentiable at every $\mathbf{x}\in D$ then f is *differentiable on D*, and f is called a *differentiable function*. The linear transformation $F\in\mathcal{L}(\mathcal{R}^n,\mathcal{R})$ is the *derivative* at $\mathbf{x}_0$ of *first order* and is also denoted by $f'(\mathbf{x}_0)$. When the limit exists, the derivative is uniquely determined. Among other generalizations of familiar properties from calculus are (6.2.1)

If $f(\mathbf{x})=c\in\mathcal{R}$ for all $\mathbf{x}\in D$, where D is open, then f is differentiable and its derivative is the zero transformation. (6.32)

and

If $f:\mathcal{R}^n\to\mathcal{R}$ is linear then f is differentiable and coincides with its derivative. (6.33)

In order to obtain the standard matrix of the derivative $F\in\mathcal{L}(\mathcal{R}^n,\mathcal{R})$ we first find that the *partial derivative* of f *with respect to* x_j *evaluated at* $\mathbf{x}_0$ is

$$\lim_{t\to 0} \frac{f(\mathbf{x}_0+t\mathbf{e}_j)-f(\mathbf{x}_0)}{t}=\partial_j f(\mathbf{x}_0) \tag{6.34}$$

for $j=1,2,\ldots,n$ which, following custom, we also write as $(\partial/\partial x_j)f(\mathbf{x}_0)$. Second, we recall that unity serves as the standard basis (2.6) in the range $\mathcal{R}$. Then

The standard matrix of $F=f'(\mathbf{x}_0)$ is

$$\mathbf{f}'(\mathbf{x}_0)=\left[\,\partial_1 f(\mathbf{x}_0),\partial_2 f(\mathbf{x}_0),\ldots,\partial_n f(\mathbf{x}_0)\right] \tag{6.35}$$

PROOF. Because $|g(t)|t|^{-1}|=|g(t)t^{-1}|$ for any g, also

$$\lim_{t\to 0}\frac{g(t)}{|t|}=0 \quad\text{iff}\quad \lim_{t\to 0}\frac{g(t)}{t}=0$$

Consequently, if the limit (6.31) exists then by (6.7) for $\mathbf{h}=t\mathbf{e}_j$

$$\lim_{t\to 0}\frac{f(\mathbf{x}_0+t\mathbf{e}_j)-f(\mathbf{x}_0)-Ft\mathbf{e}_j}{t}=0$$

Since F is linear, $Ft\mathbf{e}_j=tF\mathbf{e}_j$ and we must have

$$\lim_{t\to 0}\frac{f(\mathbf{x}_0+t\mathbf{e}_j)-f(\mathbf{x}_0)}{t}=F\mathbf{e}_j$$

That is, $F\mathbf{e}_j=\partial_j f(\mathbf{x}_0)$ for $j=1,2,\ldots,n$. (In words, the partial derivative $\partial_j f(\mathbf{x}_0)$ is the column vector of dimension $m=1$ that is the image of the jth basis vector in $\mathcal{R}^n$.) By (3.21) the standard matrix is the displayed $\mathbf{f}'(\mathbf{x}_0)$. ■

We repeatedly use primed boldfaced notation such as $\mathbf{f}'(\mathbf{x}_0)$ to denote standard matrices of derivatives. In the present case we also use the name *gradient vector*, briefly *gradient*, and the notation

$$\nabla f(\mathbf{x}_0)=\left[\partial_1 f(\mathbf{x}_0),\partial_2 f(\mathbf{x}_0),\ldots,\partial_n f(\mathbf{x}_0)\right]^T \qquad (6.36)$$

for the transpose of $\mathbf{f}'(\mathbf{x}_0)$. The symbol ∇ is called "del." (See (7.13).)

If we reconsider the argument producing (6.35) we see that the jth coordinate vector $\mathbf{e}_j$ can be replaced by any nonzero vector. In particular, if $\|\mathbf{u}\|=1$, then whenever f is differentiable at $\mathbf{x}_0$ we have the following where now the RHS is the product of a matrix times a vector.

$$\lim_{t\to 0}\frac{f(\mathbf{x}_0+t\mathbf{u})-f(\mathbf{x}_0)}{t}=\mathbf{f}'(\mathbf{x}_0)\mathbf{u} \qquad (6.37)$$

This number is the *directional derivative of f at $\mathbf{x}_0$ in the direction* $\mathbf{u}$. In terms of the gradient the directional derivative equals

$$\langle \nabla f(\mathbf{x}_0),\mathbf{u}\rangle$$

If f is differentiable then the proof of (6.35) shows that f has a directional derivative in every direction at $\mathbf{x}_0$. We can show that the converse does not hold as soon as we establish the

THEOREM. If the function f is differentiable at $\mathbf{x}_0$ then it is continuous at $\mathbf{x}_0$. (6.38)

PROOF. If we add and subtract $\langle \nabla f(\mathbf{x}_0),\mathbf{h}\rangle$ then the triangle inequality yields

$$|f(\mathbf{x}_0+\mathbf{h})-f(\mathbf{x}_0)|\leqq|f(\mathbf{x}_0+\mathbf{h})-f(\mathbf{x}_0)-\langle\nabla f(\mathbf{x}_0),\mathbf{h}\rangle|+|\langle\nabla f(\mathbf{x}_0),\mathbf{h}\rangle|$$

Using $\varepsilon=1$ in (6.31) we find $\delta>0$ such that the first term on the RHS is less than $\|\mathbf{h}\|$ while the second cannot exceed $\|\nabla f(\mathbf{x}_0)\|\cdot\|\mathbf{h}\|$ by the Cauchy–Schwarz inequality of (3.45). Consequently, for any $\varepsilon>0$ the LHS is less than ε whenever

$$0<\|\mathbf{h}\|<\min\left\{\delta, \frac{\varepsilon}{1+\|\nabla f(\mathbf{x}_0)\|}\right\}$$

and we conclude $\lim_{h\to 0} f(\mathbf{x}_0+\mathbf{h})=f(\mathbf{x}_0)$. ■

Now let us consider the following example on $\mathfrak{R}^2$

$$f(\mathbf{x})=\frac{x_1{}^2x_2}{x_1{}^4+x_2{}^2}, \qquad \mathbf{x}\neq\mathbf{0}$$

$$=0, \qquad \mathbf{x}=\mathbf{0}$$

If $\mathbf{x}=t\mathbf{u}$ where $t\neq 0$ and $\|\mathbf{u}\|=1$, then the quotient on the LHS of (6.37) is

$$\frac{u_1^2u_2}{t^2u_1^4+u_2^2}$$

The limit as $t\to 0$ in (6.37) exists: It equals $u_1^2u_2^{-1}$ if $u_2\neq 0$, and zero otherwise. However, $f(\mathbf{x})\to 1/2$ as $\mathbf{x}\to\mathbf{0}$ subject to $x_2=x_1^2$, which shows that f is not continuous at the origin. By (6.38), f cannot be differentiable there and we conclude that differentiability is a stronger property than possession of a directional derivative in every direction.

Let us develop a geometric interpretation of differentiability. First we extend the small o notation by replacing (1.10) by

$$\frac{u}{\|\mathbf{v}\|^p}\to 0$$

so, for example, $g(\mathbf{h})=o(\|\mathbf{h}\|)$ means $\|\mathbf{h}\|^{-1}g(\mathbf{h})\to 0$ as $\mathbf{h}\to\mathbf{0}$. With this notation we can express (6.31) as

The function $f: D\subset\mathfrak{R}^n\to\mathfrak{R}$ is differentiable at $\mathbf{x}_0\in D$, where D is open, iff

$$f(\mathbf{x}_0+\mathbf{h})=f(\mathbf{x}_0)+\mathbf{f}'(\mathbf{x}_0)\mathbf{h}+o(\|\mathbf{h}\|)$$

for all $\mathbf{h}$ such that D contains all points on the line segment connecting $\mathbf{x}_0$ and $\mathbf{x}_0+\mathbf{h}$. (6.39)

The preceding equation is an example of a *linear Taylor formula*. Another is the equivalent

$$f(\mathbf{x}_0+\mathbf{h})=f(\mathbf{x}_0)+\langle\nabla f(\mathbf{x}_0),\mathbf{h}\rangle+E(\mathbf{x}_0,\mathbf{h})\cdot\|\mathbf{h}\| \qquad (6.40)$$

where $E(\mathbf{x}_0, \mathbf{h})$ is the quotient in (6.31) and $E(\mathbf{x}_0, \mathbf{h}) \to 0$ as $\mathbf{h} \to \mathbf{0}$. If

$$S = \left\{ \begin{bmatrix} \mathbf{x} \\ z \end{bmatrix} : z = f(\mathbf{x}) \right\} \subset \mathcal{R}^{n+1} \tag{6.41}$$

is called a *surface*, then when f is differentiable at $\mathbf{x}_0$ there is the *tangent hyperplane*

$$H = \left\{ \begin{bmatrix} \mathbf{x} \\ z \end{bmatrix} : z = f(\mathbf{x}_0) + \langle \nabla f(\mathbf{x}_0), \mathbf{x} - \mathbf{x}_0 \rangle \right\} \tag{6.42}$$

at the point corresponding to $\mathbf{x}_0$ and $z_0 = f(\mathbf{x}_0)$. The set (6.41) is a natural generalization of two-dimensional curves and three-dimensional surfaces treated in analytic geometry. Certainly (6.42) defines a hyperplane according to (4.8); actually,

$$H = \{ \mathbf{u} : \langle \mathbf{c}, \mathbf{u} \rangle = d \}$$

where

$$\mathbf{u} = \begin{bmatrix} \mathbf{x} \\ z \end{bmatrix} \quad \text{and} \quad \mathbf{c} = \begin{bmatrix} \nabla f(\mathbf{x}_0) \\ -1 \end{bmatrix}$$

and where $d = \langle \nabla f(\mathbf{x}_0), \mathbf{x}_0 \rangle - f(\mathbf{x}_0)$ because

$$\mathbf{u}_0 = \begin{bmatrix} \mathbf{x}_0 \\ z_0 \end{bmatrix} \in H$$

In the terminology introduced for (4.8) the vector $\mathbf{c}$ is normal to H because $\langle \mathbf{c}, \mathbf{u} - \mathbf{v} \rangle = 0$ whenever $\mathbf{u}, \mathbf{v} \in H$. When f is differentiable at $\mathbf{x}_0$ and we write $\mathbf{x}_0 + \mathbf{h} = \mathbf{x}$ in (6.39), and also use the gradient, we find

$$f(\mathbf{x}) - f(\mathbf{x}_0) \approx \langle \nabla f(\mathbf{x}_0), \mathbf{x} - \mathbf{x}_0 \rangle$$

when $\mathbf{x}$ is near $\mathbf{x}_0$. That is, the RHS furnishes a linear approximation to the LHS. The numerator in (6.31) equals the error and it is small compared to the denominator $\mathbf{x} - \mathbf{x}_0$ when $\mathbf{x}$ is sufficiently near $\mathbf{x}_0$. This is the intuitive justification for (6.42) being the tangent hyperplane to (6.41) at $\mathbf{u}_0 \in \mathcal{R}^{n+1}$ and in this way we can say that f differentiable at $\mathbf{x}_0$ means f has a tangent hyperplane at that point.

The next result extends the familiar (1.12) to functions on $\mathcal{R}^n$.

MEAN VALUE THEOREM. If the function f is differentiable on the line segment joining $\mathbf{x}_0$ and $\mathbf{x}_0 + \mathbf{h}$ then

$$f(\mathbf{x}_0 + \mathbf{h}) = f(\mathbf{x}_0) + \mathbf{f}'(\mathbf{u})\mathbf{h}$$

for some $\mathbf{u} = \mathbf{x}_0 + t_0 \mathbf{h}$ where $t_0 \in (0, 1)$. (6.43)

PROOF. The result certainly holds for $\mathbf{h}=\mathbf{0}$. Let us first show that when $\mathbf{h}\neq\mathbf{0}$ the function defined on $\mathcal{R}$ by

$$g(t)=f(\mathbf{x}_0+t\mathbf{h})$$

has $g'(t)=\langle\nabla f(\mathbf{x}_0+t\mathbf{h}),\mathbf{h}\rangle$. Looking at the place $\mathbf{x}_0+t\mathbf{h}$ in (6.31) we have

$$\lim_{\mathbf{k}\to\mathbf{0}}\frac{f(\mathbf{x}_0+t\mathbf{h}+\mathbf{k})-f(\mathbf{x}_0+t\mathbf{h})-\langle\nabla f(\mathbf{x}_0+t\mathbf{h}),\mathbf{k}\rangle}{\|\mathbf{k}\|}=0$$

and then letting $\mathbf{k}=s\mathbf{h}$ produces $g'(t)$ because

$$\lim_{s\to 0}\frac{g(t+s)-g(t)-s\langle\nabla f(\mathbf{x}_0+t\mathbf{h}),\mathbf{h}\rangle}{s}=0$$

Second, we see that (1.12) applies to g and yields the stated result. ■

A useful consequence is

If f' is the zero transformation on D, where D is connected as well as open, then f is a constant function. (6.44)

PROOF. Consider a polygonal path connecting any pair of distinct points $\mathbf{x},\mathbf{y}\in D$. For any pair of successive endpoints $\mathbf{p},\mathbf{q}$ of a line segment

$$f(\mathbf{p})-f(\mathbf{q})=\mathbf{f}'(\mathbf{u})(\mathbf{p}-\mathbf{q})$$

by (6.43) and the RHS is zero since $\mathbf{f}'(\mathbf{u})=\mathbf{O}$. Thus, $f(\mathbf{p})=f(\mathbf{q})$, and by considering successive line segments, $f(\mathbf{x})=f(\mathbf{y})$. ■

Thus far, although we have shown that existence of directional derivatives in every direction will not guarantee differentiability, we have given no special sufficient conditions for differentiability. Conditions that are adequate for our needs are contained in the

THEOREM. Let $f: D\subset\mathcal{R}^n\to\mathcal{R}$ be continuous on the open set D, let the partial derivatives $\partial_1 f(\mathbf{x}),\ldots,\partial_n f(\mathbf{x})$ exist for every $\mathbf{x}\in D$, and let $\partial_1 f,\ldots,\partial_n f$ be continuous on D. Then f is differentiable on D. (6.45)

PROOF. Given $\mathbf{h}\in\mathcal{R}^n$ sufficiently small, and any $\mathbf{x}_0\in D$, let $\mathbf{h}_0=\mathbf{0},\mathbf{h}_1=h_1\mathbf{e}_1,\mathbf{h}_2=h_1\mathbf{e}_1+h_2\mathbf{e}_2,\ldots,\mathbf{h}_n=\mathbf{h}$ so that

$$f(\mathbf{x}_0+\mathbf{h})-f(\mathbf{x}_0)=\sum_{i=1}^{n}\left[f(\mathbf{x}_0+\mathbf{h}_i)-f(\mathbf{x}_0+\mathbf{h}_{i-1})\right]$$

because of telescoping summands. For each i, the two arguments $\mathbf{x}_0+\mathbf{h}_i$ and $\mathbf{x}_0+\mathbf{h}_{i-1}$ differ only in their ith coordinates. Consequently

$$f(\mathbf{x}_0+\mathbf{h})-f(\mathbf{x}_0)=\sum_{i=1}^{n}\partial_i f(\mathbf{u}_i)h_i$$

where $\mathbf{u}_i$ is a point located by (1.12) for the ith summand and satisfying $\|\mathbf{u}_i-\mathbf{x}_0\| \leqq \|\mathbf{h}\|$. If we now add and subtract the following, which we know exists even though we have not completed showing that it is in the value of the derivative,

$$\langle \nabla f(\mathbf{x}_0), \mathbf{h}\rangle = \sum_{i=1}^{n} \partial_i f(\mathbf{x}_0) h_i$$

we obtain an expression for the numerator in (6.31)

$$f(\mathbf{x}_0+\mathbf{h})-f(\mathbf{x}_0)-\langle \nabla f(\mathbf{x}_0), \mathbf{h}\rangle = \sum_{i=1}^{n} \left[\partial_i f(\mathbf{u}_i)-\partial_i f(\mathbf{x}_0)\right] h_i$$

The RHS can be viewed as the value of an inner product, so from the Cauchy–Schwarz inequality of (3.45)

$$|f(\mathbf{x}_0+\mathbf{h})-f(\mathbf{x}_0)-\langle \nabla f(\mathbf{x}_0), \mathbf{h}\rangle| \leqq \|\mathbf{h}\| \left(\sum_{i=1}^{n} \left[\partial_i f(\mathbf{u}_i)-\partial_i f(\mathbf{x}_0)\right]^2 \right)^{\frac{1}{2}}$$

As $\mathbf{h}\to\mathbf{0}$ the terms $\partial_i f(\mathbf{u}_i)-\partial_i f(\mathbf{x}_0)\to 0$ because $\mathbf{u}_i\to\mathbf{x}_0$ and each $\partial_i f$ is continuous at $\mathbf{x}_0$. In other words, the limit (6.31) is zero for $F\mathbf{h}=\langle \nabla f(\mathbf{x}_0), \mathbf{h}\rangle$ for each $\mathbf{x}_0 \in D$. ■

If f satisfies the hypothesis of (6.45) it is called a *function of Class $C^{(1)}$* on D and we write $f\in C^{(1)}$. See 6.2.4 for additional terminology.

Let us now return to the consideration of transformations with values in $\mathcal{R}^m$ for any $m \geqq 1$. A transformation $f: D \subset \mathcal{R}^n \to \mathcal{R}^m$, where D is open, is *differentiable* at $\mathbf{x}_0 \in D$ if there exists a linear transformation $A: \mathcal{R}^n \to \mathcal{R}^m$ such that

$$\lim_{\mathbf{h}\to\mathbf{0}} \frac{f(\mathbf{x}_0+\mathbf{h})-f(\mathbf{x}_0)-A\mathbf{h}}{\|\mathbf{h}\|} = \mathbf{0} \tag{6.46}$$

If f is differentiable at every $\mathbf{x}\in D$ then f is *differentiable on D* and is a *differentiable transformation*. The linear transformation $A \in \mathcal{L}(\mathcal{R}^n, \mathcal{R}^m)$ is the *derivative* of f at $\mathbf{x}_0$ of *first order* and is also denoted by $f'(\mathbf{x}_0)$. If $m=1$ then (6.46) coincides with (6.31).

With the notable exception of (6.43), which does not extend without change to transformations (because it cannot in general be guaranteed that the same $\mathbf{u}$ will work for all m coordinate functions), most properties of differentiable functions do extend without change to differentiable transformations. For example (6.2.1)

A linear transformation is differentiable and coincides with its derivative. (6.47)

In particular, the preceding holds for the zero transformation. By (6.6) the limit (6.46) is **0** iff the limit of each coordinate function is zero and

The transformation $f: D \subset \mathcal{R}^n \to \mathcal{R}^m$, where D is open, is differentiable at $\mathbf{x}_0 \in D$ iff the coordinate functions f_i are differentiable at $\mathbf{x}_0$ for $i = 1,2,\ldots,m$. (6.48)

A transformation whose coordinate functions are functions of Class $C^{(1)}$ is called a *transformation of Class* $C^{(1)}$ on D and we write $f \in C^{(1)}$. From (6.45) we obtain the corresponding sufficient conditions

Transformations of Class $C^{(1)}$ are differentiable. (6.49)

These suffice for differentiability in this book. See 6.2.4 for additional terminology.

When we work with any particular differentiable transformation we generally use

The standard matrix of $A = f'(\mathbf{x}_0)$ is

$$\mathbf{f}'(\mathbf{x}_0) = \begin{bmatrix} \partial_1 f_1(\mathbf{x}_0) & \partial_2 f_1(\mathbf{x}_0) & \cdots & \partial_n f_1(\mathbf{x}_0) \\ \partial_1 f_2(\mathbf{x}_0) & \partial_2 f_2(\mathbf{x}_0) & \cdots & \partial_n f_2(\mathbf{x}_0) \\ \vdots & \vdots & & \vdots \\ \partial_1 f_m(\mathbf{x}_0) & \partial_2 f_m(\mathbf{x}_0) & \cdots & \partial_n f_m(\mathbf{x}_0) \end{bmatrix}$$

which is the *Jacobian matrix*, briefly *Jacobian*, of f. (6.50)

PROOF. If f is differentiable and $\mathbf{h} = t\mathbf{e}_j$ in (6.46), then as in (6.35) we must have

$$\lim_{t \to 0} \frac{f(\mathbf{x}_0 + t\mathbf{e}_j) - f(\mathbf{x}_0)}{t} = A\mathbf{e}_j$$

for $j = 1,2,\ldots,n$. The LHS is the vector of partial derivatives of the coordinate functions with respect to x_j, that is

$$A\mathbf{e}_j = \begin{bmatrix} \partial_j f_1(\mathbf{x}_0) \\ \partial_j f_2(\mathbf{x}_0) \\ \vdots \\ \partial_j f_m(\mathbf{x}_0) \end{bmatrix}$$

By (3.21) the standard matrix is the Jacobian $\mathbf{f}'(\mathbf{x}_0)$ and

$$\mathbf{h} \mapsto \mathbf{f}'(\mathbf{x}_0)\mathbf{h}$$

is the linear transformation A at $\mathbf{x}_0$. ■

In formulas for changes of variables in integration, the name Jacobian is often used for the determinant of $\mathbf{f}'(\mathbf{x}_0)$. Other authors sometimes give the name to the transpose of $\mathbf{f}'(\mathbf{x}_0)$. Notice however, even though the subscripts are undeniably "reversed" in our

$$\mathbf{f}'(\mathbf{x}_0) = [\,\partial_j f_i(\mathbf{x}_0)\,]$$

this is the matrix to which we are led by (3.21). If $n=1$ and $m>n$, then f specializes to a "vector-valued function of a real variable" which represents a "curve" in $\mathcal{R}^m$ and the Jacobian becomes the "tangent vector."

We next establish four results in preparation for (6.90). The first is

Suppose f is a differentiable function on an open convex set D and $\|\nabla f(\mathbf{x})\| \leqq M$ for $\mathbf{x} \in D$. Then

$$|f(\mathbf{x}) - f(\mathbf{y})| \leqq M\|\mathbf{x}-\mathbf{y}\|$$

whenever $\mathbf{x}, \mathbf{y} \in D$. (6.51)

PROOF. We can use the mean value theorem (6.43) to write

$$f(\mathbf{x}) - f(\mathbf{y}) = \langle \nabla f(\mathbf{y} + t(\mathbf{x}-\mathbf{y})), \mathbf{x}-\mathbf{y}\rangle$$

where $t \in (0,1)$ and the result follows from the Cauchy–Schwarz inequality of (3.45). ■

We need to use a matrix norm in the next three results and by agreement it is the standard natural norm.

Suppose $f: D \subset \mathcal{R}^n \to \mathcal{R}^m$ is differentiable at $\mathbf{x}_0 \in D$ where D is open. Then given $\varepsilon > 0$ there exists $\delta > 0$ such that $N_\delta(\mathbf{x}_0) \subset D$ and

$$\|f(\mathbf{x}_0+\mathbf{h}) - f(\mathbf{x}_0)\| \leqq (\|\mathbf{f}'(\mathbf{x}_0)\| + 1)\|\mathbf{h}\|$$

whenever $\|\mathbf{h}\| < \delta$. (6.52)

PROOF. We use $\varepsilon = \|\mathbf{h}\| > 0$ to find δ based on (6.46) such that

$$\|f(\mathbf{x}_0+\mathbf{h}) - f(\mathbf{x}_0) - \mathbf{f}'(\mathbf{x}_0)\mathbf{h}\| < \|\mathbf{h}\|$$

whenever $\|\mathbf{h}\| < \delta$. Then the result follows from starting with the triangle inequality on

$$\|f(\mathbf{x}_0+\mathbf{h}) - f(\mathbf{x}_0) - \mathbf{f}'(\mathbf{x}_0)\mathbf{h} + \mathbf{f}'(\mathbf{x}_0)\mathbf{h}\|$$

and ending with

$$\|\mathbf{f}'(\mathbf{x}_0)\mathbf{h}\| + \|\mathbf{h}\| \leqq (\|\mathbf{f}'(\mathbf{x}_0)\| + 1)\|\mathbf{h}\|$$

where the compatibility property (3.54) has been used. ■

Suppose $f \in C^{(1)}$ in (6.52). Then to $\varepsilon > 0$ there corresponds $N = N_\delta(\mathbf{x}_0) \subset D$ such that

$$\|f(\mathbf{x}) - f(\mathbf{z})\| \leqq (\|f'(\mathbf{x}_0)\| + \varepsilon)\|\mathbf{x} - \mathbf{z}\|$$

whenever $\mathbf{x}, \mathbf{z} \in N$. (6.53)

PROOF. First, let F denote the derivative $f'(\mathbf{x}_0)$, define $g = f - F$, which is differentiable and has the zero transformation as its derivative, and note the identity

$$f(\mathbf{x}) - f(\mathbf{z}) = [F\mathbf{x} - F\mathbf{z}] + [g(\mathbf{x}) - g(\mathbf{z})]$$

Second, use the continuity of the partial derivatives of g to determine N (which is convex by 6.1.2) such that the intermediate result

$$\|g(\mathbf{x}) - g(\mathbf{z})\| \leqq \sum_{i=1}^{n} \|g_i(\mathbf{x}) - g_i(\mathbf{z})\| \leqq \varepsilon\|\mathbf{x} - \mathbf{z}\|$$

whenever $\mathbf{x}, \mathbf{z} \in N$, results from (6.51) applied to the coordinate functions of g. Third, use 6.1.10(3) on the final term in the identity. ■

Suppose $m = n$ and $F = f'(\mathbf{x}_0)$ is nonsingular in (6.53). Then

$$(\|F^{-1}\|^{-1} - \varepsilon)\|\mathbf{x} - \mathbf{z}\| \leqq \|f(\mathbf{x}) - f(\mathbf{z})\|$$

whenever $\mathbf{x}, \mathbf{z} \in N$. (6.54)

PROOF. First, notice that $F(\mathbf{x} - \mathbf{z}) = F\mathbf{x} - F\mathbf{z}$ implies

$$\mathbf{x} - \mathbf{z} = F^{-1}(F\mathbf{x} - F\mathbf{z})$$

and then take the norm of each side. Second, consider the difference of the two RH members of the identity in the proof of (6.53) and then use the intermediate result in that proof. ■

In order to work with derivatives of higher order than the first we need the

Definition. For linear transformations

$$B \in \mathfrak{L}(\mathfrak{R}^n, \mathfrak{L}(\mathfrak{R}^n, \mathfrak{R}))$$

$$C \in \mathfrak{L}(\mathfrak{R}^n, \mathfrak{L}(\mathfrak{R}^n, \mathfrak{R}^m))$$

$$G \in \mathfrak{L}(\mathfrak{R}^n, \mathfrak{L}(\mathfrak{R}^n, \mathfrak{L}(\mathfrak{R}^n, \mathfrak{R})))$$

we define *standard tensors* to be designated partitioned matrices of orders $n\times n, mn\times n, n^2\times n$, respectively, for which rules can be found that correctly define linear transformations for $\mathbf{h},\mathbf{k}\in\mathcal{R}^n$

$$B\mathbf{h} \text{ and } B;\ C\mathbf{h} \text{ and } C;\ G\mathbf{h},\ G\mathbf{hk}, \text{ and } G$$

respectively. The *orders* of these tensors are the orders of the partitioned matrices. These tensors are *symmetric* if each successive block of order n—numbering in all $1, m, n$, respectively—is a symmetric matrix. The term *standard representation* for a linear transformation henceforth means standard matrix or standard tensor, whichever applies. (6.55)

Any partitioned matrix of order $p\times q$ represents a linear transformation from $\mathcal{R}^q$ to $\mathcal{R}^p$ according to the rules for (2.26); we find below that the rules in question for tensors are different. But in manipulations—transpositions, sums, scalar multiples, matrix products, and so on—tensors are treated as matrices.

Suppose $f: D\subset\mathcal{R}^n\to\mathcal{R}$ is differentiable on its open domain D. Then f is *twice-differentiable at* $\mathbf{x}_0\in D$ if there exists a linear transformation $\mathbf{h}\mapsto B\mathbf{h}$ $\in\mathcal{L}(\mathcal{R}^n,\mathcal{R})$ on $\mathcal{R}^n$ such that

$$\lim_{\mathbf{h}\to\mathbf{0}}\frac{f'(\mathbf{x}_0+\mathbf{h})-f'(\mathbf{x}_0)-B\mathbf{h}}{\|\mathbf{h}\|}=O \tag{6.56}$$

In words, the condition is that f' be differentiable at $\mathbf{x}_0$. Notice that each of the three terms in the numerator is represented by an $1\times n$ matrix and the same holds for the zero transformation on the RHS. We call $B\in\mathcal{L}(\mathcal{R}^n,\mathcal{L}(\mathcal{R}^n,\mathcal{R}))$ the *second derivative* of f or derivative of *second order* and it is also denoted by $f''(x_0)$. If f is twice-differentiable at every $\mathbf{x}\in D$ then f is *twice-differentiable on D* and is a *twice-differentiable function*. We write

$$\partial_j\partial_i f(\mathbf{x}_0)=\partial_j\left(\partial_i f(\mathbf{x}_0)\right)$$

for the familiar *partial derivatives of second order* which appear in the following result.

The standard tensor for $B=f''(\mathbf{x}_0)$ is of order $n\times n$

$$\mathbf{f}''(\mathbf{x}_0)=\begin{bmatrix}\partial_1\partial_1 f(\mathbf{x}_0) & \partial_2\partial_1 f(\mathbf{x}_0) & \cdots & \partial_n\partial_1 f(\mathbf{x}_0)\\ \partial_1\partial_2 f(\mathbf{x}_0) & \partial_2\partial_2 f(\mathbf{x}_0) & \cdots & \partial_n\partial_2 f(\mathbf{x}_0)\\ \vdots & \vdots & & \vdots\\ \partial_1\partial_n f(\mathbf{x}_0) & \partial_2\partial_n f(\mathbf{x}_0) & \cdots & \partial_n\partial_n f(\mathbf{x}_0)\end{bmatrix}$$

and is called the *Hessian* of f. The rules for $\mathbf{h},\mathbf{k}\in\mathfrak{R}^n$ are
(1) The standard matrix of $B\mathbf{h}\in\mathfrak{L}(\mathfrak{R}^n,\mathfrak{R})$ is $(\mathbf{f}''(\mathbf{x}_0)\mathbf{h})^T$.
(2) $B\mathbf{hk}=\mathbf{k}^T\mathbf{f}''(\mathbf{x}_0)\mathbf{h}$. (6.57)

(Since the Hessian can be identified with a matrix of order n in all respects except for the representation of the pertinent linear transformation, it is traditional to refer to it as a matrix.)

PROOF. If f is twice-differentiable and $\mathbf{h}=t\mathbf{e}_i$ in (6.56), then proceeding as for (6.50) we find the standard matrix of $B\mathbf{e}_i$ is

$$\left[\partial_i\partial_1 f(\mathbf{x}_0),\partial_i\partial_2 f(\mathbf{x}_0),\ldots,\partial_i\partial_n f(\mathbf{x}_0)\right]$$

and by linearity of B, the standard matrix of $B\mathbf{h}$ is

$$Bh_1\mathbf{e}_1+\cdots+Bh_n\mathbf{e}_n=\mathbf{h}^T(\mathbf{f}''(\mathbf{x}_0))^T$$

which verifies (1). Finally

$$(\mathbf{f}''(\mathbf{x}_0)\mathbf{h})^T\mathbf{k}=\mathbf{k}^T\mathbf{f}''(\mathbf{x}_0)\mathbf{h}$$

because it is a scalar. ■

The name Hessian is used by some authors for the transpose of $\mathbf{f}''(\mathbf{x}_0)$. In the next section it is advantageous to work with the Hessian as a partitioned matrix given in terms of its rows $\boldsymbol{\phi}_j(\mathbf{x})$ where $\phi_j(\mathbf{x})=\partial_j f(\mathbf{x})$ for $j=1,2,\ldots,n$.

In order to work with second derivatives of functions we need the next result (which holds under weaker assumptions). Calling $\partial_i\partial_j f$ a *mixed*, or *cross, partial derivative* when $i\neq j$ we have the

THEOREM. If $f:D\subset\mathfrak{R}^n\to\mathfrak{R}$ is of Class $C^{(1)}$ on its open domain D, and if the mixed partials $\partial_i\partial_j f$ and $\partial_j\partial_i f$ are continuous on D then $\partial_i\partial_j f=\partial_j\partial_i f$. (6.58)

PROOF. For convenience we consider $\mathbf{x}_0=\mathbf{0}\in D$. It is efficient to start with $n=2$ and it is also convenient to use the familiar notation (x,y) for points in $\mathfrak{R}^2$. Suppose $N_\delta(\mathbf{0})\subset D$ and for $0<h<\delta/\sqrt{2}$ consider the second difference quotient

$$u(h)=\frac{\left[f(h,h)-f(0,h)-f(h,0)+f(0,0)\right]}{h^2}$$

Then $v(h)=f(x,h)-f(x,0)$ is defined and differentiable for $(x,0)\in D$ and $u(h)=h^{-2}[v(h)-v(0)]$. Consequently, the mean value theorem (6.43) guarantees

$$u(h)=\frac{v'(x_1)}{h}=\frac{[\partial_1 f(x_1,h)-\partial_1 f(x_1,0)]}{h}$$

for some $x_1\in(0,h)$. Similarly,

$$u(h)=\partial_2\partial_1 f(x_1,y_1)$$

for some $y_1\in(0,h)$. Clearly, we can reverse the roles of x and y to show $u(h)=\partial_1\partial_2 f(x_2,y_2)$ where $x_2,y_2\in(0,h)$. Since both mixed partials are continuous, given any $\varepsilon>0$ we can find $\delta'\in(0,\delta)$ such that if $h\in(0,\delta'/\sqrt{2}\,)$ then

$$|\partial_2\partial_1 f(x_1,y_1)-\partial_2\partial_1 f(0,0)|<\varepsilon$$

and also

$$|\partial_1\partial_2 f(x_2,y_2)-\partial_1\partial_2 f(0,0)|<\varepsilon$$

We conclude $u(h)\to\partial_2\partial_1 f(0,0)=\partial_1\partial_2 f(0,0)$ as $h\to 0$. For $n>2$ we fix our attention on $i<j$ and apply what we have just proved to the restriction of f to $\mathcal{R}^2$, say

$$g(x,y)=f(\mathbf{t})$$

where $\mathbf{t}\leftrightarrow[x_{10},x_{20},\ldots,x,\ldots,y,\ldots,x_{n0}]^T$ has components x and y. ■

We can use the preceding theorem to establish equality of higher order mixed partial derivatives. Proceeding inductively from first and second orders, we define pth *order partial derivatives* to be the partial derivatives of the $(p-1)$st order partial derivatives for $p=2,3,\ldots.$ We use the notation

$$\partial_i\partial_j\cdots\partial_p f(x)$$

for $\partial_i(\partial_j\cdots\partial_p f(x_p))$ where $\partial_j\cdots\partial_p f(x_0)$ is the $(p-1)$st order partial derivative. Results that are in practice used for higher order differentiation often involve functions and transformations as follows.

Definition. Consider $f: D\subset\mathcal{R}^n\to\mathcal{R}$ where D is open. If all sth order partial derivatives exist, where $s\geqq 1$, for every $\mathbf{x}\in D$, and if each is continuous on D, then f is a *function of Class* $C^{(s)}$ and we write $f\in C^{(s)}$. A *transformation of Class* $C^{(s)}$ is a transformation, say g, all of whose coordinate functions are functions of Class $C^{(s)}$, and we write $g\in C^{(s)}$. (6.59)

See 6.2.4 for additional terminology and properties. One of the most commonly used results is the

THEOREM. If f is a function of Class $C^{(2)}$ then it is twice-differentiable on D and its Hessian is symmetric. (6.60)

PROOF. Differentiability of f' follows from (6.49) and the Hessian is as stated by (6.58). ■

A transformation $f: D \subset \mathfrak{R}^n \to \mathfrak{R}^m$, where D is open, is *twice-differentiable* at $\mathbf{x}_0 \in D$ if it is differentiable on D and there exists a linear transformation $\mathbf{h} \mapsto C\mathbf{h} \in \mathcal{L}(\mathfrak{R}^n, \mathfrak{R}^m)$ on $\mathfrak{R}^n$ such that

$$\lim_{\mathbf{h}\to\mathbf{0}} \frac{f'(\mathbf{x}_0+\mathbf{h}) - f'(\mathbf{x}_0) - C\mathbf{h}}{\|\mathbf{h}\|} = O \tag{6.61}$$

If $m=1$ the preceding coincides with (6.56). We call $C \in \mathcal{L}(\mathfrak{R}^n, \mathcal{L}(\mathfrak{R}^n, \mathfrak{R}^m))$ the *second derivative* of f or derivative of *second order*, and it is also denoted by $f''(\mathbf{x}_0)$. If f is twice-differentiable at every $\mathbf{x} \in D$ then f is a *twice-differentiable transformation*. In this case

The standard tensor for $C = f''(\mathbf{x}_0)$ is of order $mn \times n$

$$\mathbf{f}''(\mathbf{x}_0) = \begin{bmatrix} \mathbf{f}_1''(\mathbf{x}_0) \\ \mathbf{f}_2''(\mathbf{x}_0) \\ \vdots \\ \mathbf{f}_m''(\mathbf{x}_0) \end{bmatrix}$$

where $\mathbf{f}_i''(\mathbf{x}_0)$ is the Hessian of the ith coordinate function for $i = 1, 2, \ldots, m$. The rules for $\mathbf{h}, \mathbf{k} \in \mathfrak{R}^n$ are

(1) The standard matrix of $C\mathbf{h} \in \mathcal{L}(\mathfrak{R}^n, \mathfrak{R}^m)$ is $\begin{bmatrix} (\mathbf{f}_1''(\mathbf{x}_0)\mathbf{h})^T \\ (\mathbf{f}_2''(\mathbf{x}_0)\mathbf{h})^T \\ \vdots \\ (\mathbf{f}_m''(\mathbf{x}_0)\mathbf{h})^T \end{bmatrix}$

(2) $$C\mathbf{hk} = \begin{bmatrix} \mathbf{k}^T\mathbf{f}_1''(\mathbf{x}_0)\mathbf{h} \\ \mathbf{k}^T\mathbf{f}_2''(\mathbf{x}_0)\mathbf{h} \\ \vdots \\ \mathbf{k}^T\mathbf{f}_m''(\mathbf{x}_0)\mathbf{h} \end{bmatrix} \tag{6.62}$$

PROOF. Proceeding as for (6.57), the standard matrix of $C\mathbf{e}_i$ is

$$\partial_i \mathbf{f}'(\mathbf{x}_0)$$

which is the $m \times n$ matrix formed by taking the partial derivative $\partial/\partial x_i$ of each element in the $m \times n$ Jacobian $\mathbf{f}'(\mathbf{x}_0)$. By linearity of C, $C\mathbf{h}$ has the standard representation

$$h_1 \partial_1 \mathbf{f}'(\mathbf{x}_0) + \cdots + h_n \partial_n \mathbf{f}'(\mathbf{x}_0)$$

which coincides with the matrix in (1). Then (2) follows by multiplying the matrix in (1) times $\mathbf{k}$ and then transposing each scalar element. ■

Clearly we can proceed to successively higher derivatives following the established pattern. Let us introduce one more case. Suppose $f: D \subset \mathcal{R}^n \to \mathcal{R}$ is twice-differentiable on its open domain D. Then f is *thrice-differentiable* at $\mathbf{x}_0 \in D$ if there exists a linear transformation $G \in \mathcal{L}(\mathcal{R}^n, \mathcal{L}(\mathcal{R}^n, \mathcal{L}(\mathcal{R}^n, \mathcal{R})))$ such that

$$\lim_{\mathbf{h} \to \mathbf{0}} \frac{f''(\mathbf{x}_0 + \mathbf{h}) - f''(\mathbf{x}_0) - G\mathbf{h}}{\|\mathbf{h}\|} = O \tag{6.63}$$

We call $G = f'''(\mathbf{x}_0)$ the *third derivative* of f or derivative of *third order*. If f is thrice-differentiable at every $\mathbf{x} \in D$ then f is *thrice-differentiable* on D and is a *thrice-differentiable function*. We can continue the processes used to prove (6.35) and (6.57) to find a standard tensor for $f'''(\mathbf{x}_0)$ (6.2.6). We find the real number

$$((f'''(\mathbf{x}_0)\mathbf{h})\mathbf{h})\mathbf{h} = f'''(\mathbf{x}_0)\mathbf{hhh}$$

by a different method in (6.89) where it is used in a cubic Taylor formula.

Let us use the notation introduced for (6.10) and shown in Figure 6.1 where U and V are understood to be open. Then the next theorem states that the composition of two differentiable transformations is differentiable. Because of relations between the associated matrices this is called the

CHAIN RULE. Let β be differentiable at $\mathbf{u}_0 \in U$ with derivative B and let α be differentiable at $\mathbf{v}_0 = \beta(\mathbf{u}_0)$ with derivative A. Then the composite transformation $\gamma = \alpha(\beta)$ is differentiable at $\mathbf{u}_0$ with derivative $C = AB$ and the matrix representatives satisfy

$$\boldsymbol{\gamma}'(\mathbf{u}_0) = \boldsymbol{\alpha}'(\mathbf{v}_0)\, \boldsymbol{\beta}'(\mathbf{u}_0) \tag{6.64}$$

(In words, the composite transformation $\alpha(\beta(\cdot))$ has as its derivative the product of transformations $C = AB$ and its matrix is the corresponding matrix product.)

PROOF. We adopt the following notation for the numerators in the definitions (6.46)

$$b(\mathbf{h}) = \beta(\mathbf{u}_0 + \mathbf{h}) - \beta(\mathbf{u}_0) - B\mathbf{h}$$

$$a(\mathbf{k}) = \alpha(\mathbf{v}_0 + \mathbf{k}) - \alpha(\mathbf{v}_0) - A\mathbf{k}$$

and we write the corresponding expression for γ using AB in the derivative's place as follows

$$c(\mathbf{h}) = \gamma(\mathbf{u}_0 + \mathbf{h}) - \gamma(\mathbf{u}_0) - AB\mathbf{h}$$

We will prove $\|\mathbf{h}\|^{-1}c(\mathbf{h}) \to \mathbf{0}$ as $\mathbf{h} \to \mathbf{0}$ which establishes AB as the derivative according to (6.46). Adding and subtracting $\beta(\mathbf{u}_0 + \mathbf{h}) + \beta(\mathbf{u}_0)$, rearranging terms, substituting $\mathbf{v}_0 = \beta(\mathbf{u}_0)$, and using the linearity of A, we find

$$c(\mathbf{h}) = a(\beta(\mathbf{u}_0 + \mathbf{h}) - \beta(\mathbf{u}_0)) + Ab(\mathbf{h})$$

From (6.52) the first argument on the RHS satisfies

$$\|\beta(\mathbf{u}_0 + \mathbf{h}) - \beta(\mathbf{u}_0)\| \leqq (\|B\| + 1)\|\mathbf{h}\|$$

when $\|\mathbf{h}\|$ is small. Because α is differentiable,

$$\frac{a(\beta(\mathbf{u}_0 + \mathbf{h}) - \beta(\mathbf{u}_0))}{\|\beta(\mathbf{u}_0 + \mathbf{h}) - \beta(\mathbf{u}_0)\|} \to \mathbf{0} \text{ as } \mathbf{h} \to \mathbf{0}$$

Consequently, given $\varepsilon > 0$ then for $\|\mathbf{h}\|$ sufficiently small

$$\|a(\beta(\mathbf{u}_0 + \mathbf{h}) - \beta(\mathbf{u}_0))\| \leqq \varepsilon\|\beta(\mathbf{u}_0 + \mathbf{h}) - \beta(\mathbf{u}_0)\| \leqq \varepsilon(\|B\| + 1)\|\mathbf{h}\|$$

We conclude that the first part of $c(\mathbf{h})$ satisfies

$$\lim_{\mathbf{h} \to \mathbf{0}} \frac{a(\beta(\mathbf{u}_0 + \mathbf{h}) - \beta(\mathbf{u}_0))}{\|\mathbf{h}\|} = \mathbf{0}$$

and it remains to show

$$\lim_{\mathbf{h} \to \mathbf{0}} \frac{Ab(\mathbf{h})}{\|\mathbf{h}\|} = \mathbf{0}$$

This follows from (3.54) whereby

$$\|Ab(\mathbf{h})\| \leqq \|A\| \|b(\mathbf{h})\|$$

Thus, because b is differentiable, given $\varepsilon > 0$ then $\|b(\mathbf{h})\| < \varepsilon\|\mathbf{h}\|$ for $\|\mathbf{h}\|$ sufficiently small. We conclude that the derivative of γ is AB and then the matrix representation is the stated product. ■

As a simple application of the chain rule let us show that the derivative must be zero at any interior point where a differentiable function has a minimum.

If f is differentiable at $\mathbf{x}_0$ and for some $\delta > 0$

$$f(\mathbf{x}_0) = \min\{f(\mathbf{x}) : \mathbf{x} \in N_\delta(\mathbf{x}_0)\}$$

then $\mathbf{f}'(\mathbf{x}_0) = \mathbf{O}$. (6.65)

PROOF. The function defined on $\mathcal{R}$ by

$$g(t) = f(\mathbf{x}_0 + t\mathbf{h})$$

where $\|\mathbf{h}\| = 1$ but is otherwise arbitrary, has a relative minimum at $t = 0$. By the chain rule

$$g'(0) = \langle \nabla f(\mathbf{x}_0), \mathbf{h} \rangle$$

and it must equal zero because it cannot be positive or negative. ■

EXAMPLE 6.2. Two Particular Cases of the Chain Rule. The first case can aptly be referred to as a

(1) Function of a Transformation. Consider

$$h : U \subset \mathcal{R}^n \to \mathcal{R}$$

with $h(\mathbf{x}) = f(g(\mathbf{x}))$ where $g : U \subset \mathcal{R}^n \to V$ and $f : V \subset \mathcal{R}^m \to \mathcal{R}$ are the transformation and function, respectively.

We use (6.64) by associating α, β, γ, $\mathbf{u}$, $\mathbf{v}$, $\mathbf{w}$, p, q, and r in Figure 6.1 with

$$f,\ g,\ h,\ \mathbf{x},\ \mathbf{y},\ \mathbf{z},\ n,\ m, \text{ and } 1$$

respectively. When the chain rule applies, the derivative

$$h'(\mathbf{x}_0) = f'(\mathbf{y}_0)\, g'(\mathbf{x}_0)$$

is represented by the $1 \times n$ matrix

$$\left[\partial_1 h(\mathbf{x}_0), \ldots, \partial_n h(\mathbf{x}_0)\right]$$
$$= \left[\partial_1 f(\mathbf{y}_0), \partial_2 f(\mathbf{y}_0), \ldots, \partial_m f(\mathbf{y}_0)\right] \begin{bmatrix} \partial_1 g_1(\mathbf{x}_0) & \cdots & \partial_n g_1(\mathbf{x}_0) \\ \partial_1 g_2(\mathbf{x}_0) & \cdots & \partial_n g_2(\mathbf{x}_0) \\ \vdots & & \vdots \\ \partial_1 g_m(\mathbf{x}_0) & \cdots & \partial_n g_m(\mathbf{x}_0) \end{bmatrix}$$

The jth entry, that is, the element in the $(1,j)$th position is

$$\partial_j h(\mathbf{x}_0) = \partial_1 f(\mathbf{y}_0)\partial_j g_1(\mathbf{x}_0) + \partial_2 f(\mathbf{y}_0)\partial_j g_2(\mathbf{x}_0) + \cdots + \partial_m f(\mathbf{y}_0)\partial_j g_m(\mathbf{x}_0)$$

In the notation $(\partial/\partial x_j)h = \partial_j h$, this becomes

$$\frac{\partial h}{\partial x_j} = \frac{\partial f}{\partial y_1}\frac{\partial g_1}{\partial x_j} + \frac{\partial f}{\partial y_2}\frac{\partial g_2}{\partial x_j} + \cdots + \frac{\partial f}{\partial y_m}\frac{\partial g_m}{\partial x_j}$$

which we recognize as the familiar "telescoping" chain rule from calculus because $g(\mathbf{x}) = \mathbf{y}$.

If we retain the associations for α, β, γ but consider $p=1, q=n, r=m$ then we obtain the second case which is a particular kind of

(2) Transformation of a Transformation. Consider

$$h: U \subset \mathfrak{R} \to \mathfrak{R}^m$$

with $h(x) = f(g(x))$ where $g: U \subset \mathfrak{R} \to V$ and $f: V \subset \mathfrak{R}^n \to \mathfrak{R}^m$ are the two transformations.

When the chain rule applies, the derivative

$$h'(x_0) = f'(\mathbf{y}_0)\, g'(x_0)$$

is represented by the $m \times 1$ matrix

$$\begin{bmatrix} h_1'(x_0) \\ h_2'(x_0) \\ \vdots \\ h_m'(x_0) \end{bmatrix} = \begin{bmatrix} \partial_1 f_1(\mathbf{y}_0) & \cdots & \partial_n f_1(\mathbf{y}_0) \\ \partial_1 f_2(\mathbf{y}_0) & \cdots & \partial_n f_2(\mathbf{y}_0) \\ \vdots & & \vdots \\ \partial_1 f_m(\mathbf{y}_0) & & \partial_n f_m(\mathbf{y}_0) \end{bmatrix} \begin{bmatrix} g_1'(x_0) \\ g_2'(x_0) \\ \vdots \\ g_n'(x_0) \end{bmatrix}$$

where because $x_0 \in \mathfrak{R}$, the matrices for h' and g' can be associated with vectors.

Use of the chain rule for higher derivatives is illustrated in Example 6.4.

PROBLEMS FOR SOLUTION

6.2.1 Fundamental Properties. Establish for $f: D \subset \mathfrak{R}^n \to \mathfrak{R}^m$ where D is open:

(1) If f is differentiable at $\mathbf{x}_0$ then its derivative is unique.

(2) A constant transformation has zero derivative.

(3) A linear transformation is differentiable and coincides with its derivative.

(4) Sums and inner products of differentiable transformations are differentiable.

(5) A differentiable transformation is continuous.

(6) The Jacobian (6.50) can be readily established from first principles without appeal to (3.21).

(7) The particular norms used for defining the limits in (6.31), (6.46), (6.56), (6.61), and (6.63) are as follows: ________.

(8) The function $f: D \subset \mathcal{R}^n \to \mathcal{R}$ is twice-differentiable at $\mathbf{x}_0 \in D$, where D is open, iff the *quadratic Taylor formula*

$$f(\mathbf{x}_0 + \mathbf{h}) = f(\mathbf{x}_0) + \mathbf{f}'(\mathbf{x}_0)\mathbf{h} + \frac{1}{2}\mathbf{h}^T\mathbf{f}''(\mathbf{x}_0)\mathbf{h} + o(\|\mathbf{h}\|^2)$$

holds for all $\mathbf{h}$ of sufficiently small norm.

(9) If $f' = O$ where D is also connected, then f is a constant transformation.

6.2.2 Particular Functions. Find the directional derivative at $\mathbf{0}$ in the direction $\mathbf{u}$, determine if the function is differentiable, and find the Hessians, tangent hyperplanes and normals, where such exist, for the following.

(1) $\mathbf{x} \mapsto x_1 x_2^2 x_3^3$ on $\mathcal{R}^3$.

(2) Repeat for $\mathcal{R}^4$.

(3) $\mathbf{x} \mapsto x_1^2 - 3x_1x_2 + x_2^2$ on $\mathcal{R}^2$.

(4) $\mathbf{x} \mapsto 1 - \exp[-(x_1 + x_2)^2]$ on $\mathcal{R}^2$.

(5) $\mathbf{x} \mapsto 1 - \exp[x_1x_2x_3]$ on $\mathcal{R}^3$.

(6) $\mathbf{x} \mapsto \langle \mathbf{a}, \mathbf{x} \rangle$ on $\mathcal{R}^n$.

(7) $\mathbf{x} \mapsto \langle \mathbf{x}, \mathbf{x} \rangle$ on $\mathcal{R}^n$.

(8) $\mathbf{x} \mapsto \mathbf{x}^T\mathbf{A}\mathbf{x}$ on $\mathcal{R}^n$ where $\mathbf{A} = \mathbf{A}^T$.

(9) $\mathbf{x} \mapsto x_1x_2(x_1^2 + x_2^2)^{-1}$ for $\mathbf{x} \neq \mathbf{0} \in \mathcal{R}^2$ and $\mathbf{0} \mapsto 0$.

6.2.3 Particular Transformations. Find Jacobians and standard tensors of the following.

(1) $\mathbf{x} \mapsto [x_1 + x_2, x_1 - x_2]^T$.

(2) $\mathbf{x} \mapsto [x_1 + x_2, x_1x_3]^T$.

(3) $\mathbf{x} \mapsto [x_1x_3, x_1 + x_2 + x_3]^T$.

(4) $\mathbf{x} \mapsto [(x_1^2 - x_2^2)/2, x_2x_3]^T$.

(5) $\mathbf{x} \mapsto [x_1 + x_2 + x_3, x_1x_2 + x_1x_3 + x_2x_3, x_1x_2x_3]^T$.

(6) $\mathbf{x} \mapsto [x_1r^{-1}, x_2r^{-1}, x_3r^{-1}]^T$ where $r = \langle \mathbf{x}, \mathbf{x} \rangle$.

(7) $\mathbf{x} \mapsto [e^{x_1}\cos x_2, e^{x_1}\sin x_2]^T$.

6.2.4 Continuous Differentiability. Consider a differentiable transformation $f: D \subset \mathcal{R}^n \to \mathcal{R}^m$ where D is open. If $\mathbf{x}_0 \in D$ then the particular derivative $f'(x_0)$ is continuous on $\mathcal{R}^n$ because it is a linear transformation. A different concept is

The differentiable transformation f is *continuously differentiable* if the associated transformation

$$\mathbf{x} \mapsto f'(\mathbf{x}) \in \mathfrak{L}(\mathfrak{R}^n, \mathfrak{R}^m)$$

is continuous on D.

By definition, f is continuously differentiable iff

For each $\mathbf{x} \in D$ and $\varepsilon > 0$ there exists $\delta > 0$ such that

$$\|f'(\mathbf{y}) - f'(\mathbf{x})\| < \varepsilon$$

whenever $\mathbf{y} \in N_\delta(\mathbf{x}) \subset D$.

Below (6.38) we find that existence of directional derivatives in every direction (thus including all directions $\mathbf{e}_i$ to which correspond $\partial_i f$) is not sufficient to guarantee differentiability. However, the following shows that the hypotheses of (6.45) guarantee more than differentiability. Establish:

(1) A necessary and sufficient condition that f be continuously differentiable is $f \in C^{(1)}$.

(This justifies the common practice of defining f to be continuously differentiable if $f \in C^{(1)}$.)

(2) The preceding extends to twice-continuously differentiable functions and, in general, to s-times continuously differentiable functions which can be identified with Classes $C^{(s)}$ for $s = 1, 2, 3, \ldots$.

(3) Show that $f \in C^{(s)}$ implies $f \in C^{(s-1)}$ for $s = 1, 2, 3, \ldots$, where $f \in C^{(0)}$ means f is continuous.

6.2.5 Gradients, Jacobians, and Hessians. Establish the following where, for brevity, it is supposed that $f \in C^{(2)}$ on the open set D.

(1) Gradients, Jacobians, and Hessians of f can be used to define linear transformations on D.

(2) The Hessian is the "Jacobian of the gradient."

(3) The vector ∇f and its transpose $\mathbf{f}'$ "have the same derivative."

(4) Define $\nabla^2 f(\mathbf{x}) = \nabla(\nabla f(\mathbf{x}))$ and relate to $\mathbf{f}''(\mathbf{x})$.

(5) Write the scalars $f'(\mathbf{x}_0)\mathbf{h}$ and $(f''(\mathbf{x}_0)\mathbf{h})\mathbf{h}$ as inner products using ∇ and ∇^2.

6.2.6 Thrice-Differentiable Functions. Establish: (1) A standard tensor for G in (6.63). (2) An $o(\|\mathbf{h}\|^3)$ Taylor formula for $f(\mathbf{x} + \mathbf{h})$.

6.2.7 Chain Rule Theorem. Establish: If α and β belong to their respective classes $C^{(s)}$ in (6.63) where $s \geqq 1$, then $\gamma \in C^{(s)}$.

6.3 DERIVATIVES OF VECTORS AND MATRICES

A special kind of matrix multiplication can be used to find derivatives. The main result is (6.87) which is a systematic procedure for using the chain rule. Results obtained in this section are widely used—for example, in Chapter 7 and in statistics—and are difficult to obtain without special methods.

If $\mathbf{x} \mapsto z \in \mathfrak{R}$ is a differentiable function on some open $D \subset \mathfrak{R}^n$ we use (6.35) to write

$$\frac{dz}{d\mathbf{x}} = [\partial_1 z, \partial_2 z, \ldots, \partial_n z] \tag{6.66}$$

and call it the *derivative of the scalar z with respect to* $\mathbf{x} \in \mathfrak{R}^n$, or the result of *differentiating z*. If $\mathbf{x} \mapsto \mathbf{y} \in \mathfrak{R}^r$ is a differentiable transformation we similarly use (6.50) and our notation for components to write

$$\frac{d\mathbf{y}}{d\mathbf{x}} = \begin{bmatrix} \partial_1 y_1 & \partial_2 y_1 & \cdots & \partial_n y_1 \\ \partial_1 y_2 & \partial_2 y_2 & \cdots & \partial_n y_2 \\ \vdots & \vdots & & \vdots \\ \partial_1 y_r & \partial_2 y_r & \cdots & \partial_n y_r \end{bmatrix} \tag{6.67}$$

and call it the *derivative of the vector* $\mathbf{y}$ *with respect to* $\mathbf{x} \in \mathfrak{R}^n$, or the result of *differentiating* $\mathbf{y}$. In the same way, if we associate

$$\mathbf{x} \mapsto \mathbf{A}$$

where $\mathbf{A}$ is $r \times s$, with a differentiable transformation

$$\mathbf{x} \mapsto A \in \mathcal{L}(\mathfrak{R}^s, \mathfrak{R}^r)$$

then we use (6.62) to write

$$\frac{d\mathbf{A}}{d\mathbf{x}} = \begin{bmatrix} \partial_1 a_{11} & \partial_2 a_{11} & \cdots & \partial_n a_{11} \\ \partial_1 a_{12} & \partial_2 a_{12} & \cdots & \partial_n a_{12} \\ \vdots & \vdots & & \vdots \\ \partial_1 a_{1s} & \partial_2 a_{1s} & \cdots & \partial_n a_{1s} \\ \vdots & \vdots & & \vdots \\ \partial_1 a_{r1} & \partial_2 a_{r1} & \cdots & \partial_n a_{r1} \\ \partial_1 a_{r2} & \partial_2 a_{r2} & \cdots & \partial_n a_{r2} \\ \vdots & \vdots & & \vdots \\ \partial_1 a_{rs} & \partial_2 a_{rs} & \cdots & \partial_n a_{rs} \end{bmatrix} \tag{6.68}$$

and call it the *derivative of the matrix* **A** *with respect to* $\mathbf{x} \in \mathfrak{R}^n$, or the result of *differentiating* **A**. We also use the notation $z'(\mathbf{x})$, $\mathbf{y}'(\mathbf{x})$, and $\mathbf{A}'(\mathbf{x})$ for the derivatives just introduced, and for f: $D \subset \mathfrak{R}^n \to \mathfrak{R}$ we sometimes find it convenient to use

$$\frac{df}{d\mathbf{x}}$$

to denote the standard matrix $\mathbf{f}'$ of the linear transformation f'.

The rules are straightforward. In (6.66) we successively differentiate z with respect to $x_1, x_2, \ldots, x_n$ and insert the partial derivatives as successive components of a single row. In (6.67) we successively differentiate scalars

$$y_1, y_2, \ldots, y_r$$

and insert the $1 \times n$ derivatives as successive rows of $d\mathbf{y}/d\mathbf{x}$. In (6.68) we also successively differentiate scalars moving from left to right and downward

$$a_{11}, a_{12}, \ldots, a_{1s}, \ldots, a_{r1}, a_{r2}, \ldots, a_{rs}$$

and again insert the $1 \times n$ derivatives as successive rows of $d\mathbf{A}/d\mathbf{x}$. We can also describe (6.68) as differentiating successive rows of **A** and inserting the $s \times n$ derivatives as successive blocks in $d\mathbf{A}/d\mathbf{x}$. These rules make clear the phenomena (6.2.5).

$$\frac{d\mathbf{y}}{d\mathbf{x}} = \frac{d\mathbf{y}^T}{d\mathbf{x}} \quad \text{but} \quad \frac{d\mathbf{A}}{d\mathbf{x}} \neq \frac{d\mathbf{A}^T}{d\mathbf{x}}$$

unless, for example, $\mathbf{A} = \mathbf{A}^T$ so that the same succession of objects is encountered in "traversing" the two matrices.

Higher derivatives are defined as in elementary calculus. For example, if $\mathbf{x} \mapsto z$ is twice-differentiable we can differentiate the matrix (6.66) using the rule for (6.68) to obtain

$$\frac{d^2 z}{d\mathbf{x}^2} = \begin{bmatrix} \partial_1 \partial_1 z & \partial_2 \partial_1 z & \cdots & \partial_n \partial_1 z \\ \partial_1 \partial_2 z & \partial_2 \partial_2 z & \cdots & \partial_n \partial_2 z \\ \vdots & \vdots & & \vdots \\ \partial_1 \partial_n z & \partial_2 \partial_n z & \cdots & \partial_n \partial_n z \end{bmatrix}$$

as the *second derivative of the scalar* z *with respect to* $\mathbf{x} \in \mathfrak{R}^n$.

Work with derivatives with respect to $\mathbf{x} \in \mathfrak{R}^n$ is greatly facilitated by the use of a special kind of matrix multiplication. If $\mathbf{A} = [a_{ij}]$ is $r \times s$ and $\mathbf{B} = [b_{ij}]$ is $t \times u$ then the *Kronecker product* $\mathbf{A} \otimes \mathbf{B}$ is the matrix of order

$rt \times su$ formed by replacing each a_{ij} by $a_{ij}\mathbf{B}$

$$\mathbf{A}\otimes\mathbf{B}=\begin{bmatrix} a_{11}\mathbf{B} & a_{12}\mathbf{B} & \cdots & a_{1s}\mathbf{B} \\ a_{21}\mathbf{B} & a_{22}\mathbf{B} & \cdots & a_{2s}\mathbf{B} \\ \vdots & \vdots & & \vdots \\ a_{r1}\mathbf{B} & a_{r2}\mathbf{B} & \cdots & a_{rs}\mathbf{B} \end{bmatrix} \tag{6.69}$$

There are no conformability requirements. In general, $\mathbf{A}\otimes\mathbf{B}\neq\mathbf{B}\otimes\mathbf{A}$ as shown by the following examples.

$$\mathbf{A}=\begin{bmatrix} 1 & 2 \\ 3 & 4 \end{bmatrix}, \qquad \mathbf{B}=[5,6]$$

$$\mathbf{A}\otimes\mathbf{B}=\begin{bmatrix} 5 & 6 & 10 & 12 \\ 15 & 18 & 20 & 24 \end{bmatrix}, \qquad \mathbf{B}\otimes\mathbf{A}=\begin{bmatrix} 5 & 10 & 6 & 12 \\ 15 & 20 & 18 & 24 \end{bmatrix}$$

Table 6.1 lists fundamental properties (6.3.1).

We use $\mathbf{A}\otimes\mathbf{B}$ as a tool for "escalating" vectors to matrices and matrices to tensors. Suppose $\mathbf{A}$ is $r\times s$ and specified in terms of its columns $\mathbf{a}_j\in\mathcal{R}^r$ as

$$\mathbf{A}=[\mathbf{a}_1,\mathbf{a}_2,\ldots,\mathbf{a}_s] \tag{6.70}$$

Then since, for example, $[\mathbf{a}_1,\mathbf{0},\ldots,\mathbf{0}]=\mathbf{a}_1\otimes\mathbf{e}_1^T$ where $\mathbf{e}_1\in\mathcal{R}^s$, we have

$$\mathbf{A}=\sum_{j=1}^{s}(\mathbf{a}_j\otimes\mathbf{e}_j^T) \tag{6.71}$$

which is a special case of (2.18) because the jth summand also equals

$$\mathbf{e}_j^T\otimes\mathbf{a}_j=\mathbf{a}_j\mathbf{e}_j^T$$

TABLE 6.1 PROPERTIES OF KRONECKER PRODUCTS

	Definition (6.69) $\mathbf{A}\otimes\mathbf{B}=[a_{ij}\mathbf{B}]$
(1)	$c\mathbf{A}\otimes\mathbf{B}=\mathbf{A}\otimes c\mathbf{B}=c(\mathbf{A}\otimes\mathbf{B})$
(2)	$\mathbf{A}\otimes(\mathbf{B}\otimes\mathbf{C})=(\mathbf{A}\otimes\mathbf{B})\otimes\mathbf{C}$
(3)	$(\mathbf{A}\otimes\mathbf{B})^T=\mathbf{A}^T\otimes\mathbf{B}^T$
(4)	If $\mathbf{A}+\mathbf{B}$ and $\mathbf{C}+\mathbf{D}$ are defined, then $(\mathbf{A}+\mathbf{B})\otimes(\mathbf{C}+\mathbf{D})=(\mathbf{A}\otimes\mathbf{C})+(\mathbf{A}\otimes\mathbf{D})+(\mathbf{B}\otimes\mathbf{C})+(\mathbf{B}\otimes\mathbf{D})$
(5)	If $\mathbf{AC}$ and $\mathbf{BD}$ are defined, then $(\mathbf{A}\otimes\mathbf{B})(\mathbf{C}\otimes\mathbf{D})=\mathbf{AC}\otimes\mathbf{BD}$
(6)	$\mathbf{a}^T\otimes\mathbf{b}=\mathbf{b}\otimes\mathbf{a}^T=\mathbf{ba}^T$

Using (6.71) and (3) in Table 6.1 we obtain

$$\mathbf{A}^T = \sum_{j=1}^{s} (\mathbf{a}_j^T \otimes \mathbf{e}_j) \tag{6.72}$$

in which we see that the Kronecker product inserts $\mathbf{a}_j^T$ as the jth row in $\mathbf{A}^T$.

These formulas make it clear how we can write derivatives of matrices in terms of derivatives of rows or columns. Our usual notation for the jth column of $\mathbf{A}$ is

$$\mathbf{a}_j = [a_{1j}, a_{2j}, \ldots, a_{rj}]^T$$

and its derivative can therefore be written

$$\frac{d\mathbf{a}_j}{d\mathbf{x}} = \sum_{i=1}^{r} \left(\frac{da_{ij}}{d\mathbf{x}} \otimes \mathbf{e}_i \right) \tag{6.73}$$

where $\mathbf{e}_i \in \mathfrak{R}^r$. In much the same fashion, using $\mathbf{e}_j \in \mathfrak{R}^s$,

$$\frac{d\mathbf{A}}{d\mathbf{x}} = \sum_{j=1}^{s} \left(\frac{d\mathbf{a}_j}{d\mathbf{x}} \otimes \mathbf{e}_j \right) \tag{6.74}$$

because the r rows of the derivative of $\mathbf{a}_j$ occupy rows $j, s+j, \ldots, (r-1)s+j$ in the $rs \times n$ expression for $(d/d\mathbf{x})\mathbf{A}$. The next result covers

SCALAR MULTIPLES. Suppose $c \in \mathfrak{R}$, $\mathbf{y} \in \mathfrak{R}^r$, and $\mathbf{A}$ is $r \times s$ as specified in (6.70). Then

(1) $\dfrac{d}{d\mathbf{x}}(c\mathbf{y}) = c\dfrac{d\mathbf{y}}{d\mathbf{x}} + \mathbf{y}\dfrac{dc}{d\mathbf{x}}$

(2) $\dfrac{d}{d\mathbf{x}}(c\mathbf{A}) = c\dfrac{d\mathbf{A}}{d\mathbf{x}} + \left\{ \dfrac{dc}{d\mathbf{x}} \otimes \displaystyle\sum_{j=1}^{s} (\mathbf{a}_j \otimes \mathbf{e}_j) \right\}$

where $\mathbf{e}_j \in \mathfrak{R}^s$. (6.75)

PROOF. Formula (1) results directly from the scalar relation

$$\partial_k (ca_{ij}) = c\partial_k a_{ij} + a_{ij}\partial_k c$$

Using (1), and (6.74) applied to $c\mathbf{A} = [c\mathbf{a}_1, \ldots, c\mathbf{a}_s]$ as well as to $\mathbf{A}$, we obtain (2) with relations in Table 6.1. ■

Let us now turn to matrix products. The first result we need is the following where **A** continues to be $r\times s$ and $\mathbf{b}_j\in\mathfrak{R}^s$.

$$\frac{d\mathbf{A}\mathbf{b}_j}{d\mathbf{x}}=\mathbf{A}\frac{d\mathbf{b}_j}{d\mathbf{x}}+\left(\mathbf{I}_r\otimes\mathbf{b}_j^T\right)\frac{d\mathbf{A}}{d\mathbf{x}} \tag{6.76}$$

PROOF. If we write $\mathbf{A}\mathbf{b}_j=b_{1j}\mathbf{a}_1+\cdots+b_{sj}\mathbf{a}_s$ and differentiate the RHS we obtain

$$\frac{d\mathbf{A}\mathbf{b}_j}{d\mathbf{x}}=\sum_{k=1}^{s}\left(\mathbf{a}_k\frac{db_{kj}}{d\mathbf{x}}+b_{kj}\frac{d\mathbf{a}_k}{d\mathbf{x}}\right)$$

From the first summand we obtain $\mathbf{A}(d/d\mathbf{x})\mathbf{b}_j$ according to (2.18). Using (5) in Table 6.1 with $\mathbf{e}_k\in\mathfrak{R}^s$, the second summand equals

$$\frac{d\mathbf{a}_k}{d\mathbf{x}}\otimes\mathbf{b}_j^T\mathbf{e}_k=\left(\mathbf{I}_r\otimes\mathbf{b}_j'\right)\left(\frac{d\mathbf{a}_k}{d\mathbf{x}}\otimes\mathbf{e}_k\right)$$

whereupon we verify the second term in the result by extending (4) in Table 6.1 and using (6.73). ■

We need one more preliminary result, namely

If $\mathbf{c}_k\in\mathfrak{R}^r$ and $\mathbf{e}_k\in\mathfrak{R}^t$ then

$$\frac{d}{d\mathbf{x}}\left(\mathbf{c}_k\otimes\mathbf{e}_k^T\right)=\frac{d\mathbf{c}_k}{d\mathbf{x}}\otimes\mathbf{e}_k \tag{6.77}$$

PROOF. Use (6.74) together with the fact that every column of $\mathbf{c}_k\otimes\mathbf{e}_k^T$ is zero except the kth. ■

Now we can prove the main result which, as explained below, covers essentially all derivatives of interest in this book.

MATRIX PRODUCTS. If **A** is $r\times s$ and **B** is $s\times t$ then

$$\frac{d}{d\mathbf{x}}\mathbf{A}\mathbf{B}=(\mathbf{A}\otimes\mathbf{I}_t)\frac{d\mathbf{B}}{d\mathbf{x}}+\left(\mathbf{I}_r\otimes\mathbf{B}^T\right)\frac{d\mathbf{A}}{d\mathbf{x}} \tag{6.78}$$

PROOF. The derivative of an appropriate expression for **AB** is

$$\frac{d}{d\mathbf{x}}\sum_{k=1}^{t}\left(\mathbf{A}\mathbf{b}_k\otimes\mathbf{e}_k^T\right)=\sum_{k=1}^{t}\left(\frac{d\mathbf{A}\mathbf{b}_k}{d\mathbf{x}}\otimes\mathbf{e}_k\right)$$

where $\mathbf{b}_k\in\mathfrak{R}^s$ is the kth column of **B** and $\mathbf{e}_k\in\mathfrak{R}^t$. Using (6.77) and (6.76) to write the kth summand on the RHS and then expanding using mainly

(4) and (5) of Table 6.1, we obtain

$$\left(\mathbf{A}\frac{d\mathbf{b}_k}{d\mathbf{x}}\otimes\mathbf{e}_k\right)+\left\{\left[(\mathbf{I}_r\otimes\mathbf{b}_k^T)\frac{d\mathbf{A}}{d\mathbf{x}}\right]\otimes\mathbf{e}_k\right\}$$

Summing the first term after rewriting again with (5) we have

$$\sum_{k=1}^{t}(\mathbf{A}\otimes\mathbf{I}_t)\left(\frac{d\mathbf{b}_k}{d\mathbf{x}}\otimes\mathbf{e}_k\right)=(\mathbf{A}\otimes\mathbf{I}_t)\frac{d\mathbf{B}}{d\mathbf{x}}$$

according to (6.74). In much the same way, the second summation is

$$\sum_{k=1}^{t}\left\{(\mathbf{I}_r\otimes\mathbf{b}_k^T)\otimes\mathbf{e}_k\right\}\frac{d\mathbf{A}}{d\mathbf{x}}=\mathbf{I}_r\otimes\sum_{k=1}^{t}(\mathbf{b}_k^T\otimes\mathbf{e}_k)\frac{d\mathbf{A}}{d\mathbf{x}}$$

and the second term in the stated result follows using (6.72) applied to **B**. ■

Given that we know how to differentiate scalars, vectors, and matrices, (6.78) suffices for our purposes because everything we need to differentiate can be expressed as products of matrices. But there is one caution. The formula in (6.78) does not apply to scalar multiples

$$c\mathbf{y} \text{ and } c\mathbf{A}$$

because the factors c and $\mathbf{y}$, and also c and $\mathbf{A}$ are not conformable. We can rewrite these expressions so (6.78) does apply, but usually (6.75) is more convenient (6.3.6).

Table 6.2 displays derivatives of expressions involving $\mathbf{a}\in\mathfrak{R}^n$ and $\mathbf{A}$ $m\times n$ such that $d\mathbf{a}/d\mathbf{x}$ and $d\mathbf{A}/d\mathbf{x}$ are zero matrices, that is, $\mathbf{a}$ and $\mathbf{A}$, and also $\mathbf{B}$ and $\mathbf{C}$, are constants (6.3.4).

If $\mathbf{A}_k$ is of order n for $k=1,2,\ldots,m$ we can identify the partitioned matrix

$$\mathbf{A}=\begin{bmatrix}\mathbf{A}_1\\ \mathbf{A}_2\\ \vdots\\ \mathbf{A}_m\end{bmatrix} \tag{6.79}$$

with a *tensor* of order $mn\times n$, or $n^2\times n$ if $m=n$. Then we can associate $\mathbf{x}\mapsto\mathbf{A}$ with a differentiable transformation

$$\mathbf{x}\mapsto A\in\mathcal{L}(\mathfrak{R}^n,\mathcal{L}(\mathfrak{R}^n,\mathfrak{R}^m)) \tag{6.80}$$

TABLE 6.2 DERIVATIVES INVOLVING CONSTANTS

Object		Derivative
(1)	$\mathbf{a}$ or $\mathbf{a}^T$	$\mathbf{O}_n$
(2)	$\mathbf{A}$ or $\mathbf{A}^T$	$\mathbf{O}_{mn,n}$
(3)	$\mathbf{x}$ or $\mathbf{x}^T$	$\mathbf{I}_n$
(4)	$\mathbf{a}^T\mathbf{x}$	$\mathbf{a}^T$
(5)	$\mathbf{A}\mathbf{x}$	$\mathbf{A}$
(6)	$\mathbf{x}^T\mathbf{B}$	$\mathbf{B}^T$
(7)	$\mathbf{x}^T\mathbf{x}$	$2\mathbf{x}^T$
(8)	$\mathbf{x}^T\mathbf{C}\mathbf{x}$	$\mathbf{x}^T\mathbf{C}+\mathbf{x}^T\mathbf{C}^T$
(9)	$\mathbf{x}\mathbf{x}^T$	$(\mathbf{x}\otimes\mathbf{I}_n)+(\mathbf{I}_n\otimes\mathbf{x})$
(10)	$\mathbf{x}\mathbf{a}^T$	$\mathbf{I}_n\otimes\mathbf{a}$
(11)	$\mathbf{a}\mathbf{x}^T$	$\mathbf{a}\otimes\mathbf{I}_n$

$\mathbf{a},\mathbf{x}\in\mathfrak{R}^n$, $\mathbf{A}$ is $m\times n$, $\mathbf{B}$ is $n\times p$, $\mathbf{C}$ is $n\times n$.

to define the *derivative* of a tensor

$$\frac{d\mathbf{A}}{d\mathbf{x}} \tag{6.81}$$

that continues the sequence of objects (6.66), (6.67), (6.68). In this book it suffices to find derivatives of certain expressions involving symmetric tensors defined in (6.55). According to (6.60), if f is a function of Class $C^{(2)}$ then its Hessian can be identified with a symmetric tensor.

Table 6.3 displays derivatives involving symmetric tensors (6.79). The expressions in the first column can be written as matrix products. First, we use Kronecker products in typical fashion to escalate the set of symmetric matrices $\{\mathbf{A}_1,\mathbf{A}_2,\ldots,\mathbf{A}_m\}$ to the symmetric tensor

$$\mathbf{A}=\sum_{i=1}^{m}(\mathbf{e}_i\otimes\mathbf{A}_i) \tag{6.82}$$

where $\mathbf{e}_i\in\mathfrak{R}^m$ for $i=1,2,\ldots,m$. Second, we express the object in (1) as

$$\sum_{i=1}^{m}(\mathbf{e}_i\otimes\mathbf{x}^T\mathbf{A}_i)=\sum_{i=1}^{m}\mathbf{e}_i\mathbf{x}^T\mathbf{A}_i \tag{6.83}$$

Third, the object in (2) results from multiplication on the right by $\mathbf{x}$. Finally, if $m=n$ we can multiply on the left by $\mathbf{x}^T$ to obtain the object in

TABLE 6.3 DERIVATIVES INVOLVING A SYMMETRIC TENSOR

Object	Derivative
(1) $\begin{bmatrix} \mathbf{x}^T\mathbf{A}_1 \\ \mathbf{x}^T\mathbf{A}_2 \\ \vdots \\ \mathbf{x}^T\mathbf{A}_m \end{bmatrix}$	$\begin{bmatrix} \mathbf{A}_1 \\ \mathbf{A}_2 \\ \vdots \\ \mathbf{A}_m \end{bmatrix} = \mathbf{A}$
(2) $\begin{bmatrix} \mathbf{x}^T\mathbf{A}_1\mathbf{x} \\ \mathbf{x}^T\mathbf{A}_2\mathbf{x} \\ \vdots \\ \mathbf{x}^T\mathbf{A}_m\mathbf{x} \end{bmatrix}$	$2\begin{bmatrix} \mathbf{x}^T\mathbf{A}_1 \\ \mathbf{x}^T\mathbf{A}_2 \\ \vdots \\ \mathbf{x}^T\mathbf{A}_m \end{bmatrix}$
(3) $\mathbf{x}^T\begin{bmatrix} \mathbf{x}^T\mathbf{A}_1\mathbf{x} \\ \mathbf{x}^T\mathbf{A}_2\mathbf{x} \\ \vdots \\ \mathbf{x}^T\mathbf{A}_n\mathbf{x} \end{bmatrix}$	$2\mathbf{x}^T\begin{bmatrix} \mathbf{x}^T\mathbf{A}_1 \\ \mathbf{x}^T\mathbf{A}_2 \\ \vdots \\ \mathbf{x}^T\mathbf{A}_n \end{bmatrix} + \begin{bmatrix} \mathbf{x}^T\mathbf{A}_1\mathbf{x} \\ \mathbf{x}^T\mathbf{A}_2\mathbf{x} \\ \vdots \\ \mathbf{x}^T\mathbf{A}_n\mathbf{x} \end{bmatrix}^T$

Each $\mathbf{A}_k$ is a symmetric matrix of order n.

(3). All derivatives in Table 6.3 can therefore be obtained using (6.78) (6.3.5).

It has perhaps been noticed in Table 6.3 that the symmetric tensor **A** nowhere appears in any product. In order to rewrite the objects and derivatives directly in terms of **A** we generally must "escalate" vectors and matrices. The exception is

$$\mathbf{A}\mathbf{x}$$

which is defined because **A** is $mn \times n$ and **x** is $n \times 1$. Where we start is with

$$\begin{bmatrix} \mathbf{x}^T\mathbf{A}_1 \\ \mathbf{x}^T\mathbf{A}_2 \\ \vdots \\ \mathbf{x}^T\mathbf{A}_m \end{bmatrix} = (\mathbf{I}_m \otimes \mathbf{x}^T)\mathbf{A} \tag{6.84}$$

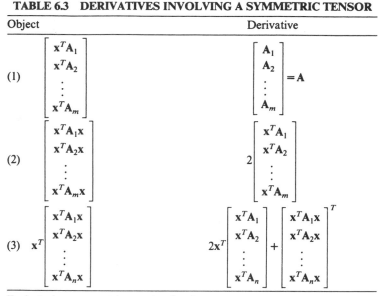

which is readily verified. Once we have this formula we can multiply by **x** on the right to obtain

$$\begin{bmatrix} \mathbf{x}^T\mathbf{A}_1\mathbf{x} \\ \mathbf{x}^T\mathbf{A}_2\mathbf{x} \\ \vdots \\ \mathbf{x}^T\mathbf{A}_m\mathbf{x} \end{bmatrix} = (\mathbf{I}_m \otimes \mathbf{x}^T)\mathbf{A}\mathbf{x} \tag{6.85}$$

If $m=n$ we can multiply on the left by $\mathbf{x}$ and then use Table 6.1 to find

$$\mathbf{x}^T\begin{bmatrix} \mathbf{x}^T\mathbf{A}_1\mathbf{x} \\ \mathbf{x}^T\mathbf{A}_2\mathbf{x} \\ \vdots \\ \mathbf{x}^T\mathbf{A}_n\mathbf{x} \end{bmatrix} = (\mathbf{x}^T \otimes \mathbf{x}^T)\mathbf{A}\mathbf{x} \tag{6.86}$$

which is the formula involving **A** that we most commonly use.

Table 6.4 displays derivatives of general expressions involving vectors and matrices. All entries follow from (6.75) and (6.78) (whose formulas also appear as entries); some are established in Problems for Solution where additional results may also be found.

TABLE 6.4 DERIVATIVES OF GENERAL EXPRESSIONS

Object	Derivative	Reference
(1) $c\mathbf{y}$ or $c\mathbf{y}^T$	$c\dfrac{d\mathbf{y}}{d\mathbf{x}} + \mathbf{y}\dfrac{dc}{d\mathbf{x}}$	(6.75)
(2) $\mathbf{y}^T\mathbf{w}$	$\mathbf{y}^T\dfrac{d\mathbf{w}}{d\mathbf{x}} + \mathbf{w}^T\dfrac{d\mathbf{y}}{d\mathbf{x}}$	(6.78)
(3) $\mathbf{F}\mathbf{z}$	$\mathbf{F}\dfrac{d\mathbf{z}}{d\mathbf{x}} + (\mathbf{I}_r \otimes \mathbf{z}^T)\dfrac{d\mathbf{F}}{d\mathbf{x}}$	(6.76)
(4) $\mathbf{y}^T\mathbf{F}$	$(\mathbf{y}^T \otimes \mathbf{I}_s)\dfrac{d\mathbf{F}}{d\mathbf{x}} + \mathbf{F}^T\dfrac{d\mathbf{y}}{d\mathbf{x}}$	(6.78)
(5) $c\mathbf{F}$	$c\dfrac{d\mathbf{F}}{d\mathbf{x}} + \left\{\dfrac{dc}{d\mathbf{x}} \otimes \sum_{j=1}^{s} (\mathbf{f}_j \otimes \mathbf{e}_j)\right\}$	(6.75)
(6) $\mathbf{y}^T\mathbf{D}\mathbf{y}$	$2\mathbf{y}^T\mathbf{D}\dfrac{d\mathbf{y}}{d\mathbf{x}} + (\mathbf{y}^T \otimes \mathbf{y}^T)\dfrac{d\mathbf{D}}{d\mathbf{x}}$	6.3.6
(7) $\mathbf{y}\mathbf{z}^T$	$(\mathbf{y} \otimes \mathbf{I}_s)\dfrac{d\mathbf{z}}{d\mathbf{x}} + (\mathbf{I}_r \otimes \mathbf{z})\dfrac{d\mathbf{y}}{d\mathbf{x}}$	(6.78)
(8) $\mathbf{F}\mathbf{G}$	$(\mathbf{F} \otimes \mathbf{I}_t)\dfrac{d\mathbf{G}}{d\mathbf{x}} + (\mathbf{I}_r \otimes \mathbf{G}^T)\dfrac{d\mathbf{F}}{d\mathbf{x}}$	(6.78)
(9) $\mathbf{F}\mathbf{G}\mathbf{H}$	$(\mathbf{F}\mathbf{G} \otimes \mathbf{I}_u)\dfrac{d\mathbf{H}}{d\mathbf{x}} + (\mathbf{F} \otimes \mathbf{H}^T)\dfrac{d\mathbf{G}}{d\mathbf{x}} + (\mathbf{I}_r \otimes \mathbf{H}^T\mathbf{G}^T)\dfrac{d\mathbf{F}}{d\mathbf{x}}$	6.3.6

$\mathbf{y}, \mathbf{w} \in \mathcal{R}^r$, $\mathbf{z}, \mathbf{e}_j \in \mathcal{R}^s$, $\mathbf{F}$ is $r \times s$, $\mathbf{G}$ is $s \times t$, $\mathbf{H}$ is $t \times u$, $\mathbf{D}=\mathbf{D}^T$ is $r \times r$.

EXAMPLE 6.3. Linear, Quadratic, and Cubic Functions on $\mathcal{R}^n$. It is straightforward to use the preceding tables to find all derivatives of the subject functions.

$$\mathbf{x} \mapsto f(\mathbf{x})$$

Results for the first case can be recalled as

(1) Linear functions on $\mathcal{R}^n$ are in one-to-one correspondence with vectors $\mathbf{a} \in \mathcal{R}^n$ where for all $\mathbf{x} \in \mathcal{R}^n$

$$f(\mathbf{x}) = \mathbf{a}^T\mathbf{x}$$

$$\mathbf{f}'(\mathbf{x}) = \mathbf{a}^T$$

$$\mathbf{f}''(\mathbf{x}) = \mathbf{O}_n$$

Note that boldface $\mathbf{f}'(\mathbf{x})$ and $\mathbf{f}''(\mathbf{x})$ denote standard matrix representations.

Quadratic functions are in one-to-one correspondence with symmetric tensors of the type represented by the Hessian (6.57) of a function of Class $C^{(2)}$. However, it is traditional to identify them in the manner of Section 5.3. This is done in the following where derivatives are covered by Table 6.3.

(2) Quadratic functions on $\mathcal{R}^n$ are in one-to-one correspondence with symmetric matrices $\mathbf{A}$ of order n where for all $\mathbf{x} \in \mathcal{R}^n$

$$f(\mathbf{x}) = \mathbf{x}^T\mathbf{A}\mathbf{x}$$

$$\mathbf{f}'(\mathbf{x}) = 2\mathbf{x}^T\mathbf{A}$$

$$\mathbf{f}''(\mathbf{x}) = 2\mathbf{A}$$

$$\mathbf{f}'''(\mathbf{x}) = \mathbf{O}_{n^2,n}$$

In words, a quadratic function can be identified with a real quadratic form where we lose no generality by requiring that the matrix of the form be symmetric.

The preceding sentence suggests how we can go to the next stage. Generalizing a quadratic form we start with the summation

(3) $$\sum_{i=1}^{n} \sum_{j=1}^{n} \sum_{k=1}^{n} a_{ijk} x_i x_j x_k$$

as the value at $\mathbf{x}$ of a *cubic function*. Then we see that there is no loss of generality in requiring

(4) $$a_{ijk} = a_{uvw}$$

whenever the associated sets of subscripts (which are not ordered sets in the sense introduced for (2.4)) satisfy

$$\{i,j,k\} = \{u,v,w\}$$

The reason is that, for example,

$$x_1x_2x_1 = x_2x_1^2 = x_1^2x_2$$

in the field $\mathcal{R}$ so that the sum of the three associated coefficients can be divided by three and assigned as the common value

$$a_{121} = a_{211} = a_{112}$$

Some consideration of possibilities leads to the matrices of order n

$$\mathbf{A}_k = [a_{ijk}] \tag{5}$$

for $k = 1,2,\ldots,n$; that is, a_{ijk} is the element in row i and column j of matrix k. First, the scalar (3) equals

$$x_1\mathbf{x}^T\mathbf{A}_1\mathbf{x} + x_2\mathbf{x}^T\mathbf{A}_2\mathbf{x} + \cdots + x_n\mathbf{x}^T\mathbf{A}_n\mathbf{x}$$

Second, each matrix (5) is a real symmetric matrix so the tensor

$$\mathbf{A} = \begin{bmatrix} \mathbf{A}_1 \\ \mathbf{A}_2 \\ \vdots \\ \mathbf{A}_n \end{bmatrix} \tag{6}$$

is symmetric of order $n^2 \times n$. Third, because of (4), each matrix (5) also has the special property that certain of its elements coincide with others in different matrices. For example, when $n=2$ there are at most four distinct elements in $\mathbf{A}$ and when $n=3$ there are at most 10, as follows:

$$\begin{bmatrix} a & b \\ b & c \\ \hdashline b & c \\ c & d \end{bmatrix} \qquad \begin{bmatrix} a & b & c \\ b & d & e \\ c & e & f \\ \hdashline b & d & e \\ d & g & h \\ e & h & i \\ \hdashline c & e & f \\ e & h & i \\ f & i & j \end{bmatrix} \tag{7}$$

Such pairs of numbers—(2,4) and (3,10)—can be characterized in various (combinatorial) ways but the significance of the associated symmetry property for the matrices (4) in the present context is that it makes possible $\mathbf{f}'''(\mathbf{x})=6\mathbf{A}$ to continue the pattern in (1) and (2) (6.3.8).

(8) Cubic functions on $\mathcal{R}^n$ are in one-to-one correspondence with symmetric tensors $\mathbf{A}$ of order $n^2\times n$ satisfying (4) where for all $\mathbf{x}\in\mathcal{R}^n$

$$f(\mathbf{x})=\mathbf{x}^T\begin{bmatrix}\mathbf{x}^T\mathbf{A}_1\mathbf{x}\\ \mathbf{x}^T\mathbf{A}_2\mathbf{x}\\ \vdots\\ \mathbf{x}^T\mathbf{A}_n\mathbf{x}\end{bmatrix}$$

$$\mathbf{f}'(\mathbf{x})=2\mathbf{x}^T\begin{bmatrix}\mathbf{x}^T\mathbf{A}_1\\ \mathbf{x}^T\mathbf{A}_2\\ \vdots\\ \mathbf{x}^T\mathbf{A}_n\end{bmatrix}+\left[\mathbf{x}^T\mathbf{A}_1\mathbf{x},\ldots,\mathbf{x}^T\mathbf{A}_n\mathbf{x}\right]$$

$$\mathbf{f}''(\mathbf{x})=2\left[x_1\mathbf{A}_1+\cdots+x_n\mathbf{A}_n\right]+2\left[\mathbf{A}_1\mathbf{x},\ldots,\mathbf{A}_n\mathbf{x}\right]+2\begin{bmatrix}\mathbf{x}^T\mathbf{A}_1\\ \mathbf{x}^T\mathbf{A}_2\\ \vdots\\ \mathbf{x}^T\mathbf{A}_n\end{bmatrix}$$

$$\mathbf{f}'''(\mathbf{x})=6\mathbf{A}$$

$$\mathbf{f}''''(\mathbf{x})=\mathbf{O}_{n^3,n}$$

PROOF. Certainly (3) represents all cubic functions and its one-to-one correspondence with (6) is established previously. The first derivative appears as (3) in Table 6.3. The second follows from Tables 6.3 and 6.4, starting with (4) in the latter and also using entry (1) (whereby a vector and its transpose "have the same derivative"). The special symmetry property (4) implies (6.3.8)

(9) $$\frac{d}{d\mathbf{x}}(x_1\mathbf{A}_1+\cdots+x_n\mathbf{A}_n)=\mathbf{A}$$

which is crucial for $\mathbf{f}'''(\mathbf{x})=6\mathbf{A}$. Finally, the zero matrix follows from considering the derivative of $f'''(\mathbf{x})=G$ in (6.63). ■

The formulas can be written in more compact form using (6.84) and following.

In summary, the preceding results enable us to write the derivatives of any third degree function on $\mathcal{R}^n$

$$\mathbf{x} \mapsto a + \mathbf{b}^T\mathbf{x} + \mathbf{x}^T\mathbf{C}\mathbf{x} + (\mathbf{x}^T \otimes \mathbf{x}^T)\mathbf{D}\mathbf{x} \tag{10}$$

where $a \in \mathcal{R}, \mathbf{b} \in \mathcal{R}^n, \mathbf{C}$ is symmetric of order n, $\mathbf{D}$ is a symmetric tensor of order $n^2 \times n$ satisfying (4), and where we have used (6.86) to write the final term.

The next procedure constitutes the most important use of derivatives of vectors and matrices in this book. It readily produces results that are otherwise difficult to obtain and, because of its usefulness for working in $\mathcal{R}^n$, it can be taken to be major justification for the substantial body of technique developed in this section.

DIFFERENTIATING COMPOSITE TRANSFORMATIONS. Replace the given notation by the standard notation of Figure 6.1. Suppose α and β belong to their respective Classes $C^{(s)}$. Proceed as follows where the standard representation of the kth derivative of $\gamma = \alpha(\beta)$ is found in Step k for $k = 1, 2, \ldots, s$.

Step 1. Use the chain rule (6.64) to write the standard representation

$$\frac{d\gamma}{d\mathbf{u}} = \frac{d\alpha}{d\mathbf{v}} \frac{d\beta}{d\mathbf{u}}$$

and note its order. Observe how this step is depicted by Figure 6.2(a).
Step 2. Differentiate the object found in Step 1 where now the chain rule is used to find the factor

$$\frac{d}{d\mathbf{u}}\left(\frac{d\alpha}{d\mathbf{v}}\right) = \frac{d^2\alpha}{d\mathbf{v}^2} \frac{d\beta}{d\mathbf{u}}$$

as depicted in Figure 6.2(b). Check conformability and verify that orders are correct for all resulting factors.
Step 3 and following. Differentiate the object found in the preceding step. Use the chain rule as necessary to differentiate all composite transformations of $\mathbf{u}$. Check conformability and verify that orders are correct for all resulting factors. (For example, to find $(d^3/d\mathbf{u}^3)\gamma$, use the chain rule to find the factor

$$\frac{d}{d\mathbf{u}}\left(\frac{d^2\alpha}{d\mathbf{v}^2}\right) = \frac{d^3\alpha}{d\mathbf{v}^3} \frac{d\beta}{d\mathbf{u}}$$

as depicted in Figure 6.2(c).) (6.87)

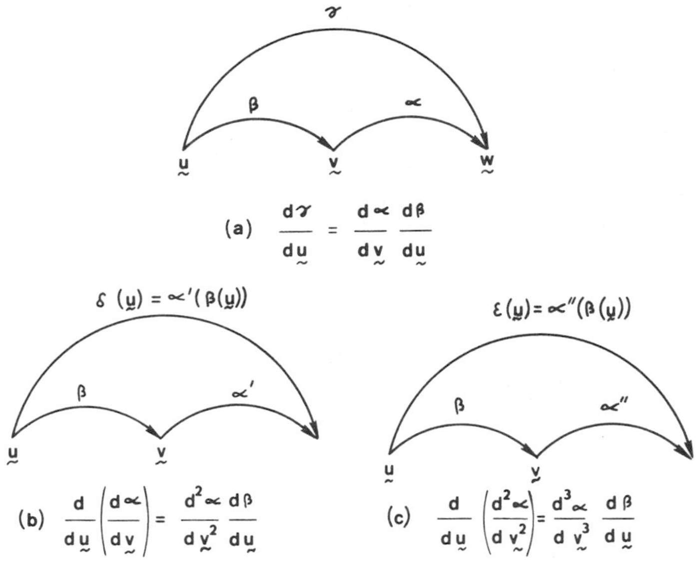

Figure 6.2. Chain rule differentiations.

PROOF. Classes $C^{(s)}$ assure that the chain rule, briefly Rule, applies throughout successive steps. Step 1 translates the Rule into the terminology and notation of the present section. The object found in Step 1 is a matrix or a tensor that can be differentiated in Step 2 using (6.75) or (6.78), or else an entry from one of Tables 6.2, 6.3, 6.4, or an extension of these tables. Where the standard representation $(d/d\mathbf{v})\alpha$—the first factor displayed on the RHS in Step 1—is to be differentiated in the process of using, say, (6.78), the Rule applies as displayed in Step 2. (This is apparent from associating δ of Figure 6.2(b) with γ of (a), and also α' with α in the figures.) Step 3 and subsequent steps are established by the same reasoning. ■

Derivatives of order three and higher of functions on $\mathcal{R}^n$ appear to have been rarely used in practice. Sufficient conditions for their existence, membership in Classes $C^{(s)}$ for $s=3,4,\ldots$, become significantly more and more restrictive as s increases. Higher order derivatives also become more and more complicated although the scheme for their generation is straightforward. In order to use (6.87) for $s=3,4,\ldots$, it would be effective to extend Table 6.4 with entries for derivatives of Kronecker products $\mathbf{A}\otimes\mathbf{B}$.

EXAMPLE 6.4. Successive Chain Rule Differentiations. This example illustrates the use of (6.87) for three important cases. We include sketches corresponding to those in the general Figure 6.2 and we recommend that such sketches be used as standard practice.

The first case is a

(1) Function of a Linear Transformation. Consider the use of a linear transformation A on the argument of a function $f: V \subset \mathcal{R}^n \to \mathcal{R}$ from $\mathbf{x}$ to $\mathbf{y} \in \mathcal{R}^n$ via

$$\mathbf{y} = \mathbf{A}\mathbf{x}$$

where ordinarily $\mathbf{A}$ is nonsingular.

Figure 6.3 contains the sketches. Matching notations as directed, we find that $\alpha, \beta, \gamma, \mathbf{u}, \mathbf{v}$ correspond to

$$g, A, f, \mathbf{x}, \mathbf{y}$$

respectively. Step 1 of the procedure yields the matrix

$$\mathbf{f}'(\mathbf{x}) = \mathbf{g}'(\mathbf{y})\mathbf{A}$$

which is of order $1 \times n$. If in Step 2 we use Table 6.4(4) to differentiate $\mathbf{f}'(\mathbf{x})$ we obtain

$$\mathbf{f}''(\mathbf{x}) = \mathbf{A}^T\left(\frac{d}{d\mathbf{x}}\mathbf{g}'(\mathbf{y})\right)$$

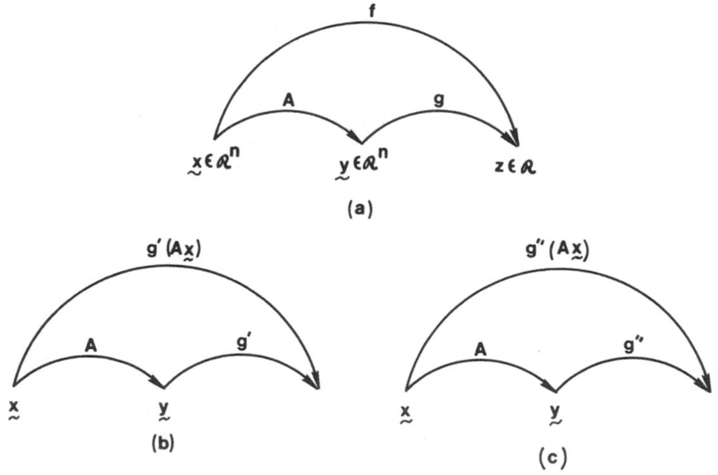

Figure 6.3. Differentiating a function of a linear transformation.

because $(d/d\mathbf{x})\mathbf{A}=\mathbf{O}$. Using the chain rule as depicted in Figure 6.3(b)

$$\frac{d}{d\mathbf{x}}\mathbf{g}'(\mathbf{y})=\mathbf{g}''(\mathbf{y})\mathbf{A}$$

and the result of completing Step 2 is

$$\mathbf{f}''(\mathbf{x})=\mathbf{A}^T\mathbf{g}''(\mathbf{y})\mathbf{A}$$

which correctly is of order n. Step 3 is completed using Table 6.4(9) and Figure 6.3(c) with result

$$\mathbf{f}'''(\mathbf{x})=(\mathbf{A}^T\otimes\mathbf{A}^T)\mathbf{g}'''(\mathbf{y})\mathbf{A}$$

which has correct order $n^2\times n$. In conclusion, we have found the formulas for the standard representations of f', f'', and f''' in terms of the new variable $\mathbf{y}=\mathbf{A}\mathbf{x}$.

The second case for demonstrating the procedure (6.87) is a

(2) Function of a Transformation. Suppose D is open and consider

$$h:D\subset\mathcal{R}^n\to\mathcal{R}$$

with $h(\mathbf{x})=g(f(\mathbf{x}))$ where $f:D\subset\mathcal{R}^n\to\mathcal{R}^m$ and $g:f(D)\subset\mathcal{R}^m\to\mathcal{R}$. This coincides with (1) in Example 6.2 except that the roles of f and g are interchanged as shown in Figure 6.4. The present notation matches the standard under

$$\alpha,\beta,\gamma,\mathbf{u},\mathbf{v}\leftrightarrow g,f,h,\mathbf{x},\mathbf{y}$$

respectively. From Step 1, $h'=g'f'$ is represented by a matrix of order $1\times n$. In Step 2 we can use Table 6.4(4) to find

$$\frac{d^2h}{d\mathbf{x}^2}=\left(\frac{dg}{d\mathbf{y}}\otimes\mathbf{I}_n\right)\frac{d^2f}{d\mathbf{x}^2}+\left(\frac{df}{d\mathbf{x}}\right)^T\frac{d}{d\mathbf{x}}\left(\frac{dg}{d\mathbf{y}}\right)$$

where for clarity we use both $d/d\mathbf{x}$ and $d/d\mathbf{y}$ without arguments rather than arguments together with primes. Applying the chain rule to the final

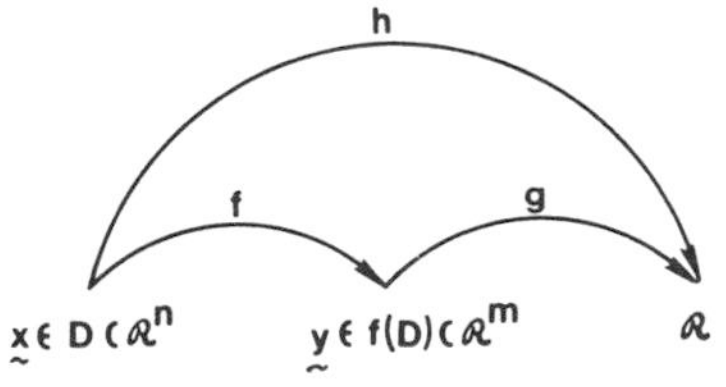

Figure 6.4. Function of a transformation.

factor on the RHS and substituting the result yields

$$\frac{d^2h}{d\mathbf{x}^2}=\left(\frac{dg}{d\mathbf{y}}\otimes\mathbf{I}_n\right)\frac{d^2f}{d\mathbf{x}^2}+\left(\frac{df}{d\mathbf{x}}\right)^T\frac{d^2g}{d\mathbf{y}^2}\frac{df}{d\mathbf{x}}$$

whose order is n as required.

The third case is the specialization of (2) corresponding to $m=1$ so it is called a

(3) Function of a Function. Suppose D is open and consider

$$h:D\subset\mathcal{R}^n\to\mathcal{R}$$

with $h(x)=g(f(x))$ where $f:D\subset\mathcal{R}^n\to\mathcal{R}$ and $g:f(D)\subset\mathcal{R}\to\mathcal{R}$. Figure 6.4 continues to apply. Here the first and second derivatives are obtained by substituting $m=1$ in those for (2). However, g' and g'' can be identified with scalars so we can rewrite

$$\frac{d^2h}{d\mathbf{x}^2}=\frac{dg}{dy}\frac{d^2f}{d\mathbf{x}^2}+\frac{d^2g}{dy^2}\left(\frac{df}{d\mathbf{x}}\right)^T\frac{df}{d\mathbf{x}}$$

whose order remains n as required. Each of the two matrices on the RHS is the scalar multiple of a matrix where, it may be noticed, the second is an outer product matrix. Each member of the RHS can therefore be differentiated using Table 6.4 to find h'''.

We can use procedure (6.87) to extend Taylor's theorem (1.13) to functions on $\mathcal{R}^n$. While the result can be established for a function f that is p-times differentiable, it suffices in this book to have the following

TAYLOR'S THEOREM. Consider $f:D\subset\mathcal{R}^n\to\mathcal{R}$, where D is open, and $\mathbf{x}_0$ and $\mathbf{h}$ are such that $\mathbf{x}_0+t\mathbf{h}\in D$ for $t\in[0,1]$. If $f\in C^{(2)}$ then the equation

$$f(\mathbf{x}_0+\mathbf{h})=f(\mathbf{x}_0)+\mathbf{f}'(\mathbf{x}_0)\mathbf{h}+\frac{1}{2!}\mathbf{h}^T\mathbf{f}''(\mathbf{u})\mathbf{h}$$

holds for some $\mathbf{u}=\mathbf{x}_0+s_0\mathbf{h}$ where $s_0\in(0,1)$. If $f\in C^{(3)}$ then the RHS of the equation has $\mathbf{u}=\mathbf{x}_0$ and fourth term

$$\frac{1}{3!}(\mathbf{h}^T\otimes\mathbf{h}^T)\mathbf{f}'''(\mathbf{w})\mathbf{h}$$

for some $\mathbf{w}=\mathbf{x}_0+t_0\mathbf{h}$ where $t_0\in(0,1)$. (6.88)

PROOF. Suppose $f\in C^{(3)}$. In this case f and its partial derivatives through the second order are differentiable, and since those of third order exist and are continuous on D we can use (6.87) to find the first three derivatives of g where

$$g(t)=f(\mathbf{x}_0+t\mathbf{h})$$

Matching notations as directed we find

$$\alpha,\beta,\gamma,\mathbf{u},\mathbf{v}\leftrightarrow f,t\mapsto \mathbf{x}_0+t\mathbf{h},g,t,\mathbf{x}$$

respectively, where in particular $\beta(t)=\mathbf{x}_0+t\mathbf{h}$. Step 1 of the procedure yields

$$g'(t)=\mathbf{f}'(\mathbf{x})\frac{d\beta}{dt}=\mathbf{f}'(\mathbf{x})\mathbf{h}$$

Using Table 6.4(2) in Step 2 to differentiate the above scalar yields

$$g''(t)=\mathbf{h}^T\frac{d}{dt}\mathbf{f}'(\mathbf{x})=\mathbf{h}^T\mathbf{f}''(\mathbf{x})\mathbf{h}$$

where the third scalar expression results from the chain rule. Next, using Table 6.4(6) and the chain rule in Step 3 yields

$$g'''(t)=(\mathbf{h}^T\otimes\mathbf{h}^T)\frac{d}{dt}\mathbf{f}''(\mathbf{x})=(\mathbf{h}^T\otimes\mathbf{h}^T)\mathbf{f}'''(\mathbf{x})\mathbf{h}$$

Finally, we use (1.13) for g with $b=1$ and $a=0$ to find t_0 and $\mathbf{w}=\mathbf{x}_0+t_0\mathbf{h}$. If $f\in C^{(2)}$ we stop at $g''(t)$ and find s_0 to which corresponds $\mathbf{u}$. ■

From the preceding explicit Taylor's theorem we can write the *cubic Taylor formula*

$$f(\mathbf{x}_0+\mathbf{h})=f(\mathbf{x}_0)+\mathbf{f}'(\mathbf{x}_0)\mathbf{h}+\frac{1}{2}\mathbf{h}^T\mathbf{f}''(\mathbf{x}_0)\mathbf{h}+\frac{1}{6}(\mathbf{h}^T\otimes\mathbf{h}^T)\mathbf{f}'''(\mathbf{x}_0)\mathbf{h}+o(\|\mathbf{h}\|^3) \tag{6.89}$$

For example, if D is convex then the preceding holds for $f\in C^{(3)}$; if moreover $f\in C^{(4)}$ and

$$\|\mathbf{f}^{(4)}(\mathbf{x})\|\leqq(4!)M$$

for $\mathbf{x}\in D$ then we can show that the absolute value of the error committed by using the cubic is bounded by $M\|\mathbf{h}\|^{-4}$.

PROBLEMS FOR SOLUTION

6.3.1 Table 6.1. Establish:

(1) All table entries.

(2) The extensions

(i) $\mathbf{A}\otimes\left(\sum_{i=1}^{p}\mathbf{X}_i\right)=\sum_{i=1}^{p}(\mathbf{A}\otimes\mathbf{X}_i)$

(ii) $\left(\sum_{i=1}^{q}\mathbf{Y}_i\right)\otimes\mathbf{C}=\sum_{i=1}^{q}(\mathbf{Y}_i\otimes\mathbf{C})$

(iii) $(\mathbf{A}\otimes\mathbf{B})\,(\mathbf{C}\otimes\mathbf{D})\,(\mathbf{E}\otimes\mathbf{F})=\mathbf{ACE}\otimes\mathbf{BDF}$.

6.3.2 Kronecker Products. Establish under specifications for conformability as necessary.

(1) $\mathbf{A}\otimes\mathbf{B}=(\mathbf{A}\otimes\mathbf{I}_t)(\mathbf{I}_s\otimes\mathbf{B})$.
(2) $(\mathbf{a}\otimes\mathbf{B})\mathbf{C}=\mathbf{a}\otimes\mathbf{BC}$.
(3) $(\mathbf{A}\otimes\mathbf{b})\mathbf{C}=\mathbf{AC}\otimes\mathbf{b}$.
(4) $\mathbf{A}(\mathbf{b}^T\otimes\mathbf{C})=\mathbf{b}^T\otimes\mathbf{AC}$.
(5) $\mathbf{A}(\mathbf{B}\otimes\mathbf{c}^T)=\mathbf{AB}\otimes\mathbf{c}^T$.
(6) $\mathbf{a}(\mathbf{b}^T\otimes\mathbf{c}^T)=\mathbf{ab}^T\otimes\mathbf{c}^T=\mathbf{b}^T\otimes\mathbf{ac}^T$.

6.3.3 Kronecker Products of Square Matrices. Establish for **A** of order r and **B** of order s.

(1) If **A** and **B** are nonsingular, then so is $\mathbf{A}\otimes\mathbf{B}$ and $(\mathbf{A}\otimes\mathbf{B})^{-1}=\mathbf{A}^{-1}\otimes\mathbf{B}^{-1}$.
(2) $\det(\mathbf{A}\otimes\mathbf{B})=(\det\mathbf{A})^r(\det\mathbf{B})^s$.
(3) $\operatorname{rank}(\mathbf{A}\otimes\mathbf{B})=(\operatorname{rank}\mathbf{A})(\operatorname{rank}\mathbf{B})$.
(4) $\operatorname{trace}(\mathbf{A}\otimes\mathbf{B})=(\operatorname{trace}\mathbf{A})(\operatorname{trace}\mathbf{B})$.
(5) If **A** has eigenvalues $\lambda_1,\lambda_2,\ldots,\lambda_r$ and **B** has $\mu_1,\mu_2,\ldots,\mu_s$, then the eigenvalues of $\mathbf{A}\otimes\mathbf{B}$ are the rs numbers $\lambda_1\mu_1,\lambda_1\mu_2,\ldots,\lambda_r\mu_s$.
(6) The eigenvalues of the *Kronecker sum* of **A** and **B**, $(\mathbf{A}\otimes\mathbf{I}_s)+(\mathbf{I}_r\otimes\mathbf{B})$, are the rs numbers $\lambda_1+\mu_1,\lambda_1+\mu_2,\ldots,\lambda_r+\mu_s$.

6.3.4 Table 6.2. Use (6.78) to derive all table entries.

6.3.5 Table 6.3. Derive all table entries using (6.83) and (6.78) as specified in the text.

6.3.6 Table 6.4. Establish:

(1) All entries attributed to (6.75).
(2) Repeat for (6.76).
(3) Repeat for (6.78).
(4) Entry (6).
(5) Entry (9).
(6) If **A** is nonsingular of order n

(i) $$\frac{d\mathbf{A}^{-1}}{d\mathbf{x}}=-\{\mathbf{A}^{-1}\otimes(\mathbf{A}^{-1})^T\}\frac{d\mathbf{A}}{d\mathbf{x}}$$

(ii) $$\frac{d\mathbf{A}}{d\mathbf{x}}=-\{\mathbf{A}\otimes\mathbf{A}^T\}\frac{d\mathbf{A}^{-1}}{d\mathbf{x}}$$

where it is to be noticed that each Kronecker product is the inverse of the other by 6.3.3(1). (Of course the derivatives are not nonsingular if $n>1$ because then they are not square.)

(7) Derive (6.75) from (6.78).

6.3.7 Second Derivatives. Verify or find second derivatives as follows.

(1) Derivatives of entries (7), (8), (9) in Table 6.2.

(2) Derivative of entry (3) in Table 6.3.

(3) Derivatives of entries (2), (3), (4), (6), and (7) in Table 6.4 given constant **y**, **F**, **F**, **D**, and **y**, respectively.

(4) $\dfrac{d}{d\mathbf{x}}(\mathbf{y}^T\otimes\mathbf{e}_i)=\mathbf{e}_i\otimes\dfrac{d\mathbf{y}}{d\mathbf{x}}$.

(5) $\dfrac{d^2}{d\mathbf{x}^2}\mathbf{a}_j=\sum_{i=1}^{r}\left(\mathbf{e}_i\otimes\dfrac{d^2}{d\mathbf{x}^2}a_{ij}\right)$.

(6) $\dfrac{d}{d\mathbf{x}}(\mathbf{Y}^T\otimes\mathbf{e}_i)=\mathbf{e}_i\otimes\dfrac{d\mathbf{Y}}{d\mathbf{x}}$.

(7) $\dfrac{d^2}{d\mathbf{x}^2}\mathbf{A}=?$

6.3.8 Example 6.3. Establish:

(1) The tensor for $n=4$ corresponding to those in (7) of the example.

(2) Property (9) is necessary and sufficient for (4).

(3) The derivatives of (10).

6.3.9 Chain Rule Differentiation. Use (6.87) to find or establish:

(1) First and second derivatives of
   (i) $(f(\mathbf{x}))^2$.
   (ii) $(f(\mathbf{x}))^{-1}$.
   (iii) $(f(\mathbf{x}))^p$ where $p\neq-1$ is an integer.
   (iv) $\exp f(\mathbf{x})$.

(2) First derivatives where **A** and **B** are real symmetric matrices of order n
   (i) $\langle\mathbf{x},\mathbf{x}\rangle^{-1}$
   (ii) $(\mathbf{x}^T\mathbf{A}\mathbf{x})^{-1}$
   (iii) $(\mathbf{x}^T\mathbf{A}\mathbf{x})\langle\mathbf{x},\mathbf{x}\rangle^{-1}$
   (iv) $(\mathbf{x}^T\mathbf{A}\mathbf{x})(\mathbf{x}^T\mathbf{B}\mathbf{x})^{-1}$
   (v) $\langle\mathbf{a},\mathbf{x}\rangle^2$.

(3) Third derivative of (3) in Example 6.4.

(4) Consider the differentiable functions on $\mathfrak{R}^2$ corresponding to those given in conventional notation as follows: $u=f(x,y)$, $v=g(x,y)$, and $z=h(u,v)$. Find the following for z as a function of (x,y) and write in conventional notation.
   (i) First derivative.
   (ii) Second derivative.

6.3.10 Cubic Taylor Formulas. Establish (6.89) for the following where $\mathbf{a} \in \mathcal{R}^n$ and $\mathbf{A} = \mathbf{A}^T$ is of order n.

(1) $\mathbf{x} \mapsto \langle \mathbf{a}, \mathbf{x} \rangle^{-1}$.
(2) $\mathbf{x} \mapsto \exp\langle \mathbf{a}, \mathbf{x} \rangle^{-1}$.
(3) $\mathbf{x} \mapsto (\mathbf{x}^T \mathbf{A} \mathbf{x})^{-1}$.
(4) $\mathbf{x} \mapsto \exp(\mathbf{x}^T \mathbf{A} \mathbf{x})^{-1}$.

6.4 THREE THEOREMS

In this section we establish theorems that are needed in Chapter 7. The results are relatively abstract and some proofs are intricate; see Problems for Solution for elementary cases. While several works contain the following approach, for example Apostol (1974) and Taylor and Manne (1972), our presentation has been influenced mainly by the more general treatment in Fleming (1965).

Let us consider $f: D \subset \mathcal{R}^n \to \mathcal{R}^n$ so it makes sense to ask if there is an inverse f^{-1}. We recall from (1.1) that the necessary and sufficient condition is that f be one-to-one and onto. We also recall that the second condition can be satisfied by confining attention to $f(D)$ the set of images of the domain. Whenever f is a linear transformation, f^{-1} exists iff f is nonsingular and there are systematic methods. At times when f is not linear we can find f^{-1} by solving

$$\begin{gathered} y_1 = f_1(\mathbf{x}) \\ y_2 = f_2(\mathbf{x}) \\ \vdots \\ y_n = f_n(\mathbf{x}) \end{gathered}$$

for the coordinates $x_i = f^{-1}(\mathbf{y})$ for $i = 1, 2, \ldots, n$, and the rule for

$$\mathbf{y} \mapsto f^{-1}(\mathbf{y})$$

But the more usual situation in practice is that we cannot find a solution, or we do not wish to do so because what we actually need to know is whether or not the inverse exists. And then when f^{-1} does exist we are generally interested in being able to find its derivative without the necessity of finding f^{-1} itself.

The next theorem resolves these matters, insofar as they can be resolved, in $\mathcal{R}^n$ for $n > 1$. Roughly speaking, it states that when the derivative f' has an inverse then so does the function f. We must suppose more than differentiability, namely that f is at least of Class $C^{(1)}$. Then $f'(\mathbf{x})$ nonzero guarantees that f has a *local inverse* g on $f(N)$ for a certain neighborhood N of $\mathbf{x}$; that is, the restriction of f to N has an inverse g defined on the set of images $f(N)$. In general it is not possible to extend such an inverse to cover the entire set $f(D)$ (6.4.1).

INVERSE TRANSFORMATION THEOREM. Consider $f: D \subset \mathfrak{R}^n \to \mathfrak{R}^n$, where D is open and $f \in C^{(p)}$ for some $p \geqq 1$. If $f'(\mathbf{x})$ is nonsingular on D then for each $\mathbf{x}_0 \in D$ there exists a neighborhood $N \subset D$ such that

(1) The set of images $f(N)$ is open.

(2) The restriction of f to N has a unique inverse $g: f(N) \to N$ of Class $C^{(p)}$ and

$$g'(\mathbf{y}) = \left[f'(g(\mathbf{y})) \right]^{-1}$$

where $\mathbf{y} = f(\mathbf{x})$. (6.90)

PROOF. A neighborhood N of $\mathbf{x}_0$ can be determined according to (6.53) for $\varepsilon = (1/2)\|\mathbf{F}^{-1}\|^{-1}$, where $\mathbf{F} = \mathbf{f}'(\mathbf{x}_0)$, so that by (6.54)

$$\|f(\mathbf{x}) - f(\mathbf{z})\| \geqq \varepsilon \|\mathbf{x} - \mathbf{z}\|$$

for all $\mathbf{x}, \mathbf{z} \in N$. Consequently, f is one-to-one on N, and by (1.1) there exists a local inverse g defined on $f(N)$ whose uniqueness follows from its definition.

If $\mathbf{y}^* \in f(N)$, then from what we have just shown, there is a unique $\mathbf{x}^* \in N$ such that $\mathbf{y}^* = f(\mathbf{x}^*)$. There is also a neighborhood $N_\delta(\mathbf{x}^*) \subset N$ whose boundary

$$B = \{\mathbf{x} : \|\mathbf{x} - \mathbf{x}^*\| = \delta\}$$

and whose closure

$$C = \{\mathbf{x} : \|\mathbf{x} - \mathbf{x}^*\| \leqq \delta\}$$

are also contained in N. Since $\mathbf{x}^* \notin B$ we also know (6.1.8(5))

$$2\gamma = d(\mathbf{y}^*, f(B))$$

which is positive. Let us prove (1) by showing

$$N_\gamma(\mathbf{y}^*) \subset f(N_\delta(\mathbf{x}^*))$$

If $\mathbf{y}^{**} \in N_\gamma(\mathbf{y}^*)$ then again by 6.1.8(5)

$$d(\mathbf{y}^{**}, f(C)) = \|\mathbf{y}^{**} - f(\mathbf{x}^{**})\|$$

for some $\mathbf{x}^{**} \in C$. Now if $\mathbf{x} \in B$ then by "subtracting and adding" $\mathbf{y}^*$ inside $\|\cdot\|$

$$\|\mathbf{y}^{**} - f(\mathbf{x})\| \geqq \|\mathbf{y}^* - f(\mathbf{x})\| - \|\mathbf{y}^{**} - \mathbf{y}^*\|$$

The first term on the RHS cannot be less than 2γ while the second must be less than γ. Consequently the RHS cannot be less than γ. But reconsidering the second term we conclude $\mathbf{x}^{**} \in \operatorname{int} C$ because $\mathbf{x}^*$ is an interior point producing a distance to $\mathbf{y}^{**}$ strictly less than that for any $\mathbf{x} \in B$. Now $\|\mathbf{y}^{**} - f(\mathbf{x})\|^2$ also assumes a minimum value at x^{**} so by (6.65) and (2) in Table 6.4

$$-2(\mathbf{y}^{**} - f(\mathbf{x}^{**}))^T f'(\mathbf{x}^{**}) = \mathbf{0}^T$$

By hypothesis the coefficient matrix is nonsingular so that $\mathbf{y}^* = f(\mathbf{x}^{**})$ from (2.21) and proof of (1) is complete.

The next step is to show that g is differentiable at $\mathbf{y}^* = f(\mathbf{x}^*)$. First,

$$g(f(\mathbf{x})) = \mathbf{x} = \mathbf{I}\mathbf{x}$$

so, by (6.47), if $g(f)$ is differentiable then its derivative must equal the identity transformation. Second, if the chain rule (6.64) applies, the derivative of $g(f)$ must equal the product $g'f'$ and we would conclude

$$g'f' = I$$

That is, the derivative of g would have to equal the inverse of the derivative of f. Let us accordingly prove that g is differentiable by showing that the linear transformation A^{-1} satisfies (6.46) where $A = f'(\mathbf{x}^*)$. Since f is differentiable we can associate $\mathbf{x}^*$ with $\mathbf{x}_0$ and $\mathbf{x} - \mathbf{x}^*$ with $\mathbf{h}$ in (6.46) to write

$$\|f(\mathbf{x}) - f(\mathbf{x}^*) - A(\mathbf{x} - \mathbf{x}^*)\| \leqq k_1 \varepsilon \|\mathbf{x} - \mathbf{x}^*\|$$

for $\|\mathbf{x} - \mathbf{x}^*\|$ sufficiently small and where $k_1 > 0$ is to be chosen later. We can rewrite the expression within $\|\cdot\|$ on the LHS as

$$A\{A^{-1}\mathbf{y} - A^{-1}\mathbf{y}^* - [g(\mathbf{y}) - g(\mathbf{y}^*)]\}$$

and then use 6.1.10(2) to obtain

$$\|g(\mathbf{y}) - g(\mathbf{y}^*) - A^{-1}(\mathbf{y} - \mathbf{y}^*)\| \leqq \|A^{-1}\| \|f(\mathbf{x}) - f(\mathbf{x}^*) - A(\mathbf{x} - \mathbf{x}^*)\|$$

If we now return to the first paragraph of this proof we see that we can use

(6.54) to assure

$$(\|A^{-1}\|^{-1}-\varepsilon)\|\mathbf{x}-\mathbf{x}^*\| \leqq \|\mathbf{y}-\mathbf{y}^*\|$$

provided $\|\mathbf{x}-\mathbf{x}^*\|$ is sufficiently small. Finally, we can combine the last three displayed inequalities to obtain

$$\| g(\mathbf{y})-g(\mathbf{y}^*)-A^{-1}(\mathbf{y}-\mathbf{y}^*)\| \leqq \varepsilon\|\mathbf{y}-\mathbf{y}^*\|$$

provided we now specify $k_1=(1-\varepsilon\|A^{-1}\|)\|A^{-1}\|^{-2}$, a quantity that is positive by (6.54). Thus A^{-1} satisfies (6.46), g is differentiable at $\mathbf{y}^*$, and proof of (2) is complete.

The final requirement is to show that if $f\in C^{(p)}$ for $p\geqq 1$ then $g\in C^{(p)}$. Suppose $f\in C^{(1)}$. We have shown $g'f'=I$ and we can rewrite this in terms of matrices as

$$(\mathbf{f}')^T(\mathbf{g}')^T=\mathbf{I}$$

This equation represents n systems of linear equations in transposes of standard matrices of derivatives $\mathbf{g}_1', \mathbf{g}_2', \ldots, \mathbf{g}_n'$. The elements of the coefficient matrix $[\partial_i f_j(g(\mathbf{y}))]$ are continuous functions of $\mathbf{y}$ by (6.10) and therefore so are the solutions by Example 6.1.(5). This establishes $g\in C^{(1)}$. In case $f\in C^{(2)}$ then $\partial_i f_j\in C^{(1)}$ and $\partial_i f_j(g(\mathbf{y}))\in C^{(1)}$ by 6.2.7. We then find $\partial_i g_j\in C^{(1)}$ so that $g\in C^{(2)}$, and we can keep repeating this process to establish $g\in C^{(p)}$. ■

The next theorem deals with m functions $u_1, u_2, \ldots, u_m$ that are conveniently presented as the coordinate functions of a special kind of transformation.

Definition. The following hold for a *standard implicit function transformation* $u: D\subset \mathfrak{R}^n \to \mathfrak{R}^m$, where D is open.

(1) $1\leqq m<n$.
(2) $u\in C^{(p)}$ where $p\geqq 1$.
(3) There exists a special point $\mathbf{x}_0\in D$ such that $u(\mathbf{x}_0)=\mathbf{0}$.
(4) $\operatorname{rank}\mathbf{u}'(\mathbf{x}_0)=m$.
(5) The first m columns of $\mathbf{u}'(\mathbf{x}_0)$ are LI. (6.91)

The first four are hypotheses for (6.93). Property (5) achieves standardization that is convenient; we know there must exist some set of m LI columns out of the full set of n. In case the first m were LD then we could duplicate all of what follows replacing the first m components $x_1, \ldots, x_m$ and their associated objects by any designated set of m components out of $x_1, \ldots, x_n$, again ordered by increasing subscripts.

The conclusion of (6.93) is that there exist functions $w_1,\ldots,w_m$ defined on $\mathcal{R}^k$ where $m+k=n$ such that

$$\begin{aligned}
x_1 &= w_1(x_{m+1},x_{m+2},\ldots,x_n)\\
x_2 &= w_2(x_{m+1},x_{m+2},\ldots,x_n)\\
&\vdots\\
x_m &= w_m(x_{m+1},x_{m+2},\ldots,x_n)\\
x_{m+1} &= x_{m+1}\\
&\vdots\\
x_n &= x_{m+k}
\end{aligned}$$

is equivalent to $u(\mathbf{x})=\mathbf{0}$ for all $\mathbf{x}$ in some neighborhood of $\mathbf{x}_0$; this is called *local equivalence* at $\mathbf{x}_0$. It should be noticed how this generalizes the result of solving $\mathbf{Ax}=\mathbf{b}$ in (3.18); if there the first m columns of $\mathbf{A}$ are LI, then $x_1,\ldots,x_m$ are expressed in terms of the nonbasic variables $x_{m+1},\ldots,x_n$.

Whenever $k=n-m$ is positive we define the special linear transformation

$$K:\mathcal{R}^n\to\mathcal{R}^k \tag{6.92}$$

where $K\mathbf{x}=[x_{m+1},x_{m+2},\ldots,x_n]^T$. With this special notation and the terminology (6.91) we have the

IMPLICIT FUNCTION THEOREM. The following exist at a special point $\mathbf{x}_0$ of a standard implicit function transformation $u:D\subset\mathcal{R}^n\to\mathcal{R}^m$.

a neighborhood A of $\mathbf{x}_0$
an open set $B\subset\mathcal{R}^k$ containing $K\mathbf{x}_0$
a unique transformation $w:B\subset\mathcal{R}^k\to\mathcal{R}^m$ of Class $C^{(p)}$

such that for all $\mathbf{x}\in A$, $u(\mathbf{x})=\mathbf{0}$ iff

$$K\mathbf{x}\in B \quad\text{and}\quad w(K\mathbf{x})=\begin{bmatrix}x_1\\x_2\\\vdots\\x_m\end{bmatrix} \tag{6.93}$$

PROOF. By (6.8), $\mathbf{x}_0$ has a neighborhood $A\subset D$ on which u' is nonsingular. Let us show that the special transformation $f:D\subset\mathcal{R}^n\to\mathcal{R}^n$ where

$$f(x)=\left[u_1(\mathbf{x}),u_2(\mathbf{x}),\ldots,u_m(\mathbf{x}),x_{m+1},\ldots,x_n\right]^T$$

has a unique local inverse $g: f(A) \to A$, where $f(A)$ is open; that is,

$$g(f(\mathbf{x})) = \mathbf{x}$$

whenever $\mathbf{x} \in A$. This follows because all hypotheses of (6.90) hold for the restriction of f to $A : A$ is open; $f \in C^{(p)}$, where $p \geqq 1$ comes from (6.91), because its coordinate functions are of Class $C^{(p)}$; and f' is nonsingular on A because, as readily verified, the matrices $\mathbf{f}'(\mathbf{x})$ and $\mathbf{u}'(\mathbf{x})$ have the same determinant. Next we define

$$B = \left\{ \mathbf{z} : \begin{bmatrix} \mathbf{0} \\ \mathbf{z} \end{bmatrix} \in f(N) \right\} \subset \mathcal{R}^k$$

which is open by (6.9) because $f(N)$ is open by (6.90). All that remains is to define w via

$$w_i\left(\begin{bmatrix} z_1 \\ \vdots \\ z_k \end{bmatrix}\right) = g_i\left(\begin{bmatrix} 0 \\ \vdots \\ 0 \\ z_1 \\ \vdots \\ z_k \end{bmatrix}\right), \qquad i = 1, 2, \ldots, m$$

and verify that it has the required properties. If $\mathbf{x} \in A$ then clearly $u(\mathbf{x}) = \mathbf{0}$ iff

$$K\mathbf{x} \in B \quad \text{and} \quad f(\mathbf{x}) = \begin{bmatrix} \mathbf{0} \\ K\mathbf{x} \end{bmatrix} \quad \text{iff} \quad g\left(\begin{bmatrix} \mathbf{0} \\ K\mathbf{x} \end{bmatrix}\right) = \mathbf{x}$$

Finally, suppose there were a second transformation Ω such that whenever $K\mathbf{x} \in B$ then

$$w(K\mathbf{x}) = \begin{bmatrix} x_1 \\ \vdots \\ x_m \end{bmatrix} \quad \text{iff} \quad \Omega(K\mathbf{x}) = \begin{bmatrix} x_1 \\ \vdots \\ x_m \end{bmatrix}$$

This would imply

$$f\left(\begin{bmatrix} w(K\mathbf{x}) \\ K\mathbf{x} \end{bmatrix}\right) = f\left(\begin{bmatrix} \Omega(K\mathbf{x}) \\ K\mathbf{x} \end{bmatrix}\right)$$

for all $\mathbf{x} \in A$ so that necessarily $w(\mathbf{z}) = \Omega(\mathbf{z})$ for all $\mathbf{z} \in B$. ■

As we noted at the beginning of this section, the present collection of theorems is used for optimization theory in Chapter 7. Specifically, they permit us to work with the following, which unlike the geometric figures of Chapter 4, are not necessarily defined by linear expressions.

Definition. Consider $u: D \subset \mathcal{R}^n \to \mathcal{R}^m$ and $k = n - m$. If
(1) $k \geqq 1$,
and on the open set D
(2) $u \in C^{(p)}$ where $p \geqq 1$,
(3) $\operatorname{rank} u' = m$,
then $S = \{\mathbf{x} : u(\mathbf{x}) = \mathbf{0}\}$ is called a *smooth hypersurface* in $\mathcal{R}^n$. (6.94)

The three imposed conditions guarantee that we are able to use the inverse transformation theorem in working with S. The integer k is called the *dimension* of S, $\dim S$, and it must be positive if S is to be an infinite set. Condition (2) justifies the label "smooth." Some authors impose (2) and (3) separately (rather than requiring that they hold for all $\mathbf{x} \in S$) and refer to points where (3) holds as "regular" or "nonsingular" points of S.

In order to work near a given point on a smooth hypersurface we need some special concepts. Consider a transformation (on a closed set!)

$$\phi : [a, b] \to \mathcal{R}^n \tag{6.95}$$

If for some $\delta > 0$ the set of images

$$C = \{\mathbf{x} : \mathbf{x} = \phi(t) \quad \text{for some } t \in (-\delta, \delta)\}$$

is a nonempty subset of S then we say that ϕ *represents* the *curve* C *in* S. We define continuity and differentiability of ϕ in terms of familiar one-sided limits from elementary calculus. A coordinate function ϕ_i, where $i = 1, 2, \ldots, n$, is *continuous* at $t = a$ if $\phi_i(a)$ equals the RH limit of ϕ_i at that point and it is *differentiable* there if the RH derivative exists and is assigned as the value of $\phi_i'(a)$. At $t = b$ we similarly assign LH objects whenever they exist. Then ϕ is *continuous* if all of its coordinate functions are continuous on $[a, b]$, and ϕ is a *transformation* of *Class* $C^{(s)}$, written $\phi \in C^{(s)}$ if all of its coordinate functions have sth derivatives that are continuous on $[a, b]$. By 6.1.7(1) and 6.2.4, $\phi \in C^{(s)}$ iff ϕ is s-times *continuously differentiable* and it is customary to use this terminology interchangeably with $\phi \in C^{(s)}$. Usually we are interested in $\phi \in C^{(p)}$ on (6.94) so that we can use the chain rule to differentiate $u(\phi(t))$.

A vector $\mathbf{h} \in \mathcal{R}^n$ is a *tangent vector to* S at $\mathbf{x}_0 \in S$ if

$$\phi(0) = \mathbf{x}_0 \quad \text{and} \quad \phi'(0) = \mathbf{h}$$

for some ϕ representing a curve in S. The set of all tangent vectors to S at

$\mathbf{x}_0$ is called the *tangent space* T of S at $\mathbf{x}_0$. It is of great significance that T is a subspace of $\mathcal{R}^n$, actually a null space of dimension $k = \dim S$, as stated in the

TANGENT SPACE THEOREM. The tangent space of the smooth hypersurface (6.94) at $\mathbf{x}_0$ is the subspace

$$T = \operatorname{Null} u'(\mathbf{x}_0)$$

and $\dim T = k$. (6.96)

PROOF. If $\mathbf{h} \in T$ then $u(\phi(t)) = 0$ for the associated ϕ and $t \in (-\delta, \delta)$ because then $\mathbf{x} = \phi(t) \in S$. By (6.32) the derivative of u with respect to t is the zero transformation, and calculating that derivative by the chain rule (6.64)

$$\mathbf{u}'(\mathbf{x}_0)\phi'(0) = \mathbf{u}'(\mathbf{x}_0)\mathbf{h} = \mathbf{0}$$

shows $\mathbf{h} \in \operatorname{Null} u'(\mathbf{x}_0)$. Conversely, if

$$\mathbf{h} \in \operatorname{Null} u'(\mathbf{x}_0)$$

we define $\psi : (-\varepsilon, \varepsilon) \to D$ with values

$$\psi(t) = g\left(\left[0, \ldots, 0, x_{m+1,0} + th_{m+1}, \ldots, x_{n0} + th_n\right]^T\right)$$

where g is the local inverse of the special transformation f found in the proof of (6.93), where $\varepsilon > 0$ is sufficiently small so that

$$t \in (-\varepsilon, \varepsilon) \quad \text{implies} \quad \mathbf{x}_0 + t\mathbf{h} \in A$$

and where we have used the notation introduced below (2.13) for components of $\mathbf{x}_0$. Clearly $\psi(0) = \mathbf{x}_0$ and ψ represents a curve in S because

$$\psi(t) = \mathbf{x}_0 + t\mathbf{h}$$

and $u(\mathbf{x}_0 + t\mathbf{h}) = \mathbf{0}$ from (6.93) by construction. By the chain rule (compare (2) for $m = n$ in Example 6.2)

$$\psi'(0) = g'\left(\left[0, \ldots, 0, x_{m+1,0}, \ldots, x_{n0}\right]^T\right)\left[0, \ldots, 0, h_{m+1}, \ldots, h_n\right]^T$$

and to verify $\mathbf{h} \in T$ it remains to show $\psi'(0) = \mathbf{h}$. First, we can differentiate the coordinates of f to find

$$f'(\mathbf{x}_0)\mathbf{h} = \left[u_1'(\mathbf{x}_0)\mathbf{h}, \ldots, u'_m(\mathbf{x}_0)\mathbf{h}, h_{m+1}, \ldots, h_n\right]^T$$

and consequently

$$f'(\mathbf{x}_0)\mathbf{h} = [0,\ldots,0,h_{m+1},\ldots,h_n]^T$$

because $\mathbf{h} \in \text{Null}\, u'(\mathbf{x}_0)$. Second, from 6.90(2)

$$g'([0,\ldots,0,x_{m+1,0},0,\ldots,x_n]^T)f'(\mathbf{x}_0) = I$$

If we now use the LHS to replace I in $I\mathbf{h} = \mathbf{h}$ and then substitute the final expression for $f'(\mathbf{x}_0)\mathbf{h}$ we obtain $\psi'(0) = \mathbf{h}$. Finally,

$$\text{nullity}\, u'(\mathbf{x}_0) = n - m = k$$

by (2.28) because $\text{rank}\, u'(\mathbf{x}_0) = m$ by (6.94). ■

PROBLEMS FOR SOLUTION

6.4.1 Inverse Transformation Theorem. Find or establish:

(1) The largest possible open sets D for which (6.90) applies in 6.2.3(1), (5), and (6).

(2) Transformation 6.2.3(7) shows that (6.90) cannot be used to establish a *global inverse* which is an inverse on the image of the entire domain.

6.4.2 An Open Transformation Theorem. A transformation that maps open sets into open sets is called an *open transformation*. (Recall from the discussion following (6.9) that continuity is in general not a sufficient condition.) Establish: If f satisfies the hypotheses of (6.90), then it is an open transformation.

6.4.3 Regular Transformations. The transformation $f: D \subset \mathcal{R}^n \to \mathcal{R}^n$ is *regular* if the following hold on the open set D: (i) f is one-to-one, (ii) f is of Class $C^{(1)}$, and (iii) f' is nonsingular. Establish

(1) A regular transformation has an inverse that is regular.

(2) A regular transformation maps an open connected set into an open connected set.

(3) The composition of two regular transformations is regular.

6.4.4 Functional Dependence. This generalization of linear dependence entails the existence of a functional (not necessarily linear) relation linking a number of functions. Prove the following basic result (which can be generalized to apply to the situation where the determinant is of rank m so that $n-m$ of the functions can be expressed in terms of the remaining m).

THEOREM. Let $f: D \subset \mathcal{R}^n \to \mathcal{R}^n$ be a transformation of Class $C^{(1)}$ on the open set D. Then given $\mathbf{x}_0 \in D$ a necessary and sufficient condition that there exists a function α on $\mathcal{R}^n$ such that $\alpha\,(\mathbf{g}) \neq 0$ for at least one $\mathbf{g}$ in every open $G \subset \mathrm{R}^n$ and

$$\alpha(f_1(\mathbf{x}),\ldots,f_n(\mathbf{x}))=0$$

for all $\mathbf{x}$ in some neighborhood of $\mathbf{x}_0$ is that $f'(\mathbf{x})$ be singular for all $\mathbf{x} \in D$.

6.4.5 Implicit Function Theorem. Establish for (6.93):

(1) The theorem does not guarantee that we can solve for any m of the x_i in terms of the remaining $n-m$.

(2) The example $(x_1-x_2)^2=0$ shows that the conditions are not necessary.

(3) Whether or not $x^5+y^5+xy=3$ has a unique solution for y near $x=y=1$.

(4) The two variable case can be stated as follows in conventional notation.

THEOREM. Let $f(\cdot,\cdot)$ be a function of Class $C^{(1)}$ in some neighborhood of (x_0, y_0) and suppose

$$f(x_0,y_0)=0 \quad \text{and} \quad \partial_2 f(x,y)\neq 0$$

Then there is a $\delta>0$ and a unique function g of Class $C^{(1)}$ on $N_\delta(x_0)$ such that $y_0=g(x_0)$ and $f(x,g(x))=0$ for all $x\in N_\delta(x_0)$. Furthermore,

$$g'(x)=-\frac{\partial f/\partial x}{\partial f/\partial y}$$

on this neighborhood.

(5) The counterpart theorem for three variables using $f(\cdot,\cdot,\cdot)$, (x_0, y_0, z_0), $N_\delta(x_0, y_0)$, and $g(\cdot,\cdot)$, and show how to find $\partial z/\partial x$ and $\partial z/\partial y$

(6) The following is an equivalent form.

THEOREM. Let f be a transformation on the open set $D\subset \mathcal{R}^n \times \mathcal{R}^m$ to $\mathcal{R}^m$ of Class $C^{(p)}$, where $p\geqq 1$, at $(\mathbf{x}_0, \mathbf{y}_0)\in D$. Suppose $f(\mathbf{x}_0, \mathbf{y}_0)=\mathbf{0}$ and $\operatorname{rank} f'(\mathbf{x}_0, \mathbf{y}_0)=m$ where in fact

$$\begin{bmatrix} \dfrac{\partial f_1}{\partial y_1} & \dfrac{\partial f_1}{\partial y_2} & \cdots & \dfrac{\partial f_1}{\partial y_m} \\ \dfrac{\partial f_2}{\partial y_1} & \dfrac{\partial f_2}{\partial y_2} & \cdots & \dfrac{\partial f_2}{\partial y_m} \\ \vdots & \vdots & & \vdots \\ \dfrac{\partial f_m}{\partial y_1} & \dfrac{\partial f_m}{\partial y_2} & \cdots & \dfrac{\partial f_m}{\partial y_m} \end{bmatrix}$$

is nonsingular at $(\mathbf{x}_0, \mathbf{y}_0)$. Then there exists $\delta_1>0$ such that the preceding matrix is nonsingular on $N_{\delta_1}(\mathbf{x}_0, \mathbf{y}_0)$. Furthermore, there exist $\delta_2 > 0$ and $g: N_{\delta_2}(\mathbf{x}_0) \subset \mathcal{R}^n \to \mathcal{R}^m$ of Class $C^{(p)}$ such that $\mathbf{y}_0 = g(\mathbf{x}_0)$ and $f(\mathbf{x}, g(\mathbf{x})) = \mathbf{0}$ for all $\mathbf{x} \in N_{\delta_2}(\mathbf{x}_0)$.

6.4.6 Tangent Space Theorem. Establish:

(1) Correspondence between the two concepts of surfaces: (6.41) and (6.94).

(2) The set of vectors *normal* to a smooth hypersurface S at $\mathbf{x}_0$, that is, the set of vectors orthogonal to every vector in the tangent space T, forms a vector space of dimension m.

(3) The following is an acceptable definition for the k-plane that is tangent to a smooth hypersurface S at $\mathbf{x}_0$

$$P = \{\mathbf{x} : \mathbf{x} = \mathbf{x}_0 + \mathbf{h}, \mathbf{h} \in T\}$$

6.5 CONVEX FUNCTIONS

As shown in Table 4.1 there are many special kinds of convex sets in $\mathcal{R}^n$. In this section we develop properties of a special kind of associated function that is used in Section 7.5. Our ultimate objective is Procedure (6.107).

A function $f: D \subset \mathcal{R}^n \to \mathcal{R}$, where D is convex, is a *convex function* if

$$f(t\mathbf{x}_1 + (1-t)\mathbf{x}_2) \leqq tf(\mathbf{x}_1) + (1-t)f(\mathbf{x}_2) \tag{6.97}$$

for every $\mathbf{x}_1, \mathbf{x}_2 \in D$ and every $t \in [0, 1]$. The geometric interpretation of (6.97) is that all points on the line segment on the RHS lie on or above the corresponding point on the "surface" $z = f(\mathbf{x})$ on the LHS. Authors have defined convex functions using $t = 1/2$ only but these are now called *midconvex functions*. The function f is *strictly convex* if strict inequality holds whenever $\mathbf{x}_1 \neq \mathbf{x}_2$ and $t \in (0, 1)$. The function f is *concave*, or *strictly concave*, if its negative $-f$ is convex, or strictly convex, respectively. Since we can obtain results for concave functions by replacing f by $-f$ in results for convex functions, it generally suffices to confine our attention to the latter. Notable exceptions occur in optimization where, for example, minima of the two kinds of functions have different properties.

The definition of a convex function can be based on

JENSEN'S INEQUALITY. If f is a convex function on D and

$$\mathbf{x} = t_1\mathbf{x}_1 + \cdots + t_r\mathbf{x}_r$$

is a convex combination of points $\mathbf{x}_i \in D$ then

$$f(\mathbf{x}) \leqq t_1 f(\mathbf{x}_1) + \cdots + t_r f(\mathbf{x}_r) \tag{6.98}$$

PROOF. The conclusion is trivial for $r=1$ and it holds for $r=2$ by (6.97). It follows by induction for any r using the technique in the proof for (4.19). ■

The next result is but one of several boundedness properties that hold for convex functions; we need it in (6.100) to establish continuity.

If f is convex on D and $A \subset D$ is compact then f is bounded above on A. (6.99)

PROOF. Since A is compact it can be covered with a finite number of closed simplexes, each of which is contained in D. Now every $\mathbf{x} \in A$ can be written in terms of its barycentric coordinates

$$\mathbf{x} = t_1\mathbf{p}_1 + \cdots + t_{k+1}\mathbf{p}_{k+1}$$

The conclusion follows with Jensen's inequality which shows that $f(\mathbf{x})$ cannot exceed the largest value it assumes on the $k+1$ vertexes $\mathbf{p}_i$. ■

Next we present the elegant proof of Bourbaki (1953) for the

THEOREM. A convex function on an open domain is continuous. (6.100)

PROOF. We lose no generality, but we gain considerable simplicity, by showing that f is continuous at $\mathbf{0}$ where $f(\mathbf{0})=0$. Suppose that $\varepsilon \in (0,1)$ is specified and we select some specific $\delta > 0$ so that we can appeal to (6.99) to write

$$f(\mathbf{x}) \leqq M \quad \text{whenever } \|\mathbf{x}\| < \delta$$

We then define $t = \varepsilon M^{-1}$ and consider an arbitrary $\mathbf{y}$ satisfying $\|\mathbf{y}\| < t\delta$. Since f is convex,

$$\mathbf{y} = (1-t)\mathbf{0} + t(t^{-1}\mathbf{y})$$

implies $f(\mathbf{y}) \leqq tf(t^{-1}\mathbf{y}) \leqq \varepsilon$. In the same way

$$\mathbf{0} = \frac{t}{1+t}(-t^{-1}\mathbf{y}) + \frac{1}{1+t}\mathbf{y}$$

implies $0 \leqq \varepsilon + f(\mathbf{y})$ so $f(\mathbf{y}) \in N_\varepsilon(\mathbf{0})$ whenever $\mathbf{y} \in N_{t\delta}(\mathbf{0})$. ■

Let us consider a way of characterizing a convex function that leads to criteria for differentiability and to the useful result (6.103) which holds whenever f is differentiable. We say that any function $f: D \subset \mathcal{R}^n \to \mathcal{R}$ has *support* at $\mathbf{x} \in D$ if there exists at least one $\mathbf{p} \in \mathcal{R}^n$ such that either of the following two equivalent relations holds for every $\mathbf{u} \in D$.

$$(1) \quad f(\mathbf{u}) \geqq f(\mathbf{x}) + \langle \mathbf{p}, \mathbf{u}-\mathbf{x} \rangle \qquad (6.101)$$

$$(2) \quad f(\mathbf{x}) \leqq f(\mathbf{u}) + \langle \mathbf{p}, \mathbf{x}-\mathbf{u} \rangle$$

The (geometric) *supports* are the affine RHSs which equal $f(\mathbf{x})$ for $\mathbf{u}=\mathbf{x}$. Given that D is open as well as convex, the existence of support (unique or not) at each $\mathbf{x}\in D$ is necessary and sufficient for convexity of f; see Roberts and Varberg (1973, Section 43). For our purposes it suffices to establish

If f has support at each point in its convex domain D then f is convex. (6.102)

PROOF. Let $\mathbf{x}=t\mathbf{x}_1+(1-t)\mathbf{x}_2$ where $\mathbf{x}_1,\mathbf{x}_2\in D$ and $t\in(0,1)$. Then $\mathbf{x}=t\mathbf{x}+(1-t)\mathbf{x}$ and

$$\mathbf{0}=t(\mathbf{x}-\mathbf{x}_1)+(1-t)(\mathbf{x}-\mathbf{x}_2)$$

If we substitute this expression in

$$f(\mathbf{x})=f(\mathbf{x})-\langle\mathbf{p},\mathbf{0}\rangle$$

and "switch signs," then its RHS becomes

$$t\left[f(\mathbf{x})+\langle\mathbf{p},\mathbf{x}_1-\mathbf{x}\rangle\right]+(1-t)\left[f(\mathbf{x})+\langle\mathbf{p},\mathbf{x}_2-\mathbf{x}\rangle\right]$$

Using 6.101(1) for $\mathbf{u}=\mathbf{x}_1$ and $\mathbf{u}=\mathbf{x}_2$ shows that the preceding expression cannot exceed $t[f(\mathbf{x}_1)]+(1-t)[f(\mathbf{x}_2)]$. ■

When f is differentiable, there are useful characterizations of convexity that generalize the familiar situation for $n=1$. The first appears as (1) in (6.103) and states that $f(\mathbf{x})$ lies on or above its tangent hyperplane (6.42) at each point $[x_1,\ldots,x_n,f(\mathbf{x})]^T\in\mathfrak{R}^{n+1}$. That is, whenever $\mathbf{x},\mathbf{x}+\mathbf{h}\in D$

$$f(\mathbf{x}+\mathbf{h})\geqq f(\mathbf{x})+\mathbf{f}'(\mathbf{x})\mathbf{h}$$

where the RHS is the "height" z to the tangent hyperplane at the stated point in $\mathfrak{R}^{n+1}$. Equivalent ways to express this are: the tangent hyperplane never overestimates $f(\mathbf{x}+\mathbf{h})$; and, f has support at each $\mathbf{x}_1$ with $\mathbf{p}=\nabla f(\mathbf{x}_1)$. Condition (2) of (6.103) generalizes the familiar situation in elementary calculus where f' nondecreasing means

$$x_2>x_1 \quad \text{implies} \quad f'(x_2)\geqq f'(x_1)$$

This involves four real numbers and is equivalent to

$$\left[f'(x_2)-f'(x_1)\right](x_2-x_1)\geqq 0$$

The counterpart makes sense on $\mathfrak{R}^n$ because the first term is a matrix, the second is a vector, and the two are conformable. If $f:D\subset\mathfrak{R}^n\to\mathfrak{R}$, where D is open, is differentiable, we therefore call its derivative f' *nondecreasing* if

$$[f'(\mathbf{x}_2)-f'(\mathbf{x}_1)](\mathbf{x}_2-\mathbf{x}_1)\geqq 0$$

for every $\mathbf{x}_1,\mathbf{x}_2\in D$; if the strict inequality holds whenever $\mathbf{x}_1\neq\mathbf{x}_2$ we call f' *increasing*. Usually we use the equivalent inequality

$$\langle\nabla f(\mathbf{x}_2)-\nabla f(\mathbf{x}_1),\mathbf{x}_2-\mathbf{x}_1\rangle\geqq 0$$

as in the

THEOREM. If f is differentiable on its convex open domain D then f is convex iff each of the following two equivalent conditions holds for all $\mathbf{x}_1,\mathbf{x}_2\in D$

(1) $f(\mathbf{x}_2)\geqq f(\mathbf{x}_1)+\langle\nabla f(\mathbf{x}_1),\mathbf{x}_2-\mathbf{x}_1\rangle$

(2) $\langle\nabla f(\mathbf{x}_2)-\nabla f(\mathbf{x}_1),\mathbf{x}_2-\mathbf{x}_1\rangle\geqq 0$ (6.103)

PROOF. Let us first show that convexity implies (1). Since $\mathbf{x}_1+t\mathbf{h}=t\mathbf{x}_2+(1-t)\mathbf{x}_1$ where $\mathbf{h}=\mathbf{x}_2-\mathbf{x}_1$ and $t\in(0,1)$, we have

$$f(\mathbf{x}_1+t\mathbf{h})\leqq tf(\mathbf{x}_2)+(1-t)f(\mathbf{x}_1)$$

and adding $-t\mathbf{f}'(\mathbf{x}_1)\mathbf{h}$ to both sides yields

$$\frac{1}{t}\left[f(\mathbf{x}_1+t\mathbf{h})-f(\mathbf{x}_1)-t\mathbf{f}'(\mathbf{x}_1)\mathbf{h}\right]\leqq f(\mathbf{x}_2)-f(\mathbf{x}_1)-\mathbf{f}'(\mathbf{x}_1)\mathbf{h}$$

By (6.37) the LHS goes to zero as $t\in(0,1)$ goes to zero so that the RHS, which does not involve t, must be nonnegative. Second, (1) implies (2) as we see by combining (1) as written with

$$f(\mathbf{x}_1)-f(\mathbf{x}_2)\geqq\langle\nabla f(\mathbf{x}_2),\mathbf{x}_1-\mathbf{x}_2\rangle$$

obtained by interchanging $\mathbf{x}_1$ and $\mathbf{x}_2$. Third, (2) implies (1) because we can use the mean value theorem (6.43) with $\mathbf{u}=(1-t)\mathbf{x}_1+t\mathbf{x}_2$ to write

$$f(\mathbf{x}_2)-f(\mathbf{x}_1)=\langle\nabla f(\mathbf{x}_1),\mathbf{x}_2-\mathbf{x}_1\rangle+t\langle\nabla f(\mathbf{x}_2)-\nabla f(\mathbf{x}_1),\mathbf{x}_2-\mathbf{x}_1\rangle$$

Finally, (1) guarantees that f has support at each point in D and so by (6.102) it implies that f is convex on D. ■

As presented in 6.5.3, we can adapt almost the same proof technique (there need to be some alterations) to establish

Conditions (1) and (2) of (6.103) are each necessary and sufficient for strict convexity if strict inequalities hold for all $\mathbf{x}_1\neq\mathbf{x}_2$. (6.104)

Applying the previous two results to $-f$ yields necessary and sufficient conditions for f to be concave and strictly concave.

Whenever (5.28) can be used, as for example when the elements of the Hessian are constants, the following guarantees that it can be applied to the problem of resolving convexity.

THEOREM. If $f \in C^{(2)}$ on its convex open domain D then f is convex iff $\mathbf{f}''(\mathbf{x})$ is positive semidefinite at each $\mathbf{x} \in D$. (6.105)

PROOF. If f is convex then so is g, where $g(t) = f(\mathbf{x} + t\mathbf{h})$, for all $\mathbf{x} \in D$ and all $t \in T$ where T is some real interval containing zero, and where $\mathbf{h} \in D$ is arbitrary. By the chain rule (6.64) $g''(t) = \mathbf{h}^T\mathbf{f}''(\mathbf{x} + t\mathbf{h})\mathbf{h}$, and $g''(0) = \mathbf{h}^T\mathbf{f}''(\mathbf{x})\mathbf{h} \geqq 0$ by (1.14). To prove the converse we use Taylor's theorem (6.88) to write

$$f(\mathbf{x}+\mathbf{h}) = f(\mathbf{x}) + \mathbf{f}'(\mathbf{x})\mathbf{h} + \frac{1}{2}\mathbf{h}^T\mathbf{f}''(\mathbf{x}+t\mathbf{h})\mathbf{h}$$

for some $t \in (0, 1)$. Then clearly if $\mathbf{f}''(\mathbf{x})$ is positive semidefinite

$$f(\mathbf{x}+\mathbf{h}) \geqq f(\mathbf{x}) + \mathbf{f}'(\mathbf{x})\mathbf{h}$$

and f is convex on D by (6.103). ■

From this theorem we see

$$\mathbf{x} \mapsto \mathbf{x}^T\mathbf{A}^T\mathbf{A}\mathbf{x}$$

is convex for any $m \times n$ $\mathbf{A}$. Any "sum of squares"

$$\mathbf{x} \to \mathbf{x}^T\,\mathbf{diag}(\lambda_1, \lambda_2, \ldots, \lambda_n)\mathbf{x}$$

is convex iff all $\lambda_i \geqq 0$. Next we have

If $f \in C^{(2)}$ on its convex open domain D then positive definiteness of $\mathbf{f}''(\mathbf{x})$ for all $\mathbf{x} \in D$ is sufficient, but not necessary, for f to be strictly convex on D. (6.106)

PROOF. Sufficiency follows from the corresponding proof of (6.105) with appeal to (6.104). A simple counterexample disproves necessity. ■

Thus

$$\mathbf{x} \mapsto \mathbf{x}^T\mathbf{I}\mathbf{x} = \|\mathbf{x}\|^2$$

is strictly convex on $\mathscr{R}^n$ but $\mathbf{x} \mapsto \|\mathbf{x}\|$ is not strictly convex except on highly restricted $D \subset \mathscr{R}^n$.

How do we determine whether or not a particular function is convex? Usually it is difficult to rely on the definition (6.97) alone and much the same can be said for the other criteria from (6.98) through (6.104). A reasonable approach is the following.

ESTABLISHING CONVEXITY. The convexity of a substantial number of functions can be established using the particular functions and results in Example 6.5. Most of these latter do not involve differentiation; when the function is sufficiently differentiable, use of (6.105), (6.87), and (5.28) should be considered. (6.107)

Experience confirms that use of (6.105) is generally not a preferred method, given that some other result can be used.

EXAMPLE 6.5. Particular Convex Functions. A constant function is trivially convex, and a function is both convex and concave iff it is affine (6.5.4). By (6.105) the quadratic form

$$\mathbf{x} \mapsto \mathbf{x}^T \mathbf{A} \mathbf{x}$$

is convex on $\mathcal{R}^n$ iff $\mathbf{A}$ is positive semidefinite. As noted in the text, $\mathbf{x} \to \|\mathbf{x}\|^2$ is strictly convex on $\mathcal{R}^n$ while $\mathbf{x} \to \|\mathbf{x}\|$ is not. In summary, a model convex function is

$$\mathbf{x} \mapsto a + \mathbf{b}^T \mathbf{x} + \mathbf{x}^T \mathbf{C} \mathbf{x}$$

where $a \in \mathcal{R}$, $\mathbf{b} \in \mathcal{R}^n$, and $\mathbf{C}$ is positive semidefinite of order n.

A number of convex functions can be constructed using the result

(1) Nonnegative multiples of convex functions, the sum of convex functions, and consequently LCs with nonnegative coefficients of convex functions, all on a common convex domain, are convex.

which is readily established.

Many additional examples on $\mathcal{R}^n$ can be established using composite functions involving convex functions on $\mathcal{R}$. Important instances of these latter are the strictly convex functions corresponding to the following.

(2) e^u and $|u+a|^p$, $p>1$, on $\mathcal{R}$
(3) $u \log u$ and $-\log u$ on $(0, \infty)$
(4) u^p, $p>1$, on $[0, \infty)$
(5) $-u^p$, $p \in (0, 1)$, on $[0, \infty]$

A result that is particularly easy to combine with the above is "a convex function of a linear function is convex." More precisely,

(6) Suppose the linear transformation $\mathbf{x} \mapsto \mathbf{A}\mathbf{x}$ maps the convex set $D \subset \mathcal{R}^n$ into $E \subset \mathcal{R}^m$. If $g: E \to \mathcal{R}$ is convex on E, then the composite function $g(\mathbf{A}\mathbf{x})$ is convex on D.

PROOF. It can easily be shown that E is convex. Continuing, suppose

$\mathbf{x}_1, \mathbf{x}_2 \in D$ and $t \in (0,1)$. If $\mathbf{x} = t\mathbf{x}_1 + (1-t)\mathbf{x}_2$ and $\mathbf{u}_i = \mathbf{A}\mathbf{x}_i$, then also

$$\mathbf{u} = \mathbf{A}\mathbf{x} = t\mathbf{u}_1 + (1-t)\mathbf{u}_2$$

so clearly $g(\mathbf{A}\mathbf{x}) \leqq tg(\mathbf{A}\mathbf{x}_1) + (1-t)g(\mathbf{A}\mathbf{x}_2)$ and g is convex. ■

For example, $\mathbf{x} \mapsto \exp \mathbf{a}^T\mathbf{x}$ and $\mathbf{x} \mapsto \exp \mathbf{x}^T\mathbf{A}\mathbf{x}$ are convex on $\mathfrak{R}^n$ for $\mathbf{a} \in \mathfrak{R}^n$ and $\mathbf{A}$ positive semidefinite, respectively. For $\mathbf{b} > \mathbf{0}$, $\mathbf{x} \mapsto \log \mathbf{b}^T\mathbf{x}$ is concave on the positive orthant.

Another useful result states "a nondecreasing convex function of a convex function is convex," namely

(7) Let $f: D \subset \mathfrak{R}^n \to \mathfrak{R}$ be convex on D while $g: I \subset \mathfrak{R} \to \mathfrak{R}$ is both convex and nondecreasing on I. If $f(D) \subset I$ then the composite function $g(f(\mathbf{x}))$ is convex on D.

PROOF. If $\mathbf{x} = t\mathbf{x}_1 + (1-t)\mathbf{x}_2$ where $\mathbf{x}_1, \mathbf{x}_2 \in D$ and $t \in (0,1)$, then $f(\mathbf{x}) \leqq tf(\mathbf{x}_1) + (1-t)f(\mathbf{x}_2)$. Because g is nondecreasing

$$g(f(\mathbf{x})) \leqq g(tf(\mathbf{x}_1) + (1-t)f(\mathbf{x}_2))$$

and the RHS cannot exceed

$$tg(f(\mathbf{x}_1)) + (1-t)g(f(\mathbf{x}_2))$$

because g is convex. ■

Thus $\mathbf{x} \mapsto \|\mathbf{x} - \mathbf{a}\|^p$ is convex on $\mathfrak{R}^n$ for all $\mathbf{a} \in \mathfrak{R}^n$ and $p \geqq 1$. If f is positive and $\log f$ is convex then f is convex since $f(x) = \exp(\log f(\mathbf{x}))$. Such functions are called *log-convex*.

Many extensions of elementary inequalities can be established using various convex functions f and numbers t in the definition (6.97); see, for example, Roberts and Varberg (1973, Chapter VI). As an example of the reverse process let us establish

(8) If $f: D \subset \mathfrak{R}^n \to \mathfrak{R}$ is convex on D then $\mathbf{x} \mapsto 1/f(\mathbf{x})$ is concave on $N = \{\mathbf{x} : f(\mathbf{x}) < 0\}$.

PROOF. Since f is convex, for $\mathbf{x}_1, \mathbf{x}_2 \in N$ and $t \in (0,1)$ we have

$$\left[-f(t\mathbf{x}_1 + (1-t)\mathbf{x}_2)\right]^{-1} \leqq -\left[tf(\mathbf{x}_1) + (1-t)f(\mathbf{x}_2)\right]^{-1}$$

and the RHS cannot exceed

$$t\left[-f(\mathbf{x}_1)\right]^{-1} + (1-t)\left[-f(\mathbf{x}_2)\right]^{-1}$$

by 1.2.3(2). ■

PROBLEMS FOR SOLUTION

6.5.1 Criteria. Show that each of the following is necessary and sufficient for f to be convex on its convex domain D.

(1) Jensen's Inequality (6.98) holds for every set of r points from D.

(2) If $\mathbf{x}_1, \mathbf{x}_2 \in D$ and $t \in (0,1)$ then

$$f(t\mathbf{x}_1 + (1-t)\mathbf{x}_2) < ta + (1-t)b$$

whenever $f(\mathbf{x}_1) < a$ and $f(\mathbf{x}_2) < b$.

(3) The *epigraph of f*

$$E = \left\{ \begin{bmatrix} \mathbf{x} \\ z \end{bmatrix} : f(\mathbf{x}) \leqq z \right\}$$

is a convex set in $\mathcal{R}^{n+1}$.

(4) Consider $c \in \mathcal{R}$ and the associated set

$$C = \{\mathbf{x} : f(\mathbf{x}) \leqq c\}.$$

Show:

(i) If f is a convex function then C is a convex set.

(ii) The converse need not hold; that is, for every $c \in \mathcal{R}$ the associated C can be a convex set and yet f is not a convex function (6.5.8).

6.5.2 Examples. Furnish simple examples where the conclusions in (6.99) and (6.100) fail because the special conditions are not met.

6.5.3 Strict Convexity. Let (1)′ and (2)′ denote the respective conditions in (6.103) with $\geqq$ replaced by $>$. Establish:

(1) Strict convexity implies (1)′.

(2) Condition (1)′ implies (2)′.

(3) Condition (2)′ implies that f is strictly convex.

(4) A simple example where f is strictly convex at $\mathbf{x}$ but $\mathbf{f}''(\mathbf{x})$ is not positive definite.

6.5.4 Verifications for Example 6.5. Establish:

(1) A function f is both convex and concave iff $f(\mathbf{x}) = a + \langle \mathbf{b}, \mathbf{x} \rangle$ for some $a \in \mathcal{R}$ and $\mathbf{b} \in \mathcal{R}^n$, that is, iff f is affine.

(2) $\mathbf{x} \mapsto \|\mathbf{x}\|$ is not strictly convex on any domain containing two distinct points lying on a line through the origin.

(3) Statements (1)–(5).

(4) The set E in (6) is convex.

6.5.5 Additional Convex Functions. Establish:

(1) If f and g are convex functions on a common domain D then

$$\mathbf{x} \mapsto \max\{f(\mathbf{x}), g(\mathbf{x})\}$$

is convex on D.

(2) Suppose $A \subset \mathfrak{R}^n$ is nonempty and closed. Then the distance 6.1.8(5) from $\mathbf{x}$ to A is a convex function iff A is a convex set.

(3) The convexity or nonconvexity of monic polynomials of degrees 0 through 5 on $\mathfrak{R}$.

(4) The projection $\mathbf{x} \mapsto x_i$ is convex on $\mathfrak{R}^n$ for $i = 1, 2, \ldots, n$.

(5) The product of two nonnegative, nondecreasing, convex functions on $I \subset \mathfrak{R}$ is nonnegative, nondecreasing, and convex on I.

(6) The generalization of the preceding for functions on $D \subset \mathfrak{R}^n$.

(7) Suppose $f(x) = xg(x)$ where g is twice-differentiable on $P = \{x : x > 0\}$. Then f is convex on P iff $g(1/x)$ is convex on P.

(8) If f is strictly convex, $f(\mathbf{0}) = 0$, and $\mathbf{x} \neq \mathbf{0}$ then $t \mapsto t^{-1}f(\mathbf{x})$ is increasing on $\{t : t > 0\}$.

6.5.6 Log-Convex Functions. Establish for convex D:

(1) f is log-convex iff $f(\mathbf{x}) > 0$ for all $\mathbf{x} \in D$ and

$$f(t\mathbf{x}_1 + (1-t)\mathbf{x}_2) \leqq f^t(\mathbf{x}_1) f^{1-t}(\mathbf{x}_2)$$

whenever $\mathbf{x}_1, \mathbf{x}_2 \in D$ and $t \in (0, 1)$.

(2) The product of two log-convex functions on a common domain is log-convex on that domain.

(3) The counterpart for the sum.

6.5.7 Midconvex Functions. Show that if f is midconvex and continuous on its convex open domain then it is convex.

6.5.8 Quasiconvex Functions. Among the many weaker properties that are used in practice, let us cite; $f: D \subset \mathfrak{R}^n \to \mathfrak{R}$, where D is convex, is a *quasiconvex function* if the sets C of 6.5.1 are convex for all $c \in \mathfrak{R}$. Establish:

(1) f is quasiconvex on its convex domain D iff

$$f(t\mathbf{x}_1 + (1-t)\mathbf{x}_2) \leqq \max\{f(\mathbf{x}_1), f(\mathbf{x}_2)\}$$

whenever $\mathbf{x}_1, \mathbf{x}_2 \in D$ and $t \in [0, 1]$.

(2) If f is differentiable on its convex open domain D then f is quasiconvex iff

$$f(\mathbf{x}_2) \leqq f(\mathbf{x}_1) \quad \text{implies} \quad f'(\mathbf{x}_1)(\mathbf{x}_2 - \mathbf{x}_1) \leqq 0$$

whenever $\mathbf{x}_1, \mathbf{x}_2 \in D$.

7
OPTIMIZATION THEORY ON $\mathcal{R}^n$

In this book we use *optimization* to refer to matters connected with minima and maxima—in applied contexts they are called *optima*—of functions. In this chapter we use results from Section 5.3 and Chapter 6 to find conditions—necessary or sufficient, or both—that hold at solution points for problems of minimization subject to equality and inequality constraints. We delve deeply into conditions that are verifiable using systematic procedures. But we must omit minimization on infinite-dimensional spaces (such as in the calculus of variations) and on specialized subsets of $\mathcal{R}^n$ (such as sets of points with integer components) where results different from those in this book are required. For methods of finding solution points we cite Avriel (1976) from among the many works that have appeared since the pioneering Fiacco and McCormick (1968).

7.1 GENERAL CONSIDERATIONS

Let us introduce the subject and obtain results for what is sometimes called unconstrained minimization on an open set.

We work with an *objective function* $f: D \subset \mathcal{R}^n \to \mathcal{R}$ where D is an open set. The *constrained minimization problem* is to find $\mathbf{x}_0$ such that

$$f(\mathbf{x}_0) = \min\{f(\mathbf{x}) : \mathbf{x} \in K\}$$

where $K \subset D$ is specified. Such a point $\mathbf{x}_0$ is a *solution point* and K is the *constraint set*. In standard format, the preceding is the *general constraint problem*

$$\begin{aligned} &\text{minimize } f(\mathbf{x}) \\ &\text{s.t. } \mathbf{x} \in K \end{aligned} \tag{7.1}$$

where s.t. is read *subject to*. When $K=D$, (7.1) is the *unconstrained minimization problem*.

Requiring D to be open simplifies matters, and rules out few cases of general interest in this chapter where derivatives are used for nearly all results. Generality is achieved in the present section by permitting K to be any kind of subset of D; but we should point out that results are significant only when $\operatorname{int} K \neq \varnothing$. In particular cases where D cannot be open, it might happen that an equivalent problem can be found; for example, in place of $f(x)=\sqrt{x}$ and $D=K=[0,\infty)$ in (7.1) we can use $f(x)=x^2$ and $D=K=\mathcal{R}$.

It is conventional to refer to K as the *feasible region* or set of *feasible points*. Thus, $\mathbf{x}$ is *feasible* for (7.1) iff $\mathbf{x}\in K$. Points $\mathbf{x}\in\operatorname{int} K$ are *feasible interior points* and points $\mathbf{x}\in K\cap\operatorname{bdry} K$ are *feasible boundary points*. A limit point of K is a *feasible limit point* and an isolated point of K is a *feasible isolated point*; often we need to use the fact that each feasible point can be called one and only one of these two latter names.

An *extremum* of f *over* K is any one of the following types of minima or maxima. There is a *global minimum* at $\mathbf{x}_0\in K$ if $f(\mathbf{x})\geqq f(\mathbf{x}_0)$ holds for all $\mathbf{x}\neq\mathbf{x}_0$ in K; there is a *strict global minimum* there if $f(\mathbf{x})>f(\mathbf{x}_0)$ for all such $\mathbf{x}$. Global minima constitute absolute, or unconditional minima over the entire set K and in practice we must be prepared to deal with less universal phenomena. There is a *local minimum* at $\mathbf{x}_0\in K$ if $f(\mathbf{x})\geqq f(\mathbf{x}_0)$ holds for all

$$\mathbf{x}\in N\cap K$$

where $N=N_\delta(\mathbf{x}_0)$ is some neighborhood (6.3) of $\mathbf{x}_0$; there is a *strict local minimum* there if $f(\mathbf{x})>f(\mathbf{x}_0)$ for all such $\mathbf{x}$. Thus in both cases strict minima are unique minima in the sense that the value $f(\mathbf{x}_0)$ is not duplicated elsewhere in K or in $N\cap K$. We define *global maximum, strict global maximum, local maximum,* and *strict local maximum* by replacing $>$ by $<$ in the respective definitions for minima.

To each type of minimum or maximum there corresponds a similarly designated solution point. Thus we have *global solution points* which are *global minimum points*, or *global maximum points*, or *strict global minimum points*, and so on, for all of the cases introduced. Throughout the chapter we are interested in both necessary and sufficient conditions for $\mathbf{x}_0$ to be a solution point. We use derivatives and call conditions *first-order* or *second-order* to identify the highest order derivative used.

Figure 7.1 shows parts of the graphs of two parabolas where

$$f(x)=1+(x-1)^2 \quad \text{and} \quad f(x)=1+(x+1)^2$$

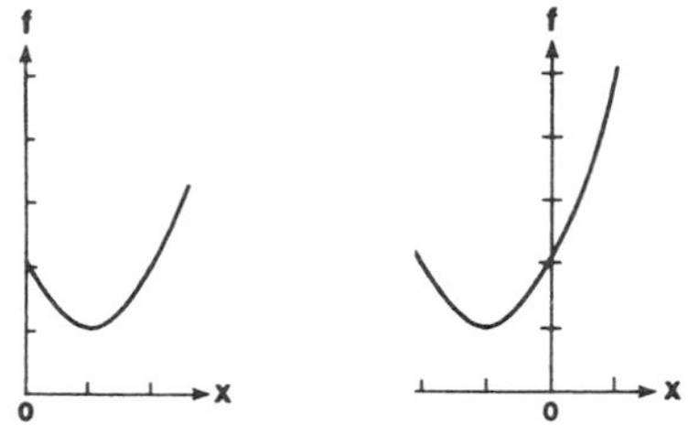

(a) (b) **Figure 7.1.** Global minima.

in (a) and (b), respectively. Let us use the associated functions on $D=\mathfrak{R}$ to illustrate how different choices for K lead to different solution points. First, if $K=D$ there are strict global minimum points $x_0=1$ and $x_0=-1$ in (a) and (b), respectively. Second, if $K=(0,\infty)$ the situation is unchanged in (a) but there is no solution point of any kind in (b); but if $K=[0,\infty)$ then the boundary point $x_0=0$ is the strict global solution point in (b). Third, if

$$K=\left[0,\tfrac{1}{2}\right]\cup\left[\tfrac{3}{2},2\right]$$

then each of $x_0=1/2$ and $x_0=3/4$ is a strict local solution point in (a) while $x_0=0$ is the strict global solution point in (b). In order to illustrate solution points that are not strict solution points consider the change from (b)

$$\begin{aligned} f(x)&=2, && x\leqq 0\\ &=1+(x+1)^2, && x>0 \end{aligned}$$

where for $D=\mathfrak{R}$ every point $x\leqq 0$ is a global minimum point.

In the following sections we work with five cases of Problem (7.1). The *equality constraint problem* is

$$\begin{aligned} &\text{minimize } f(\mathbf{x})\\ &\quad\text{s.t. } h(\mathbf{x})=\mathbf{0} \end{aligned} \tag{7.2}$$

where $h: D\subset\mathfrak{R}^n\to\mathfrak{R}^k$. The *inequality constraint problem* is

$$\begin{aligned} &\text{minimize } f(\mathbf{x})\\ &\quad\text{s.t. } g(\mathbf{x})\leqq\mathbf{0} \end{aligned} \tag{7.3}$$

where $g: D \subset \mathcal{R}^n \to \mathcal{R}^m$. The *mixed constraint problem* is

$$\begin{aligned} &\text{minimize } f(\mathbf{x}) \\ &\text{s.t. } g(\mathbf{x}) \leqq \mathbf{0} \\ &\qquad h(\mathbf{x}) = \mathbf{0} \end{aligned} \tag{7.4}$$

The *convex programming problem* is

$$\begin{aligned} &\text{minimize } f(\mathbf{x}) \\ &\text{s.t. } g(\mathbf{x}) \leqq \mathbf{0} \\ &\qquad \mathbf{A}\mathbf{x} = \mathbf{b} \end{aligned} \tag{7.5}$$

where f, and each coordinate function of g, is a convex function, $\mathbf{A}$ is $k \times n$, and $\mathbf{b} \in \mathcal{R}^k$. The *standard linear programming problem* is

$$\begin{aligned} &\text{minimize } \mathbf{c}^T\mathbf{x} \\ &\text{s.t. } \mathbf{A}\mathbf{x} = \mathbf{b} \\ &\qquad \mathbf{x} \geqq \mathbf{0} \end{aligned} \tag{7.6}$$

where now $\mathbf{A}$ is $m \times n$ and satisfies certain conditions, $\mathbf{b} \in \mathcal{R}^m$, and $\mathbf{c} \in \mathcal{R}^n$. With the exception of Problem (7.5) (where $-f$ will not be convex unless f is affine), it suffices to consider the stated cases because

$$\max\{f(\mathbf{x}) : \mathbf{x} \in K\} = -\min\{-f(\mathbf{x}) : \mathbf{x} \in K\} \tag{7.7}$$

can be used to translate results for minimization into corresponding ones for maximization. The relations that are to hold for coordinate functions in Problems (7.2)–(7.6) are called *individual constraints*. A *constraint qualification* for a case of (7.1) is a set of special conditions imposed on K; these appear as conditions on g, h, and $\mathbf{A}$.

If the constraint set K in (7.1) is closed and bounded then the extreme value theorem (6.19) guarantees that there is a solution point whenever f is continuous. If $\operatorname{int} K \neq \varnothing$ then we can work there in the same way as in any other open subset of D; in particular cases, we can consider any remaining points of K separately. But in all cases for K we can consider the

Definition. A nonzero vector $\mathbf{h} \in \mathcal{R}^n$ is called a *feasible direction* at $\mathbf{x} \in K$ if there exists $\tau > 0$ such that

$$\mathbf{x} + t\mathbf{h} \in K$$

whenever $t \in [0, \tau]$. (7.8)

In words, $\mathbf{x}$ must be the endpoint of a line segment in the direction $\mathbf{h}$ that lies entirely in K. If $\mathbf{x} \in \operatorname{int} K$ then every direction is feasible at $\mathbf{x}$ but counterexamples show that the converse does not hold for every K (7.1.2). There are no feasible directions at a feasible isolated point.

Feasible directions make it possible to work at feasible limit points in a general constraint set K using the associated function on $\mathfrak{R}$

$$t \mapsto f(\mathbf{x}_0 + t\mathbf{h})$$

that figures so prominently in Chapter 6. (See, for example, the proofs of (6.7), (6.43), (6.50), (6.65), (6.88), (6.96), (6.103), and (6.105)). The following is a *first-order necessary condition*.

If $f \in C^{(1)}$ and $\mathbf{h}$ is a feasible direction at the local solution point $\mathbf{x}_0$ of Problem (7.1) then

$$\mathbf{f}'(\mathbf{x}_0)\mathbf{h} \geqq 0 \tag{7.9}$$

PROOF. If ϕ denotes the associated function on $\mathfrak{R}$ then by the chain rule (6.64)

$$\phi'(0) = \mathbf{f}'(\mathbf{x}_0)\mathbf{h}$$

If now $\phi'(0) < 0$ then by (6.8) actually $\phi'(t) < 0$ on some $(0, \delta)$. This is impossible because by (1.12)

$$\phi(t) - \phi(0) = t\phi'(u)$$

where $0 < u < t < \delta$ and the negative RHS contradicts the fact that $\phi(0)$ is a local minimum. ■

Let us reconsider $K = [0, \infty)$ for the objective function sketched in Figure 7.1(b). The feasible direction at $x_0 = 0$ is the positive direction and $f'(x_0)h = 2h$ for any $h > 0$. The surface of revolution with

$$f(\mathbf{x}) = (x_1 + 1)^2 + (x_2 - 1)^2 + 1 \tag{7.10}$$

provides a similar example when K is the nonnegative quadrant in $\mathfrak{R}^2$. The global solution point for Problem (7.1) is $\mathbf{x}_0 = [0, 1]^T$ on the boundary of K. Here the standard matrix is 1×2, namely

$$\mathbf{f}'(\mathbf{x}) = \left[2(x_1 + 1), 2(x_2 - 1)\right]$$

and feasible directions $\mathbf{h} = [h_1, h_2]^T$ have $h_1 \geqq 0$ so the condition of (7.9) is satisfied as

$$\mathbf{f}'(\mathbf{x}_0)\mathbf{h} = 2h_1 \geqq 0$$

If we change to $K=\mathcal{R}^2$ then the global solution point of the unconstrained minimization problem is the interior point $[-1,1]^T$ at which every direction is a feasible direction.

The *second-order necessary condition* corresponding to (7.9) is included in

If $f\in C^{(2)}$ and $\mathbf{h}$ is a feasible direction at the local solution point $\mathbf{x}_0$ of Problem (7.1) then
(1) $\mathbf{f}'(\mathbf{x}_0)\mathbf{h}\geqq 0$,
(2) $\mathbf{h}^T\mathbf{f}''(\mathbf{x}_0)\mathbf{h}\geqq 0$ whenever $\mathbf{f}'(\mathbf{x}_0)\mathbf{h}=0$. (7.11)

(Condition (2) can be restated as: the Hessian $\mathbf{f}''(\mathbf{x}_0)$ is positive semidefinite on $S\cap \operatorname{Null}\mathbf{f}'(\mathbf{x}_0)$ where S is the set of feasible directions at $\mathbf{x}_0$.)

PROOF. Statement (1) comes from (7.9). For (2) we use Taylor's theorem (6.68) to write

$$f(\mathbf{x}_0+t\mathbf{h})-f(\mathbf{x}_0)=\frac{t^2}{2}\mathbf{h}^T\mathbf{f}''(\mathbf{u})\mathbf{h}$$

where $\mathbf{u}=\mathbf{x}_0+t'\mathbf{h}$ for some $t'<t$. Suppose $\mathbf{h}^T\mathbf{f}''(\mathbf{x}_0)\mathbf{h}<0$ and we rewrite the preceding as

$$f(\mathbf{x}_0+t\mathbf{h})-f(\mathbf{x}_0)=\frac{t^2}{2}\mathbf{h}^T\mathbf{f}''(\mathbf{x}_0)\mathbf{h}+\frac{t^2}{2}\mathbf{h}^T\left[\mathbf{f}''(\mathbf{u})-\mathbf{f}''(\mathbf{x}_0)\right]\mathbf{h}$$

Since $f\in C^{(2)}$ the term within $[\cdot]$ goes to zero as $t\to 0$ so that for $t>0$ sufficiently small the entire RHS is negative; this is impossible since $\mathbf{x}_0$ is a local minimum point. ■

The Hessian for (7.10)

$$\mathbf{f}''(\mathbf{x})=\begin{bmatrix}2 & 0\\ 0 & 2\end{bmatrix}$$

is actually positive definite on all of $\mathcal{R}^2$.

Suppose f is differentiable. Then $\mathbf{x}_0$ is called a *critical point* of Problem (7.1) if $\mathbf{x}_0\in\operatorname{int}K$ and

$$\mathbf{f}'(\mathbf{x}_0)=\mathbf{O}$$

or equivalently, $\nabla f(\mathbf{x}_0)=\mathbf{0}$. With this terminology we can restate (6.65) as

If f is differentiable and $\mathbf{x}_0\in K$ is a local solution point of Problem (7.1), then $\mathbf{x}_0$ is a critical point. (7.12)

Let us use the analysis thus far to explain the idea behind *methods of steepest descent* for locating solution points. Suppose an interior point $\mathbf{y}_0$ is not a minimum point because $\mathbf{f}'(\mathbf{y}_0)\neq\mathbf{O}$ and we seek a direction $\mathbf{h}$ such that the linear approximation

$$f(\mathbf{y}_0+\mathbf{h})\approx f(\mathbf{y}_0)+\mathbf{f}'(\mathbf{y}_0)\mathbf{h}$$

will be as small as possible. That is, we seek a direction $\mathbf{h}$ of fastest decrease of f per unit of distance as measured by the standard norm. In terms of the gradient, we wish to solve for fixed $\mathbf{y}_0$

$$\text{minimize } \langle \nabla f(\mathbf{y}_0), \mathbf{h}\rangle$$
$$\text{s. t. } \|\mathbf{h}\|=1$$

The extreme value theorem (6.19) guarantees that there exists a solution point $\mathbf{h}_0$. According to the Cauchy–Schwarz inequality of (3.45)

$$-\|\nabla f(\mathbf{y}_0)\|\cdot\|\mathbf{h}\|\leqq\langle\nabla f(\mathbf{y}_0),\mathbf{h}\rangle\leqq\|\nabla \mathbf{f}(\mathbf{y}_0)\|\cdot\|\mathbf{h}\|$$

for all $\mathbf{h}$ and so we conclude that the minimum occurs at

$$\mathbf{h}_0=-\|\nabla f(\mathbf{y}_0)\|^{-1}\nabla f(\mathbf{y}_0)$$

where equality holds for the first two members. In words,

When distance is measured by the standard norm, the negative of the gradient vector $\nabla f(\mathbf{x})$ gives the direction of steepest descent from $f(\mathbf{x})$, and the gradient vector itself points in the direction of steepest ascent. (7.13)

The preceding determines no direction at a critical point but this does not mean that there must be an extremum there; a counterexample is $x_0=0$ for $x\mapsto x^3$ on $\mathfrak{R}$. A critical point $\mathbf{x}_0$ is called a *saddle point* of f if every neighborhood of $\mathbf{x}_0$ contains points $\mathbf{y}$ such that $f(\mathbf{y})<f(\mathbf{x}_0)$ and other points $\mathbf{z}$ such that $f(\mathbf{x}_0)<f(\mathbf{z})$.

Let us next give the classical necessary conditions for interior points.

THEOREM. If $f\in C^{(2)}$ and $\mathbf{x}_0\in\operatorname{int}K$ is a local solution point of Problem (7.1) then
(1) $\mathbf{f}'(\mathbf{x}_0)=\mathbf{O}$,
(2) $\mathbf{f}''(\mathbf{x}_0)$ is positive semidefinite. (7.14)

PROOF. This follows from (7.11) because in the present case every direction is feasible and $\mathbf{f}'(\mathbf{x}_0)\mathbf{h}=0$ for every $\mathbf{h}\in\mathfrak{R}^n$. ■

For sufficient conditions, we recall procedures from elementary calculus and consider the quadratic approximation

$$f(\mathbf{x}_0+\mathbf{h})\approx f(\mathbf{x}_0)+\mathbf{f}'(\mathbf{x}_0)\mathbf{h}+\frac{1}{2}\mathbf{h}^T\mathbf{f}''(\mathbf{x}_0)\mathbf{h}$$

We know that $\mathbf{x}_0$ must be a critical point so the RHS of

$$f(\mathbf{x}_0+\mathbf{h})-f(\mathbf{x}_0)\approx\frac{1}{2}\mathbf{h}^T\mathbf{f}''(\mathbf{x}_0)\mathbf{h}$$

must be nonnegative. The earlier example $x\mapsto x^3$ indicates that we must impose positivity and the result is the classical

THEOREM. If $f\in C^{(2)}$ the following are sufficient for a feasible interior point $\mathbf{x}_0$ to be a strict local solution point of Problem (7.1).
(1) $\mathbf{f}'(\mathbf{x}_0)=\mathbf{O}$,
(2) $\mathbf{f}''(\mathbf{x}_0)$ is positive definite. (7.15)

PROOF. Again by (6.19), $\mathbf{h}\mapsto\mathbf{h}^T\mathbf{f}''(\mathbf{x}_0)\mathbf{h}$ has a minimum $m_0>0$ on the compact set $\{\mathbf{h}:\|\mathbf{h}\|=1\}$. Consequently

$$\mathbf{h}^T\mathbf{f}''(\mathbf{x}_0)\mathbf{h}=\|\mathbf{h}\|^2\frac{\mathbf{h}^T}{\|\mathbf{h}\|}\mathbf{f}''(\mathbf{x}_0)\frac{\mathbf{h}}{\|\mathbf{h}\|}\geqq\|\mathbf{h}\|^2m_0$$

for all nonzero $\mathbf{h}\in\mathcal{R}^n$. Next, let us use Taylor's theorem (6.88) to write

$$f(\mathbf{x}_0+\mathbf{h})-f(\mathbf{x}_0)=\frac{1}{2}\mathbf{h}^T\mathbf{f}''(\mathbf{x}_0)\mathbf{h}+\frac{1}{2}\mathbf{h}^T\left[\mathbf{f}''(\mathbf{u})-\mathbf{f}''(\mathbf{x}_0)\right]\mathbf{h}$$

and then rewrite the RHS as

$$\frac{1}{2}\|\mathbf{h}\|^2\left\{\mathbf{d}^T\mathbf{f}''(\mathbf{x}_0)\mathbf{d}+\mathbf{d}^T\left[\mathbf{f}''(\mathbf{u})-\mathbf{f}''(\mathbf{x}_0)\right]\mathbf{d}\right\}$$

where $\mathbf{d}=\|\mathbf{h}\|^{-1}\mathbf{h}$ is the unit vector in the direction of the arbitrary nonzero $\mathbf{h}$. By the continuity of the matrix elements $\partial_i\partial_j f$ the second term within $\{\cdot\}$ goes to zero as $\mathbf{h}\to\mathbf{0}$ so there exists a neighborhood N of $\mathbf{0}$ such that

$$|\mathbf{d}^T\left[\mathbf{f}''(\mathbf{u})-\mathbf{f}''(\mathbf{x}_0)\right]\mathbf{d}|<\frac{m_0}{2}$$

whenever $\mathbf{h}\in N$. For such $\mathbf{h}$ we have

$$f(\mathbf{x}_0+\mathbf{h})-f(\mathbf{x}_0)>\frac{1}{2}\|\mathbf{h}\|^2\left(m_0-\frac{m_0}{2}\right)$$

which shows that a strict local minimum occurs at $\mathbf{x}_0$. ■
If $\mathbf{f}''(\mathbf{x}_0)$ is indefinite in (2) then $\mathbf{x}_0$ is a saddle point (7.1.8).

EXAMPLE 7.1. Least Squares Inverse. This example illustrates the power of the techniques of Section 6.3 and establishes a result used in Example 7.2. Suppose initially that the $m\times n$ system $\mathbf{Ax}=\mathbf{b}$ is inconsistent because rank $\mathbf{A}=n<m$ and rank$[\mathbf{A},\mathbf{b}]=n+1$. Even though there is no solution we can calculate the *residual*

$$r(\mathbf{x})=\mathbf{b}-\mathbf{A}\mathbf{x}$$

for each $\mathbf{x}\in\mathfrak{R}^n$. The norm of the residual measures the squared error associated with $\mathbf{x}$ as an approximate solution of $\mathbf{A}\mathbf{x}=\mathbf{b}$; indeed, $\mathbf{x}$ is a solution iff $\|r(\mathbf{x})\|=0$. We are thus led to the unconstrained minimization problem

$$\text{minimize } \|\mathbf{b}-\mathbf{A}\mathbf{x}\|$$
$$\text{s. t. } \mathbf{x}\in\mathfrak{R}^n$$

whose solution is called the *least squares solution* of $\mathbf{A}\mathbf{x}=\mathbf{b}$. Let us solve this problem using (7.15). Clearly it is equivalent to minimize the square of the norm $\langle\mathbf{b}-\mathbf{A}\mathbf{x},\mathbf{b}-\mathbf{A}\mathbf{x}\rangle$ rewritten as

$$f(\mathbf{x})=\langle\mathbf{b},\mathbf{b}\rangle-2\langle\mathbf{b},\mathbf{A}\mathbf{x}\rangle+\langle\mathbf{A}\mathbf{x},\mathbf{A}\mathbf{x}\rangle$$

The first derivative can be found from (4) and (8) in Table 6.2

$$\mathbf{f}'(\mathbf{x})=-2\mathbf{b}^T\mathbf{A}+\mathbf{x}^T\mathbf{A}^T\mathbf{A}+\mathbf{x}^T\mathbf{A}^T\mathbf{A}$$

so critical points satisfy

$$-2\mathbf{b}^T\mathbf{A}+2\mathbf{x}^T\mathbf{A}^T\mathbf{A}=\mathbf{O}$$

This can be written as

(1) $$(\mathbf{A}^T\mathbf{A})\mathbf{x}=\mathbf{A}^T\mathbf{b}$$

and since $\mathbf{A}^T\mathbf{A}$ is nonsingular on account of $\mathbf{A}$ having full column rank

(2) $$\mathbf{x}_0=(\mathbf{A}^T\mathbf{A})^{-1}\mathbf{A}^T\mathbf{b}$$

is the unique critical point. The second derivative is

$$\mathbf{f}''(\mathbf{x}_0)=2\mathbf{A}^T\mathbf{A}$$

which is positive definite by 5.3.3 and so $\mathbf{x}_0$ is the unique least squares solution by (7.15).

Whenever $\mathbf{A}$ has full column rank, that is without regard to $\mathbf{A}\mathbf{x}=\mathbf{b}$ for any particular $\mathbf{b}$, we call (the left inverse)

(3) $$\mathbf{A}^L=(\mathbf{A}^T\mathbf{A})^{-1}\mathbf{A}^T$$

The *least squares inverse* of **A**; if **A** is furthermore square then it is nonsingular and $\mathbf{A}^L = \mathbf{A}^{-1}$.

As an example, if we denote the first factor on the RHS of (3.38) by **B** then

(4) $$\mathbf{B}^L = \frac{1}{2}\begin{bmatrix} 2 & 2 & -2 \\ -1 & -1 & 2 \end{bmatrix}$$

PROBLEMS FOR SOLUTION

7.1.1 Convex Constraint Sets. Specify the types of convex sets from Table 4.1 that appear as the constraint sets K in (7.2)–(7.6) when g and h are affine.

7.1.2 Feasible Directions. Establish:

(1) If $\mathbf{x} \in \operatorname{int} K$ then every direction is feasible at $\mathbf{x}$.
(2) The converse does not hold for every K; that is, find an example where every direction is feasible at $\mathbf{x} \in K$ and yet $\mathbf{x} \notin \operatorname{int} K$.
(3) Conditions such that the set S of feasible directions at $\mathbf{x}_0$ is (i) $\mathcal{R}^n$, (ii) a proper subset of $\mathcal{R}^n$, (iii) a convex cone, (iv) a polyhedral convex cone, (v) $\varnothing$.

7.1.3 Numerical Exercises. Apply second-order conditions to classify critical points for $f(\mathbf{x})$ as follows.

(1) $(x^2 - 4x + 6)$ s.t. $x \in [0, 1]$.
(2) $(x_1^2 - x_2)^2 + (x_1 - 1)^2$ s.t. $\mathbf{x} \in \mathcal{R}^2$.
(3) $x_1^2 - 3x_1x_2 + x_2^2$ s.t. $\mathbf{x} \in \mathcal{R}^2$.
(4) $ax_1x_2 - x_1^2x_2 - x_1x_2^2$ s.t. $x_1 > 0$, $x_2 > 0$ and given $a > 0$.
(5) $(x_1 - 1)(x_2 + 2)$ s.t. $\mathbf{x} \in \mathcal{R}^2$.
(6) $(x_1^2 + 2x_2^2)\exp[-(x_1^2 + x_2^2)]$ s.t. $\mathbf{x} \in \mathcal{R}^2$.
(7) $x_1^4 + x_2^4 - 4x_1 - 32x_2 + 1$ s.t. $\mathbf{x} \in \mathcal{R}^2$.
(8) $a + 2\mathbf{b}^T\mathbf{x} + \mathbf{x}^T\mathbf{C}\mathbf{x}$ s.t. $\mathbf{x} \in \mathcal{R}^n$ where $\mathbf{C} = \mathbf{C}^T$ is nonsingular.
(9) $\mathbf{x}^T\mathbf{A}\mathbf{x}/\mathbf{x}^T\mathbf{x}$ s.t. $\mathbf{x} \neq \mathbf{0}$ where $\mathbf{A} = \mathbf{A}^T$.
(10) $\mathbf{x}^T\mathbf{A}\mathbf{x}/\mathbf{x}^T\mathbf{B}\mathbf{x}$ s.t. $\mathbf{x} \neq \mathbf{0}$ where **A** and **B** are symmetric and **B** is positive definite.

7.1.4 Example of Peano. Show that $f(\mathbf{x}) = (x_2 - x_1^2)(x_2 - 2x_1^2)$ has a minimum at the origin in $\mathcal{R}^2$ along every line through the origin; however, f does not have a two-dimensional minimum or maximum at the origin.

7.1.5 Gradients and Norms. Find the counterpart to (7.13), that is, the direction of steepest descent from $f(\mathbf{x})$, when distance is measured by

(1) The absolute norm.
(2) The maximum norm.

7.1.6 Inapplicability of Differential Calculus. Furnish a good example of Problem (7.1) where a solution point can be established from first principles but not from use of derivatives.

7.1.7 Differentiating Components. Find $\mathbf{A}^L$ by solving the minimization problem by differentiating components in Σ-summation notation.

7.1.8 Summary. Discuss possibilities for Problem (7.1) when $f \in C^{(2)}$ and

(1) No critical points exist.

(2) Critical points exist and all are known .

7.2 EQUALITY CONSTRAINTS

In this section we establish necessary and sufficient conditions for solution points $\mathbf{x}_0$ of

$$\text{minimize } f(\mathbf{x})$$
$$\text{s.t. } h(\mathbf{x}) = \mathbf{0}$$

which is Problem (7.2).

At times it is possible to "incorporate" the relation $h(\mathbf{x}) = \mathbf{0}$ into f so as to obtain an equivalent but unconstrained problem in a new objective function. Having done so, the methods of Section 7.1 can be applied. For example, consider the problem in $\mathfrak{R}^3$ of dividing a length $a > 0$ into three parts so that the product of the three lengths will be a maximum.

$$\text{maximize } x_1 x_2 x_3$$
$$\text{s.t. } x_1 + x_2 + x_3 = a$$
$$x_1, x_2, x_3 > 0$$

If we solve for $x_3 = a - x_1 - x_2$ and use (7.7) we obtain the equivalent unconstrained minimization problem

$$\text{minimize } -a x_1 x_2 + x_1^2 x_2 + x_1 x_2^2$$
$$\text{s.t. } \mathbf{x} \in K$$

where K is the interior of the first quadrant in $\mathfrak{R}^2$. The new problem can readily be solved by considering the critical points that belong to K.

What if we are not able, or we do not choose, to incorporate the equality constraints into the objective function? Based on what follows, one answer is that we can consider the related problem

$$\text{minimize } f(\mathbf{x}) + \boldsymbol{\lambda}^T h(\mathbf{x})$$
$$\text{s.t. } \mathbf{x} \in D \tag{7.16}$$

where minimization is with respect to $\mathbf{x}$ alone, and the vector $\boldsymbol{\lambda} \in \mathfrak{R}^k$ is a *parameter* in the specific sense that to each such $\boldsymbol{\lambda}$ there corresponds one instance of Problem (7.1). The objective function of (7.16)

$$\psi(\mathbf{x},\boldsymbol{\lambda}) = f(\mathbf{x}) + \boldsymbol{\lambda}^T h(\mathbf{x}) \tag{7.17}$$

is the *Lagrangian* and the components $\lambda_1, \lambda_2, \ldots, \lambda_k$ are *Lagrange multipliers* for Problem (7.2). Critical points $\mathbf{x}_0$ of (7.16) satisfying $h(\mathbf{x}) = \mathbf{0}$ are called *critical points* of Problem (7.2).

Notice that $k = \operatorname{rank} \mathbf{h}'(\mathbf{x}_0)$ must satisfy $k < n$ in the

LAGRANGE MULTIPLIER THEOREM. Suppose $f \in C^{(1)}$ while $h \in C^{(1)}$ determines a smooth hypersurface (6.94). If $\mathbf{x}_0$ is a local solution point of Problem (7.2) then there exists a unique $\boldsymbol{\lambda}_0 \in \mathfrak{R}^k$ such that

$$\mathbf{f}'(\mathbf{x}_0) + \boldsymbol{\lambda}_0^T \mathbf{h}'(\mathbf{x}_0) = \mathbf{O} \tag{7.18}$$

(In words, there must exist a unique parameter $\boldsymbol{\lambda}_0$ and Lagrange multipliers $\lambda_{10}, \lambda_{20}, \ldots, \lambda_{k0}$ such that $\mathbf{x}_0$ is a critical point of the associated Problem (7.16).)

PROOF. Let $\mathbf{y}$ be any vector in the tangent space of the hypersurface S, determined by h, at $\mathbf{x}_0$ and let the associated function be ϕ. That is, $\mathbf{x} = \phi(t)$ is a curve in S for $t \in [-\delta, \delta]$, $\phi(0) = \mathbf{x}_0$, and $\boldsymbol{\phi}'(0) = \mathbf{y}$. Since $\mathbf{x}_0$ is a local solution point of Problem (7.2), $\theta(t) = f(\phi(t))$ has a local minimum at $t = 0$. By the chain rule (6.64)

$$\theta'(0) = \mathbf{f}'(\mathbf{x}_0)\mathbf{y} = 0$$

Now let us formulate Gale's problems in (4.30).

I. Find $\mathbf{w} \in \mathfrak{R}^k$ such that $[\mathbf{h}'(\mathbf{x}_0)]^T \mathbf{w} = [\mathbf{f}'(\mathbf{x}_0)]^T$
II. Find $\mathbf{y} \in \mathfrak{R}^n$ such that $\mathbf{h}'(\mathbf{x}_0)\mathbf{y} = 0$ and $\mathbf{f}'(\mathbf{x}_0)\mathbf{y} = 1$

By (6.96) $\mathbf{h}'(\mathbf{x}_0)\mathbf{y} = 0$ iff $\mathbf{y}$ is in the tangent space of S; by what we have just shown, $\mathbf{f}'(\mathbf{x}_0)\mathbf{y} = 0$ for every $\mathbf{y}$ in that space and so Problem II has no solution. Gale's theorem (4.30) thus assures us that Problem I must have a solution, say $\mathbf{w} = -\boldsymbol{\lambda}_0 \in \mathfrak{R}^k$. The coefficient matrix has full column rank and so, given that the equations are consistent, there must be a unique solution by (3.18). ■

The Lagrangian for the earlier example of dividing a into three parts is

$$\psi(\mathbf{x},\lambda) = -x_1 x_2 x_3 + \lambda(x_1 + x_2 + x_3)$$

and (7.18) applies. Critical points for this case of Problem (7.2) are determined from

$$-x_2x_3+\lambda=0$$

$$-x_1x_3+\lambda=0$$

$$-x_1x_2+\lambda=0$$

$$x_1+x_2+x_3=a$$

Since we deal with positive x's we may write $x_2=\lambda x_3^{-1}=x_1$ from the first two equations and conclude $x_1=x_2$. In the same way $x_2=x_3$ and

$$\mathbf{x}_0=\left[\frac{a}{3},\frac{a}{3},\frac{a}{3}\right]^T$$

without needing to find λ_0.

Values of both $\boldsymbol{\lambda}$ and $\mathbf{x}$ are generally produced in the process of implementing computer programs for solving (7.2). As illustrated in Example 7.3, knowing $\boldsymbol{\lambda}$ is helpful in practice.

In handwork with small problems we must determine critical points for Problem (7.2) by solving equations which are generally nonlinear in $\mathbf{x}$

$$\mathbf{f}'(\mathbf{x})+\boldsymbol{\lambda}^T\mathbf{h}'(\mathbf{x})=\mathbf{O}$$

$$h(\mathbf{x})=\mathbf{0} \qquad (7.19)$$

for pairs $(\mathbf{x},\boldsymbol{\lambda})$. Sometimes this last equation is written in forms equivalent to

$$\frac{d}{d\boldsymbol{\lambda}}\psi(\mathbf{x},\boldsymbol{\lambda})=\mathbf{O}$$

because (Table 6.2(4)) it is the same as $[h(\mathbf{x})]^T=\mathbf{0}^T$. The LHS of the first equation in (7.19) can also be written as the sum of $1\times n$ matrices

$$\mathbf{f}'(\mathbf{x})+\lambda_1\mathbf{h}_1'(\mathbf{x})+\cdots+\lambda_k\mathbf{h}_k'(\mathbf{x})$$

Similarly, we note that the second derivative with respect to $\mathbf{x}$ of the Lagrangian can be written as either of the members in

$$\mathbf{f}''(\mathbf{x})+(\boldsymbol{\lambda}^T\otimes\mathbf{I}_n)\mathbf{h}''(\mathbf{x})=\mathbf{f}''(\mathbf{x})+\lambda_1\mathbf{h}_1''(\mathbf{x})+\cdots+\lambda_k\mathbf{h}_k''(\mathbf{x}) \qquad (7.20)$$

The LHS is appropriate for use with the derivatives of Section 6.3 ($\mathbf{h}''(\mathbf{x})$ is a tensor of order $kn\times n$) while the RHS is expressed in terms of second derivatives of functions. A felicitous expression for the common value in (7.20) is

$$\frac{d^2}{d\mathbf{x}^2}\psi(\mathbf{x},\boldsymbol{\lambda})$$

in the notation used throughout Section 6.3

First- and second-order necessary conditions are contained in the

THEOREM. Suppose $f \in C^{(2)}$ while $h \in C^{(2)}$ determines a smooth hypersurface (6.94). If $\mathbf{x}_0$ is a local solution point of Problem (7.2) then there exists a unique $\boldsymbol{\lambda}_0 \in \mathcal{R}^k$ such that the Lagrangian $\psi(\mathbf{x}, \boldsymbol{\lambda})$ satisfies

(1) $\frac{d}{d\mathbf{x}}\psi(\mathbf{x}_0, \boldsymbol{\lambda}_0) = \mathbf{O}$

(2) $\frac{d^2}{d\mathbf{x}^2}\psi(\mathbf{x}_0, \boldsymbol{\lambda}_0)$ is positive semidefinite on Null $\mathbf{h}'(\mathbf{x}_0)$. (7.21)

PROOF. Statement (1) comes from (7.18). If $\mathbf{y}$ is any vector in the tangent space Null $\mathbf{h}'(\mathbf{x}_0)$ as in the proof of (7.18), we can write the quadratic form in (2) as

$$\mathbf{y}^T\{\mathbf{f}''(\mathbf{x}_0) + [\boldsymbol{\lambda}_0^T \otimes \mathbf{I}_n]\mathbf{h}''(\mathbf{x}_0)\}\mathbf{y} = \mathbf{y}^T\mathbf{f}''(\mathbf{x}_0)\mathbf{y} + \boldsymbol{\lambda}_0^T(\mathbf{I}_k \otimes \mathbf{y}^T)\mathbf{h}''(\mathbf{x}_0)\mathbf{y}$$

The LHS uses (7.20) and the RHS follows from properties of Kronecker products (7.2.1). If we add the RHS to

$$[\mathbf{f}'(\mathbf{x}_0) + \boldsymbol{\lambda}_0^T\mathbf{h}'(\mathbf{x}_0)]\phi''(0) = 0$$

produced from (1) we obtain

$$\{\mathbf{y}^T\mathbf{f}''(\mathbf{x}_0)\mathbf{y} + \mathbf{f}'(\mathbf{x}_0)\phi''(0)\} + \{\boldsymbol{\lambda}_0^T\mathbf{h}'(\mathbf{x}_0)\phi''(0) + \boldsymbol{\lambda}_0^T(\mathbf{I}_k \otimes \mathbf{y}^T)\mathbf{h}''(\mathbf{x}_0)\mathbf{y}\}$$

The first expression within $\{\cdot\}$ is nonnegative because it equals $\theta''(0)$ where $\theta(t) = f(\phi(t))$ is the function that has a local minimum at $t = 0$. The second expression within $\{\cdot\}$ is zero because (7.2.1) it is the second derivative of $\boldsymbol{\lambda}^T h(\mathbf{x})$ at $\mathbf{x}_0$ and $\boldsymbol{\lambda}^T h(\phi(t)) = 0$ for $t \in [-\delta, \delta]$. ■

In the example of subdividing a, the seond derivative of the Lagrangian with respect to $\mathbf{x}$ is

$$\mathbf{A} = \frac{a}{3}\begin{bmatrix} 0 & -1 & -1 \\ -1 & 0 & -1 \\ -1 & -1 & 0 \end{bmatrix}$$

which by (5.21) is indefinite. However, using (3.27) the matrix of the standard basis for Null $[1, 1, 1]$ is found to be

$$\mathbf{U} = \begin{bmatrix} 1 & 0 \\ 0 & 1 \\ -1 & -1 \end{bmatrix}$$

Resolving the definiteness of **A** on Range **U** using (5.29) we arrive at

$$\mathbf{U}^T\mathbf{A}\mathbf{U}=\begin{bmatrix} 2 & 1 \\ 1 & 2 \end{bmatrix}$$

which is actually positive definite on $\mathfrak{R}^2$. This completes the verification of the necessary conditions of (7.21).

The stronger property of positive definiteness of $d^2\psi/d\mathbf{x}^2$ on Null $\mathbf{h}'$ that holds in this example is shown to be a sufficient condition in the next theorem. (Just as in unconstrained problems, positive definiteness is not necessary at local solution points.) Notice that h need not necessarily determine a smooth hypersurface in the

THEOREM. Suppose $f\in C^{(2)}$ and $h\in C^{(2)}$. Then the following are sufficient for a feasible point $\mathbf{x}_0$ to be a strict local solution point of Problem (7.2): there exists $\boldsymbol{\lambda}_0\in\mathfrak{R}^k$ such that

(1) $\dfrac{d}{d\mathbf{x}}\psi(\mathbf{x}_0,\boldsymbol{\lambda}_0)=\mathbf{O}$

(2) $\dfrac{d^2}{d\mathbf{x}^2}\psi(\mathbf{x}_0,\boldsymbol{\lambda}_0)$ is positive definite on Null $\mathbf{h}'(\mathbf{x}_0)$ (7.22)

PROOF. Simultaneous minima and maxima occur at feasible isolated points so it suffices to suppose that $\mathbf{x}_0$ is a feasible limit point. If $\mathbf{x}_0$ is not a strict local solution then $f(\mathbf{z})\leqq f(\mathbf{x}_0)$ for some $\mathbf{z}\neq\mathbf{x}_0$ in every neighborhood of $\mathbf{x}_0$ and there must exist a sequence $\{\mathbf{x}_i\}$ converging to $\mathbf{x}_0$ with the following additional properties: $h(\mathbf{x}_i)=\mathbf{0}$, $\mathbf{x}_i\neq\mathbf{x}_0$, and $f(\mathbf{x}_i)\leqq f(\mathbf{x}_0)$ for $i=1,2,\ldots$. Furthermore we can define a second sequence $\{\mathbf{y}_i\}$ where $\mathbf{x}_i=\mathbf{x}_0+t_i\mathbf{y}_i$ with $\|\mathbf{y}_i\|=1, t_i>0$, and a real (third) sequence $\{t_i\}$ converging to zero. Since the $\mathbf{y}_i$ lie in a compact set there must by (6.16) be a limit point $\mathbf{y}_0$ with $\|\mathbf{y}_0\|=1$, so $\mathbf{y}_0\neq\mathbf{0}$. Rather than specifying a convergent subsequence we take advantage of (6.4) and suppose $\{\mathbf{y}_i\}\rightarrow\mathbf{y}_0$. Two tasks remain. First, we show $\mathbf{y}_0\in$ Null $\mathbf{h}'(\mathbf{x}_0)$. By the mean value theorem (6.43) we can write

$$h_j(\mathbf{x}_i)-h_j(\mathbf{x}_0)=\mathbf{h}_j'(\mathbf{u}_{ij})(t_i\mathbf{y}_i)$$

for the jth coordinate function. Rewriting this as

$$\frac{h_j(\mathbf{x}_i)-h_j(\mathbf{x}_0)}{t_i}=\mathbf{h}_j'(\mathbf{u}_{ij})\mathbf{y}_i$$

we have zero on the LHS because $\mathbf{x}_0$ and all $\mathbf{x}_i$ are feasible points; in the limit, the RHS goes to $\mathbf{h}_j'(\mathbf{x}_0)\mathbf{y}$ because $h\in C^{(1)}$ as well as $C^{(2)}$. We conclude $\mathbf{h}'(\mathbf{x}_0)\mathbf{y}_0=0$. Second, we show that the quadratic form in (2) is nonpositive at $\mathbf{y}_0$. Now $h(\mathbf{x}_0)=\mathbf{0}$ so we can apply Taylor's theorem (6.88) to

$f(\mathbf{x}_0)=\psi(\mathbf{x}_0,\boldsymbol{\lambda}_0)$ to obtain

$$0\geqq\frac{f(\mathbf{x}_i)-f(\mathbf{x}_0)}{t_i^2}=\frac{1}{2}\mathbf{y}_i^T\frac{d^2\psi(\boldsymbol{\xi}_i,\boldsymbol{\lambda}_0)}{d\mathbf{x}^2}\mathbf{y}_i$$

for some $\boldsymbol{\xi}_i$ on the line segment between $\mathbf{x}_0$ and $\mathbf{x}_i=\mathbf{x}_0+t_i\mathbf{y}_i$. Because $\psi\in C^{(2)}$, the RHS goes to

$$\frac{1}{2}(\mathbf{y}_0)^T\frac{d^2\psi(\mathbf{x}_0,\boldsymbol{\lambda}_0)}{d\mathbf{x}^2}\mathbf{y}_0\leqq 0$$

which contradicts (2). ■

Sometimes Problem (7.16) is presented as an unconstrained problem which, at the cost of adding k variables, serves to replace the original constrained Problem (7.2). This cannot be done in general. As is clear from McCormick (1972, 211–212), even though we may find $\boldsymbol{\lambda}_0$ for which the associated $\mathbf{x}_0$ is a solution point for (7.2), it is not necessarily true that $\mathbf{x}_0$ is a solution to (7.16) as an unconstrained problem on D unless we impose severe restrictions on f and h. See Falk (1967) and (1969) for results on cases of (7.16). What is true is that Problem (7.16) is closely related to (7.2) in several important respects as established in (7.18), (7.21), and (7.22).

EXAMPLE 7.2. Minimum Norm Inverse and Generalized Inverse. In Example 7.1 we derived one type of approximate solution, namely the least squares solution, for an inconsistent system of linear equations. There is one additional case to consider where $\mathbf{Ax}=\mathbf{b}$ fails to have a unique solution, namely, the case where there are an infinite number. If $\mathbf{A}$ is $m\times n$ and rank $\mathbf{A}=m<n$, then the *minimum norm solution* of $\mathbf{Ax}=\mathbf{b}$ is defined to be the solution point of

$$\begin{aligned}&\text{minimize } \|\mathbf{x}\|\\ &\text{s.t. } \mathbf{Ax}=\mathbf{b}\end{aligned}$$

Once again we take advantage of the fact that we can equivalently minimize $(1/2)\|\mathbf{x}\|^2$ with Lagrangian

$$\psi(\mathbf{x},\boldsymbol{\lambda})=\frac{1}{2}\langle\mathbf{x},\mathbf{x}\rangle+\boldsymbol{\lambda}^T(\mathbf{Ax}-\mathbf{b})$$

where $\boldsymbol{\lambda}\in\mathcal{R}^m$. Condition (7.18) then yields

$$\begin{aligned}\mathbf{x}&=-\mathbf{A}^T\boldsymbol{\lambda}\\ \mathbf{Ax}&=\mathbf{b}\end{aligned}$$

Since $\mathbf{A}$ has full row rank, $\mathbf{AA}^T$ is nonsingular by (4.5) and there is the unique solution pair

$$\boldsymbol{\lambda}_0 = -(\mathbf{AA}^T)^{-1}\mathbf{b}$$

$$\mathbf{x}_0 = \mathbf{A}^T(\mathbf{AA}^T)^{-1}\mathbf{b}$$

The second derivative of the Lagrangian at $(\mathbf{x}_0, \boldsymbol{\lambda}_0)$ is $\mathbf{I}$ which is positive definite. From (7.22) we conclude that $\mathbf{x}_0$ is the unique minimum norm solution.

Whenever $\mathbf{A}$ has full row rank we call (the right inverse)

(1) $$\mathbf{A}^N = \mathbf{A}^T(\mathbf{AA}^T)^{-1}$$

the *minimum norm inverse* of $\mathbf{A}$; if $\mathbf{A}$ is square as well, then it is nonsingular and $\mathbf{A}^N = \mathbf{A}^{-1}$.

As an example, if we denote the second factor on the RHS of (3.38) by $\mathbf{S}$ then

(2) $$\mathbf{S}^N = \frac{1}{8}\begin{bmatrix} 4 & 2 \\ 0 & 0 \\ 2 & 3 \\ 0 & 0 \\ -2 & 1 \\ 2 & 3 \\ -2 & 1 \end{bmatrix}$$

The next idea is to combine $\mathbf{S}^N$ and $\mathbf{B}^L$ where $\mathbf{A} = \mathbf{BS}$ is the factorization of (3.36) for an arbitrary matrix $\mathbf{A}$. But first let us explain our objective.

If $\mathbf{A}$ is any real matrix of order $m \times n$ let us seek an $n \times m$ matrix $\mathbf{A}^G$ called a *generalized inverse* of $\mathbf{A}$ such that for any $\mathbf{b}_0 \in \mathfrak{R}^m$ the vector $\mathbf{x}_0 = \mathbf{A}^G\mathbf{b}_0$ is

(3) The least squares solution of $\mathbf{Ax} = \mathbf{b}_0$ when there is no solution.
(4) The unique solution when there is but one solution.
(5) The minimum norm solution when there are an infinite number of solutions.

Suppose $\operatorname{rank} \mathbf{A} = k$ and $\mathbf{A} = \mathbf{BS}$ is the factorization of (3.36) where $\mathbf{B}$ has full column rank k and $\mathbf{S}$ has full row rank k. Then $\mathbf{A}^G = \mathbf{S}^N\mathbf{B}^L$ has properties (3)–(5) and it is the only such matrix according to the

(6) **THEOREM.** If $\mathbf{A} = \mathbf{O}$, then trivially $\mathbf{A}^G = \mathbf{O}^T$. A nonzero matrix $\mathbf{A} = \mathbf{BS}$, where the factorization is from (3.36), has

$$\mathbf{A}^G = \mathbf{S}^N\mathbf{B}^L = \mathbf{S}^T(\mathbf{S}\mathbf{S}^T)^{-1}(\mathbf{B}^T\mathbf{B})^{-1}\mathbf{B}^T$$

as its unique generalized inverse satisfying (3), (4), (5).

PROOF. The zero matrix trivially provides for (3) when $\mathbf{b}_0 \neq \mathbf{0}$ and (5) when $\mathbf{b}_0 = \mathbf{0}$. If there is no solution when $\mathbf{BS} \neq \mathbf{O}$ then the minimization problem of Example 7.1 yields

$$\mathbf{S}^T\mathbf{B}^T\mathbf{BSx} = \mathbf{S}^T\mathbf{B}^T\mathbf{b}_0$$

in place of (1) in that example. Multiplying both sides by $\mathbf{S}$ and by appropriate inverses we obtain

$$\mathbf{Sx} = \mathbf{B}^L\mathbf{b}_0$$

Multiplication by $\mathbf{S}^N$ yields $\mathbf{x} = \mathbf{S}^N\mathbf{B}^L\mathbf{b}_0$ so (3) holds. Property (4) follows at once from 3.2.1. If there are an infinite number of solutions, then just as at the beginning of the present example, (7.18) yields

$$\mathbf{x} = -\mathbf{S}^T\mathbf{B}^T\boldsymbol{\lambda}$$

$$\mathbf{BSx} = \mathbf{b}_0$$

and we find that (5) holds. To prove uniqueness we first use the four-factor expression in (6) to verify that the $\mathbf{A}^G$ we have found satisfies

(7) $$\mathbf{AA}^G \text{ and } \mathbf{A}^G\mathbf{A} \text{ are symmetric}$$

(8) $$\mathbf{A}^G\mathbf{AA}^G = \mathbf{A}^G$$

(9) $$\mathbf{AA}^G\mathbf{A} = \mathbf{A}$$

Second, we verify that if there were another matrix $\mathbf{Z}$ that satisfied (7), (8), (9) then necessarily $\mathbf{Z} = \mathbf{A}^G$ (7.2.8). ■

Concluding our examples for the matrix on the LHS of (3.38), we obtain

$$\begin{bmatrix} 1 & 0 & 1 & 0 & 0 & 1 & 0 \\ 1 & 0 & 1 & 0 & 0 & 1 & 0 \\ 1 & 0 & 2 & 0 & 1 & 2 & 1 \end{bmatrix}^G = \frac{1}{16}\begin{bmatrix} 6 & 6 & -4 \\ 0 & 0 & 0 \\ 1 & 1 & 2 \\ 0 & 0 & 0 \\ -5 & -5 & 6 \\ 1 & 1 & 2 \\ -5 & -5 & 6 \end{bmatrix}$$

by forming the product of $\mathbf{S}^N$ in (2) and $\mathbf{B}^L$ from Example 7.1 (4). Additional material on generalized inverses is contained in 7.2.8.

PROBLEMS FOR SOLUTION

7.2.1 Kronecker Products. Establish the expressions used in the proof of (7.21) for

(1) $\theta''(t)$.

(2) $\mathbf{y}^T(\boldsymbol{\lambda}^T \otimes \mathbf{I}_n)$.

(3) $\dfrac{d^2}{d\mathbf{x}^2}\boldsymbol{\lambda}^T h(\mathbf{x})$.

7.2.2 Numerical Exercises. Apply second-order conditions to classify critical points for the following.

(1) mimimize $(x_1-2)^2+x_2^2$
s.t. $x_1-x_2^2=0$

(2) minimize $x_1+x_2+x_3$
s.t. $x_1^2+x_2^2=1$
$x_3=1$

(3) maximize $-x^2-y^2$
s.t. $x+2y-z=0$
$2x-y+z=1$

(4) minimize $x^2+y^2+z^2$
s.t. $(x-y)^2-z^2=1$

(5) maximize $x_1x_2x_3$
s.t. $x_1^2+x_2^2+x_3^2=8$
$x_1, x_2, x_3>0$

(6) minimize $x_1^2+x_2^2+x_3^2$
s.t. $x_1x_2-2x_3^2+1=0$

(7) maximize $x_1x_2^2x_3^3$
s.t. $x_1+x_2+x_3=1$
$x_1, x_2, x_3>0$

(8) minimize $x_1^2+x_2^2$
s.t. $(x_1-1)^3-x_2^2=0$

7.2.3 Closest Point in a Hypersurface. Establish or furnish:

(1) Existence of solutions to
minimize $\|\mathbf{x}-\mathbf{a}\|$
s.t. $h(\mathbf{x})=\mathbf{0}$
where the transformation h determines a smooth hypersurface (6.94).

(2) Which cases in 7.2.2 can be interpreted as instances of the subject problem?

(3) An example of the subject problem to which 7.22(2) does not apply.

7.2.4 Closest Point in a k-Plane. Solve in $\mathcal{R}^n$:

(1) minimize $\|\mathbf{x}-\mathbf{a}\|$
s.t. $\mathbf{A}\mathbf{x}=\mathbf{b}$
where $\mathbf{A}$ has full row rank $n-k$.
(2) Case for a line, $k=1$.
(3) Case for a hyperplane, $k=n-1$.

7.2.5 Two Problems. Solve and relate solutions given $\mathbf{a}\neq\mathbf{0}$.

(1) minimize $\|\mathbf{x}\|$
s.t. $\langle\mathbf{a},\mathbf{x}\rangle=1$
(2) maximize $\langle\mathbf{a},\mathbf{x}\rangle$
s.t. $\|\mathbf{x}\|=1$

7.2.6 Central Quadrics. Suppose $\mathbf{A}=\mathbf{A}^T$ and show that the following three problems are equivalent: they have the same extrema and the same solution points.

(1) minimize $\|\mathbf{x}\|$
s.t. $\mathbf{x}^T\mathbf{A}\mathbf{x}=1$
(2) maximize $\mathbf{x}^T\mathbf{A}\mathbf{x}$
s.t. $\|\mathbf{x}\|=1$
(3) maximize $\|\mathbf{x}\|^{-1}$
s.t. $\mathbf{x}^T\mathbf{A}\mathbf{x}=1$

7.2.7 Constrained Minimum of Quadratic Form. Solve the problem

$$\begin{aligned}&\text{minimize } \frac{1}{2}\mathbf{x}^T\mathbf{C}\mathbf{x}\\&\text{s.t. } \mathbf{A}\mathbf{x}=\mathbf{b}\end{aligned}$$

where $\mathbf{C}=\mathbf{C}^T$ is positive definite and $\mathbf{A}$ is $m\times n$ of rank m.

7.2.8 Generalized Inverses. Establish:

(1) The matrix $\mathbf{A}^G$ satisfies (7)–(9) in Example 7.2.
(2) If $\mathbf{Z}$ also satisfies (7)–(9) then $\mathbf{Z}=\mathbf{A}^G$.
(3) If $k\neq0$ then $(k\mathbf{A})^G=k^{-1}\mathbf{A}^G$.
(4) $(\mathbf{A}^G)^G=\mathbf{A}$.
(5) $(\mathbf{A}^T)^G=(\mathbf{A}^G)^T$.
(6) If $\mathbf{A}$ is symmetric then $\mathbf{A}^G$ is symmetric.
(7) $(\mathbf{A}\mathbf{B})^G\neq\mathbf{B}^G\mathbf{A}^G$ in general unless $\mathbf{A}$ and $\mathbf{B}$ are both of full rank.
(8) $\operatorname{rank}\mathbf{A}=\operatorname{rank}\mathbf{A}^G$.
(9) Explicit formulas for $\mathbf{A}^G$ given $\mathbf{A}$ as follows
  (i) $\mathbf{a}\neq\mathbf{0}$.
  (ii) $\mathbf{b}^T\neq\mathbf{0}^T$.
  (iii) $\mathbf{a}\mathbf{b}^T$.
  (iv) $\begin{bmatrix} a & b \\ c & d \end{bmatrix}$ where $ad-bc=0$.
  (v) (3.37).

7.2.9 Conditional Inverse. Establish the following properties of a *conditional inverse* of an $m \times n$ matrix $\mathbf{A}$ which is any matrix $\mathbf{A}^C$ satisfying $\mathbf{A}\mathbf{A}^C\mathbf{A} = \mathbf{A}$:

(1) Any left inverse of $\mathbf{A}$ is a conditional inverse.

(2) If $\mathbf{A}\mathbf{x} = \mathbf{b}$ is consistent then all solutions are given by

$$\mathbf{x} = \mathbf{A}^C\mathbf{b} + (\mathbf{I} - \mathbf{A}^C\mathbf{A})\mathbf{v}$$

where $\mathbf{v} \in \mathcal{R}^n$ is arbitrary.

7.3 INEQUALITY CONSTRAINTS

Let us next establish necessary and sufficient conditions for solution points $\mathbf{x}_0$ of

$$\text{minimize } f(\mathbf{x})$$
$$\text{s.t. } g(\mathbf{x}) \leqq \mathbf{0}$$

which is Problem (7.3).

We adapt the terminology introduced for (4.66) and call any individual constraint $g_i(\mathbf{x}) \leqq 0$ for $i = 1, 2, \ldots, m$ a *binding constraint* at $\mathbf{x}_0 \in D$ if $g_i(\mathbf{x}_0) = 0$ and a *slack constraint* at $\mathbf{x}_0$ if $g_i(\mathbf{x}_0) < 0$. Then we can say that the most important feature to be noticed in the present section is

> The properties of a local solution point $\mathbf{x}_0$ of Problem (7.3) are determined by the binding constraints at $\mathbf{x}_0$.

This does not say that we can entirely disregard the slack constraints. For example, if $\mathbf{x}_0$ is a global solution point, then omitting one slack constraint might cause some $\mathbf{x}_1 \in \mathcal{R}^n$ to become feasible where $f(\mathbf{x}_1) < f(\mathbf{x}_0)$. But even if all slack constraints are omitted, it is clear that $\mathbf{x}_0$ remains a local solution point and in fact it is a local solution point of the equality constraint Problem (7.2) that corresponds to the set of binding constraints.

Figure 7.2 illustrates the displayed feature for $m = 3$ constraints. There

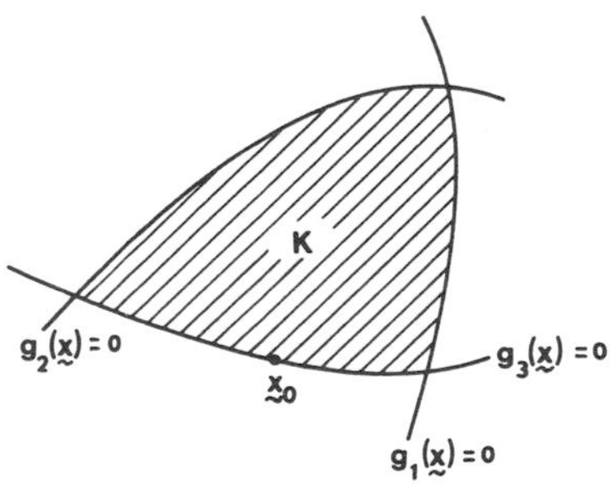

Figure 7.2. Feasible region for Problem (7.3).

$g_3(\mathbf{x})=0$ is the only binding constraint at $\mathbf{x}_0$ because both $g_1(\mathbf{x}_0)<0$ and $g_2(\mathbf{x}_0)<0$. In this example $\mathbf{x}_0$ is a local solution point of Problem (7.3) iff $\mathbf{x}_0$ is a local solution point of

$$\text{minimize } f(\mathbf{x})$$
$$\text{s.t. } g_3(\mathbf{x})=0$$

which is the associated Problem (7.2).

Of course at the start we usually do not know which are the binding constraints and there may be no practical method for sorting them out in advance so as to solve Problem (7.3) as an instance of (7.2) or (7.1). Enumeration of all possible such equality constraint problems is impractical except perhaps when g has only a few, say $m=4$ or fewer, individual constraints. But we find that important properties of solution points can be expressed in terms of the

Definition. If $\mathbf{x}$ is a feasible point for Problem (7.3) or (7.4) then

$$B(\mathbf{x})=\{i: i\in\{1,2,\ldots,m\} \text{ and } g_i(\mathbf{x})=0\}$$

is the *binding index set at* $\mathbf{x}_0$ for the designated problem

Possibilities for $B(\mathbf{x}_0)$ range from $\varnothing$ to $\{1,2,\ldots,m\}$.

Results in this section are based on differentiability assumptions that the function f and the transformation g belong to their respective Classes $C^{(1)}$, or $C^{(2)}$, on their common open domain D. The constraint set

$$K=\{\mathbf{x}: g(\mathbf{x})\leqq \mathbf{0}\}$$

is closed and, as always, $\mathbf{x}_0\in K$ is either a feasible limit point or a feasible isolated point. Since g is continuous at $\mathbf{x}_0\in D$, $B(\mathbf{x}_0)=\varnothing$ implies that $\mathbf{x}_0$ is a feasible interior point but examples show that the converse does not hold for every such g (7.3.1). When $B(x_0)\neq\varnothing$ we make use of the associated Problem (7.2) as described for Figure 7.2.

The following theorem of John (1948) is more general than (7.18) because it does not require a corresponding constraint qualification that g' have full rank $m<n$.

JOHN'S THEOREM. Suppose $f\in C^{(1)}$ and $g\in C^{(1)}$. If $\mathbf{x}_0$ is a local solution point of Problem (7.3) then there exist $\bar{u}\in\mathcal{R}$, $\mathbf{u}_0\in\mathcal{R}^m$ such that

(1) $\begin{bmatrix}\bar{u}\\ \mathbf{u}_0\end{bmatrix}$ is semipositive.

(2) $\mathbf{u}_0^T g(\mathbf{x}_0)=0$.

(3) $\bar{u}\mathbf{f}'(\mathbf{x}_0)+\mathbf{u}_0^T\mathbf{g}'(\mathbf{x}_0)=\mathbf{O}$. (7.23)

Note. The $m+1$ numbers $\bar{u}$ and $u_1, u_2, \ldots, u_m$, the components of $\mathbf{u}_0$, are called *John multipliers*. According to (1) they cannot all be zero and none can be negative. Condition (2)

$$u_1 g_1(\mathbf{x}_0) + u_2 g_2(\mathbf{x}_0) + \cdots + u_m g_m(\mathbf{x}_0) = 0$$

is called the *complementary slackness condition* because it requires for $i = 1, 2, \ldots, m$

$$\text{If } g_i(\mathbf{x}_0) < 0 \text{ then } u_i = 0.$$

$$\text{If } \quad u_i > 0 \text{ then } g_i(\mathbf{x}_0) = 0.$$

In words, at most one of $g_i(\mathbf{x}_0) \leqq 0$ and $u_i \geqq 0$ can be slack for each i. Following this theorem, we return to our usual notation where u_{10}, $u_{20}, \ldots, u_{m0}$ are the components of $\mathbf{u}_0$.

PROOF. If $B(\mathbf{x}_0) = \varnothing$ then $\mathbf{x}_0$ is an interior point and $\mathbf{f}'(\mathbf{x}_0) = \mathbf{O}$ by (7.12). Consequently (1), (2), (3) hold for $\bar{u} = 1$, $\mathbf{u}_0 = \mathbf{0}$. The remaining possibility is $B(\mathbf{x}_0) \neq \varnothing$ and it suffices to show that Gordan's Problem II in (4.34) with matrix

$$\mathbf{A} = \left[\begin{array}{c|c} 0 & \mathbf{f}'(\mathbf{x}_0) \\ \hline g(\mathbf{x}_0) & \mathbf{g}'(\mathbf{x}_0) \end{array}\right] = \left[\begin{array}{c|ccc} 0 & \partial_1 f(\mathbf{x}_0) & \cdots & \partial_n f(\mathbf{x}_0) \\ \hline g_1(\mathbf{x}_0) & \partial g_1(\mathbf{x}_0) & \cdots & \partial_n g_1(\mathbf{x}_0) \\ \vdots & \vdots & & \vdots \\ g_m(\mathbf{x}_0) & \partial_1 g_m(\mathbf{x}_0) & \cdots & \partial_n g_m(\mathbf{x}_0) \end{array}\right]$$

has no solution in $\mathfrak{R}^{n+1}$. Let us suppose that there were a solution corresponding to $\bar{z} \in \mathfrak{R}$ and $\mathbf{z}_0 \in \mathfrak{R}^n$. Then $\mathbf{z}_0$ would be a solution to the Problem II corresponding to a submatrix of $\mathbf{A}$, namely

$$\mathbf{f}'(\mathbf{x}_0)\mathbf{z}_0 < 0$$

$$\mathbf{g}_i'(\mathbf{x}_0)\mathbf{z}_0 < 0 \quad \text{for } i \in B(\mathbf{x}_0)$$

where the final equations have the stated form because $g_i(\mathbf{x}_0) = 0$ for $i \in B(\mathbf{x}_0)$. By continuity of the derivatives and g, there would exist by (6.8) a neighborhood $N = N_\delta(\mathbf{x}_0)$ such that

$$\mathbf{f}'(\mathbf{x})\mathbf{z}_0 < 0$$

$$\mathbf{g}_i'(\mathbf{x})\mathbf{z}_0 < 0 \quad \text{for } i \in B(\mathbf{x}_0)$$

$$g_i(\mathbf{x}) < 0 \quad \text{for } i \notin B(\mathbf{x}_0)$$

whenever $\mathbf{x} \in N$. We could then use the mean value theorem (6.43) to write the following for $t>0$

$$f(\mathbf{x}_0 + t\mathbf{z}_0) = f(\mathbf{x}_0) + \mathbf{f}'\big(\mathbf{x}_0 + \bar{t}\mathbf{z}_0\big) t\mathbf{z}_0$$

$$g_i(\mathbf{x}_0 + t\mathbf{z}_0) = g_i(\mathbf{x}_0) + \mathbf{g}_i'(\mathbf{x}_0 + t_i\mathbf{z}_0) t\mathbf{z}_0 \quad \text{for } i \in B(\mathbf{x}_0)$$

where $\bar{t}, t_i \in (0,t)$. For $t>0$ sufficiently small $\mathbf{x}_0 + t\mathbf{z}_0 \in N$ and we would conclude

$$f(\mathbf{x}_0 + t\mathbf{z}_0) - f(\mathbf{x}_0) = t\mathbf{f}'\big(\mathbf{x}_0 + \bar{t}\mathbf{z}_0\big)\mathbf{z}_0 < 0$$

This is a contradiction, whether $\mathbf{x}_0$ is a feasible isolated point or not, because as we now show, such points $\mathbf{x}_0 + t\mathbf{z}_0$ are feasible. First, as already noted they belong to N. Second, if $i \notin B(\mathbf{x}_0)$ then $g_i(\mathbf{x}_0 + t\mathbf{z}_0) < 0$. Third and finally, if $i \in B(\mathbf{x}_0)$ then

$$g_i(\mathbf{x}_0 + t\mathbf{z}_0) = t\mathbf{g}_i'(\mathbf{x}_0 + t_i\mathbf{z}_0) < 0$$

Thus, by (4.34) there must exist a solution to Gordan I for matrix $\mathbf{A}$, and by design its components are the multipliers $\bar{u}, u_1, \ldots, u_n$. ■

John multipliers are sometimes called generalized Lagrange multipliers since they exist in some cases where the latter do not. Consider the simple example

$$\text{minimize } x_1 + x_2$$
$$\text{s.t. } x_1^2 + x_2^2 \leqq 0$$

Here $\bar{u} = 0$ and $u_1 = 1$ are John multipliers and yet there are no Lagrange multipliers for the equivalent problem

$$\text{minimize } x_1 + x_2$$
$$\text{s.t. } x_1^2 + x_2^2 = 0$$

This happens because $x_1 + x_2 + \lambda(x_1^2 + x_2^2)$ has no critical point at the solution point $\mathbf{x}_0 = \mathbf{0}$. Of course (7.18) does not apply to the second problem because $\mathbf{h}'(\mathbf{x}) = [2x_1, 2x_2]$ does not have full rank 1 at $\mathbf{x}_0$ as required.

Whenever $\bar{u} = 0$, as in the preceding example, we have a pathological case in the sense that the objective function f plays no role in the problem: Any function of Class $C^{(1)}$ on D can replace f and there will be no change in the necessary conditions of (7.23).

Whenever $\bar{u} \neq 0$ we lose no generality by assigning $\bar{u}=1$. Then the remaining John multipliers $u_1, u_2, \ldots, u_m$ are nonnegative by (1), they satisfy the complementary slackness condition (2), they are interpretable as Lagrange multipliers because of (3), and they are called *Kuhn–Tucker multipliers* because they were first obtained (in a different manner) in Kuhn and Tucker (1951).

One of the most direct ways of guaranteeing that $\bar{u}=1$ is possible in (7.23) is to impose the following

Constraint Qualification. The transformation g has the property

$$\{\mathbf{g}_i'(\mathbf{x}_0) : i \in B(\mathbf{x}_0)\} \text{ is LI}$$

That is, the submatrix of $\mathbf{g}'(\mathbf{x}_0)$ corresponding to the binding constraints at $\mathbf{x}_0$ has full row rank. (7.24)

In contrast to many constraint qualifications, it is straightforward to verify (7.24) using, for example, (6.87) and (3.22). Let us show that (7.24) makes $\bar{u}=1$ possible by establishing the fundamental

KUHN–TUCKER CONDITIONS FOR INEQUALITY CONSTRAINTS. Suppose $f \in C^{(1)}$, $g \in C^{(1)}$, and $\mathbf{x}_0$ is a local solution point of Problem (7.3) at which the constraint qualification (7.24) holds. Then there exists a unique $\mathbf{u}_0 \in \mathcal{R}^m$ such that

(1) $\mathbf{u}_0$ is nonnegative.
(2) $\mathbf{u}_0^T g(\mathbf{x}_0) = 0$.
(3) $\mathbf{f}'(\mathbf{x}_0) + \mathbf{u}_0^T \mathbf{g}'(\mathbf{x}_0) = \mathbf{O}$. (7.25)

Note. Here we return to our usual notation and use a second subscript for components of $\mathbf{u}_0$ to stress that they go with $\mathbf{x}_0$. Conditions (1), (2), (3) are called the *Kuhn–Tucker conditions* for Problem (7.3).

PROOF. If $\bar{u}=0$ then by (1) $\mathbf{u}_0$ has at least one positive component. But each such $u_{j0} > 0$ implies $j \in B(\mathbf{x}_0)$ by (2) so we have a nontrivial LC

$$\sum_{j \in B(\mathbf{x}_0)} u_{j0} g_j(\mathbf{x}_0) \mathbf{0}^T$$

that we can differentiate to obtain a contradiction to (7.24). Thus $\bar{u} > 0$, and replacing $\mathbf{u}_0$ in (7.23) by $\bar{u}^{-1}\mathbf{u}_0$ produces the unique vector in $\mathcal{R}^m$ that has the stated properties. ■

In the simple example

$$\text{maximize} \quad x_1^2 + 4x_2^2$$

$$\text{s.t.} \quad x_1^2 + x_2^2 \leqq 1$$

$$x_1^2 - x_2 \leqq 0$$

the feasible region lies inside the unit circle and inside the parabola $x_2 = x_1^2$. Only the first constraint is binding at the global solution point where $x_1 = 0$, $x_2 = 1$. Condition (7.24) is satisfied because

$$\mathbf{g}_1'(\mathbf{x}_0) = [0, 2]$$

is nonzero. The Kuhn–Tucker conditions hold with multipliers $u_1 = 4$, $u_2 = 0$.

The Kuhn–Tucker conditions are surprisingly strong. By adapting the proof of (7.25) we find that its hypotheses guarantee

$$\mathbf{u}_0 = \mathbf{0} \quad \text{iff} \quad \mathbf{f}'(\mathbf{x}_0) = \mathbf{O}$$

When $\mathbf{u}_0 = \mathbf{0}$, examples show that $\mathbf{x}_0$ need not be an interior point of K (7.3.1). If $\mathbf{u}_0 \neq \mathbf{0}$ then $\mathbf{x}_0$ is a local solution point of an associated Problem (7.2) as illustrated in Figure 7.2. Further, as observed by James E. Falk, and as presented in 7.3.6, the Kuhn-Tucker conditions distinguish minima from maxima and consequently can be said to be more powerful than their Lagrange counterparts when both are considered apart from their accompanying second-order conditions. Expressing this another way, the Kuhn–Tucker conditions are "harder" to satisfy for Problem (7.3) than (7.18) is for Problem (7.2): When a pair $(\mathbf{x}_0, \mathbf{u}_0)$ is found, the chances heuristically are better that $\mathbf{x}_0$ is indeed a local solution point of (7.3) than they are in the corresponding situation for (7.2).

Let us note that earlier general remarks on Lagrange multipliers apply to Kuhn–Tucker multipliers. In handwork on small problems or in cases of problems possessing special characteristics such as convexity, we may be able to solve for $\mathbf{x}$ and $\mathbf{u}$ from the equations of (7.25). Whenever we find $\mathbf{x}_0$ we can solve linear equations for $\mathbf{u}_0$, but in most cases of interest we can say that an algorithm would simultaneously deliver $\mathbf{x}_0$ and $\mathbf{u}_0$. In Example 7.3 we illustrate uses of multipliers for sensitivity analyses.

First- and second-order necessary conditions for the inequality constrained Problem (7.3) follow at once if we merely restrict our attention to the binding constraints.

THEOREM. Suppose $f \in C^{(2)}$, $g \in C^{(2)}$, and all hypotheses of (7.25) hold. Then the Kuhn–Tucker conditions (1), (2), (3) remain necessary for $\mathbf{x}_0$ to be a local solution point of Problem (7.3), and furthermore

(4) $\dfrac{d^2}{d\mathbf{x}^2}[f(\mathbf{x}_0) + \mathbf{u}_0^T g(\mathbf{x}_0)]$ is positive semidefinite on the subspace

$$\{\mathbf{y} : \mathbf{g}_i'(\mathbf{x}_0)\mathbf{y} = 0 \quad \text{for all } i \in B(\mathbf{x}_0)\} \tag{7.26}$$

Note. The subspace is the null space of the derivative of the transformation determined by the binding constraints at $\mathbf{x}_0$. By (6.96) it also is the tangent space of the hypersurface determined by those same constraints.

PROOF. The conclusions of (7.25) certainly follow. Furthermore, $\mathbf{x}_0$ is a local solution point of the new problem

$$\text{minimize } f(\mathbf{x})$$
$$\text{s.t. } g_0(\mathbf{x}) = \mathbf{0}$$

where $g_0(\mathbf{x}) = \mathbf{0}$ iff $g_i(\mathbf{x}) = 0$ for all $i \in B(\mathbf{x}_0)$. Since the constraint qualification (7.24) holds, (7.21) applies to the new problem and yields (4). ■

Second-order sufficient conditions for a strict local solution can be established with much the same approach used to prove (7.22). We start by supposing that we can find a feasible point $\mathbf{x}_0$ and a vector $\mathbf{u}_0$ such that the Kuhn–Tucker conditions in (7.25) are satisfied. Then we would expect to require the Hessian

$$\mathbf{f}''(\mathbf{x}_0) + u_{10}\mathbf{g}_1''(\mathbf{x}_0) + \cdots + u_{m0}\mathbf{g}_m''(\mathbf{x}_0) \tag{7.27}$$

to be positive definite on some null space. Now there are two important observations. First, a matrix $\mathbf{g}_j''(\mathbf{x}_0)$ drops out of (7.27) whenever $u_{j0} = 0$. Second, $u_{j0} = 0$ can occur in either of the two ways

(i) $g_j(\mathbf{x}_0) < 0$ and $u_{j0} = 0$ by complementary slackness,
(ii) $g_j(\mathbf{x}_0) = 0$ and $u_{j0} = 0$ also happens.

When (ii) holds the associated constraint is binding but it is also degenerate insofar as its effect on the definiteness of (7.27). In this way we are led to consideration of

$$B^+(\mathbf{x}_0, \mathbf{u}_0) = \{i : g_i(\mathbf{x}_0) = 0 \quad \text{and} \quad u_{i0} > 0\}$$

which is a subset of $B(\mathbf{x}_0)$ corresponding to the given Kuhn–Tucker vector $\mathbf{u}_0$. We call $B^+(\mathbf{x}_0, \mathbf{u}_0)$ the *nondegenerate binding index set* corresponding to $\mathbf{u}_0$ at the feasible point $\mathbf{x}_0$. While $B(\mathbf{x}_0)$ is uniquely determined for any feasible $\mathbf{x}_0$, such cannot be said for $B^+(\mathbf{x}_0, \mathbf{u}_0)$ which is a function of both $\mathbf{x}_0$ and $\mathbf{u}_0$.

With the preceding as background and motivation we now state and prove the

THEOREM. Suppose $f \in C^{(2)}$ and $g \in C^{(2)}$. Then the following are sufficient for a feasible point $\mathbf{x}_0$ to be a strict local solution point of Problem (7.3): there exists $\mathbf{u}_0 \in \mathfrak{R}^m$ such that the Kuhn–Tucker conditions (1), (2), (3) of (7.25) hold, and furthermore

(4) The matrix (7.27) is positive definite on the subspace

$$P=\{\mathbf{y}:\mathbf{g}_j'(\mathbf{x}_0)\mathbf{y}=0 \quad \text{for} \quad j\in B^+(\mathbf{x}_0,\mathbf{u}_0)\} \tag{7.28}$$

Note. The subspace P is determined by the binding constraints at $\mathbf{x}_0$ that have positive Kuhn–Tucker multipliers.

PROOF. As in the proof of (7.22) we suppose that $\mathbf{x}_0$ is a limit point but not a strict local solution point. Then we can let $\{\mathbf{x}_i\}$ be a sequence of feasible points with associated objects as used in that proof. By the mean value theorem (6.43)

$$f'(\mathbf{z}_i)\mathbf{y}_i=\frac{f(\mathbf{x}_i)-f(\mathbf{x}_0)}{t_i}$$

for some $\mathbf{z}_i$ so that in the limit $\mathbf{f}'(\mathbf{x}_0)\mathbf{y}_0\leqq 0$. In the same way, if $j\in B^+(\mathbf{x}_0,\mathbf{u}_0)$ then for some $\mathbf{w}_{ij}$

$$\mathbf{g}_j'(\mathbf{w}_{ij})\mathbf{y}_i=\frac{g_j(\mathbf{x}_i)-g_j(\mathbf{x}_0)}{t_i}\leqq 0$$

where the inequality holds because $g_j(\mathbf{x}_i)$ determines the sign of the fraction for such j and it is nonpositive by feasibility. We conclude $\mathbf{g}_j'(\mathbf{x}_0)\mathbf{y}_0\leqq 0$. Now we can use the Kuhn–Tucker condition (3) to write

$$\mathbf{f}'(\mathbf{x}_0)\mathbf{y}_0=-\sum_{j\in B^+(\mathbf{x}_0,\mathbf{u}_0)} u_{j0}\mathbf{g}_j'(\mathbf{x}_0)\mathbf{y}_0$$

and use the preceding conclusion to deduce $\mathbf{f}'(\mathbf{x}_0)\mathbf{y}_0\geqq 0$. Consequently, $\mathbf{f}'(\mathbf{x}_0)\mathbf{y}_0=0$ and we conclude $\mathbf{y}_0\in P$ from the displayed summation. This completes the counterpart of the first task for proving (7.22). In order to finish we proceed exactly as in the second task for (7.22) and show that the quadratic form with matrix (7.27) would assume a nonpositive value at y_0. ■

Let us note two matters for P. First, in case the constraint qualification (7.24) does not hold then we must of course exercise care in finding a basis for P in preparation for using the procedure (5.29) for resolving definiteness on a subspace. Second, it is possible that $P=\varnothing$ in which case we cannot use (7.28) as stated; however, as shown in Fiacco and McCormick (1968, 32) the cited (1), (2), (3) are then sufficient by themselves.

EXAMPLE 7.3. Interpreting Multipliers as Partial Derivatives. The subject interpretations are useful for studying effects on solution points due to changes in definition of the feasible region. Such studies are included in what is called *sensitivity analysis* in optimization theory.

Before introducing such analysis let us consider a special problem in which it is particularly easy to relate multipliers to partial derivatives, namely,

$$\text{minimize } f(\mathbf{x})$$
$$\text{s.t. } -\mathbf{x} \leqq \mathbf{0}$$

which is Problem (7.3) with the nonnegative orthant as feasible region. Constraint qualification (7.24) holds since any nonempty collection of rows of $\mathbf{g}'(\mathbf{x}) = -\mathbf{I}_m$ must be LI. If $\mathbf{x}_0$ is a local solution point then Kuhn–Tucker condition (3) becomes

$$\mathbf{f}'(\mathbf{x}_0) = \mathbf{u}_0^T$$

so multiplier u_{i0} equals partial derivative $\partial_i f(\mathbf{x}_0)$. For the surface of revolution considered in (7.10), the solution point is on the boundary, $\mathbf{x}_0 = [0, 1]^T$, and the partial derivatives $\partial_1 f(\mathbf{x}_0) = 2, \partial_2 f(\mathbf{x}_0) = 0$ are the Kuhn–Tucker multipliers.

Now let us proceed to a sensitivity analysis of Problem (7.2) by considering

(1)
$$\text{minimize } f(\mathbf{x})$$
$$\text{s.t. } h(\mathbf{x}) = \mathbf{b}$$

We suppose that for $\mathbf{b} = \mathbf{0}$ we have a strict solution point $\mathbf{x}_0$ and associated vector $\boldsymbol{\lambda}_0$ of Lagrange multipliers satisfying the conditions of (7.22). Let us furthermore suppose that h determines a smooth hypersurface (6.94). These are strong conditions which actually imply that (1) has a solution for each $b \in \mathfrak{R}^k$ sufficiently near $\mathbf{0}$ and (7.22) applies at the solution points $\mathbf{x}(\mathbf{b})$; see Fiacco (1976). Given that this is the situation, we can apply the implicit function theorem (6.93) to a certain transformation ϕ in $\mathfrak{R}^{n+k}$ so as to guarantee solutions of $n+k$ equations in $n+k$ unknowns $x_1, \dots, x_n, \lambda_1, \dots, \lambda_k$ in a neighborhood of $\mathbf{b} = \mathbf{0}$ in $\mathfrak{R}^k$. These are the coordinates of the solution point $\mathbf{x}$ and the vector of multipliers $\boldsymbol{\lambda}$ for (1) also according to Fiacco (1976). The transformation ϕ is defined for each $\mathbf{b}$ sufficiently near $\mathbf{0}$ by the $n+k$ scalar LHSs in

(2)
$$\mathbf{f}'(\mathbf{x}) + \boldsymbol{\lambda}^T \mathbf{h}'(\mathbf{x}) = \mathbf{O}$$
$$h(\mathbf{x}) - \mathbf{b} = \mathbf{0}$$

Notice that the first vector LHS is the first derivative $(d/d\mathbf{x})\psi(\mathbf{x}, \boldsymbol{\lambda})$. The only nonobvious condition in (6.91) is that ϕ' be nonsingular at the point in $\mathfrak{R}^{n+k}$ corresponding to $\mathbf{x}_0$ and $\boldsymbol{\lambda}_0$. This derivative has the $n+k$ order matrix

$$(3)\qquad \begin{bmatrix} \dfrac{d^2}{d\mathbf{x}^2}\left[f(\mathbf{x})+\boldsymbol{\lambda}^T h(\mathbf{x})\right] & \left[\mathbf{h}'(\mathbf{x})\right]^T \\ \mathbf{h}'(\mathbf{x}) & \mathbf{O}_k \end{bmatrix} = \begin{bmatrix} \mathbf{A} & \mathbf{B}^T \\ \mathbf{B} & \mathbf{O} \end{bmatrix}$$

where: $\mathbf{A}$ is given by (7.20); the (1,2) element is $(d/d\boldsymbol{\lambda})(d\psi/d\mathbf{x})$ found by Table 6.2(6); the (2,1) element is $(d/d\mathbf{x})h(\mathbf{x})$; and, the (2,2) element is $(d/d\boldsymbol{\lambda})h(\mathbf{x})$. Let us suppose matrix (3) is singular for $\mathbf{x}=\mathbf{x}_0$ and $\boldsymbol{\lambda}=\boldsymbol{\lambda}_0$. This means there is a nontrivial LC of its columns equalling $\mathbf{0}$, say

$$(4)\qquad \begin{aligned} \mathbf{Ay}+\mathbf{B}^T\mathbf{z}&=\mathbf{0} \\ \mathbf{By}&=\mathbf{0} \end{aligned}$$

where not both $\mathbf{y}\in\mathcal{R}^n$ and $\mathbf{z}\in\mathcal{R}^k$ are zero. But if $\mathbf{y}\neq\mathbf{0}$, then $\mathbf{y}\in\operatorname{Null}\mathbf{B}$ and $\mathbf{y}^T\mathbf{Ay}+\mathbf{y}^T\mathbf{B}^T\mathbf{z}=0$ which is impossible because it states $\mathbf{y}^T\mathbf{Ay}=-\langle\mathbf{By},\mathbf{z}\rangle=0$ contradicting (2) in (7.22). Thus $\mathbf{y}=\mathbf{0}$ so we must have $\mathbf{z}\neq\mathbf{0}$. But this too is impossible: From the first equation in (4), $\mathbf{B}^T\mathbf{z}=\mathbf{0}$ and this contradicts LI of the columns of $\mathbf{B}^T$ which must hold because (6.94) has been imposed as a constraint qualification. Therefore ϕ' is nonsingular, the implicit function theorem applies, and for sufficiently small $\|\mathbf{b}\|$ there exist transformations of Classes $C^{(1)}$ (and not necessarily $C^{(2)}$)

$$(5)\qquad \begin{aligned} \mathbf{x}&=\alpha(\mathbf{b}) & \qquad \mathbf{x}_0&=\alpha(\mathbf{0}) \\ \boldsymbol{\lambda}&=\beta(\mathbf{b}) & \qquad \boldsymbol{\lambda}_0&=\beta(\mathbf{0}) \end{aligned}$$

satisfying the first-order necessary conditions (2) for Problem (1). As stated previously, $\mathbf{x}$ is the solution point and $\boldsymbol{\lambda}$ is the vector of Lagrange multipliers for Problem (1) as a function of $\mathbf{b}$.

A principal concern in sensitivity analysis is how the minimum value $f(\mathbf{x})$ changes in (1) as the RHS $\mathbf{b}$ changes. Now that we are assured of the existence and differentiability of the transformations in (5) we can use the chain rule (6.64) to find the derivative of $f(\alpha(\mathbf{b}))$ with respect to $\mathbf{b}$. Using that theorem together with the first equation in (2) we find (writing $\mathbf{x}_0(\mathbf{b})$ only in the first term)

$$\frac{df(\mathbf{x}_0(\mathbf{b}))}{d\mathbf{b}}=\frac{df(\mathbf{x}_0)}{d\mathbf{x}}\frac{d\mathbf{x}_0}{d\mathbf{b}}=\left(-\boldsymbol{\lambda}_0^T\frac{dh(\mathbf{x}_0)}{d\mathbf{x}}\right)\frac{d\mathbf{x}_0}{d\mathbf{b}}$$

But with the second equation in (2) the product of the last two factors is $\mathbf{I}_k$

so

$$\frac{df(\mathbf{x}_0(\mathbf{b}))}{d\mathbf{b}} = -\boldsymbol{\lambda}_0^T \tag{6}$$

Thus $\partial_i f(\mathbf{x}_0)$ is the negative of the ith Lagrange multiplier, which means that $-\lambda_{i0}$ equals the incremental rate of change of $f(\mathbf{x}_0(\mathbf{b}))$ per unit change in the ith constraint component. It is also possible to calculate $(d/d\mathbf{b})\boldsymbol{\lambda} = (d^2/d\mathbf{b}^2)f(\mathbf{x}_0(\mathbf{b}))$ in order, for example, to test for convexity.

In applied work the preceding has considerable significance because it means, especially in resource allocation problems, that multipliers play the roles of implicit "prices" for the objects represented by the constraints. For example, suppose $f(\mathbf{x})$ represents total monetary cost of producing something according to "plan" $\mathbf{x}$ while $g_i(\mathbf{x})$ is associated with the available amount of the ith resource. Then Problem (7.3), rather than (7.2), will be the preferable formulation given that we do not know in advance the amounts of individual resources to be consumed by the most advantageous "plan."

Analysis comparable to the preceding can be performed for Problem (7.3) with the result that for small $\|\mathbf{c}\|$ in

$$\begin{aligned} &\text{minimize } f(\mathbf{x}) \\ &\text{s.t. } g(\mathbf{x}) \leqq \mathbf{c} \end{aligned} \tag{7}$$

we have the counterpart to (6), namely

$$\frac{df(\mathbf{x}_0(\mathbf{c}))}{d\mathbf{c}} = -\mathbf{u}_0^T \tag{8}$$

Now Kuhn–Tucker multipliers are nonnegative, and their interpretations as implicit prices turn out as follows. If $u_{i0} = 0$ then the ith resource is in plentiful supply, it is a free good, and it does not affect the minimum cost $f(\mathbf{x}_0)$. If $u_{j0} > 0$ then the jth resource is in scarce supply, it is priced by u_{j0}, and all available units are consumed in the solution "plan" $\mathbf{x}_0$. In economic terms this is expressed by saying that the Kuhn–Tucker multipliers correspond to prices associated with small changes in resource requirements, that is, they are what are called marginal prices.

PROBLEMS FOR SOLUTION

7.3.1 Feasible Region. Establish for continuous g:

(1) $B(\mathbf{x}_0) = \varnothing$ implies $\mathbf{x}_0 \in \operatorname{int} K$.

(2) $\mathbf{x}_0 \in \operatorname{int} K$ does not imply $B(\mathbf{x}_0) = \varnothing$.

(3) In (7.25), $\mathbf{x}_0 \in \operatorname{int} K$ implies $\mathbf{u}_0 = \mathbf{0}$.
(4) In (7.25), $\mathbf{u}_0 = \mathbf{0}$ does not imply $\mathbf{x}_0 \in \operatorname{int} K$.

7.3.2 Binding Constraints. Let $\mathbf{x}_0$ be a local solution point of Problem (7.3) and consider the associated Problem (7.2)

$$\text{minimize } f(\mathbf{x})$$
$$\text{s.t. } \mathbf{x} \in K(\mathbf{x}_0)$$

where $K(\mathbf{x}_0) = \{\mathbf{x} : g_i(\mathbf{x}) = 0 \text{ for all } i \in B(\mathbf{x}_0)\}$. Establish conditions such that the following apply.
(1) (7.14).
(2) (7.21).

7.3.3 Other Problems. Consider the following.

$$\begin{array}{lll} \text{(I)} & \text{minimize} & f(\mathbf{x}) \\ & \text{s.t.} & g(\mathbf{x}) \geqq \mathbf{0} \\ \text{(II)} & \text{maximize} & f(\mathbf{x}) \\ & \text{s.t.} & g(\mathbf{x}) \leqq \mathbf{0} \\ \text{(III)} & \text{maximize} & f(\mathbf{x}) \\ & \text{s.t.} & g(\mathbf{x}) \geqq \mathbf{0} \\ \text{(IV)} & \text{minimize} & f(\mathbf{x}) \\ & \text{s.t.} & g(\mathbf{x}) \leqq \mathbf{b} \\ & & \mathbf{x} \geqq \mathbf{0} \end{array}$$

Establish the four counterparts to each of the following results for Problem (7.3):
(1) John's theorem (7.23).
(2) Necessary conditions (7.26).
(3) Sufficient conditions (7.28).

7.3.4 Bootstrap? Can John's theorem be used to produce useful conditions for Problem (7.2) by consideration of the following?

$$\text{minimize } f(\mathbf{x})$$
$$\text{s.t. } h(\mathbf{x}) \leqq \mathbf{0}$$
$$-h(\mathbf{x}) \leqq \mathbf{0}$$

7.3.5 Strict Complementary Slackness. Kuhn–Tucker multipliers are said to have the subject property if condition (2) in (7.25) is replaced by: $u_{i0} = 0$ iff $g_i(\mathbf{x}_0) < 0$. What effect does this have on (1) (7.26), (2) (7.28)?

7.3.6 Lagrange Multipliers and Kuhn–Tucker Multipliers. (James E. Falk.) Establish:

(1) A precise statement covering "Lagrange multipliers in (7.18) do not distinguish between minima and maxima of $f(\mathbf{x})$."
(2) A corresponding statement for pairs $\mathbf{x}_0$ and $\mathbf{u}_0$ in (7.25) considering both Problem (7.3) and (II) in 7.3.3.
(3) What the preceding suggests concerning the relative merits of (7.18) and (7.25) for their problems given that we consider them apart from their accompanying second-order conditions?

7.3.7 Interpreting Kuhn–Tucker Multipliers. Suppose we have $\mathbf{x}_0$ and $\mathbf{u}_0$ such that all conditions for a strict local solution in (7.28) hold. Suppose further that $B^+(\mathbf{x}_0,\mathbf{u}_0)=B(\mathbf{x}_0)$ so all binding constraints are nondegenerate. Perform the analysis for Problem (7) in Example 7.3 leading to (8).

7.4 MIXED CONSTRAINTS

Here we quickly deal with criteria for solution points of Problem (7.4), namely

$$\begin{aligned} &\text{minimize } f(\mathbf{x}) \\ &\text{s.t. } g(\mathbf{x}) \leqq \mathbf{0} \\ &\quad h(\mathbf{x}) = \mathbf{0} \end{aligned}$$

which is also called the *general nonlinear programming problem.*

Criteria are readily forthcoming under the following verifiable

Constraint Qualification. The transformations g and h have the property

$$\{\mathbf{g}_i'(\mathbf{x}_0) : i \in B(\mathbf{x}_0)\} \cup \{\mathbf{h}_j'(\mathbf{x}_0) : j = 1,2,\ldots,k\} \text{ is LI}$$

where the first set is omitted whenever there are no binding inequalities at $\mathbf{x}_0$. (7.29)

Whenever g does not appear in Problem (7.4) and $h \in C^{(1)}$ determines a smooth hypersurface, then (7.29) holds. Whenever h does not appear, (7.29) coincides with (7.24). In this way the results of the present section specialize to those in the preceding two.

Results for local solution points $\mathbf{x}_0$ of the inequality constraint Problem (7.3) are derived in the preceding section from those for Problem (7.2) following the observation that they depend on the binding constraints at $\mathbf{x}_0$. The corresponding feature of the mixed constraint problem is

Adjoining $h(\mathbf{x})=\mathbf{0}$ to Problem (7.3) merely increases the set of binding constraints at a local solution point.

In what follows we find very little remains to be done after noting this fact.

Let us establish the generalization of (7.25), namely the

KUHN–TUCKER CONDITIONS FOR MIXED CONSTRAINTS. Suppose f,g,h belong to their respective Classes $C^{(1)}$ and $\mathbf{x}_0$ is a local solution point of Problem (7.4) at which the constraint qualification (7.29) holds. Then there exist unique $\mathbf{u}_0 \in \mathcal{R}^m, \boldsymbol{\lambda}_0 \in \mathcal{R}^k$ such that

(1) $\mathbf{u}_0$ is nonnegative.

(2) $\mathbf{u}_0^T g(\mathbf{x}_0)=0$.

(3) $\mathbf{f}'(\mathbf{x}_0)+\mathbf{u}_0^T \mathbf{g}'(\mathbf{x}_0)+\boldsymbol{\lambda}_0^T \mathbf{h}'(\mathbf{x}_0)=\mathbf{O}$. (7.30)

PROOF. If we consider the constraints involved in (7.29) then the Lagrange multiplier theorem (7.18) assures us that there exist associated multipliers. In order to obtain a full set we assign zero multipliers to the remaining constraints so that (2) and (3) then hold. Suppose (1) fails because $u_{\alpha 0}<0$. Then the tangent space at $\mathbf{x}_0$ defined by the set of $1 \times n$ matrices in (7.29) with the one for $i=\alpha$ deleted must contain a vector $\mathbf{y}$ such that $\mathbf{g}'_{\alpha 0}(\mathbf{x}_0)\mathbf{y} \neq 0$; otherwise $\mathbf{g}'_{\alpha 0}(\mathbf{x}_0)$ would be a LC of the specified matrices in violation of (7.29). We are free to suppose $\mathbf{g}'_{\alpha 0}(\mathbf{x}_0)\mathbf{y}<0$ and use (3) to conclude

$$\mathbf{f}'(\mathbf{x}_0)\mathbf{y} = -u_{\alpha 0}\mathbf{g}'_{\alpha 0}(\mathbf{x}_0)\mathbf{y}<0$$

This is impossible because it implies

$$f(\mathbf{x}_0+t\mathbf{y})<f(\mathbf{x}_0)$$

for sufficiently small $t>0$. ■

PROBLEMS FOR SOLUTION

7.4.1 Extensions. Establish counterparts to the following for Problem (7.4):

(1) John's theorem (7.23).

(2) Second-order necessary conditions (7.26).

(3) First- and second-order sufficient conditions (7.28).

(4) Sensitivity analysis in Example 7.3.

7.4.2 Kuhn–Tucker Conditions. Extend 7.3.3 by establishing 7.4.1 (1), (2), (3) for each of the problems:

(I′) minimize $f(\mathbf{x})$

s.t. $g(\mathbf{x}) \geqq \mathbf{0}$

$h(\mathbf{x}) = \mathbf{0}$

(II′) maximize $f(\mathbf{x})$

s.t. $g(\mathbf{x}) \leqq \mathbf{0}$

$h(\mathbf{x}) = \mathbf{0}$

(III′) maximize $f(\mathbf{x})$

s.t. $g(\mathbf{x}) \geqq \mathbf{0}$

$h(\mathbf{x}) = \mathbf{0}$

(IV′) minimize $f(\mathbf{x})$

s.t. $g(\mathbf{x}) \leqq \mathbf{b}$

$h(\mathbf{x}) = \mathbf{c}$

$\mathbf{x} \geqq \mathbf{0}$

7.5 CONVEX PROGRAMMING

In this section we work with an important set of problems that includes the linear programming problem (7.6) which is introduced in Example 7.4 using results from Section 4.3.

Let us start with the most remarkable result that holds for all instances of the *convex problem*

$$\begin{aligned} &\text{minimize } f(\mathbf{x}) \\ &\text{s.t. } \mathbf{x} \in D \end{aligned} \tag{7.31}$$

where f is convex on the convex open set D.

THEOREM. Any local solution point of a convex problem (7.31) is a global solution point. (7.32)

PROOF. Suppose f has a local minimum at $\mathbf{x}_0$ that is not global because $f(\mathbf{u}) < f(\mathbf{x}_0)$ for some $\mathbf{u} \in D$. By (6.97)

$$f(t\mathbf{x}_0 + (1-t)\mathbf{u}) \leqq tf(\mathbf{x}_0) + (1-t)f(\mathbf{u})$$

for $t \in (0,1)$ and consequently the RHS is less than $f(\mathbf{x}_0)$. Considering t near unity we see that this contradicts the assumption that $\mathbf{x}_0$ locates a local minimum. ■

We can establish a few additional general results before we consider the special case of the convex programming problem (7.5). Let us start with

If S is the set of solution points of the convex problem (7.31) then
(1) S is a convex set.
(2) $S=\varnothing$ or $S=\{\mathbf{x}_0\}$ whenever f is strictly convex. (7.33)

PROOF. The null set is convex and, otherwise, if $\mathbf{x}_0$ is a solution point then

$$S=\{\mathbf{x}: f(\mathbf{x}) \leqq f(\mathbf{x}_0)\}$$

is convex by 6.5.1 (4). Statement (2) holds because if $\mathbf{x}_0$ and $\mathbf{u}$ are distinct points in S, then the inequality displayed in the proof of (7.32) shows that f is not strictly convex. ■

There are three possibilities for the number of elements in S: 0, 1, or ∞. There are none for the (strictly convex) case $f(x)=x^2$ on $(0,1)$, exactly one if D is changed to $(-1,1)$, and an infinite number for

$$\begin{aligned} f(x) &= x^2, \qquad x \in (-2,-1] \cup [1,2) \\ &= 1, \qquad x \in (-1,1) \end{aligned}$$

on $(-2,2)$. When f is differentiable, then solution points in D coincide with critical points; that is, the condition in (7.12) is sufficient as well as necessary. Another way of stating this is

If f is differentiable on its convex open domain D then the convex problem (7.31) has a global solution at $\mathbf{x}_0$ iff $\mathbf{f}'(\mathbf{x}_0)=\mathbf{O}$. (7.34)

PROOF. Necessity results from (7.12) and sufficiency from (1) in (6.103). ■

If f is also strictly convex then it can have at most one critical point by (7.33).

The phenomenon of necessary conditions becoming sufficient also occurs for the convex programming problem (7.5), namely

$$\begin{aligned} &\text{minimize } f(\mathbf{x}) \\ &\text{s.t. } g(\mathbf{x}) \leqq \mathbf{0} \\ &\mathbf{A}\mathbf{x} = \mathbf{b} \end{aligned}$$

where f, and each coordinate function of g, is a convex function.

In our usual notation, the derivative of the rewritten equality constraint is

$$\frac{d}{d\mathbf{x}}(\mathbf{Ax}-\mathbf{b})=\mathbf{A}=\begin{bmatrix}\mathbf{a}_1^T\\ \mathbf{a}_2^T\\ \vdots\\ \mathbf{a}_k^T\end{bmatrix}$$

When both $\mathbf{Ax}=\mathbf{b}$ and $g(\mathbf{x})\leqq\mathbf{0}$ actually appear in Problem (7.5) then the constraint qualification (7.29) is as follows for any feasible point $\mathbf{x}$.

$$\{\mathbf{g}_i'(\mathbf{x}): i\in B(\mathbf{x})\}\cup\{\mathbf{a}_1^T,\ldots,\mathbf{a}_k^T\}\text{ is LI} \tag{7.35}$$

If $\mathbf{Ax}=\mathbf{b}$ does not appear then the above coincides with (7.24) while if g does not appear in the problem then it means $\operatorname{rank}\mathbf{A}=k$.

KUHN–TUCKER THEOREM FOR CONVEX PROGRAMMING. Suppose f and g belong to their respective Classes $C^{(1)}$ and the constraint qualification (7.35) holds for the feasible point $\mathbf{x}_0$. Then $\mathbf{x}_0$ is a global solution point of Problem (7.5) iff there exist unique $\mathbf{u}_0\in\mathcal{R}^m, \boldsymbol{\lambda}_0\in\mathcal{R}^k$ such that

(1) $\mathbf{u}_0$ is nonnegative.

(2) $\mathbf{u}_0^T g(\mathbf{x}_0)=0$.

(3) $\mathbf{f}'(\mathbf{x}_0)+\mathbf{u}_0^T\mathbf{g}'(\mathbf{x}_0)+\boldsymbol{\lambda}_0^T\mathbf{A}=\mathbf{O}$. (7.36)

PROOF. Necessity holds by (7.30). Conversely, if $\mathbf{x}_0+\mathbf{h}$ is a feasible point then by (6.103)

$$f(\mathbf{x}_0+\mathbf{h})-f(\mathbf{x}_0)\geqq\mathbf{f}'(\mathbf{x}_0)\mathbf{h}$$

and we complete the proof by showing that the RHS is nonnegative. By (3),

$$\mathbf{f}'(\mathbf{x}_0)\mathbf{h}=-\mathbf{u}_0^T\mathbf{g}'(\mathbf{x}_0)\mathbf{h}-\boldsymbol{\lambda}_0^T\mathbf{Ah}$$

The first term on this RHS is nonnegative because, again by (6.103),

$$\mathbf{g}_i'(\mathbf{x}_0)\mathbf{h}\leqq g_i(\mathbf{x}_0+\mathbf{h})-g_i(\mathbf{x}_0)$$

for $i=1,2,\ldots,m$; consequently, since $\mathbf{x}_0+\mathbf{h}$ is feasible

$$-\mathbf{g}_i'(\mathbf{x}_0)\mathbf{h}\geqq 0$$

for $i \in B(\mathbf{x}_0)$, that is, for $g_i(\mathbf{x}_0)=0$. The second term $\boldsymbol{\lambda}_0^T\mathbf{Ah}$ is zero by feasibility: $\mathbf{Ax}_0+\mathbf{Ah}=\mathbf{b}$ and $\mathbf{Ax}_0=\mathbf{b}$ imply $\mathbf{Ah}=\mathbf{0}$. ■

There are many ways of weakening the above hypotheses; see, for example, Mangasarian (1969) and Stoer and Witzgall (1970).

If we wish to use (7.7) to translate the results of this section into results for maximization then we address the *concave problem*

$$\text{maximize } f(\mathbf{x})$$
$$\text{s.t. } \mathbf{x} \in D \tag{7.37}$$

where $f: D \subset \mathfrak{R}^n \to \mathfrak{R}$ is a concave function (7.5.2). A linear function is both convex and concave so it is not surprising that linear programming problems (7.6) have many special properties.

EXAMPLE 7.4. Linear Programming. In this example we obtain results by using (7.36) and material from Chapter 4. We impose a constraint qualification that we would not need if we were starting from scratch, and we stop far short of anything like a complete treatment of linear programming.

Let us relabel Problem (7.6) as the *primal problem*

(1)
$$\text{minimize } \mathbf{c}^T\mathbf{x}$$
$$\text{s.t. } \mathbf{Ax}=\mathbf{b}$$
$$\mathbf{x} \geqq \mathbf{0}$$

and introduce the *dual problem*

(2)
$$\text{maximize } \mathbf{b}^T\mathbf{y}$$
$$\text{s.t. } \mathbf{A}^T\mathbf{y} \leqq \mathbf{c}$$

It is customary to refer to (1) and (2) as linear programming problems in *standard form*. (*Canonical forms* appear in 7.5.5.) It is also conventional to let $\mathbf{A}$ be $m \times n$ whereby $\mathbf{x}, \mathbf{c} \in \mathfrak{R}^n$ and $\mathbf{y}, \mathbf{b} \in \mathfrak{R}^m$, and to understand that $\mathbf{x}$ and $\mathbf{y}$ always go with (1) and (2), respectively; for example, "$\mathbf{x}$ is feasible" is understood to refer to (1).

If $\mathbf{x}$ and $\mathbf{y}$ are feasible then $\mathbf{b}=\mathbf{Ax}$ implies $\mathbf{b}^T\mathbf{y}=\mathbf{x}^T\mathbf{A}^T\mathbf{y}$. But $\mathbf{x} \geqq \mathbf{0}$ and $\mathbf{A}^T\mathbf{y} \leqq \mathbf{c}$ imply

$$\mathbf{x}^T\mathbf{A}^T\mathbf{y} \leqq \mathbf{x}^T\mathbf{c}=\mathbf{c}^T\mathbf{x}$$

and we conclude

(3) If $\mathbf{x}$ and $\mathbf{y}$ are feasible points, then $\mathbf{b}^T\mathbf{y} \leqq \mathbf{c}^T\mathbf{x}$.

If equality holds for feasible $\mathbf{x}_0$ and $\mathbf{y}_0$, then since $\mathbf{b}^T\mathbf{y} \leqq \mathbf{b}^T\mathbf{y}_0$ for all feasible $\mathbf{y}$, and $\mathbf{c}^T\mathbf{x} \geqq \mathbf{c}^T\mathbf{x}_0$ for all feasible $\mathbf{x}$, we have

(4) If $\mathbf{x}_0$ and $\mathbf{y}_0$ are feasible points, then $\mathbf{b}^T\mathbf{y}_0 = \mathbf{c}^T\mathbf{x}_0$ implies that both are solution points.

We find in (7) that the converse is also valid. Result (4) figures prominently in the development below and it is important for sensitivity analysis (7.5.6).

Linear programming problems are special cases of the convex programming problem (7.5). In (1) we have $f(\mathbf{x}) = \mathbf{c}^T\mathbf{x}$ and for (2) we use $f(\mathbf{y}) = -\mathbf{b}^T\mathbf{y}$. In (1) the inequality constraint has $g(\mathbf{x}) = -\mathbf{I}_n\mathbf{x}$ and in (2) we must assign

$$g(\mathbf{y}) = \mathbf{A}^T\mathbf{y} - \mathbf{c}$$

In order to apply (7.36) we impose (7.35) as the

(5) Constraint Qualification. The matrix $\mathbf{A}$ has the following properties

(i) If $\mathbf{x}$ is a feasible point then the rows of

$$\begin{bmatrix} \mathbf{A} \\ \mathbf{E} \end{bmatrix}$$

are LI where $\mathbf{E}$ is the submatrix of $\mathbf{I}_n$ corresponding to the binding inequalities in $\mathbf{x} \geqq \mathbf{0}$.

(ii) If $\mathbf{y}$ is a feasible point then the rows of $\mathbf{A}^T$ corresponding to the binding inequalities in $\mathbf{A}^T\mathbf{y} \leqq \mathbf{c}$ are LI.

It is sufficient that the preceding hold at solution points but it simplifies the wording of results if we impose it on the feasible regions. In this way we are led to a complementary slackness condition that furnishes

(6) **CRITERIA FOR SOLUTION POINTS.** Suppose constraint qualification (5) holds. Then a feasible point is a solution point of (1) or (2) iff there exists a feasible point for the other problem such that

$$(\mathbf{x}_0)^T(\mathbf{A}^T\mathbf{y}_0 - \mathbf{c}) = 0$$

PROOF. We show that this is a corollary of (7.36) by starting with problem (2) where $\mathbf{f}'(\mathbf{y}) = -\mathbf{b}^T$ and $\mathbf{g}'(\mathbf{y}) = \mathbf{A}^T$. A feasible point $\mathbf{y}_0$ is a solution point iff there exists $\mathbf{x}_0$ such that the correspondingly numbered conditions of (7.36) hold

(i) $\mathbf{x}_0 \geqq \mathbf{0}$.

(ii) $\mathbf{x}_0^T(\mathbf{A}^T\mathbf{y}_0 - \mathbf{c}) = 0$.

(iii) $-\mathbf{b}^T + \mathbf{x}_0^T\mathbf{A}^T = \mathbf{O}$.

Next let us show that the result also holds for (1) where $\mathbf{f}'(\mathbf{x}) = \mathbf{c}^T$ and $\mathbf{g}'(\mathbf{x}) = -\mathbf{I}_n$. A feasible point $\mathbf{x}_0$ is a solution point iff there exist $\mathbf{u}_0 \in \mathfrak{R}^n, \boldsymbol{\lambda}_0 \in \mathfrak{R}^m$ such that

(i′) $\mathbf{u}_0 \geqq \mathbf{0}$.

(ii′) $\mathbf{u}_0^T \mathbf{x}_0 = 0$.

(iii′) $\mathbf{c}^T - \mathbf{u}_0^T \mathbf{I}_n + \boldsymbol{\lambda}_0^T \mathbf{A} = \mathbf{O}$.

Now (i′) and (iii′) are equivalent to $\mathbf{A}^T \boldsymbol{\lambda}_0 \geqq -\mathbf{c}$ which means $\mathbf{y}_0 = -\boldsymbol{\lambda}_0$ is a feasible point. Given $\mathbf{u}_0 = \mathbf{c}^T - \mathbf{A}^T \mathbf{y}_0$ in this new notation we see that (ii′) produces the required condition. ■

The preceding criteria show that solution points of the primal and dual problems are closely related. Their exact relationship is specified in the

(7) **DUALITY THEOREM OF LINEAR PROGRAMMING.** Suppose constraint qualification (5) holds. Then the primal problem has a solution point iff the dual problem has one. Moreover, the extreme values of the two objective functions coincide.

PROOF. Either problem has a solution point iff the other problem has a feasible point at which the complementary slackness condition in (6) holds. Then this condition together with (4) shows that there is a pair of solution points at which the objective functions have equal values. ■

More than this is established in works on linear programming. For example, both problems have solution points iff both feasible regions are nonempty, the dual of the dual is the primal, and so on.

Methods of differential calculus based on (7.12) are not useful for finding solution points for (1) because there is a critical point iff $\mathbf{c} = \mathbf{0}$. But from (7.14)

(8) If $\mathbf{c} \neq \mathbf{0}$ then any solution points of (1) must lie on the boundary of the feasible region.

Consequently, (1) can fail to have a solution point in only two different ways. First, the individual constraints can be inconsistent which means that there is no feasible point. Second, the feasible region, and hence also the objective function, can be unbounded below; in applied work this suggests that individual constraints need to be added.

Now let us turn to the main results underlying the *simplex method* for finding solution points. Relying on (4.49) and (7.7) we may suppose that we have an instance of the standard problem (1) which is the form with which the elementary simplex method is used. According to (4.50) and (4.51)

(9) If the feasible region in (1) is nonempty then it contains a vertex.

Examples show that a similar statement cannot be made for (2). This brings us to the main result which is so intuitively evident in $\mathfrak{R}^2$.

(10) **FUNDAMENTAL THEOREM OF LINEAR PROGRAMMING.** Suppose constraint qualification (5) holds. If (1) has a solution point, then one of the finite set of vertexes is a solution point.

PROOF. Given a solution point $\mathbf{x}_0$ there exists $\mathbf{y}_0 \in \mathcal{R}^m$ according to (6). We may assume that the first r components of $\mathbf{x}$ are positive and all others are zero. If the first r columns of $\mathbf{A}$ are LI then $\mathbf{x}$ is a vertex and we are done. Otherwise the columns are LD and by (4.50) there exists a vertex $\bar{\mathbf{x}}$ corresponding to some subset of the first r columns of $\mathbf{A}$. By construction $\bar{\mathbf{x}} \geqq \mathbf{0}$ and $\mathbf{A}\bar{\mathbf{x}} = \mathbf{b}$ so $\bar{\mathbf{x}}$ is feasible. Moreover $\bar{\mathbf{x}}$ is a solution point because $\bar{\mathbf{x}}$ and $\mathbf{y}_0$ satisfy (7). There are at most $C(n,m)$ vertexes since each corresponds to a selection of m LI columns from the n columns of $\mathbf{A}$. ∎

It is in general impractical to test all vertexes to find the achieved minimum in (1). There may be too many to examine on any computer and, even if their number is moderate, an exhaustive search is wasteful. The effectiveness of the simplex method is that starting at any vertex we move at each step to another so as to diminish the value of the objective function. Very few steps are commonly required before it is resolved that a solution point has been reached, or that none will ever be reached because of unboundedness of the objective function. It is even true that the simplex method can be used on a different problem to get started for (1) by

(11) **FINDING A VERTEX.** First, change signs as necessary so $\mathbf{b} \geqq \mathbf{0}$. Second, use the simplex method to solve in $\mathcal{R}^{m+n}$

$$\begin{aligned} &\text{minimize } \mathbf{0}^T\mathbf{x} + \mathbf{e}^T\mathbf{u} \\ &\text{s.t. } \mathbf{A}\mathbf{x} + \mathbf{u} = \mathbf{b} \\ &\mathbf{x} \geqq \mathbf{0},\ \mathbf{u} \geqq \mathbf{0} \end{aligned}$$

Then the feasible region for (1) is nonempty iff the preceding problem has a solution point with $\mathbf{u} = \mathbf{0}$.

Notice that an initial vertex for (11) is

$$\begin{bmatrix} \mathbf{x} \\ \mathbf{u} \end{bmatrix} = \begin{bmatrix} \mathbf{0} \\ \mathbf{b} \end{bmatrix}$$

and in the present context the components $u_1, \dots, u_m$ are called *artificial variables*.

Let us describe how the simplex method proceeds for (1). An initial vertex $\mathbf{x}_1$ is either apparent or else (11) is used as follows. If (11) has a solution point with $\mathbf{u} \neq \mathbf{0}$ then (1) has no solution because its feasible region is empty. Otherwise there is a solution point with $\mathbf{u}=\mathbf{0}$

$$\begin{bmatrix} \mathbf{x}_1 \\ \mathbf{0} \end{bmatrix}$$

and we proceed to find it. Following this, the initial data for (1) are assembled in a partitioned matrix

$$\mathbf{P}=\begin{bmatrix} \mathbf{A} & \mathbf{b} \\ \mathbf{c}^T & \mathbf{c}^T\mathbf{x}_1 \end{bmatrix}$$

All subsequent calculations in the simplex method produce matrices that are row-equivalent to **P** by means of Gauss–Jordan pivot operations in positions originally occupied by elements of **A**. That is, we never pivot in the last row or last column.

Since $\mathbf{x}_1$ is a vertex the columns of **A** corresponding to its positive components are LI. Let us assume *nondegeneracy* whereby $\mathbf{x}_1$ has m positive components. Then by Gauss–Jordan pivoting in the first m rows of **P** we can replace **P** by $\mathbf{P}_1$ where unity is the only nonzero element in each of the initial n columns corresponding to nonzero components in $\mathbf{x}_1$. In other words, we pivot in each of the designated columns, in any order, but never in the last row. The submatrix of designated columns of **A** is the matrix obtained by some permutation of the columns of $\mathbf{I}_m$; the associated components of $\mathbf{x}_1$ comprise the first set of basic variables. This completes *Stage 1* of the simplex procedure.

Entering *Stage 2* we examine the elements in the last row of $\mathbf{P}_1$ but not in the last column. If all of these elements are nonnegative, then $\mathbf{x}_1$ is a solution point. Otherwise, and in the absence of degenerate vertexes, we can diminish $\mathbf{c}^T\mathbf{x}_1$ by exchanging the associated column for one of those associated with positive components of $\mathbf{x}_1$. In practice it is convenient to select the column associated with the smallest negative quantity, although such does not ensure that the total number of simplex stages will be minimal. After selecting the next pivot column in this fashion, the pivot row is determined as the one for which there is a positive pivot such that its ratio to the element in the last column is minimum over all rows having positive pivots. If there are no nonnegative elements in the pivot column, then the objective function is unbounded below and there is no solution point. Otherwise, we perform a Gauss–Jordan pivot operation to complete *Stage 2*.

Stage 3 proceeds as did *Stage 2* and we continue successive stages until we find a solution point or find that there are none due to unboundedness.

Details on the remarkable simplex method, and many indications of its practical significance, can be found in the first five books cited in the Preface.

PROBLEMS FOR SOLUTION

7.5.1 Feasible Region. Show that the convex programming problem (7.5) is a special case of the convex problem (7.31).

7.5.2 Maxima. Establish:

(1) If a convex function achieves its global maximum at an interior point then it must be a constant function.
(2) Whether or not (1) holds for the case of a local maximum.
(3) Counterpart of (7.36) for the *concave programming problem*

$$\text{maximize } f(\mathbf{x})$$
$$\text{s.t. } g(\mathbf{x}) \leqq \mathbf{0}$$
$$\mathbf{Ax} = \mathbf{b}$$

where f is concave and where g, $\mathbf{A}$ and $\mathbf{b}$ are as in (7.5).

7.5.3 Solution Points in Linear Programming. Establish for Example 7.4:

(1) If the primal problem has an objective function that is unbounded below, then the dual problem has no feasible point.
(2) The converse in (1) need not hold.
(3) If the dual problem has an objective function that is unbounded above, what can be said for the primal?
(4) Property (8) in the example.
(5) When the primal problem can have more than one solution point.
(6) Earlier results in the text can be used to prove that if the feasible region in the primal problem is a convex polyhedron then at least one of its vertexes is a solution point.

7.5.4 Constraint Qualification for Linear Programming. Establish for (5) in Example 7.4:

(1) If the primal problem has a feasible point and (i) holds, then rank $\mathbf{A} = m$.
(2) The feasible region in the dual problem has a vertex iff rank $\mathbf{A} = m$.
(3) A simple example to show that there can be a feasible vertex for the primal problem but none for the dual.
(4) An example where the primal problem has a feasible point but no feasible vertex?
(5) An example to show that both feasible regions can have vertexes and yet that one of properties (i) and (ii) holds while the other fails.

7.5.5 Canonical Linear Programming Problems. Establish:

(1) Results from Chapter 4 can be used to show that the *standard primal*

can always be transformed into an instance of the *canonical primal*

$$\begin{aligned} &\text{minimize } \mathbf{c}^T\mathbf{x} \\ &\quad \text{s.t. } \mathbf{A}\mathbf{x} \geqq \mathbf{b} \\ &\qquad\quad \mathbf{x} \geqq \mathbf{0} \end{aligned}$$

Find the new matrix **A** and new RHS **b**.

(2) The converse holds; find the new **A**, **b** for the standard primal.

(3) The *standard dual* can always be transformed into an instance of the *canonical dual*

$$\begin{aligned} &\text{maximize } \mathbf{b}^T\mathbf{y} \\ &\quad \text{s.t. } \mathbf{A}^T\mathbf{y} \leqq \mathbf{c} \\ &\qquad\quad \mathbf{y} \geqq \mathbf{0} \end{aligned}$$

Find the new **A**, **b**.

(4) The converse holds; find the new **A**, **b**.

(5) The dual of the canonical dual is the canonical primal.

(6) The preceding can be used to show that the dual of the standard dual is the standard primal.

(7) Results from Section 4.2 can be used to show that if both canonical problems have feasible points, then both have solution points.

(8) The same conclusion holds for the standard problems.

(9) The counterpart of constraint qualification (5) for the two canonical problems.

(10) The counterpart for the canonical problems of (6) in Example 7.4.

(11) The (general) duality theorem of linear programming, namely, that each of the following is an equivalent statement.

  (i) The primal problem has a solution point.

  (ii) The dual problem has a solution point.

  (iii) Both have solution points and the extreme values of the two objective functions coincide.

  (iv) Both feasible regions are nonempty.

7.5.6 Sensitivity Analysis. Show how (4) can be used to measure the effects on the minimum in (1) due to changes in the RHS vector **b**.

7.5.7 The Simplex Method. Establish:

(1) A simple example to show that the feasible region in (2) can be nonempty and yet fail to have a vertex.

(2) Property (11) in Example 7.4.

(3) If all of the coefficients of the objective function are nonnegative upon entering Stage 2, or any later stage, then the vertex arrived at in the preceding stage is a solution point.
(4) In the absence of degenerate vertexes we can diminish the objective function when not all of the above-mentioned coefficients are nonnegative.
(5) If there are no nonnegative elements in a pivot column that has been selected in (4) just above, then there is no solution point.

7.6 SEPARATION AND REPRESENTATION

Results in this section concern convex sets and are not used in later chapters. But they are widely applied in optimization theory as, for example, to generalize (7.36) and to develop linear programming without invoking a constraint qualification. In this section, we prove some separation theorems, and two representation theorems—(7.40) and (7.43)—that generalize results on "halfspace forms" and "point forms" established by algebraic methods in Section 4.4.

The elementary theorem of the separating hyperplane (4.14) is the simplest separation theorem in this book, and by virtue of (4.33) we can say that Farkas' theorem is the separation theorem that has been the most useful. In general, a hyperplane H *separates* two convex sets C and D if C is contained in one of the closed halfspaces determined by H and D lies in the other. When one of the sets consists of a single point, say $D=\{\mathbf{d}\}$, then we say that H *separates* C and $\mathbf{d}$, or, that it *separates* $\mathbf{d}$ *from* C. If $H=\{\mathbf{x}:\langle\mathbf{a},\mathbf{x}\rangle=b\}$ does separate C and D and

$$C\subset\{\mathbf{x}:\langle\mathbf{a},\mathbf{x}\rangle\leqq b\}$$

then $D\subset\{\mathbf{x}:\langle\mathbf{a},\mathbf{x}\rangle\geqq b\}$. Since the defining equation for H can also be written $\langle-\mathbf{a},\mathbf{x}\rangle=-b$ it is clear that we can express C in the foregoing as a subset of a halfspace with $\geqq$ in its defining relation. It is also true that C and D need not be disjoint in order to be separated by a hyperplane (7.6.1).

Let us first establish a generalization of (4.14) where now we actually exhibit a special type of separating hyperplane. A hyperplane H is said to be a *supporting hyperplane* for a closed convex set K at $\mathbf{x}_0\in K$ if $\mathbf{x}_0\in H$ and K is contained in one of the closed halfspaces determined by H. In this case we also say that *H supports K at $\mathbf{x}_0$*. Notice that $K\cap H$ is a convex set that can contain points distinct from $\mathbf{x}_0$, that is, it can be larger than $\{\mathbf{x}_0\}$. Figure 7.3 illustrates the next theorem and also the objects involved in its proof.

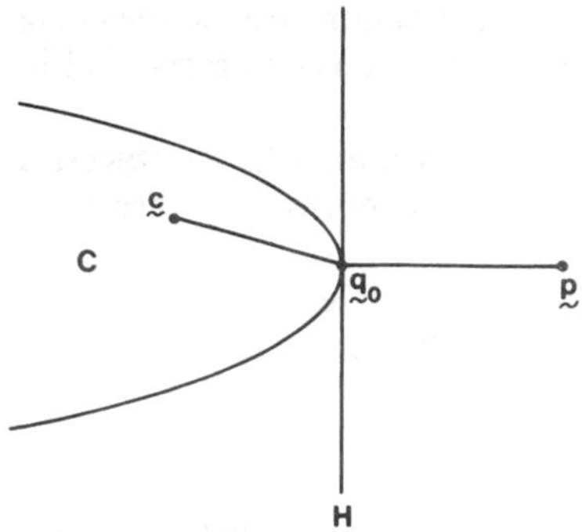

Figure 7.3. A supporting hyperplane.

Let C be a nonempty, closed, convex set and $\mathbf{p} \notin C$. Then C has a supporting hyperplane H that separates $\mathbf{p}$ from C. (7.38)

PROOF. First (7.6.3), there exists a unique solution point, say $\mathbf{q}_0$, for the problem

$$\text{minimize } \|\mathbf{p}-\mathbf{x}\|$$
$$\text{s.t. } \mathbf{x} \in C$$

Let us now show that the required hyperplane can be one through $\mathbf{q}_0$, namely

$$H = \{\mathbf{x} : \langle \mathbf{p}-\mathbf{q}_0, \mathbf{x}-\mathbf{q}_0 \rangle = 0\}$$

To this end, let $\mathbf{c} \in C$, define

$$g(t) = \|\mathbf{p} - (t\mathbf{c} + (1-t)\mathbf{q}_0)\|^2$$

and consider the problem (see Figure 7.3)

$$\text{minimize } g(t)$$
$$\text{s.t. } t \in [0,1]$$

The solution point is $t=0$ and $g'(0) \geqq 0$ since g is differentiable. Thus

$$-2\langle \mathbf{p}-\mathbf{q}_0, \mathbf{c}-\mathbf{q}_0 \rangle \geqq 0$$

which, in the notation introduced for Figure 4.3 states $C \subset H^-$. On the other hand

$$\langle \mathbf{p}-\mathbf{q}_0, \mathbf{p}-\mathbf{q}_0 \rangle > 0$$

so $\mathbf{p} \in H^+ \subset \{\mathbf{x} : \langle \mathbf{p}-\mathbf{q}_0, \mathbf{x}-\mathbf{q}_0 \rangle \geqq 0\}$. ■

It is instructive to note

1. The point $\mathbf{q}_0$ used in the preceding proof must be a boundary point, but
2. This proof does not show that there is at least one supporting hyperplane through every boundary point.

The reason for (2) is that we must show that an arbitrary $\mathbf{q}_0 \in \operatorname{bdry} C$ is the point in the closure of C that is the "nearest point" in C to some point not in C. Later (in the proof of (7.42)) we show that there does exist such a point. But next let us establish

MINKOWSKI'S SEPARATION THEOREM. Let C and D be nonempty, closed, convex sets with $C \cap D = \varnothing$ and C compact. Then there exist supporting hyperplanes H for C and H' for D such that $H \cap H' = \varnothing$. (7.39)

PROOF. Since one of the sets is compact they are at positive distance apart. That is (7.6.3) $\|\mathbf{p}_0 - \mathbf{q}_0\| > 0$ for every $\mathbf{p}_0 \in D$ and $\mathbf{q}_0 \in C$. Then we use (7.38) to obtain supporting hyperplanes

$$H = \{\mathbf{x} : \langle \mathbf{p}_0 - \mathbf{q}_0, \mathbf{x} - \mathbf{q}_0 \rangle = 0\}$$

$$H' = \{\mathbf{x} : \langle \mathbf{q}_0 - \mathbf{p}_0, \mathbf{x} - \mathbf{p}_0 \rangle = 0\}$$

and it is straightforward (7.6.3) to show $H \cap H' = \varnothing$. ■

Examples show that we need compactness even in $\mathfrak{R}^2$ (7.6.4).

Now we can readily prove an important result which is called the "kernel of many of the important applications of convexity" in Fenchel (1953). Notice how it generalizes the relationship between a convex set and its "tangent lines" in $\mathfrak{R}^2$.

FIRST REPRESENTATION THEOREM. A closed convex set C is the intersection of all closed halfspaces containing it. (7.40)

PROOF. The result is trivial for $\varnothing$ and there is a vacuous result for $C = \mathfrak{R}^n$. Otherwise, if $\mathbf{p} \notin C$ then by (7.38) there is a closed halfspace containing C but not containing $\mathbf{p}$. ■

More generally, if C is any convex set then (7.40) applies to its closure.

It is clear from the proof of (7.39), and indeed it is a special case of that result, that there exists a hyperplane H' through $\mathbf{p}$ in Figure 7.3 such that C lies in one of the open halfspaces determined by H'. Actually, C need not be closed.

If $\mathbf{p} \notin C$ where C is a nonempty convex set then there exists a hyperplane H through $\mathbf{p}$ such that C lies in one of the open halfspaces determined by H. (7.41)

PROOF. The result holds for the closure of C by (7.39) and this implies that it holds for C (7.6.3). ■

Let us now follow the method of proof attributed to E. J. McShane in Botts (1942) and use optimization theory to establish the

SUPPORTING HYPERPLANE THEOREM. If $\mathbf{q}_0$ is a boundary point of a convex set C then C has a supporting hyperplane through $\mathbf{q}_0$. (7.42)

PROOF. Again we lose no generality by supposing C is closed and it is convenient to suppose $\mathbf{q}_0=\mathbf{0}$ (7.6.3). If $S=\{\mathbf{x}:\|\mathbf{x}\|=1\}$ then the problem

$$\text{maximize } d(\mathbf{x},C)$$

$$\text{s.t. } \mathbf{x}\in S$$

has solution point $\mathbf{p}_0\notin C$. We will show that $\mathbf{q}_0=\mathbf{0}$ is the point in C closest to $\mathbf{p}_0$, that is

$$\text{minimize } \|\mathbf{p}_0-\mathbf{x}\|$$

$$\text{s.t. } \mathbf{x}\in C$$

has solution point $\mathbf{0}$. Let $\varepsilon\in(0,1)$, $\|\mathbf{y}\|<\varepsilon$ but $\mathbf{y}\notin C$ (see Figure 7.4). By (7.41) there exists a hyperplane through $\mathbf{y}$

$$H=\{\mathbf{x}:\langle\mathbf{a},\mathbf{x}\rangle=b\}$$

where we may suppose $\|\mathbf{a}\|=1$ and where C lies in one open halfspace. Now the distance $d(\mathbf{0},H)=|b|$ so $|b|<\varepsilon$. On the other hand,

$$d(\mathbf{a},H)=1-|b|>1-\varepsilon$$

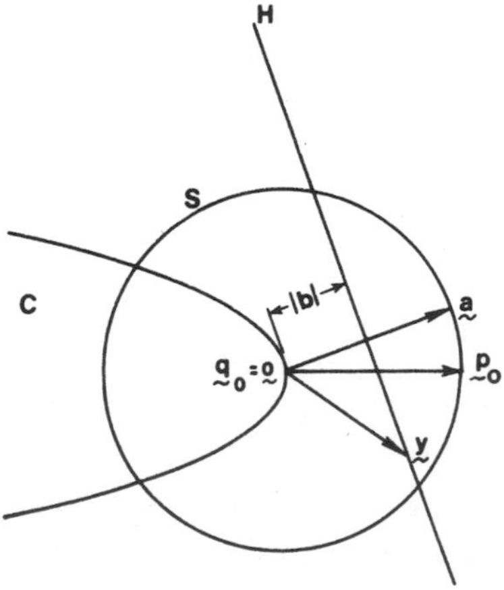

Figure 7.4. Proof of supporting hyperplane theorem.

Consequently, since $\mathbf{a}$ and C are on opposite sides of H, we conclude $d(\mathbf{a},C)>1-\varepsilon$. Thus $d(\mathbf{p}_0,C)\geqq 1$ and then $\|\mathbf{p}_0-\mathbf{0}\|=1$ shows that the bound is achieved for $\mathbf{q}_0=\mathbf{0}$. The point $\mathbf{0}$ is therefore the solution point to the last displayed problem and the desired result follows from (7.38). ■

The companion to (7.40) is the

SECOND REPRESENTATION THEOREM. A compact convex set C is the convex hull of its extreme points. (7.43)

PROOF. The conclusion holds for $\dim C=k=0,1$. Suppose that it holds for dimension $k-1\leqq n$ and let us work with C in $\mathcal{R}^k$, that is, let us use k-dimensional neighborhoods to define a topology. If $\mathbf{q}_0\in\operatorname{bdry} C$ then there exists a supporting hyperplane through $\mathbf{q}_0$ by (7.42). Any line segment not in H that contained an interior point in H would by (7.39) contain points separated by H. Consequently (7.6.3), the extreme points of $C\cap H$ are extreme points of C and the result follows by the inductive hypothesis. If $\mathbf{q}_0\in\operatorname{int} C$ then any line through $\mathbf{q}_0$ contains two boundary points and the result follows from that for the boundary. ■

The main result states that two convex sets need not be disjoint to be separated by a hyperplane. In its full generality it is not easy to prove; we follow the artful proof in Fenchel (1953) and remark on other versions below.

SEPARATION THEOREM. Let C and D be nonempty convex sets where furthermore $\operatorname{int} C\neq\varnothing$. If $D\cap\operatorname{int} C=\varnothing$ then there is a hyperplane that separates C and D. (7.44)

PROOF. We suppose that C and D are closed because the theorem holds as stated if it holds for the closures of C and D (7.6.3). Let $\mathbf{p}\in\operatorname{int} C$ and define the sequence of sets $\{C_m\}$ where (7.6.2)

$$C_m=\left\{\mathbf{x}:\mathbf{x}=\frac{1}{m}\mathbf{p}+\left(1-\frac{1}{m}\right)\mathbf{c},\ \mathbf{c}\in C\right\}\cap\{\mathbf{x}:\|\mathbf{x}-\mathbf{p}\|\leqq m\}$$

(For each m the first set is a shrunken or contracted closed subset of $\operatorname{int} C$; as $m\to\infty$ its boundary approaches more closely the boundary of C (7.6.2). The second set is a closed neighborhood of radius m which serves to keep the intersection bounded.) The displayed intersection C_m is convex, compact, and it does not intersect D. We can therefore use (7.39) to produce a separating hyperplane for each m

$$H_m=\{\mathbf{x}:\langle\mathbf{a}_m,\mathbf{x}\rangle=b_m\}$$

where for convenience we arrange $\|\mathbf{a}_m\|=1$, and

$$C_m\subset\{\mathbf{x}:\langle\mathbf{a}_m,\mathbf{x}\rangle\leqq b_m\}$$

while D is contained in the other closed halfspace. If $\mathbf{d}\in D$ then

$$\langle \mathbf{a}_m, \mathbf{p}\rangle \leqq b_m \leqq \langle \mathbf{a}_m, \mathbf{d}\rangle$$

so $\{\mathbf{a}_m\}$ and $\{b_m\}$ are bounded sequences. Consequently, we can select subsequences converging to $\mathbf{a}$ and b, respectively, and

$$\langle \mathbf{a}, \mathbf{x}\rangle \geqq b$$

for every $\mathbf{x}\in D$ because every term formed from the subsequences has this property. Also in the limit

$$\langle \mathbf{a}, \mathbf{x}\rangle \leqq b$$

for every $\mathbf{x}\in \operatorname{int} C$ and also for every $\mathbf{x}\in C$. Thus

$$H=\{\mathbf{x}:\langle \mathbf{a}, \mathbf{x}\rangle = b\}$$

separates C and D. ■

If we let $D=\{\mathbf{q}_0\}$ where $\mathbf{q}_0$ is a boundary point of C, we obtain a weaker version of (7.42) where we must have $\operatorname{int} C\neq\varnothing$. The most general version, which requires only that C and D be nonempty and have nonintersecting relative interiors (6.1.14), was established in Fenchel (1953); see Stoer and Witzall (1970). For different kinds of proof see Bourbaki (1953), Valentine (1964), and Rockafeller (1970). The Bourbaki approach involves the following geometric version of the *Hahn–Banach theorem* (7.6.3)

If C is an open convex set and D is a k-plane that does not intersect C, then there exists a hyperplane H such that $D\subset H$ and $C\cap H=\varnothing$. (7.45)

Also see Eggleston (1958), Mangasarian (1969), and Roberts and Varberg (1973) for material on this theorem.

PROBLEMS FOR SOLUTION

7.6.1 Elementary Examples. Furnish examples of:

(1) Two convex sets in $\mathcal{R}^2$ that are not disjoint but are separated by a hyperplane.

(2) Two disjoint sets, not both convex, that are not separated by any hyperplane.

7.6.2 Fenchel's Sets. Illustrate the linear contraction sets used as the first members in the definition of C_m in the proof of (7.44).

(1) $C=[0,1]$ and $p=\frac{1}{4}$.

(2) $C=\{\mathbf{x}:x_1^2+x_2^2=1\}$ and $\mathbf{p}=\mathbf{0}$.

(3) C in (2) and $\mathbf{p}=\left[\frac{1}{4},0\right]^T$.

7.6.3 Details. Establish for use in the proofs of the designated results.

(1) For (7.38): $\mathbf{q}_0$ exists.
(2) For (7.38): $g'(0) \geqq 0$.
(3) For (7.39): $\mathbf{p}_0$ and $\mathbf{q}_0$ exist.
(4) For (7.39): $H \cap H' = \varnothing$.
(5) For (7.41): If conclusion holds for the closure of C, then it holds for C.
(6) For (7.42): supposition that C is closed.
(7) For (7.42): changes if proof is carried out for $\mathbf{q}_0 \neq \mathbf{0}$.
(8) For (7.43): Extreme points of $C \cap H$ are extreme points of C.
(9) For (7.43): Any line through $\mathbf{q}_0 \in \operatorname{int} C$ contains two boundary points.
(10) For (7.44): supposition that C and D are closed.
(11) For (7.44): C_m is convex, compact, and it does not intersect D.
(12) For (7.44): Subsequences converging to $\mathbf{a}$ and b can be selected as stated.
(13) For (7.44): $\mathbf{x} \in C$ implies $\langle \mathbf{a}, \mathbf{x} \rangle \leqq b$.
(14) For (7.45): (7.44) can be used for the proof.

7.6.4 Different Types of Separation. Establish:

(1) Separation in (7.39) is *strong separation* in the sense that H lies strictly between two translates of H that separate the two sets in question.
(2) An example of two disjoint convex closed sets in $\mathfrak{R}^2$ that are not strongly separable.
(3) The concept of *strict separation*—one of the sets lies in one open halfspace determined by H and the other lies in the opposite—is distinct from strong separation.

7.6.5 Hyperplanes through the Origin. State the following using the origin as the specially designated point.

(1) (7.38).
(2) (7.41).
(3) (7.42).

8
COMPLEX VARIABLES

This chapter can be omitted because it relates to procedures—mainly (9.64) and (10.89)—that can be used in place of more elementary ones. Readers with knowledge at the level of Churchill, Brown, and Verhey (1974) can certainly proceed directly to Chapter 9. But here we give background, state theorems without proof, and present three examples devoted to useful techniques.

8.1 ANALYTIC FUNCTIONS

Let us use some results from Chapters 1 and 6 to describe a special kind of function that has remarkable infinite series representations.

The concepts of neighborhoods, open sets, limits, and so forth, in Section 6.1 are readily introduced in $\mathcal{C}$ using the one-to-one correspondence

$$z = x + iy \leftrightarrow \begin{bmatrix} x \\ y \end{bmatrix} = \mathbf{z} \tag{8.1}$$

between complex numbers $z \in \mathcal{C}$ and vectors $\mathbf{z} \in \mathcal{R}^2$ under which distances $|z_1 - z_2|$ and $\|\mathbf{z}_1 - \mathbf{z}_2\|$ are preserved. A nonempty connected open set in $\mathcal{C}$ is a *region*. A *complex function* is a function

$$f: D \subset \mathcal{C} \to \mathcal{C} \tag{8.2}$$

where D is a region. Points in D are traditionally referred to as *complex variables*. Using the classical uv notation—where $w = f(z)$, $u = \operatorname{Re} w$, and $v = \operatorname{Im} w$—complex functions are in one-to-one correspondence with transformations on $\mathcal{R}^2$

$$\begin{bmatrix} x \\ y \end{bmatrix} \mapsto \begin{bmatrix} u \\ v \end{bmatrix} \tag{8.3}$$

We can identify u and v with coordinate functions on the corresponding set $D \subset \mathcal{R}^2$ and obtain counterparts to most of the results in Section 6.1; for example, (8.2) is continuous iff u and v are continuous.

The complex function (8.2) is *differentiable* at $z_0 \in D$, and its *derivative* is denoted by $f'(z_0)$, if the corresponding transformation (8.3) is differentiable according to (6.46). Possession of a derivative in $\mathcal{C}$ is important only in relation to all points in an open set according to the

Definition. The function (8.2) is *analytic in a region G* if $f'(z)$ exists for every $z \in G$. It is *analytic on* $S \subset \mathcal{C}$ if it is analytic in some region containing S.

In particular, S can consist of a single point. A basic result is the

THEOREM. The function (8.2) is analytic in a region $G \subset D$ iff both of the following hold at each point in $G \subset \mathcal{R}^2$

(1) Each of u and v is differentiable.

(2) $$\frac{\partial u}{\partial x} = \frac{\partial v}{\partial y}, \quad \frac{\partial u}{\partial y} = -\frac{\partial v}{\partial x}. \tag{8.4}$$

The equations in (2) are the *Cauchy–Riemann* conditions; see 8.1.1 for the manner by which they are suggested from considerations of $\mathcal{C}$ and $\mathcal{R}^2$ as vector spaces.

Complex counterparts of real series are used to define the complex *exponential function* $\exp z = e^z$. The characteristic relationship involves real functions, namely

$$e^z = e^x(\cos y + i \sin y) \tag{8.5}$$

In keeping with $z = \rho e^{i\theta}$, the *principal branch* of the complex *logarithmic function* is defined for $z \neq 0$ by

$$\log z = \log \rho + i\theta, \qquad -\pi < \theta \leqq \pi$$

Trigonometric and hyperbolic functions and their inverses can be expressed in terms of exponential and logarithmic functions, and essentially all of the familiar properties—multiple angle formulas, expressions for derivatives, and so on—remain valid for complex functions. From $w = z^c$, where $z \neq 0$ and $c \in \mathcal{C}$, expressed as $w = \exp(c \log z)$, polynomials and rational functions, as well as the foregoing, are examples of the

Definition. *Elementary functions* are functions (8.2) such that $f(z)$ can be expressed in terms of a finite number of operations involving additions, subtractions, multiplications, and divisions performed on exponential and logarithmic functions.

If f and g are elementary functions, then so are $f \pm g, f \cdot g, f/g$ provided $g(\cdot) \neq 0$; $\exp f$, $\log f$, and f^c for $c \in \mathcal{C}$ provided $f(\cdot) \neq 0$; and c^f provided $c \neq 0$. But the following are not: $\operatorname{Re} z$, $\operatorname{Im} z$, $\bar{z}$, and $|z|$. Expressions involving the latter may or may not qualify as elementary functions; for example, $\operatorname{Re} z + i \operatorname{Im} z$ does while $\bar{z} + \operatorname{Re} z$ does not. In particular cases we can use

Elementary functions of z, and composite elementary functions, are analytic except at points that make
(1) A denominator vanish, or
(2) An argument for a logarithm or a power touch the nonpositive real axis. (8.6)

or we can apply (8.4).

Analytic functions can be defined in terms of *power series*

$$\sum_{j=0}^{\infty} a_j (z - z_0)^j \tag{8.7}$$

of complex numbers because of the

THEOREM. Let f be analytic in a neighborhood $N_\delta(z_0)$. Then there is a unique power series, namely, the *Taylor's series*

$$f(z_0) + f'(z_0)(z - z_0) + \cdots + \frac{f^{(n)}(z_0)}{n!}(z - z_0)^n + \cdots$$

that converges to $f(z)$ for each $z \in N_\delta(z_0)$. Conversely, any power series that converges in a neighborhood represents an analytic function in that neighborhood. (8.8)

The radius δ can be as large as the distance from z_0 to the nearest point, if any, where f fails to be analytic. Examples show that this distance, the *radius of convergence* of the Taylor's series at z_0, can be finite or infinite. When it is finite, the associated circle with center z_0 is the *circle of convergence* and its interior is the *disc of convergence* of the Taylor's series. Examples also show that no general statement can be made for convergence on the circle itself.

Series with both positive and negative powers of $(z - z_0)$ are introduced using

$$\sum_{j=-\infty}^{\infty} a_j (z - z_0)^j = \sum_{j=1}^{\infty} a_{-j} (z - z_0)^{-j} + \sum_{j=0}^{\infty} a_j (z - z_0)^j \tag{8.9}$$

where the first member on the RHS is a power series in $(z-z_0)^{-1}$. If the latter converges, it converges on the complement of some $N_\delta(z_0)$. Series such as (8.9) are important because of the following theorem where integrals are counterparts to line integrals in calculus and where *positively oriented circles* are circles traced out counterclockwise.

THEOREM. Let f be analytic in the annular region $A=\{z: 0 \leqq r_1 < |z-z_0| < r_2\}$. Then there is a unique series of the form (8.9)—namely the *Laurent's series* where

$$a_n = \frac{1}{2\pi i}\int_C \frac{f(p)}{(p-z_0)^{n+1}}\,dp$$

for $n=0, \pm 1, \pm 2, \ldots$ and C is any positively oriented circle with center z_0 and radius r, $r_1 < r < r_2$—that converges to $f(z)$ for each $z \in A$. Conversely, any series (8.9) that converges in an annular region represents an analytic function in that region. (8.10)

If $r_1 = 0$ then the Laurent's series coincides with Taylor's. If $r_1 > 0$ is required then there must be at least one point inside the circle with center z_0 and radius r_2 where f fails to be analytic.

EXAMPLE 8.1. Finding Series Representations. In this example we recall methods for finding Taylor's series and we illustrate some for finding Laurent's series for rational functions.

Procedures from elementary calculus can be used to find Taylor's series. Because such series are unique, the idea is to find any convenient method. One method is *substitution*: the series for $\exp(-z^2)$ can be found from that for $\exp t$. Another is *integration*: Integrating with $t=-z$ in (5) below yields the series for $\log(1+z)$. *Differentiation* can be used to obtain the series for $(1-z)^{-2}$. Methods of *undetermined coefficients* apply, and sometimes some of the preceding methods can be used to find series that can be multiplied to give the desired Taylor's series.

The methods we present for Laurent's series are called *division algorithms*. They are equivalent to familiar processes of "long division," but we can distinguish two procedures.

(1) Finding the quotient of two power series.
(2) Using binomial series together with substitution, especially in

$$\frac{1}{(1-z)^n} = 1 + nz + \cdots + C(n+m-1, m)z^m + \cdots$$

where $n > 0$ and $|z| < 1$.

Most often in Chapters 9 and 10 we have two polynomials in (1).

Let us use (1) for $f(z)=(z-2)^{-1}$. Familiar long division produces

$$\frac{1}{z-2}=z^{-1}+2z^{-2}+4z^{-3}+\cdots \tag{3}$$

If we change signs on the LHS and then divide, we obtain

$$-\frac{1}{2-z}=-\frac{1}{2}-\frac{1}{4}z-\frac{1}{8}z^2-\cdots \tag{4}$$

In the next paragraph we show that series (3) is the Laurent's series for $z_0=0$, $r_1=2$ and any $r_2>2$, and series (4) is the Taylor's series for $z_0=0$ and $\delta=2$.

More elegant derivations result from using (2) and they have the advantage that the form of the general term and radii for the annular region are apparent. For $n=1$ we use the familiar *geometric series*

$$\frac{1}{1-t}=1+t+t^2+\cdots \tag{5}$$

which converges for $|t|<1$ or

$$\frac{1}{1-1/t}=1+\frac{1}{t}+\frac{1}{t^2}+\cdots \tag{6}$$

which converges for $|t|>1$. To obtain (3) we first write

$$\frac{1}{z-2}=\frac{1}{z}\,\frac{1}{1-2/z}$$

and then use (6) for $t=z/2$ and multiply by $1/z$. From this it is clear that the general term is $2^{m-1}z^{-m}$ for $m=1,2,3,\ldots$ and convergence occurs for $|z/2|>1$, that is, for $|z|>2$. To obtain (4) we write $f(z)=-(1/2)(1-z/2)^{-1}$ and use (5) to conclude that the radius of convergence is 2.

The methods of the preceding paragraph can be adapted to find series in powers of $(z-z_0)$ for any z_0. For example, to obtain ascending powers of $(z-6)$, the denominator is first rewritten $4+(z-6)$ and second divided by 4; the resulting series converges inside $N_4(6)$. For descending powers, the change is to divide by $(z-6)$.

Let us find the Laurent's series for

$$\frac{z}{(z-1)(z-2)}=-\frac{1}{z-1}+\frac{2}{z-2} \tag{7}$$

in the annular region where $1<|z|<2$. Here $z_0=0$, $r_1=1$, $r_2=2$, we need descending powers of z for the first term on the RHS, and ascending powers for the second. If we use (2) for both terms and then add we obtain

(8) $$\cdots-\frac{1}{z^2}-\frac{1}{z}-1-\left(\frac{z}{2}\right)-\left(\frac{z}{2}\right)^2-\cdots$$

which represents (7) in the given region.

PROBLEMS FOR SOLUTION

8.1.1 Vector Space Considerations. Deduce the Cauchy–Riemann conditions from the fact that the linear transformations in $\mathcal{C}$ are not in one-to-one correspondence with those in $\mathcal{R}^2$.

8.1.2 Example 8.1. Establish:

(1) An example for fixed f and z_0 where different Laurent's series are obtained for different choices of r_1 and r_2.

(2) Multiplication of series and the method of undetermined coefficients can be used to find the series for $(1-z^2)^{-1}\cos z$ in powers of z.

(3) Several nonzero terms of series for the following in powers of z:
   - (i) $\sin z/\sin 2z$.
   - (ii) $e^z/(1+e^z)$.
   - (iii) $z/(e^z-1)$.

(4) Taylor's series for $f(z)=(1-z)^{-1}$
   - (i) About $z_0=3$.
   - (ii) About $z_0=i$.

(5) Laurent's series for $f(z)$ as follows,
   - (i) $[z(z-a)]^{-1}$ in $\{z:0<|z|<a\}$.
   - (ii) $(2z-7)/[(z-3)(z-4)]$ in $\{z:3<|z|<4\}$.
   - (iii) $[(z-1)(z-2)]^{-1}$ in $\{z:|z|>2\}$.
   - (iv) $[(z^2+1)(z+2)]^{-1}$ in $\{z:1<|z|<2\}$ using division algorithms with real arithmetic.
   - (v) Repeat using complex arithmetic.
   - (vi) $(z-1)/[(z-2)(z-3)^2]^2$ in $\{z:0<|z-3|<1\}$.

(6) The annular regions A where the following represent analytic functions, and identify the functions.
   - (i) $\cdots+\left(\frac{2}{z}\right)^2+\left(\frac{2}{z}\right)+\left(\frac{z}{3}\right)+\left(\frac{z}{3}\right)^2+\cdots.$
   - (ii) $\cdots+\left(\frac{1}{z}\right)^2+\left(\frac{1}{z}\right)+a+\left(\frac{z}{4}\right)+\left(\frac{z}{4}\right)^2+\cdots.$

8.2 SINGULARITIES AND RESIDUES

Let us continue to describe material that we use in Chapters 9 and 10 for optional procedures.

Suppose f is analytic in some deleted neighborhood $N_\delta(z_0)-\{z_0\}$ but not at the point z_0 itself. Then f has a Laurent's series in the annular region

$$A=\{z:\varepsilon<|z-z_0|<\delta\} \tag{8.11}$$

where $\varepsilon>0$ is the radius of the inner circle. There are three significantly different cases.

First, the Laurent's series may have no negative powers of $(z-z_0)$ appearing with nonzero coefficients. In this case we say that f has a *removable singularity* at z_0 and we agree to assign $f(z_0)=a_0$, the coefficient for $n=0$ in (8.10). This truly removes all difficulties at z_0 because the Laurent's series becomes the Taylor's series for $N_\delta(z_0)$ and f becomes analytic at z_0.

Second, the Laurent's series may have a finite number of negative powers of $(z-z_0)$ appearing because

$$a_{-k}\neq 0 \quad \text{but} \quad a_{-n}=0 \quad \text{for} \quad n>k>0$$

Then z_0 is a *pole* of *order* k of f. If $k=1$, we call z_0 a *simple pole*, and if $k>1$, it is a *multiple pole*. For example, both z^{-1} and $(\cos z)/z$ have simple poles at $z=0$ while $(z-z_0)^{-3}(z-z_1)^{-2}$ has poles of order 3 at z_0 and order 2 at z_1.

The third, and final, possibility is that the Laurent's series may have an infinite number of negative powers of $(z-z_0)$ appearing with nonzero coefficients. Then z_0 is an *essential singularity* of f. An example is $z=0$ for $\exp z^{-1}$.

If f has a pole or an essential singularity at z_0 we call z_0 an *isolated singularity of* f.

Definition. The Laurent's series for f in a deleted neighborhood of an isolated singularity z_0 is the *principal Laurent's series* for f at z_0. The sum of the negative powers of $(z-z_0)$ is called the *principal part* of f at z_0. The sum of the nonnegative powers is called the *regular part* of f at z_0. (8.12)

Such series correspond to $\varepsilon=0$ in (8.11) and they are unique because any positive δ can be used up to the distance from z to the nearest other point, if any, where f fails to be analytic.

By far the most important coefficient in any Laurent's series is a_{-1}

because for $n = -1$ in (8.10) we have

$$a_{-1} = \frac{1}{2\pi i} \int_C f(z)\, dz$$

When z_0 is an isolated singularity and the series is the principal Laurent's series, a_{-1} is called the *residue* of f at z_0 and is denoted by $\operatorname{res}(f, z_0)$. If we rewrite the preceding display as

$$\int_C f(z)\, dz = 2\pi i \operatorname{res}(f, z_0)$$

we obtain the simplest example of the use of the following theorem to calculate an integral.

RESIDUE THEOREM FOR A CIRCLE. Let f be analytic on a positively oriented circle C. If f is analytic inside C except at a finite number of isolated singularities $z_1, z_2, \ldots, z_m$, then

$$\int_C f(z)\, dz = 2\pi i \sum_{j=1}^{m} \operatorname{res}(f, z_j) \tag{8.13}$$

In works on complex variables this follows from a result called *Cauchy's integral theorem* which also ensures that if f is analytic inside C then the integral has value zero.

Let us present four methods for calculating residues at isolated singularities which, we caution, are the only points at which we define residues.

Differentiation at Poles

If for some positive integer k the function defined by

$$\phi(z) = (z - z_0)^k f(z) \tag{8.14}$$

is analytic and nonzero at z_0, then f has a pole of order k at z_0. This result, perhaps with L'Hôpital's rule, is used to determine the order of a pole and to calculate the residue. When $k = 1$ we have the formula

$$\operatorname{res}(f, z_0) = \lim_{z \to z_0} \phi(z) \tag{8.15}$$

which is the first instance of the subject method. For the positive integer k, ϕ has a Taylor's series and $\operatorname{res}(f, z_0)$ is the $(k-1)$st coefficient in that series for ϕ, that is

$$\operatorname{res}(f,z_0)=\frac{1}{(k-1)!}\lim_{z\to z_0}\phi^{(k-1)}(z) \tag{8.16}$$

Division Algorithms

The idea is to find the principal Laurent's series and then read off the coefficient of the $(z-z_0)^{-1}$ term.

Substitution in Formulas at Poles

Consider

$$f(z)=\frac{p(z)}{q(z)} \tag{8.17}$$

where p and q are analytic at z_0, $p(z_0)\neq 0$, and z_0 is a pole of order 1. If we use (8.15) and L'Hôpital's rule we obtain

$$\operatorname{res}\left(\frac{p}{q},z_0\right)=\frac{p(z_0)}{q'(z_0)} \tag{8.18}$$

A division algorithm can be used to find other formulas (8.2.2).

Partial Fraction Expansions for Rational Functions

Once we have the expansion we can read off the residues. For example, if f is given by

$$\frac{z^3+1}{z(z-1)^3}=-\frac{1}{z}+\frac{2}{(z-1)^3}+\frac{1}{(z-1)^2}+\frac{2}{z-1} \tag{8.19}$$

we can see at once $\operatorname{res}(f,0)=-1$ and $\operatorname{res}(f,1)=2$.

EXAMPLE 8.2. Calculating Residues. Usually (8.15) is effective at a simple pole. For example

$$\lim_{z\to 0}\frac{z\cos z}{\sin z}=\lim_{z\to 0}\frac{-z\sin z+\cos z}{\cos z}$$

establishes $\operatorname{res}(\cot z,0)=1$.

Formula (8.15) is particularly convenient for calculating residues at simple poles of rational functions. When the denominator is factored, the method is conveniently carried out using *Heaviside coverup*. For example, to find the residue at $z = 0$ we cover up the factor z in the denominator of

$$\frac{z+1}{z(z-1)(z-2)} \tag{1}$$

and evaluate what remains for $z=0$. This is equivalent to using (8.15) because

$$\lim_{z\to 0} z\cdot\frac{z+1}{z(z-1)(z-2)} = \lim_{z\to 0}\frac{z+1}{(z-1)(z-2)}$$

Although we deal with partial fractions at the end of this example, let us observe here that three Heaviside coverups produce the expansion of (1)

$$\frac{z+1}{z(z-1)(z-2)} = \frac{1}{2}\frac{1}{z} - 2\frac{1}{z-1} + \frac{3}{2}\frac{1}{z-2}$$

Conversely, any method of expanding the LHS in partial fractions will produce the three residues $1/2$, -2, and $3/2$.

Formula (8.15) can be used for nonreal poles. Either we factor the denominator or use L'Hôpital's rule. For example

$$\lim_{z\to i}\frac{z-i}{z^2+1} = \lim_{z\to i}\frac{1}{2z}$$

establishes $\operatorname{res}((z^2+1)^{-1}, i) = (2i)^{-1}$.

At multiple poles we can use (8.16). The function

$$f(z) = \frac{1}{z(e^z-1)} \tag{2}$$

has a pole of order 2 at $z=0$. The associated function (8.14)

$$\phi(z) = \frac{z}{e^z-1}$$

must be differentiated once since $k-1=1$. Now

$$\phi'(z) = \frac{e^z-1-ze^z}{(e^z-1)^2}$$

is indeterminate at $z=0$ and two successive applications of L'Hôpital's rule produce $\text{res}(f,0) = -1/2$.

L'Hôpital's rule is not required when (8.16) is used to find residues at multiple poles of rational functions. For example, $z=1$ is a pole of order 3 of

$$f(z) = \frac{z^3+1}{z(z-1)^3} \tag{3}$$

and the associated $\phi(z) = z^2 + z^{-1}$ must be differentiated $3-1=2$ times. We find $\phi''(z)/2 = 1 + z^{-3}$ so $\text{res}(f,1) = 2$.

Procedures (1) and (2) of Example 8.1 offer effective methods for finding residues at essential singularities as well as at poles. Even though, as we have cautioned, the residue at z_0 is by definition the a_{-1} coefficient in the principal Laurent's series at z_0, we can also use other series. For example, as readily verified

(4) The series in (3) of Example 8.1 is not a principal Laurent's series but the coefficient of z^{-1} is the residue at $z=2$, and not at $z_0=0$.

We have the same situation in (8) of Example 8.1 where the coefficient of z^{-1} is the residue at $z=1$ and not at $z_0=0$. There is another possibility not yet illustrated and that is the case where more than one isolated singularity occurs inside the inner circle of the annulus for Laurent's series. This happens if we expand (7) in Example 8.1 in powers of $(z-3)$. Then, as readily verified, the coefficient of $(z-3)^{-1}$ equals the sum of the residues at $z=1,2$. The rule is

(5) The coefficient of the $(z-z_0)^{-1}$ term in any series that can be identified as a Laurent's series is a residue if there is one and only one isolated singularity inside the inner circle of the associated annulus.

We do not say iff, for the trivial reason that there might be two isolated singularities involved and one of the residues could be zero.

Division algorithms can also be used to show that a point is a pole of a certain order, or an essential singularity. For example, (2) does have a pole of order 2 at $z=0$ because

$$\frac{1}{z(e^z-1)} = \frac{1}{z^2}\,\frac{1}{1+z/2!+z^2/3!+\cdots} \tag{6}$$

Naturally we must be certain we are using an appropriate series. For example, it is nonsense to conclude from (3) in Example 8.1 that $-(2-z)^{-1}$ has an essential singularity at $z=0$. The rule is

(7) The existence of an essential singularity, or the order of a pole, at z_0 can be established from the principal Laurent's series at z_0.

In the text we note the example $\exp(z^{-1})$ which has an essential singularity at $z=0$ and it is easily seen that the residue is unity.

In working with various kinds of linear systems in the following chapters we find that almost all of the residues with which we are concerned are connected with poles of rational functions. We lose no generality (1.2.9) in supposing that any rational function f can be expressed as

$$f(z)=r(z)+\frac{p(z)}{q(z)} \tag{8}$$

where r is a polynomial and the degree of p is strictly less than that of q. Let us suppose further that we know all zeros of q together with their multiplicities, and that p and q have no zeros in common. Because r is a polynomial it need no longer be of concern and we concentrate our attention on p/q.

Because we can divide both numerator and denominator by the leading coefficient of q, we suppose this coefficient is unity and

$$q(z)=(z-b_1)^{k_1}(z-b_2)^{k_2}\cdots(z-b_m)^{k_m} \tag{9}$$

Thus the b_j are the distinct poles of respective multiplicities k_j of the rational function p/q. The principal Laurent's series for p/q at the pole b_j consists of the sum of the principal part

$$A_j(z)=\frac{a_{1j}}{z-b_j}+\frac{a_{2j}}{(z-b_j)^2}+\cdots+\frac{a_{k_j j}}{(z-b_j)^{k_j}} \tag{10}$$

and the regular part

$$B_j(z)=\frac{p(z)}{q(z)}-A_j(z)$$

Now $B_j(z)$ has a removable singularity at b_j, and poles at the remaining poles of p/q. By induction

$$\frac{p(z)}{q(z)}=A_1(z)+\cdots+A_m(z) \tag{11}$$

except at the poles $b_1,\ldots,b_m$. The sum on the RHS of (11) is defined to be the *partial fraction expansion* of p/q and we see that it coincides with the familiar representation used in algebra and calculus.

Expanding p/q in partial fractions requires finding the k_j numbers

$a_{1j}, \ldots, a_{k_j j}$ in (10). They are also coefficients in the Taylor's series for

$$\Phi_j(z) = (z - b_j)^{k_j} \frac{p(z)}{q(z)} \tag{12}$$

in a neighborhood of b_j where $\Phi_j(z)$ is analytic. However, they appear in reverse order

$$\Phi_j(z) = a_{k_j j} + \cdots + a_{2j}(z - b_j)^{k_j - 2} + a_{1j}(z - b_j)^{k_j - 1} + \cdots$$

and so from (8.8) the formulas are

$$a_{\alpha j} = \frac{1}{(k_j - \alpha)!} \Phi_j^{(k_j - \alpha)}(b_j), \qquad \alpha = 1, 2, \ldots, k_j \tag{13}$$

and in particular, $\operatorname{res}(p/q, b_j) = a_{1j}$ so

$$\operatorname{res}\left(\frac{p}{q}, b_j\right) = \frac{1}{(k_j - 1)!} \Phi_j^{(k_j - 1)}(b_j) \tag{14}$$

which is a special case of (8.16). In using (13) to find the residues of p/q we see that we must perform all the differentiations of the different Φ_j required for finding the complete partial fraction expansion of p/q.

Using (13) for the example (8.19) we start with (11) in the form

$$\frac{z^3 + 1}{z(z-1)^3} = \frac{a_{11}}{z} + \frac{a_{12}}{z-1} + \frac{a_{22}}{(z-1)^2} + \frac{a_{32}}{(z-1)^3}$$

Then $a_{11} = -1$ by Heaviside coverup and we are done with $b_1 = 0$. Next, $a_{32} = 2$ for $\alpha = 3$ and $j = 2$, again by Heaviside coverup, and we proceed to work with

$$\Phi_2(z) = z^2 + z^{-1}$$

$$\Phi_2'(z) = 2z - z^{-2}$$

$$\Phi_2''(z) = 2 + 2z^{-3}$$

to find $a_{22} = 1$ and, finally, the residue $a_{12} = 4/2 = 2$.

Additional methods from algebra and calculus can be explained in terms of (11). If we multiply both sides of (11) by $q(z)$ in the factored form (9) we obtain polynomials in z on both sides, the resulting relation is an identity holding for all z, and the coefficients of the polynomial on the

RHS are linear expressions in the coefficients a_{ij}. If there are

$$k_1 + k_2 + \cdots + k_m = n$$

coefficients to be determined then we can proceed as follows

(15) Substitute any n distinct values of z into both sides of

$$p(z) = q(z)\,A_1(z) + \cdots + q(z)\,A_m(z)$$

or into derivatives of both sides, to obtain n linear equations, by equating coefficients of like powers of z from the two sides. The unique solution is the set of coefficients a_{ij} in (10).

For the example (8.19) treated just previously,

$$\begin{aligned} 1 &= a_{11} + a_{12} \\ 0 &= -3a_{11} - 2a_{12} + a_{22} \\ 0 &= 3a_{11} + a_{12} - a_{22} + a_{32} \\ 2 &= a_{32} \end{aligned}$$

As before, Heaviside coverup yields $a_{11} = -1$ and $a_{32} = 2$ at the start. But in any event, a substantial part of the work in this method is finding the linear equations.

Whenever $\bar{b}_j$ is also a pole of p/q, and this must happen when b_j is not real and q has real coefficients, then q has the real factor

$$\left[(z - b_j)(z - \bar{b}_j)\right]^{\beta}$$

where β is the smaller of the two multiplicities. For some purposes it suffices to permit powers of such real quadratics to appear as denominators in the RHS of (11). Expressions of this type that are equivalent to the RHS of (11) are often called partial fraction expansions and treated through "short cut" methods in textbooks on calculus or linear systems. These methods have the advantage that they require only real arithmetic whenever p and q have real coefficients. They can be established using (11).

See Henrici (1974) for more advanced methods of finding residues.

EXAMPLE 8.3. Using Rouché's Theorem. Let us illustrate one more result that is used in work with linear systems. In works on complex variables it is established for more general curves C, but here it suffices to have

ROUCHÉ'S THEOREM FOR A CIRCLE. Let f and g be analytic inside and on a circle C. If $|g(z)|<|f(z)|$ for all $z\in C$, then f and $f+g$ have the same number of zeros, counting multiplicities, inside C.

The idea is that if the number of zeros are known, say from obvious considerations, for one of f and $f+g$, then the theorem establishes that the other must have the same number inside C. A typical use is in the proof of the

(1) **THEOREM.** A polynomial of degree n with complex coefficients

$$z^n+a_{n-1}z^{n-1}+\cdots+a_1z+a_0$$

has exactly n zeros.

PROOF. Let the two analytic functions be $f(z)=z^n$ and

$$g(z)=a_{n-1}z^{n-1}+\cdots+a_1z+a_0$$

and let C be a circle about the origin of radius $M>1$ where M is to be determined later. For any $M>1$ and $z\in C$, $|f(z)|=M^n$ and

$$|g(z)|\leqq|a_{n-1}|M^{n-1}+\cdots+|a_1|M+|a_0|\leqq(|a_{n-1}|+\cdots+|a_0|)M^{n-1}$$

Thus we have the strict inequality

$$(|a_{n-1}|+\cdots+|a_0|)M^{n-1}<M^n$$

provided we select M satisfying

$$M>\max\{1,|a_{n-1}|+\cdots+|a_0|\}$$

In this way $|g(z)|<M^n=|f(z)|$ whenever $z\in C$ and Rouché's theorem applies. The desired result follows because f clearly has exactly n zeros at the origin which is inside C. ■

With this theorem we immediately obtain the fundamental theorem of algebra (1.8).

Let us next use Rouché's theorem to show that the n zeros of a polynomial of degree n are, in a certain sense, continuous functions of the n coefficients. We consider polynomials a and b where in fixed notation

$$a(z)=z^n+a_{n-1}z^{n-1}+\cdots+a_0=(z-r_1)^{m_1}\cdots(z-r_s)^{m_s}$$

$$b(z)=z^n+b_{n-1}z^{n-1}+\cdots+b_0=(z-z_1)(z-z_2)\cdots(z-z_n)$$

and where intentionally we label the zeros of $a(z)$ so as to record their multiplicities m_k but we label those of $b(z)$ without any such regard. For any $\delta>0$ we define the $\delta-$*neighborhood* of the polynomial a to be the collection of polynomials b of degree n satisfying

$$b_j \in N_\delta(a_j) \quad \text{for} \quad j=0,1,\ldots,n-1$$

and denoted by $\mathfrak{N}_\delta(a)$. Then the specific meaning we assign to *the zeros of a polynomial a are continuous functions of the coefficients* $a_0, a_1, \ldots, a_{n-1}$ is

> Given any sufficiently small $\varepsilon>0$ there exists $\delta>0$ such that b has m_k zeros in $N_\varepsilon(r_k)$ for $k=1,\ldots,s$ whenever $b\in\mathfrak{N}_\delta(a)$.

Notice that continuity defined in this manner does not mean that the zeros are "well-behaved" in ways we might desire. It does mean that the circle of radius ε about r_j always contains m_j zeros of b; but maybe r_j is real while two or more distinct zeros z_t appear in $N_\varepsilon(r_j)$. For example

$$a(z)=(z-0)^3$$

has $r_1=0$, $m_1=3$ while for real $c>0$

$$b(z)=z^3-c^2$$

has three distinct zeros $z_1=c, z_2=c\omega, z_3=c\omega^2$ where $\omega=\exp(2\pi i/3)$. This means that a change in only one of the coefficients of a, in this case a_0 which changes from 0 to c^2, has led to one real zero of multiplicity 3 being replaced by three distinct zeros. On the other hand, all three are indeed within a circle of radius $2c$ centered at the original repeated zero $r_1=0$. We are interested in specifying how large the radii ε about the r_j can be, and then finding how small the radii δ about the a-coefficients must be. If a has only one distinct zero then $\varepsilon>0$ is otherwise unrestricted. If $s>1$ then $\varepsilon>0$ *sufficiently small* means

$$0<\frac{\varepsilon}{2}<\min\{|r_\alpha-r_\beta|:\alpha,\beta=1,\ldots,s,\ \alpha\neq\beta\}$$

so that the zeros of a have disjoint neighborhoods $N_\varepsilon(r_j)$, $j=1,\ldots,s$. With these definitions we have

(2) The zeros of a polynomial are continuous functions of its coefficients.

PROOF. Given $\varepsilon>0$ sufficiently small as specified above, let C_k denote the boundary circle of $N_\varepsilon(r_k)$ for $k=1,2,\ldots,s$. By Rouché's theorem it suffices to find $\delta>0$ such that for $b\in\mathfrak{N}_\delta(a)$

$$|b(z)-a(z)|<|a(z)|$$

whenever z is a point on any one of the s boundary circles C_k. Let us continue to adapt the preceding proof by using the extreme value theorem

(6.19) to find two numbers M and m. On the one hand, for each k there exists

$$M_k = \max\{|z|^{n-1} + \cdots + |z| + 1 : z \in C_k\}$$

so for any $\delta > 0$, $b \in \mathcal{N}_\delta(a)$, and $M = \max\{M_1, \ldots, M_s\}$

$$|b(z) - a(z)| \leqq M\delta$$

whenever $z \in C_k$. On the other hand, $a(z) \neq 0$ on C_k because the neighborhoods are disjoint so there are lower bounds

$$0 < m_k \leqq |a(z)|$$

and we can select $m = \min\{m_1, \ldots, m_s\}$. Finally, we select $\delta > 0$ satisfying $M\delta < m$ because then

$$|b(z) - a(z)| \leqq M\delta < m \leqq |a(z)|$$

shows that we have established the appropriate inequality. ■

Rouché's theorem can be used for particular functions. Gross and Harris (1974) make essential uses for certain polynomials appearing in queueing theory. See Henrici (1974) for the manner in which it contributes to a systematic development of methods for locating zeros of polynomials within halfplanes, neighborhoods, and so on.

PROBLEMS FOR SOLUTION

8.2.1 Principal Laurent's Series. Which series in 8.1.2(5) are principal Laurent's series?

8.2.2 Residues. Establish or accomplish:

(1) The counterpart of (8.17) for a pole of order 2.
(2) An example of a function having a zero residue.
(3) Determine orders of all poles and find all residues.

 (i) $(e^z - 1)^{-1}$.
 (ii) $(z(z-1))^{-1}$.
 (iii) $\exp(z^{-1})$.
 (iv) $z^{-4}e^z$.
 (v) $(\cos z)^{-1}$.
 (vi) $(z-1)/z(z-2)^3$.
 (vii) $e^z/(z^2+z+1)$.
 (viii) $(z+1)(z^2+1)^{-2}$.
 (ix) $(z^5+2z^3+1)/z(z-1)^4$.
 (x) $e^z/((z^2+2)^2(z^2-1))$.

8.2.3 An Equation from Queueing Theory (Gross and Harris (1974)). Establish:

(1) Rouché's theorem can be used to show that if $k>0$ and $\operatorname{Re} s>0$, then

$$z^2-(1+k+s)z+k=0$$

has one and only one root inside the unit circle.

(2) Which of the following could describe the roots in the preceding?
   (i) Both real.
   (ii) Neither real.
   (iii) Complex conjugates.
   (iv) Only one real.

8.2.4 Counting Roots. Use Rouché's theorem to prove that if $a>e$ then $e^z-az^n=0$ has n roots inside the unit circle.

9

LINEAR DIFFERENTIAL EQUATIONS

We use the term *linear systems* for systems of linear differential and difference equations. In this chapter we extend familiar solutions for $n=1,2$ to results for arbitrary order n. We develop direct algebraic, and indirect Laplace transform, methods as alternative solution procedures. The results are widely used in diverse quantitative settings, they represent initial stages for studying variable coefficient cases, and nearly every one is a prototype for a result in Chapter 10. The prerequisite (for Section 9.4) is Section 5.4.

9.1 FUNDAMENTAL CONCEPTS

Here we introduce the subject and specify the approach that we follow in the chapter.

An *ordinary differential equation*, briefly *differential equation*, is an equation that can be written as

$$\phi(t, y, y', \ldots, y^{(n)}) = 0 \tag{9.1}$$

A *solution* of (9.1) is a continuous function

$$y : \mathcal{I} \to \mathcal{R} \tag{9.2}$$

where $\mathcal{I}$ is a real interval, such that

$$\phi(t, y(t), y'(t), \ldots, y^{(n)}(t)) = 0$$

for all $t \in \operatorname{int} \mathcal{I}$,

$$\phi(t_0+, y(t_0+), y'(t_0+), \ldots, y^{(n)}(t_0+)) = 0$$

where we use RH limits whenever $\mathcal{I}$ has a LH endpoint t_0, and

$$\phi(t_1-, y(t_1-), y'(t_1-), \ldots, y^{(n)}(t_1-))=0$$

where we use LH limits whenever $\mathcal{I}$ has a RH endpoint t_1.

Equation (9.1) is a *linear differential equation* of *order n* if it can be written

$$y^{(n)}+\alpha_{n-1}(t)y^{(n-1)}+\cdots+\alpha_1(t)y'+\alpha_0(t)y=b(t) \tag{9.3}$$

where $\alpha_{n-1},\ldots,\alpha_1,\alpha_0,b$ are continuous functions on $\mathcal{I}$ to $\mathcal{R}$. If not all of the α's are constant functions, (9.3) is said to have *variable coefficients*; otherwise, it has *constant coefficients*. If $b(t)=0$ for all $t\in\mathcal{I}$ then (9.3) is called *homogeneous*; otherwise it is *nonhomogeneous*.

In this chapter we suppose $0\in\mathcal{I}$ and seek solutions of (9.3) such that

$$\begin{aligned} y(0)&=y_0\\ y'(0)&=y_0'\\ &\vdots\\ y^{(n-1)}(0)&=y_0^{(n-1)} \end{aligned} \tag{9.4}$$

where $y_0, y_0', \ldots, y_0^{(n-1)}$ are n specified real numbers. Such problems are called *initial value problems*, and the RHSs in (9.4) are prescribed *initial values*, in keeping with the situation in applications where the domain so often corresponds to "time."

In the next two sections we work with two cases of (9.3). The *first order linear initial value problem* is

$$\begin{aligned} y'(t)&=a(t)y(t)+f(t)\\ y(0)&=y_0 \end{aligned} \tag{9.5}$$

The $n=2$ constant coefficient problem is

$$\begin{aligned} &y''(t)+\alpha_1 y'(t)+\alpha_0 y(t)=b(t)\\ &y(0)=y_0, \qquad y'(0)=y_0' \end{aligned} \tag{9.6}$$

Let us introduce the notation

$$\begin{aligned} y(t) &= y_1(t) \\ y'(t) &= y_2(t) \\ \vdots \quad & \quad \vdots \\ y^{(n-1)}(t) &= y_n(t) \end{aligned} \tag{9.7}$$

so (9.3) can be rewritten as

$$\begin{aligned} y_1'(t) &= \qquad\qquad y_2(t) \\ y_2'(t) &= \qquad y_3(t) \\ &\vdots \\ y_n'(t) &= -\alpha_{n-1}y_n(t) - \cdots - \alpha_1 y_2(t) - \alpha_0 y_1(t) + b(t) \end{aligned}$$

This suggests introducing vector-matrix notation for *data*

$$\mathbf{A} = \begin{bmatrix} 0 & 1 & 0 & \cdots & 0 \\ 0 & 0 & 1 & \cdots & 0 \\ \vdots & \vdots & \vdots & & \vdots \\ 0 & 0 & 0 & \cdots & 1 \\ -\alpha_0 & -\alpha_1 & -\alpha_2 & \cdots & -\alpha_{n-1} \end{bmatrix}, \qquad \mathbf{y}(t) = \begin{bmatrix} y_1(t) \\ y_2(t) \\ \vdots \\ y_{n-1}(t) \\ y_n(t) \end{bmatrix}$$

$$\mathbf{y}_0 = \begin{bmatrix} y_0 \\ y_0' \\ \vdots \\ y_0^{(n-1)} \end{bmatrix}, \qquad \mathbf{f}(t) = \begin{bmatrix} 0 \\ 0 \\ \vdots \\ b(t) \end{bmatrix} \tag{9.8}$$

Note. In the remainder of the book we use boldface notation—as in the preceding for $\mathbf{y}(t)$ and $\mathbf{f}(t)$—for values in $\mathfrak{R}^n$ of transformations. We continue to use the notation of (6.50) whereby $\mathbf{y}'(t)$ denotes the $n\times 1$ Jacobian of $y: \mathcal{I} \to \mathfrak{R}^n$.

We see that $\mathbf{A}$ is the companion matrix of

$$p(\lambda) = \lambda^n + \alpha_{n-1}\lambda^{n-1} + \cdots + \alpha_1\lambda + \alpha_0$$

We call $p(\lambda)$ the *characteristic polynomial*, and

$$p(\lambda) = 0 \tag{9.9}$$

the *characteristic equation*, of (9.3).

Combining the equations following (9.7) with (9.4) we obtain a special case (because **A** and f in (9.8) are special) of the *linear differential systems problem*

$$\mathbf{y}'(t) = \mathbf{A}\mathbf{y}(t) + \mathbf{f}(t)$$

$$\mathbf{y}(0) = \mathbf{y}_0 \tag{9.10}$$

where **A** is a constant matrix of order n, f is continuous on $\mathcal{I}$, and

$$\mathbf{y}_0 = [y_{10}, y_{20}, \ldots, y_{n0}]^T$$

is the prescribed *initial value vector*. If $\mathbf{f}(t) = \mathbf{0}$ for all $t \in \mathcal{I}$ then (9.10) is a *homogeneous system* and otherwise it is *nonhomogeneous*.

In this way we are led to the

THEOREM. Solving the constant coefficient case of (9.3) subject to (9.4) is equivalent to solving (9.10) for the case (9.8). (9.11)

PROOF. We have shown how to go from (9.3) to (9.10) and the process is reversible. ■

See 9.1.4 concerning the relative generality of (9.3) and (9.10).

We are interested in solving (9.10) for any continuous f and any constant **A**. We solve constant coefficient cases of (9.3) for any n by using results for (9.10). In general, we have the following standard procedure for

SOLVING INITIAL VALUE PROBLEMS. First, find at least one explicit formula that can be verified as producing a solution, either

(1) in general, or

(2) in specific cases for given functions and initial values.

Second, prove that the problem can have at most one solution. (9.12)

We find formulas satisfying (1) for all problems we consider and, using Laplace transforms in Section 9.5, we find formulas that fall under (2).

PROBLEMS FOR SOLUTION

9.1.1 Standard Intervals. Establish:

(1) No generality is lost by supposing $0 \in \mathcal{I}$.

(2) A set of nine standard intervals that cover all cases for $\mathcal{I}$.

9.1.2 Transforming Equations. Exhibit the following as instances of (9.10) by identifying **A** and **f**.

(1) $y'' + 4y' + 4y = e^{-2t} + t + 1$

$y(0) = 0,\ y'(0) = 1.$

(2) $y''-2y'-3y=2e^t+3\sin t$
$y(0)=c,\ y'(0)=d.$
(3) $y''+4y=2\cos 2t+\sin t+2e^{-t}$
$y(0)=c,\ y'(0)=d.$

9.1.3 An Extension. Generalize the preceding in order to write the following as an $n=4$ instance of (9.10).

$$z_1''+z_2'+2z_1+z_2=t$$

$$2z_1'-z_2''+z_1+z_2=1+t^2$$

9.1.4 Transforming Systems. Establish:
(1) The $n=2$ constant coefficient homogeneous system (9.10) can be solved by solving (9.3).
(2) Repeat for nonhomogeneous case where $f(t)$ is differentiable.

9.1.5 Finding Differential Equations. Find homogeneous linear differential equations of minimal order satisfied by the following.
(1) e^{at}.
(2) $a+bt$.
(3) $a+bt+ct^2$.
(4) $a\cos t+b\sin t$.
(5) $a\cos ct+b\sin ct$.

9.2 FIRST-ORDER EQUATIONS

Let us assemble some familiar results as one-dimensional cases that we can generalize for constant coefficients.

We start with a special case of (9.5), namely

$$y'(t)=f(t)$$

$$y(0)=y_0 \tag{9.13}$$

Since f is continuous, this is the familiar problem of integral calculus. Certainly one solution is

$$y(t)=y_0+\int_0^t f(u)\,du \tag{9.14}$$

and this is the one and only solution according to the

EXISTENCE AND UNIQUENESS THEOREM. There exists a unique solution to Problem (9.13) and it is given by (9.14). (9.15)

PROOF. Formula (9.14) makes sense because f is integrable on $\mathcal{I}$ and it clearly satisfies (9.13) because it is continuous, it satisfies the differential equation, and it assumes the value y_0 for $t=0$. If there were a second solution z then $w=y-z$ would be a solution to

$$w'(t)=0$$

$$w(0)=0 \tag{9.16}$$

By (6.44) this requires $w(t)=0$ for all $t \in \operatorname{int} \mathcal{I}$. Finally, w could not fail to vanish at an endpoint of $\mathcal{I}$, whenever there were such a point, because this would violate the one-sided continuity of w. ■

The simple result (9.15) serves as a model for resolution of initial value problems. It represents completion of (9.12) through provision of a formula of type (1). Nevertheless, it is important to recall that there may be no way to perform the integration in (9.14) in terms of elementary functions so that in particular cases it may be necessary to use infinite series or numerical methods.

According to (6.38) we lose no generality by requiring that solutions be continuous. If the highest derivative $y^{(n)}$ is piecewise continuous then its integral $y^{(n-1)}$ will be continuous and so also will be the lower derivatives and y itself.

EXAMPLE 9.1. Handling Piecewise Continuous Functions. Let us use Problem (9.13) to illustrate how we can use results in the text to (i) cope with piecewise continuous functions in problem statements or (ii) obtain solutions that have specified finite jumps. It suffices to consider cases where there is a single point of discontinuity.

Suppose $\mathcal{I}=[0,2], y_0=0$, and consider the function illustrated in Figure 9.1(a), namely

$$\begin{aligned} f(t)&=0, \qquad 0 \leqq t<1 \\ &=1, \qquad 1<t \leqq 2 \end{aligned}$$

Now f is integrable on $\mathcal{I}$, as is any piecewise continuous function on that interval, and all we need do is solve (9.13) separately, on $\mathcal{I}_1=[0,1]$ and $\mathcal{I}_2=[1,2]$. The solution on $\mathcal{I}_1$ is

$$y_1(t)=\int_0^t 0\,du=0$$

The next initial value is $y_1(1)=0$ and the solution on $\mathcal{I}_2$ is

$$y_2(t)=\int_1^t du=t-1$$

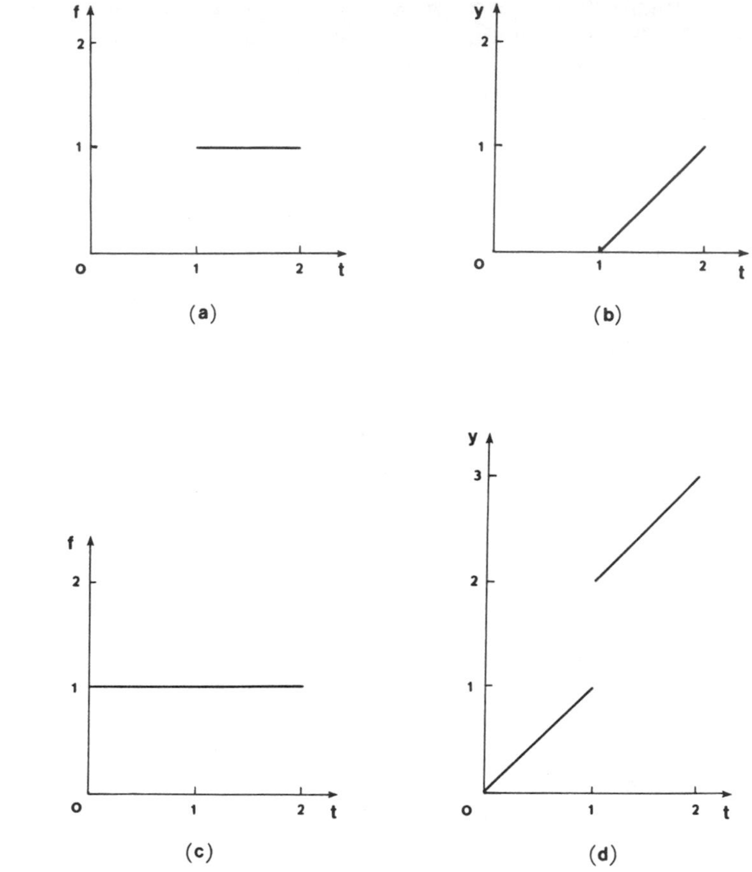

Figure 9.1. Discontinuities.

Consequently, the solution on $\mathcal{I}$ is the continuous function

$$\begin{aligned} y(t) &= 0, \qquad 0 \leqq t \leqq 1 \\ &= t-1, \qquad 1 \leqq t \leqq 2 \end{aligned}$$

shown in Figure 9.1(b).

Let us next illustrate process (ii) by solving (9.13) for the case of the function in Figure 9.1(c), namely

$$f(t) = 1, \qquad t \in \mathcal{I}$$

where $y(0)=0$ and $y(1+)-y(1-)=1$. First we solve on $\mathcal{I}_1$ obtaining

$$y_1(t)=\int_0^t du=t$$

Then since $y_1(1)=1$, we solve on $\mathcal{I}_2$ using $y_2(1)=2$ and obtain

$$y_2(t)=2+\int_1^t du=t+1$$

Putting the pieces together, the required solution is

$$\begin{aligned} y(t)&=t, \qquad 0\leqq t<1\\ &=t+1, \qquad 1<t\leqq 2 \end{aligned}$$

as drawn in Figure 9.1(d).

Elementary texts provide various rationalizations for the procedure that solves (9.5), but multiplying both sides of

$$y'(t)=a(t)y(t)+f(t)$$

by the integrating factor

$$m(t)=\exp\left(\int_0^t -a(u)\,du\right) \tag{9.17}$$

shows that we can use our results for (9.13). The reason, as readily verified, is

$$\frac{d}{dt}\left[m(t)y(t)\right]=m(t)f(t) \tag{9.18}$$

and therefore (9.15) implies the

EXISTENCE AND UNIQUENESS THEOREM. There exists a unique solution $y(t)$ to Problem (9.5) and it is given by

$$m(t)y(t)=y(0)+\int_0^t m(u)f(u)\,du$$

where $m(t)$ is given by (9.17). (9.19)

Usually it is convenient to work with the solution as written here because the result is complicated when we multiply both sides by $[m(t)]^{-1}$ and use the explicit formulas for $m(\cdot)$. However, if $a(t)=a$ is constant for all $t\in\mathcal{I}$, then it is convenient to work with

$$y(t)=e^{at}y_0+\int_0^t e^{a(t-u)}f(u)\,du \tag{9.20}$$

PROBLEMS FOR SOLUTION

9.2.1 Another Formula. Verify that another expression for (9.14) is

$$y(t)=y_0-\int_t^0 f(u)\,du, \qquad t<0$$

$$=y_0+\int_0^t f(u)\,du, \qquad t\geqq 0$$

9.2.2 Exact Equations. If $(\partial/\partial y)p=(\partial/\partial t)q$ then

$$p\,dt+q\,dy=0$$

is *exact* and has solution $u(t,y)=c$ where $(\partial/\partial t)u=p$ and $(\partial/\partial y)u=q$. Solve:

(1) $y\sin t\,dt-\cos t\,dy=0$
(2) $(t^2+y)\,dt+(t+\sin y)\,dy=0$

9.2.3 Reducible Equations. Establish:

(1) Introducing $v=y/t$ makes it possible to solve

$$\frac{dy}{dt}=\frac{y-t}{y+t}$$

(2) An *integrating factor* is a factor that transforms an equation into an exact equation. The following can be solved by finding such a factor.

$$y\,dt+2(y^4-t)\,dy=0$$

9.2.4 Power Series Solution. Find a formal series solution for (9.13) where $f(t)=\exp(-t^2)$.

9.2.5 Uniqueness in (9.15). Can we use the following argument to prove the subject result? Problem (9.16) has solution

$$w(t)=0+\int_0^t 0\,du$$

which is clearly zero for all t.

9.2.6 Linear Equations. Solve and check the solutions:
(1) $y' = y + 1 + 2t$, $y(0) = 0$.
(2) $(t^2+1)y' - ty = 1$, $y(0) = 1$.

9.3 SECOND-ORDER EQUATIONS

Here we obtain some familiar results for Problem (9.6) in special ways that we recall in Section 10.4.

Let us seek constants λ_1 and λ_2 such that under the transformation y to z via

$$z(t) = y'(t) - \lambda_1 y(t)$$

the second order homogeneous problem in y becomes

$$\begin{aligned} z'(t) &= \lambda_2 z(t) \\ z(0) &= y_0' - \lambda_1 y_0 \end{aligned} \tag{9.21}$$

This requires

$$y'' + \alpha_1 y' + \alpha_0 y = \frac{d}{dt}(y' - \lambda_1 y) - \lambda_2(y' - \lambda_1 y)$$

for all solutions y. Expanding the RHS, collecting terms, and equating corresponding coefficients we find that λ_1 and λ_2 must be the roots of the characteristic equation

$$\lambda^2 + \alpha_1\lambda + \alpha_0 = 0 \tag{9.22}$$

Using (9.19) we can find the unique solution of (9.21) and write

$$\begin{aligned} y'(t) &= \lambda_1 y(t) + (y_0' - \lambda_1 y_0)e^{\lambda_2 t} \\ y(0) &= y_0 \end{aligned} \tag{9.23}$$

which is an instance of (9.5). If $\lambda_1 \neq \lambda_2$ the solution of (9.23) is

$$y(t) = c_1 e^{\lambda_1 t} + c_2 e^{\lambda_2 t} \tag{9.24}$$

where c_1 and c_2 are the unique solutions of

$$\begin{aligned} c_1 + c_2 &= y_0 \\ \lambda_1 c_1 + \lambda_2 c_2 &= y_0' \end{aligned}$$

If $\lambda_1 = \lambda_2 = \lambda$ then the solution of (9.23) is

$$y(t) = k_1 e^{\lambda t} + k_2 t e^{\lambda t} \tag{9.25}$$

where k_1 and k_2 are the unique solutions of

$$\begin{aligned} k_1 \qquad &= y_0 \\ \lambda k_1 + k_2 &= y_0' \end{aligned}$$

In this way (9.15) and (9.19) yield

There exists a unique solution to Problem (9.6), and it is given by (9.24) when the roots of (9.22) are distinct, and by (9.25) when the roots coincide. (9.26)

When the roots are not real, say

$$\lambda_1 = \mu + i\nu, \qquad \lambda_2 = \mu - i\nu$$

then it is sometimes convenient to use the real form of (9.24)

$$y(t) = d_1 e^{\mu t} \cos \nu t + d_2 e^{\mu t} \sin \nu t \tag{9.27}$$

where d_1 and d_2 are the unique solutions of

$$\begin{aligned} d_1 \qquad &= y_0 \\ \mu d_1 + \nu d_2 &= y_0' \end{aligned}$$

Since this system is nonsingular, the numbers d_1 and d_2 can assume any pair of real values, and the same can be said for the c's and k's. It therefore makes sense to call the applicable expression (9.24) or (9.25), and we can always replace the former by (9.27), the *general solution* of the homogeneous differential equation

$$y''(t) + \alpha_1 y'(t) + \alpha_0 y(t) = 0 \tag{9.28}$$

The functions $\exp \lambda_1 t$, $\exp \lambda_2 t$, $\exp \lambda t$, and $t \exp \lambda t$ are called *fundamental solutions* of (9.28). In summary, we have obtained the

FUNDAMENTAL SOLUTIONS THEOREM. Any solution of (9.28) can be written as a LC of two fundamental solutions. (9.29)

This theorem, which in our development appears almost as an afterthought, plays an important role in the study of variable coefficient cases where concepts illustrated in the next example are used.

EXAMPLE 9.2. Vector Spaces of Solutions. When the coefficients α_1 and α_0 in Problem (9.6) are not constant there is no explicit formula for solutions comparable to that for $n=1$ in (9.19). There are only infinite series solutions, numerical solutions, and special results for special cases. In order to develop effective theory for the variable coefficient case, essential uses are made of the fact that the solutions can be interpreted as vectors in a vector space. Let us illustrate concepts using the $n=2$ constant coefficient case.

As cited in (9) of Example 2.1 the set of twice-differentiable functions on $\mathcal{I}=(-1,1)$ furnishes an example of a vector space; call this space V. Then V is an infinite dimensional space because no finite collection of functions could possibly serve as a basis. As cited in (10) of that example, the set of all solutions to

(1) $$y''(t)+\alpha_1 y'(t)+\alpha_0 y(t)=0$$

also forms a vector space; let us call this space the *solution space* S of (1). Clearly, S is a vector space because any LC of solutions is a solution; this is the important *superposition principle* for linear systems. What is not so clear is that S is a finite-dimensional space. This follows from (9.29) and we have the

(2) **THEOREM.** The solution space S has dimension $n=2$ and pairs of fundamental solutions provide bases.

Now LD of $\{y_1, y_2\}$ in S entails the existence of $c_1, c_2 \in \mathcal{R}$, not both zero, such that

(3) $$c_1 y_1(t)+c_2 y_2(t)=0$$

for all $t\in\mathcal{I}$. Equation (3) holding for all points in an interval rather than at a single point is something new. In order to work with such LD in S we make use of the *Wronskian* determinant

$$w(y_1(t),y_2(t))=\begin{vmatrix} y_1(t) & y_2(t) \\ y_1'(t) & y_2'(t) \end{vmatrix}$$

and the basic result is the

(4) **THEOREM.** A necessary and sufficient condition for $\{y_1(t), y_2(t)\}$ to be LD in S is $w(y_1(t), y_2(t))=0$ for every $t\in\mathcal{I}$.

PROOF. Necessity is obvious. Conversely, if $w=0$ on $\mathcal{I}$ there exist $c_{01}, c_{02}\in\mathcal{R}$, not both zero, satisfying

$$c_{01} y_1(0)+c_{02} y_2(0)=0$$

$$c_{01} y_1'(0)+c_{02} y_2'(0)=0$$

because the coefficient matrix is in fact singular for all $t\in\mathcal{I}$. Now define for all $t \in \mathcal{I}$

$$y(t) = c_{01} y_1(t) + c_{02} y_2(t)$$

This function is a solution of (1), both initial values are by construction zero, and therefore by (9.26)

$$y(t) = 0$$

for all $t \in \mathcal{I}$. We have thus shown the existence of a nontrivial LC of $\{y_1, y_2\}$ that vanishes on $\mathcal{I}$. ■

Examples show that the condition is not sufficient in V.

One of the most useful results for working with variable coefficient cases is that

(5) $$y \mapsto Ty = y'' + \alpha_1 y' + \alpha_0 y$$

is a linear transformation in V and $S = \text{Null}\, T$. The inner product 3.4.2(1) is used in V and the adjoint transformation of (5)

$$y \mapsto T^A y = y'' - (\alpha_1 y)' + \alpha_0 y$$

is important.

We call any solution of the nonhomogeneous equation

$$y''(t) + \alpha_1 y'(t) + \alpha_0 y(t) = b(t) \tag{9.30}$$

a *particular solution* of that equation. We are interested in finding such solutions because (recall (3.20)) of the

THEOREM. The general solution of (9.30) is given by the sum of the general solution of (9.28) and any particular solution of (9.30). (9.31)

PROOF. If z is an arbitrary solution, and p is any given particular solution, of (9.30) then $z - p$ is a solution of the homogeneous (9.28). By (9.29) there must exist constants c_1 and c_2 such that

$$z - p = c_1 y_1 + c_2 y_2$$

where y_1 and y_2 are fundamental solutions of (9.28). ■

Methods of finding particular solutions are given in 9.3.4, 9.3.5, 9.4.2, and Example 9.4.

Let us rephrase (9.26) as the

EXISTENCE AND UNIQUENESS THEOREM. There exists a unique solution to Problem (9.6) and it can be found by finding the pair of coefficients that makes the general solution in (9.31) satisfy the initial values. (9.32)

The final end product is the

DIRECT PROCEDURE FOR SOLVING PROBLEM (9.6).

(1) Find the roots of (9.22) and write the general solution of (9.28) using (9.24), (9.25), or (9.27).
(2) Find a particular solution of (9.30).
(3) Form the sum of the two solutions and determine the coefficients that satisfy the initial values.
(4) Check that the solution satisfies both the differential equation and the initial values. (9.33)

The last step is a matter of prudence.

PROBLEMS FOR SOLUTION

9.3.1 Wronskians. Establish:

(1) Two LI functions whose Wronskian vanishes for all $t \in \mathcal{I}$.
(2) Let $y_1(t)$ and $y_2(t)$ be two LI solutions to the homogeneous equation and let $w(t)$ denote their Wronskian. Show that

$$p(t)=\int_0^t \frac{\begin{vmatrix} y_1(u) & y_2(u) \\ y_1(t) & y_2(t) \end{vmatrix}}{w(u)} b(u)\,du$$

is a particular solution of the nonhomogeneous equation.
(3) Abel's Formula. The Wronskian can be written

$$w(t)=w(0)\exp\left[-\int_0^t \alpha_1(u)\,du\right]$$

9.3.2 Reduction to First-Order Equation. Establish:

(1) The variable coefficient homogeneous case can be solved explicitly in the following cases for $n=2$.
(i) $\alpha_0(t)=0$ for all $t \in \mathcal{I}$.
(ii) One solution $z(t)$ is known where $z(t) \neq 0$ for all $t \in \mathcal{I}$.
(2) The variable coefficient nonhomogeneous equation can be solved in the preceding cases (or whenever two LI solutions of the homogeneous equation are known).

9.3.3 Euler's Rationale. Establish:

(1) It is possible to substitute $y(t)=e^{\lambda t}$ into the homogeneous equation and determine conditions on λ such that two LI solutions are obtained.

(2) The preceding can be interpreted as using an "operator equation" $p(D)y(t)=0$ where $Dy(t)=y'(t)$.

9.3.4 Undetermined Coefficients. Show how to find particular solutions of the nonhomogeneous equation when $b(t)$ has the following forms. (Provide for two cases: (I) $b(t)$ is not a solution of (9.28), and (II) $b(t)$ is a solution.)

(1) Polynomial in t.

(2) $\exp kt$.

(3) LC of $\sin kt$ and $\cos kt$.

9.3.5 Variation of Parameters. Let $y_1(t)$ and $y_2(t)$ be two LI solutions of the homogeneous equation. Show how to determine $c_1(t)$ and $c_2(t)$ such that

$$y(t)=c_1(t)y_1(t)+c_2(t)y_2(t)$$

is a particular solution to the nonhomogeneous equation. (Note that 9.3.1 (2) gives the general result.)

9.3.6 Exercises. Solve and check:

(1) Problems in 9.1.2.

(2) $y''+y'-2y=2+2t-2t^2$, $y(0)=0$, $y'(0)=-3$.

(3) $y''+y=t$, $y(0)=y'(0)=0$.

(4) $y''+y=\sin t$, $y(0)=y'(0)=1$.

(5) $y''+y=t+\sin t$, $y(0)=y'(0)=0$.

(6) $y''+y'+y=t^2$, $y(0)=y'(0)=0$.

(7) $y''-y'-y=0$, $y(0)=1$, $y'(0)=0$.

9.3.7 Strict Stability. The homogeneous equation is called *strictly stable* if all solutions tend to zero at $t\to\infty$. Establish:

(1) Necessary and sufficient conditions for strict stability for $n=1,2,3$.

(2) Corresponding geometric conditions in $\alpha_0\alpha_1\alpha_2$-space.

9.3.8 Vector Space for $n=1$. Specialize (1)–(5) in Example 9.2 so as to present the solutions of $y'(t)+\alpha_0 y(t)=0$ as a finite-dimensional vector space.

9.3.9 Two-Point Problems. Establish:

(1) Conditions for the existence of a unique solution to (9.28) subject to $y(t_0)=y_0$, $y(t_1)=y_1$.

(2) The solution of $y''(t)=0$ so as to find the equation of the line joining (t_0, y_0) and (t_1, y_1).

9.4 SYSTEMS OF EQUATIONS

Let us use results from Section 5.4 and obtain direct procedures for finding solutions.

Following the familiar pattern, we start with the homogeneous case of (9.10), namely

$$\mathbf{y}'(t)=\mathbf{A}\mathbf{y}(t)$$
$$\mathbf{y}(0)=\mathbf{y}_0 \tag{9.34}$$

If $\mathbf{A}$ were a diagonal matrix we could write a solution by inspection. Let us then use the transformation

$$\mathbf{y}(t)=\mathbf{P}\mathbf{z}(t)$$

where $\mathbf{P}$ is the nonsingular matrix of order n appearing as $\mathbf{T}$ in the Jordan canonical form theorem (5.38). There are two cases.

First, if $\mathbf{A}$ has n LI eigenvectors, it is diagonalizable, $\mathbf{z}'=\mathbf{P}^{-1}\mathbf{A}\mathbf{P}\mathbf{z}$, and we obtain the new system

$$\mathbf{z}'(t)=\mathbf{diag}(\lambda_1,\lambda_2,\ldots,\lambda_n)\mathbf{z}(t)$$
$$\mathbf{z}(0)=\mathbf{z}_0 \tag{9.35}$$

where $\lambda_1,\ldots,\lambda_n$ are the n, not necessarily distinct, eigenvalues of $\mathbf{A}$ and $\mathbf{z}_0=\mathbf{P}^{-1}\mathbf{y}_0$. In our usual notation for components of vectors,

$$\mathbf{z}(t)=\begin{bmatrix} z_1(t)\\ z_2(t)\\ \vdots\\ z_n(t)\end{bmatrix},\qquad \mathbf{z}_0=\begin{bmatrix} z_{10}\\ z_{20}\\ \vdots\\ z_{n0}\end{bmatrix}$$

and we can write the new system as the following system of scalar equations.

$$\begin{aligned} z_1'(t)&=\lambda_1 z_1(t), & z_1(0)&=z_{10}\\ z_2'(t)&=\lambda_2 z_2(t), & z_2(0)&=z_{20}\\ &\vdots & &\vdots\\ z_n'(t)&=\lambda_n z_n(t), & z_n(0)&=z_{n0} \end{aligned} \tag{9.36}$$

With (9.20) we obtain

$$z_i(t)=z_{i0}e^{\lambda_i t},\qquad i=1,2,\ldots,n$$

so (9.36) has solution

$$\mathbf{z}(t)=\mathbf{diag}(e^{\lambda_1 t},\ldots,e^{\lambda_n t})\mathbf{z}_0$$

Using the inverse transformation $\mathbf{z}=\mathbf{P}^{-1}\mathbf{y}$,

$$\mathbf{y}(t)=\mathbf{P}\,\mathbf{diag}(e^{\lambda_1 t},\ldots,e^{\lambda_n t})\mathbf{P}^{-1}\mathbf{y}_0 \tag{9.37}$$

is a solution to (9.34).

Second, if $\mathbf{A}$ has fewer then n LI eigenvectors, then in place of (9.35) we have

$$\mathbf{z}'(t)=\mathbf{J}\mathbf{z}(t)$$

where $\mathbf{J}$ is the JCF from (5.38). In our present notation, using λ's for eigenvalues, let us suppose that the upper LH corner of $\mathbf{J}$ contains a block of order m

$$\mathbf{J}_1=\begin{bmatrix}\lambda_1 & 1 & 0 & \cdots & 0\\ 0 & \lambda_1 & 1 & \cdots & 0\\ \vdots & \vdots & \vdots & & \vdots\\ 0 & 0 & 0 & \cdots & \lambda_1\end{bmatrix} \tag{9.38}$$

Following the transformation $\mathbf{y}=\mathbf{P}\mathbf{z}$, the first m equations in the counterpart of (9.36) form a system

$$\mathbf{v}'(t)=\mathbf{J}_1\mathbf{v}(t)$$

$$\mathbf{v}(0)=\mathbf{v}_0 \tag{9.39}$$

where $\mathbf{v}\in\mathfrak{R}^m$ is notation for the first m components of $\mathbf{z}\in\mathfrak{R}^n$. The m individual scalar equations can be written as follows in our usual notation for components

$$\begin{aligned}
v_1'(t)&=\lambda_1 v_1(t)+v_2(t), & v_1(0)&=v_{10}\\
v_2'(t)&=\lambda_1 v_2(t)+v_3(t), & v_2(0)&=v_{20}\\
&\vdots & &\vdots\\
v_{m-1}'(t)&=\lambda_1 v_{m-1}(t)+v_m(t), & v_{m-1}(0)&=v_{m-1,0}\\
v_m'(t)&=\lambda_1 v_m(t), & v_m(0)&=v_{m0}
\end{aligned}$$

Again, (9.20) applies to each equation. We start with the mth equation, and proceed upward in order, obtaining solutions

$$v_m(t)=v_{m0}e^{\lambda t}$$

$$v_{m-1}(t)=v_{m-1,0}e^{\lambda t}+v_{m0}te^{\lambda t}$$

$$\vdots$$

$$v_2(t)=v_{20}e^{\lambda t}+v_{30}te^{\lambda t}+\cdots+v_{m0}\frac{t^{m-2}}{(m-2)!}e^{\lambda t}$$

$$v_1(t)=v_{10}e^{\lambda t}+v_{20}te^{\lambda t}+\cdots+v_{m-1,0}\frac{t^{m-2}}{(m-2)!}e^{\lambda t}+v_{m0}\frac{t^{m-1}}{(m-1)!}e^{\lambda t}$$

Let us denote the coefficient matrix

$$e^{t\mathbf{J}_1}=\begin{bmatrix} e^{\lambda t} & te^{\lambda t} & \frac{t^2}{2!}e^{\lambda t} & \cdots & \frac{t^{m-1}}{(m-1)!}e^{\lambda t} \\ 0 & e^{\lambda t} & te^{\lambda t} & \cdots & \frac{t^{m-2}}{(m-2)!}e^{\lambda t} \\ \vdots & \vdots & \vdots & & \vdots \\ 0 & 0 & 0 & \cdots & e^{\lambda t} \end{bmatrix} \tag{9.40}$$

and write the solution we have found for (9.39) as

$$\mathbf{v}(t)=e^{t\mathbf{J}_1}\mathbf{v}_0 \tag{9.41}$$

We can in the same way solve a total of q subsystems, where q is the number of Jordan blocks in $\mathbf{J}$, to solve the transformed problem

$$\mathbf{z}'(t)=\mathbf{J}\mathbf{z}(t)$$

$$\mathbf{z}(0)=\mathbf{z}_0$$

Using (9.40) for each individual block, we form

$$e^{t\mathbf{J}}=\begin{bmatrix} e^{t\mathbf{J}_1} & \mathbf{O} & \cdots & \mathbf{O} \\ \mathbf{O} & e^{t\mathbf{J}_2} & \cdots & \mathbf{O} \\ \vdots & \vdots & & \vdots \\ \mathbf{O} & \mathbf{O} & \cdots & e^{t\mathbf{J}_q} \end{bmatrix} \tag{9.42}$$

and then write

$$\mathbf{y}(t)=\mathbf{P}e^{t\mathbf{J}}\mathbf{P}^{-1}\mathbf{y}_0 \tag{9.43}$$

as the solution to (9.34). By (5.38), this includes (9.37) as a special case.

A real matrix can have nonreal eigenvalues in which case the preceding matrix $\mathbf{P}$ has nonreal eigenvectors as columns. For this reason we use the following

Definition. Suppose $\mathbf{A}=\mathbf{PJP}^{-1}$ where $\mathbf{A}$ is any complex matrix of order n and $\mathbf{J}$ is the JCF from (5.38). For any $t\in\mathfrak{R}$ the *matrix exponential* of $\mathbf{A}$ is

$$e^{t\mathbf{A}}=\mathbf{P}e^{t\mathbf{J}}\mathbf{P}^{-1}$$

where $e^{t\mathbf{J}}$ is formed by replacing the Jordan block $\mathbf{J}_\alpha$ by the corresponding block (9.40) for $\alpha=1,2,\ldots,q$. The matrix exponential is also denoted by $\exp t\mathbf{A}$. (9.44)

Various properties of $\exp t\mathbf{A}$ are developed below. Thus far we have shown

Problem (9.34) has solution

$$\mathbf{y}(t)=e^{t\mathbf{A}}\mathbf{y}_0 \tag{9.45}$$

This takes care of existence of solutions for the homogeneous case of the system (9.10). To prove uniqueness we use the following result.

Suppose $g:\mathfrak{I}\rightarrow\mathfrak{R}$ is continuous and nonnegative for all $t\in\mathfrak{I}$. If for some $M\geqq 0$ and all $t\in\mathfrak{I}$

$$g(t)\leqq M\left|\int_0^t g(u)\,du\right|$$

then $g(t)=0$ for all $t\in\mathfrak{I}$. (9.46)

PROOF. We define for $t\in\mathfrak{I}$

$$\begin{aligned} G(t)&=-\int_0^t g(u)\,du, && t<0\\ &=\int_0^t g(u)\,du, && t\geqq 0\end{aligned}$$

so $G(t)\geqq 0$ for all $t\in\mathfrak{I}$. For $t\geqq 0$, $G'(t)=g(t)$ and we can write

$$\exp(-Mt)\left[G'(t)-MG(t)\right]\leqq 0$$

Integrating this expression we obtain

$$\exp(-Mt)G(t)-G(0)\leqq 0$$

Since $G(0)=0$ we must have $G(t)=0$ and hence $g(t)=0$ for $t\geqq 0$ in $\mathcal{I}$. For $t<0$, $G'(t)=-g(t)$ and we can use the same integrating factor to write the same inequality as before. ■

Generalizing (9.26) we have

There exists a unique solution to (9.34), and it is given by (9.45). (9.47)

PROOF. It remains to prove uniqueness. If **v** were a second solution, then $\mathbf{w}=\mathbf{y}-\mathbf{v}$ would be a solution to

$$\mathbf{w}'(t)=\mathbf{A}\mathbf{w}(t)$$

$$\mathbf{w}(0)=\mathbf{w}_0$$

Letting nM denote the largest of the n^2 values $|a_{ij}|$ we find that each component of w satisfies

$$|w_i(t)|\leqq M\int_0^t(|w_1(u)|+\cdots+|w_n(u)|)\,du$$

and their sum $g(t)=|w_1(t)|+\cdots+|w_n(t)|$ satisfies

$$g(t)\leqq M\left|\int_0^t g(u)\,du\right|$$

Then $g(t)=0$ from (9.46) implies $\mathbf{w}(t)=\mathbf{0}$ for all $t\in\mathcal{I}$. ■

On the strength of this theorem we are entitled to call

$$\mathbf{y}(t)=e^{t\mathbf{A}}\mathbf{c} \tag{9.48}$$

where $\mathbf{c}\in\mathcal{R}^n$ is arbitrary, the *general solution* of

$$\mathbf{y}'(t)=\mathbf{A}\mathbf{y}(t) \tag{9.49}$$

To the columns of $\exp t\mathbf{A}$ there correspond *n fundamental solutions* $t\mapsto$ (column j) for $j=1,2,\ldots,n$, and we have the following generalization of (9.29)

FUNDAMENTAL SOLUTIONS THEOREM. Any solution of (9.49) can be written as a LC of n fundamental solutions. (9.50)

Any solution of the nonhomogeneous equation

$$\mathbf{y}'(t)=\mathbf{A}\mathbf{y}(t)+\mathbf{f}(t) \tag{9.51}$$

is called a *particular solution* of that equation. Generalizing (9.31) we have the

THEOREM. The general solution of (9.51) is given by the sum of the general solution of (9.49) and any particular solution of (9.51). (9.52)

PROOF. Using (9.50) for $\mathbf{z}$ and $\mathbf{p}$ generalizing z and p, respectively, in the proof of (9.31),

$$\mathbf{z}-\mathbf{p}=e^{t\mathbf{A}}\mathbf{c}$$

for some $\mathbf{c}\in\mathfrak{R}^n$. ■

Table 9.1 displays properties of matrix exponentials. Entry (1) is often used as the definition. Limits of sequences of matrices $\{\mathbf{X}_m\}$ are defined in the manner used for vectors in Section 6.1. According to (6.30) we can use any convenient matrix norm, and we have the additional property (3.54) to work with. In particular, using (6.6)

$$\lim\{\mathbf{X}_m\}=\mathbf{O}$$

iff all n^2 (scalar) sequences of matrix entries converge to $0\in\mathfrak{R}$. The series in Table 9.1(1) converges iff its sequences of partial sums converge iff its n^2 sequences of matrix entries converge. The general term is

$$\mathbf{P}\left(\frac{t^m}{m!}\mathbf{J}^m\right)\mathbf{P}^{-1}$$

because of "telescoping" of adjacent factors $\mathbf{P}^{-1}\mathbf{P}$. If $\mathbf{A}$ is diagonalizable,

TABLE 9.1 PROPERTIES OF MATRIX EXPONENTIALS

Definition (9.44) $e^{t\mathbf{A}}=\mathbf{P}e^{t\mathbf{J}}\mathbf{P}^{-1}$	
(1)	$e^{t\mathbf{A}}=\mathbf{I}+t\mathbf{A}+\frac{t^2}{2!}\mathbf{A}^2+\cdots+\frac{t^m}{m!}\mathbf{A}^m+\cdots$
(2)	If $\mathbf{Q}$ is nonsingular, $e^{t\mathbf{Q}^{-1}\mathbf{A}\mathbf{Q}}=\mathbf{Q}^{-1}e^{t\mathbf{A}}\mathbf{Q}$
(3)	$e^{(t_1+t_2)\mathbf{A}}=e^{t_1\mathbf{A}}e^{t_2\mathbf{A}}$
(4)	In general, $e^{t(\mathbf{A}+\mathbf{B})}\neq e^{t\mathbf{A}}e^{t\mathbf{B}}$
(5)	$(e^{t\mathbf{A}})^{-1}=e^{-t\mathbf{A}}$
(6)	$\det e^{t\mathbf{A}}=e^{t(\operatorname{trace}\mathbf{A})}$
(7)	$(e^{t\mathbf{A}})^H=\exp t\mathbf{A}^H$
(8)	For any $\mathbf{c}\in\mathfrak{R}^n$, $\frac{d}{dt}e^{t\mathbf{A}}\mathbf{c}=\mathbf{A}e^{t\mathbf{A}}\mathbf{c}=e^{t\mathbf{A}}\mathbf{A}\mathbf{c}$
(9)	$\frac{d}{dt}e^{t\mathbf{A}}\neq\mathbf{A}e^{t\mathbf{A}}$

the series in (1) is shown to converge to

$$\mathbf{P}\,\mathbf{diag}(e^{\lambda_1 t},\ldots,e^{\lambda_n t})\mathbf{P}^{-1}$$

by adapting familiar arguments from calculus to show that the series of main diagonal terms of

$$\frac{t^m}{m!}\,\mathbf{diag}(\lambda_1^m,\ldots,\lambda_n^m)$$

converge to $e^{\lambda_j t}$. If **A** is defective, essentially the same procedure can be used (5.4.2). The remaining properties in Table 9.1 follow readily from (1) using the calculus of power series of matrices whereby the series can be operated on term by term. No convergence proofs are required if we use (9.44) but we need to use induction to establish the properties for any order n.

Property (2) in Table 9.1 is immediate since any $\mathbf{B}=\mathbf{Q}^{-1}\mathbf{A}\mathbf{Q}$ has the same JCF as does **A**. Property (3) follows from the corresponding property for Jordan blocks. Any pair of matrices such that $\mathbf{AB}\neq\mathbf{BA}$ provides a counterexample sufficient to establish (4). Property (5) follows from (3); (6) and (7) are also based on easy calculations. Property (8) follows by showing that (9.48) satisfies (9.49) and also: **A** and $\exp t\mathbf{A}$ commute. We have included (9) because the formula in question does appear in some works where the derivative of a matrix $\mathbf{B}=[b_{ij}(t)]$ with respect to the scalar t is

$$\left[\frac{d}{dt}b_{ij}(t)\right]$$

But we work with derivatives with respect to vectors and according to (6.68)

$$\frac{d}{dt}e^{t\mathbf{A}}$$

is a tensor of order $n^2\times 1$ (9.4.1).

If $t\mapsto\mathbf{h}(t)\in\mathcal{R}^n$ is any continuous transformation on $\mathcal{I}$ then its component functions are also continuous on $\mathcal{I}$ and we define

$$\int_0^t \mathbf{h}(u)\,du=\begin{bmatrix}\int_0^t h_1(u)\,du\\ \vdots\\ \int_0^t h_n(u)\,du\end{bmatrix} \tag{9.53}$$

for $t \in \mathcal{I}$. In words, the integral of a transformation on $\mathcal{I}$ is the vector whose components are the integrals of the corresponding components. Properties of the component integrals extend to **h** and, in particular

$$\frac{d}{dt}\left[\int_0^t \mathbf{h}(u)\,du\right] = \mathbf{h}(t)$$

Consequently, we can write a useful general formula for the nonhomogeneous system by collecting together results for individual components. If we use the second term in (9.20) for each component we are led to

A particular solution of (9.51) is given by

$$\mathbf{p}(t) = \int_0^t e^{(t-u)\mathbf{A}}\mathbf{f}(u)\,du \tag{9.54}$$

PROOF. If we consider the integrand to be the product of $\exp t\mathbf{A}$ and the vector

$$\mathbf{c} = e^{-u\mathbf{A}}\mathbf{f}(u)$$

then it is clear that from Leibniz's rule and Table 9.1 (8)

$$\mathbf{p}'(t) = \int_0^t \mathbf{A}e^{t\mathbf{A}}e^{-u\mathbf{A}}\mathbf{f}(u)\,du + \mathbf{f}(t)$$

as required. ■

The preceding $\mathbf{p}(t)$ is the unique solution of

$$\mathbf{p}'(t) = \mathbf{A}\mathbf{p}(t) + \mathbf{f}(t)$$

$$\mathbf{p}(0) = \mathbf{0} \tag{9.55}$$

and we now have all we need to establish the

EXISTENCE AND UNIQUENESS THEOREM. There exists a unique solution to Problem (9.10) and it is given by

$$\mathbf{y}(t) = e^{t\mathbf{A}}\mathbf{y}_0 + \mathbf{p}(t)$$

where $\mathbf{p}(t)$ is given by (9.54). (9.56)

PROOF. This is an immediate consequence of (9.52), (9.54), and the fact that $\exp t\mathbf{A}$ is nonsingular by Table 9.1(5). ■

We see that (9.20) serves as a "scalar" prototype for the preceding "vector" result.

EXAMPLE 9.3. Matrix Exponentials. The problem of finding $\exp t\mathbf{A}$ for an arbitrary square matrix $\mathbf{A}$ of order n can be solved by finding the Jordan canonical form $\mathbf{J}$ and using (9.44). For n larger than 3 or 4, and even for $n=2$, this would benefit from the use of a computer. In the present example we illustrate some 2×2 cases and we show how the Cayley–Hamilton theorem (5.39) leads to a second method of finding $\exp t\mathbf{A}$.

The problem

$$(1)\qquad \begin{aligned} y_1'(t) &= y_2(t)+1, \qquad & y_1(0)&=3 \\ y_2'(t) &= y_1(t)+t, \qquad & y_2(0)&=1 \end{aligned}$$

is an instance of Problem (9.10) where

$$\mathbf{A}=\begin{bmatrix} 0 & 1 \\ 1 & 0 \end{bmatrix}, \qquad \mathbf{f}(t)=\begin{bmatrix} 1 \\ t \end{bmatrix}, \qquad \mathbf{y}_0=\begin{bmatrix} 3 \\ 1 \end{bmatrix}$$

This matrix is diagonalizable and

$$\mathbf{P}^{-1}\mathbf{A}\mathbf{P}=\frac{1}{2}\begin{bmatrix} 1 & 1 \\ 1 & -1 \end{bmatrix}\begin{bmatrix} 0 & 1 \\ 1 & 0 \end{bmatrix}\begin{bmatrix} 1 & 1 \\ 1 & -1 \end{bmatrix}$$

is the Jordan canonical form

$$\mathbf{J}=\begin{bmatrix} 1 & 0 \\ 0 & -1 \end{bmatrix}$$

Consequently $\exp t\mathbf{A}$ equals

$$(2)\qquad \mathbf{P}\begin{bmatrix} e^{t} & 0 \\ 0 & e^{-t} \end{bmatrix}\mathbf{P}^{-1}=\frac{1}{2}\begin{bmatrix} e^{t}+e^{-t} & e^{t}-e^{-t} \\ e^{t}-e^{-t} & e^{t}+e^{-t} \end{bmatrix}$$

and the unique solution to the homogeneous problem (9.34) is $(\exp t\mathbf{A})\mathbf{y}_0$ which is

$$(3)\qquad \mathbf{y}(t)=\begin{bmatrix} 2e^{t}+e^{-t} \\ 2e^{t}-e^{-t} \end{bmatrix}$$

From (9.54) we have the particular solution

$$\mathbf{p}(t)=\frac{1}{2}\begin{bmatrix} \displaystyle\int_0^t \left[e^{t}(e^{-u}+ue^{-u})+e^{-t}(e^{u}-ue^{u})\right]du \\ \displaystyle\int_0^t \left[e^{t}(e^{-u}+ue^{-u})+e^{-t}(-e^{u}+ue^{u})\right]du \end{bmatrix}$$

where we have evaluated the integrand and taken the integral signs inside [·] in accordance with (9.53). Upon integrating we obtain

(4) $$\mathbf{p}(t)=\begin{bmatrix} e^{t}-e^{-t}-t \\ e^{t}+e^{-t}-2 \end{bmatrix}$$

By (9.56) the unique solution to (1) is given by the sum of the expressions in (3) and (4)

(5) $$\mathbf{y}(t)=\begin{bmatrix} 3e^{t}-t \\ 3e^{t}-2 \end{bmatrix}$$

It is readily verified that (5) satisfies (1).

In the next example we transform a second-order equation into two first-order equations. If we apply the transformation (9.7) to the second-order problem

(6) $$y''(t)-y(t)=e^{2t}$$
$$y(0)=2, \qquad y'(0)=0$$

We obtain the following instance of Problem (9.10)

(7) $$\mathbf{y}'(t)=\begin{bmatrix} 0 & 1 \\ 1 & 0 \end{bmatrix}\mathbf{y}(t)+\begin{bmatrix} 0 \\ e^{2t} \end{bmatrix}$$
$$\mathbf{y}_0=\begin{bmatrix} 2 \\ 0 \end{bmatrix}$$

which has the same matrix as (1). Using $\exp t\mathbf{A}$ from (2), the solution to the homogeneous problem (7) is

$$\mathbf{y}(t)=\begin{bmatrix} e^{t}+e^{-t} \\ e^{t}-e^{-t} \end{bmatrix}$$

and the particular solution to the nonhomogeneous problem is found from (9.54) as

$$\mathbf{p}(t)=\begin{bmatrix} -\frac{1}{2}e^{t}+\frac{1}{6}e^{-t}+\frac{1}{3}e^{2t} \\ -\frac{1}{2}e^{t}-\frac{1}{6}e^{-t}+\frac{2}{3}e^{2t} \end{bmatrix}$$

Then the solution of (7) is the sum

$$\mathbf{y}(t)=\begin{bmatrix} \frac{1}{2}e^{t}+\frac{7}{6}e^{-t}+\frac{1}{3}e^{2t} \\ \frac{1}{2}e^{t}-\frac{7}{6}e^{-t}+\frac{2}{3}e^{2t} \end{bmatrix} \tag{8}$$

in which, according to (9.7), the first component is the solution of (6)

$$y(t)=\frac{1}{2}e^{t}+\frac{7}{6}e^{-t}+\frac{1}{3}e^{2t} \tag{9}$$

There are easier ways to find (9) (see Example 9.4) but the preceding illustrates the method of solving an nth order problem by solving a system of n first-order equations.

For an example of a nondiagonalizable case we have the homogeneous Problem (9.10) with

$$\mathbf{A}=\begin{bmatrix} 1 & 0 \\ 2 & 1 \end{bmatrix}, \qquad \mathbf{y}_0=\begin{bmatrix} 1 \\ 1 \end{bmatrix} \tag{10}$$

Here we have eigenvalues $\lambda_1=\lambda_2=1$,

$$\mathbf{P}=\begin{bmatrix} 0 & \frac{1}{2} \\ 1 & 0 \end{bmatrix}, \qquad \mathbf{J}=\begin{bmatrix} 1 & 1 \\ 0 & 1 \end{bmatrix}$$

the matrix exponential is

$$e^{t\mathbf{A}}=\begin{bmatrix} e^{t} & 0 \\ 2te^{t} & e^{t} \end{bmatrix} \tag{11}$$

and the solution is

$$\mathbf{y}(t)=\begin{bmatrix} e^{t} \\ e^{t}+2te^{t} \end{bmatrix} \tag{12}$$

Next we describe an alternative method for finding $\exp t\mathbf{A}$. According to the Cayley–Hamilton theorem (5.39), $\mathbf{A}$ and its eigenvalues satisfy the same polynomial equation. This suggests that if all eigenvalues satisfy

$$e^{\lambda t}=h_0(t)+h_1(t)\lambda+\cdots+h_{n-1}(t)\lambda^{n-1} \tag{13}$$

then so also should $\exp t\mathbf{A}$. When there are n distinct eigenvalues there is a unique solution for the $h_j(t)$ because the determinant of the coefficient matrix is Vandermonde's determinant. When there are repeated eigenvalues, we can differentiate (13) with respect to λ one time less than the multiplicity of a repeated λ_α and obtain a full set of equations for λ_α. Again we can show that we obtain a nonsingular coefficient matrix. Let us illustrate for $n=2$.

If **A** is of order 2 with distinct eigenvalues then the equations to be solved are

$$e^{\lambda_1 t}=h_0(t)+h_1(t)\lambda_1 \qquad (14)$$

$$e^{\lambda_2 t}=h_0(t)+h_1(t)\lambda_2$$

If we substitute the unique solutions h_0 and h_1 in (13), written for $\exp t\mathbf{A}$, we obtain

$$e^{t\mathbf{A}}=\frac{\lambda_1 e^{\lambda_2 t}-\lambda_2 e^{\lambda_1 t}}{\lambda_1-\lambda_2}\mathbf{I}+\frac{e^{\lambda_1 t}-e^{\lambda_2 t}}{\lambda_1-\lambda_2}\mathbf{A} \qquad (15)$$

which is readily shown to be a correct formula (9.4.3).

When **A** is of order 2 with repeated eigenvalue $\lambda_1=\lambda_2=\lambda$ the equations are

$$e^{\lambda t}=h_0(t)+h_1(t)\lambda \qquad (16)$$

$$te^{\lambda t}= \qquad h_1(t)$$

and the formula is

$$e^{t\mathbf{A}}=(1-\lambda t)e^{\lambda t}\mathbf{I}+te^{\lambda t}\mathbf{A} \qquad (17)$$

which is also readily verified.

We can now state the principal

DIRECT PROCEDURE FOR SOLVING PROBLEM (9.10).

(1) Find the JCF of **A**.
(2) Evaluate the formula in (9.56).
(3) Check the result. (9.57)

On the strength of this procedure and (9.11), we can solve any constant coefficient linear differential equation of order n with prescribed initial values as follows.

DIRECT PROCEDURE FOR SOLVING (9.3) SUBJECT TO (9.4).

(1) Solve Problem (9.10) for data (9.8).
(2) Extract the solution $y(t)=y_1(t)$.
(3) Check the result. (9.58)

PROBLEMS FOR SOLUTION

9.4.1 Table 9.1. Establish:

(1) Entry (2) using (9.44).
(2) Entry (3).
(3) A counterexample for (4).
(4) If $\mathbf{AB}=\mathbf{BA}$ then equality holds in (4).
(5) Entry (5) using any of the preceding properties derived from (9.44).
(6) Repeat for Entry (6).
(7) Repeat for Entry (7).
(8) Repeat for Entry (8).
(9) An expression for $(d/dt)\exp t\mathbf{A}$ using Table 6.4.
(10) Entries (2) to (8) using (1) without appeal to (9.44).

9.4.2 Particular Solution (9.54). Establish:

(1) The unique solution to (9.55) is given by $\mathbf{p}(t)$.
(2) Result (9.54) yields a particular solution for use for $n=2$ in (9.33).

9.4.3 Explicit Formulas for $\exp t\mathbf{A}$. Establish:

(1) Formulas (15) and (17) in Example 9.3.
(2) Corresponding formulas in the three possible cases for $n=3$.
(3) The number of formulas required for $n=4$.

9.4.4 Exercises. Solve the problems in 9.3.6 as 2×2 systems.

9.4.5 Strict Stability. Find a necessary and sufficient condition that all solutions of (9.34) tend to 0 as $t\to\infty$.

9.4.6 Functions of Matrices. Establish:

(1) Definition of $g(\mathbf{A})$ where $\mathbf{A}$ is square and $g(\cdot)$ denotes the value of a scalar function.
(2) Closed form expressions for $\sin\mathbf{A}$ and $\cos\mathbf{A}$ for $\mathbf{A}$ of order n.
(3) $\sin^2\mathbf{A}+\cos^2\mathbf{A}=\mathbf{I}$.

9.5 FORMAL LAPLACE TRANSFORMS

In this section we present additional methods for solving initial value problems. These methods are generally effective, they serve as prototypes for those in Section 10.6, and they are especially helpful in Section 10.7.

Broader treatments are to be found in Churchill (1972) and in works cited therein.

The basic formula for the *Laplace transformation* y to η is

$$\eta(s)=\int_0^\infty e^{-st}y(t)\,dt \tag{9.59}$$

where, in this book, s is real. We call the function η the *Laplace transform* of y if there is some $x_0\in\mathfrak{R}$ such that (9.59) exists whenever $s>x_0$; in such cases, y is the *inverse Laplace transform* of η. We write

$$\eta(s)=\mathfrak{L}\{y(t)\},\qquad y(t)=\mathfrak{L}^{-1}\{\eta(s)\}$$

and we denote Laplace transforms of functions $b,f,z,\ldots$ by the generally corresponding Greek letters $\beta,\phi,\zeta,\ldots$. Elementary properties of improper integrals imply the following criteria for

EXISTENCE OF LAPLACE TRANSFORMS. If $y:\mathfrak{R}\to\mathfrak{R}$ satisfies
(1) $y(t)=0$ for $t<0$,
(2) $y(t)$ is piecewise continuous,
(3) $y(t)=O(e^{x_0t})$ for some $x_0\in\mathfrak{R}$,
then $y(t)$ has a Laplace transform. (9.60)

Condition (1) can be omitted but we include it, and impose it whenever we use (9.59), to provide for what amount to unique inverse Laplace transforms; see Churchill (1972), and works cited therein, for precise statements on uniqueness of inverse Laplace transforms. Table 9.2 contains a basic collection of pairs of functions that correspond under (9.59).

Let us consider the entries in Table 9.2. If a is real, we can say that all entries are found by the methods of calculus for evaluating (9.59). If $a=c+id$ is nonreal in (2), we use (8.5) to write $\mathfrak{L}\{\exp at\}$ as

$$\mathfrak{L}\{e^{ct}\cos dt\}+i\mathfrak{L}\{e^{ct}\sin dt\}$$

and then obtain $\eta(s)=(s-a)^{-1}$ using (7) and (8). In words, Entry (2) holds whether a is real or nonreal, and the same can be said for (4). It is convenient to use $a\in\mathcal{C}$ in working with partial fraction expansions for $\eta(s)$.

We work with $\mathcal{I}=[0,\infty)$ in most cases and for brevity we write $y(0)$ in place of $y(0+)$. Whenever we need to work with an interval $J\subset\mathcal{I}$, we gain that effect by defining $y(t)=0$ for $t\notin J$.

Table 9.3 contains properties that we use in this book for $a,b,c\in\mathfrak{R}$. Entries are established by using changes of variable, differentiation under the integral sign, and so on, for (9.59).

TABLE 9.2 A BRIEF TABLE OF LAPLACE TRANSFORMS

	Inverse $y(t)$	Laplace Transform $\eta(s)$	Valid $s > x_0$ x_0
(1)	1	$\frac{1}{s}$	0
(2)	e^{at}	$\frac{1}{s-a},\ a\in\mathcal{C}$	$\operatorname{Re} a$
(3)	$t^m,\ m=1,2,\ldots$	$\frac{m!}{s^{m+1}}$	0
(4)	$t^m e^{at},\ m=1,2,\ldots$	$\frac{m!}{(s-a)^{m+1}},\ a\in\mathcal{C}$	$\operatorname{Re} a$
(5)	$\sin bt$	$\frac{b}{s^2+b^2}$	0
(6)	$\cos bt$	$\frac{s}{s^2+b^2}$	0
(7)	$e^{ct}\sin dt$	$\frac{d}{(s-c)^2+d^2}$	c
(8)	$e^{ct}\cos dt$	$\frac{s-c}{(s-c)^2+d^2}$	c

TABLE 9.3 PROPERTIES OF LAPLACE TRANSFORMS

	Inverse	Laplace Transform
(1)	$y(t)$	$\eta(s)$
(2)	$ay(t)+bz(t)$	$a\eta(s)+b\zeta(s)$
(3)	$y'(t)$	$s\eta(s)-y(0)$
(4)	$y^{(n)}(t)$	$s^n\eta(s)-s^{n-1}y(0)$ $-\cdots-y^{(n-1)}(0)$
(5)	$y_c(t)=0,\quad t<c$ where $c>0$ $=y(t-c),\ t\geqq c$	$e^{-cs}\eta(s)$
(6)	$\frac{1}{a}\exp\left(-\frac{bt}{a}\right)y\left(\frac{t}{a}\right), a>0$	$\eta(as+b)$
(7)	$t^m y(t),\ m=1,2,\ldots$	$(-1)^m\eta^{(m)}(s)$
(8)	$t^{-1}y(t)$	$\int_s^\infty \eta(u)\,du$
(9)	$\int_0^t y(t-u)z(u)\,du$	$\eta(s)\zeta(s)$

It is easy to state what we mean by the method of

USING FORMAL LAPLACE TRANSFORMS. Proceed as follows to solve an initial value problem, based on a linear differential equation with constant coefficients, to which an existence and uniqueness theorem applies.

(1) Transform the differential equation in $y(t)$ into an equation in $\eta(s)$ using Tables 9.2 and 9.3 to transform individual terms.
(2) Solve the resulting linear algebraic equation, and call the solution $\eta(s)$ the *formal Laplace transform* of $y(t)$.
(3) Use Tables 9.2 and 9.3 to find the *formal inverse Laplace transform* $y(t)$.
(4) Verify that $y(t)$ is a solution, and hence that it must be the unique solution, to the initial value problem. (9.61)

Use of the modifier *formal* is appropriate because we do not justify the several interchanges of order for limits in (1), and in (3) we do not show that the correspondence $\eta(s)$ to $y(t)$ is unique. On the strength of (4) we lose no effectiveness by these omissions. For brevity, we do not always write the modifier, but we always proceed in accordance with (9.61) so that it does apply. Procedure (9.61) provides an example of following (2) in (9.12).

Tables of Laplace transforms generally suffice when $y(t)$ is given and we need to find $\eta(s)$. When no entry for $y(t)$ can be found, we turn to (9.59) or to Table 9.3. For example, for the unit step function

$$\begin{aligned} f(t) &= 0, \qquad t<1 \\ &= 1, \qquad t \geqq 1 \end{aligned} \tag{9.62}$$

it is a simple matter to use (9.59) to calculate $\phi(s)$ as

$$\int_1^{\infty} e^{-st}\,dt = \frac{e^{-s}}{s} \tag{9.63}$$

This result can also be written at once from $\mathcal{L}\{1\}$ in Table 9.2(1) and the translation operation in Table 9.3(5). Figure 9.2 illustrates a simple example of the translation of a function $f(t)$ by replacing it by $f_c(t)$ according to this entry (5). Such translations are of use in working with piecewise continuous functions, for example, in Section 10.7.

We also turn to Tables 9.2 and 9.3 when $\eta(s)$ is known and we need to find $y(t)$. There are different ways to proceed in the important case when $\eta(s)$ is a rational function. We can find the partial fraction expansion (11) in Example 8.2, where all denominators are powers of linear factors, and then use Table 9.2(2) and (4). We can avoid working with nonreal numbers

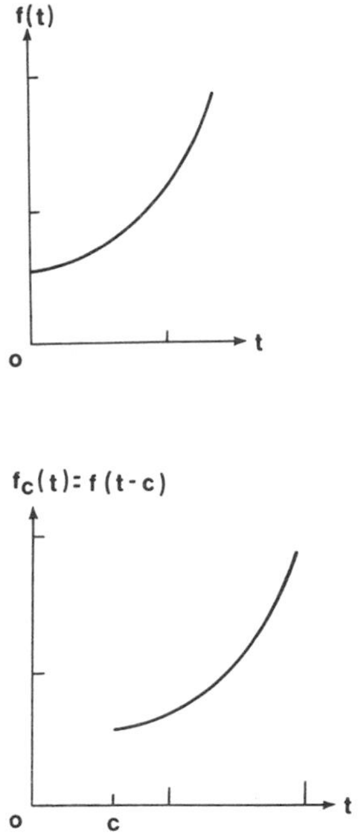

Figure 9.2. Translation of $f(t)$.

in the preceding by using Table 9.2(7) and (8) for quadratic denominators that correspond to conjugate pairs of nonreal zeros. There is also the optional

INVERSION FORMULA FOR RATIONAL FUNCTIONS. Let $\eta(s)$ be a rational function such that (1) its numerator has degree strictly less than that of its denominator, and (2) its numerator and denominator have no zeros in common. Then $\eta(s)$ is the formal Laplace transform of

$$y(t) = \sum_{j=1}^{m} \operatorname{res}\big(e^{st}\eta(s), b_j\big)$$

where $b_1, b_2, \ldots, b_m$ are the poles of $\eta(s)$. (9.64)

PROOF. Using the partial fraction expansion we have for each sum (10) from Example 8.2

$$\mathfrak{L}^{-1}\big\{A_j(s)\big\} = a_{1j}e^{b_jt} + \cdots + a_{k_jj}\frac{t^{k_j-1}}{(k_j-1)!}e^{b_jt}$$

by Table 9.2 (4) for $m=1,\ldots,k_j$. We see from (8.16) that each term on the RHS is the residue of e^{st} times the corresponding term in (10). ■
The same inversion formula can be established for a larger class of functions. But it cannot be used for (9.63) because it would yield $y(t)=1$ instead of (9.62).

There are two apparent reasons for the power of methods that use Laplace transforms. First, and as revealed by (9.60), the presence of e^{-st} in the integrand of (9.59) serves to make the integral converge for a wide variety of functions. Second, as stated in (2) of Table 9.3 the transformation is linear so that with the important property (4) we can be sure that a linear algebraic equation for $\eta(s)$ will result from (2) of (9.61). Directly in terms of elementary differential equations, we can express the reasons as follows. Using (9.61) we can effectively handle

1. Piecewise continuous RHSs as easily as continuous ones (without needing to solve successive problems as in Example 9.1).
2. Nonhomogeneous equations in the same manner as homogeneous ones (without needing first to find a general solution of the homogeneous case and second to find a particular solution of the full equation).
3. Prescribed initial values directly (without needing to solve at the end for the appropriate coefficients).

These are of great significance in practice.

EXAMPLE 9.4. Solving Low-Order Differential Equations. Let us first illustrate (9.61) by solving the instance of (9.5)

(1)
$$y'(t)=-4y(t)+f(t)$$
$$y(0)=0$$

where $f(t)$ is the unit step function (9.62) and $\mathcal{I}=[0,\infty)$. Formally transforming both sides of the differential equation, and incorporating the initial value, we obtain

$$s\eta(s)=-4\eta(s)+\frac{e^{-s}}{s}$$

Step (2) of (9.61) is trivial, which is typical, and

$$\eta(s)=e^{-s}\frac{1}{s(s+4)}$$

As is also typical, we next expand the RHS in partial fractions so we can use the basic Table 9.2. By Heaviside coverup,

$$\eta(s)=\frac{1}{4}e^{-s}\left(\frac{1}{s}-\frac{1}{s+4}\right)$$

and so, with Table 9.3(5) we find

$$\text{(2)} \qquad \begin{aligned} y(t)&=0, & t&<1 \\ &=\frac{1}{4}(1-e^{-4(t-1)}), & t&\geqq 1 \end{aligned}$$

Completion of the final step (4) is immediate so (2) is the unique solution to (1).

Nothing substantially different is involved for $n=2$ as we may see by solving (6) in Example 9.3. We first obtain

$$s^2\eta(s)-2s-\eta(s)=\frac{1}{s-2}$$

and then solve for η (s) and expand in partial fractions

$$\text{(3)} \qquad \eta(s)=\frac{1}{2}\frac{1}{s-1}+\frac{7}{6}\frac{1}{s+1}+\frac{1}{3}\frac{1}{s-2}$$

The inverse transform is (9) in Example 9.3.

In case we wish to find a particular solution for a differential equation of order 2 or higher, Laplace transforms provide an effective method. We use zero initial values to obtain the transform in simple form. For (6) in Example 9.3 we solve

$$\text{(4)} \qquad p''(t)-p(t)=e^{2t}$$

$$p(0)=0, \qquad p'(0)=0$$

and find the particular solution

$$p(t)=-\frac{1}{2}e^{t}+\frac{1}{6}e^{-t}+\frac{1}{3}e^{2t}$$

Let us now solve (9.5). In the special case (9.13),

$$\eta(s)=y_0\frac{1}{s}+\frac{1}{s}\phi(s) \tag{9.65}$$

Using the *convolution formula* (9) of Table 9.3, we see that the inverse

Laplace transform is given by (9.14). For (9.5)

$$\eta(s)=y_0\frac{1}{s-a}+\frac{1}{s-a}\phi(s) \tag{9.66}$$

so we use convolution once again to obtain the solution (9.20).

In order to solve the systems problem (9.10) we find the formal Laplace transforms of the component functions of $\mathbf{y}(t)$ and then write the results as

$$(s\mathbf{I}-\mathbf{A})\boldsymbol{\eta}(s)=\mathbf{y}_0+\boldsymbol{\phi}(s) \tag{9.67}$$

where $\boldsymbol{\eta}(s)=[\eta_1(s),\ldots,\eta_n(s)]^T$ is the vector of Laplace transforms of the components of $\mathbf{y}$, and we similarly use the corresponding notation $\boldsymbol{\phi}$ for the Laplace transform of $\mathbf{f}(t)$. If s is not an eigenvalue of $\mathbf{A}$, the coefficient matrix in (9.67) is nonsingular. Thus, for s sufficiently large

$$\boldsymbol{\eta}(s)=(s\mathbf{I}-\mathbf{A})^{-1}\mathbf{y}_0+(s\mathbf{I}-\mathbf{A})^{-1}\boldsymbol{\phi}(s) \tag{9.68}$$

The matrix $(s\mathbf{I}-\mathbf{A})^{-1}$ is called the *resolvent matrix* of $\mathbf{A}$ and

$$\mathcal{L}\{e^{t\mathbf{A}}\}=(s\mathbf{I}-\mathbf{A})^{-1} \tag{9.69}$$

by considering (9.56) for $\mathbf{f}(t)=\mathbf{0}$. Let us summarize by stating the

FORMAL LAPLACE TRANSFORM PROCEDURE FOR SOLVING PROBLEM (9.10). Applying (9.61) for component functions yields (9.67) which must be solved for $\boldsymbol{\eta}(s)$ as an explicit function of s. The remaining Steps (3) and (4) of (9.61) are carried out for the components of $\boldsymbol{\eta}(s)$ using appropriate vector-matrix methods.

In order to find a particular solution (say, to avoid using (9.54)), solve

$$(s\mathbf{I}-\mathbf{A})\mathbf{p}(s)=\boldsymbol{\phi}(s)$$

for $\mathbf{p}(s)$ and then find its formal inverse Laplace transform. (9.70)

The difficult task is expressing the RHS of (9.69) as an explicit function of s. For $n>3$ it is in general necessary to use special results from numerical analysis and a computer.

EXAMPLE 9.5. Resolvent Matrices. The result of transforming (1) in Example 9.3 is

$$s\begin{bmatrix}\eta_1(s)\\ \eta_2(s)\end{bmatrix}-\begin{bmatrix}3\\ 1\end{bmatrix}=\begin{bmatrix}0 & 1\\ 1 & 0\end{bmatrix}\begin{bmatrix}\eta_1(s)\\ \eta_2(s)\end{bmatrix}+\begin{bmatrix}\dfrac{1}{s}\\ \dfrac{1}{s^2}\end{bmatrix}$$

and collecting terms yields the form (9.67)

$$(1)\qquad \begin{bmatrix} s & -1 \\ -1 & s \end{bmatrix}\begin{bmatrix} \eta_1(s) \\ \eta_2(s) \end{bmatrix} = \begin{bmatrix} 3 \\ 1 \end{bmatrix} + \begin{bmatrix} \frac{1}{s} \\ \frac{1}{s^2} \end{bmatrix}$$

With a 2×2 system we can at once write the inverse matrix using 3.1.8. The resolvent matrix of the coefficient matrix $\mathbf{A}$ is

$$(2)\qquad (s\mathbf{I}-\mathbf{A})^{-1} = \frac{1}{(s-1)(s+1)}\begin{bmatrix} s & 1 \\ 1 & s \end{bmatrix}$$

Let us first suppose that we are interested in proceeding directly to the solution $\mathbf{y}(t)$ without pausing to find $\exp t\mathbf{A}$. Multiplying both sides of (1) by (2) we obtain

$$\eta(s) = \frac{1}{(s-1)(s+1)}\begin{bmatrix} 3s+1 \\ s+3 \end{bmatrix} + \frac{1}{s^2(s-1)(s+1)}\begin{bmatrix} s^2+1 \\ 2s \end{bmatrix}$$

Just as in working with scalar equations (and all we are actually doing is working with two such equations simultaneously) we need to expand the RHS in partial fractions. In place of working individually with the components we write

$$(3)\qquad \boldsymbol{\eta}(s) = \frac{1}{s-1}\begin{bmatrix} 3 \\ 3 \end{bmatrix} + \frac{1}{s+1}\begin{bmatrix} 0 \\ 0 \end{bmatrix} + \frac{1}{s^2}\begin{bmatrix} -1 \\ 0 \end{bmatrix} + \frac{1}{s}\begin{bmatrix} 0 \\ -2 \end{bmatrix}$$

and say we have expanded the RHS in *vector partial fractions*. We see that the inverse transform of the RHS is the solution (5) found in Example 9.3.

In order to find $\exp t\mathbf{A}$ we expand the RHS of (2) in *matrix partial fractions*

$$\boldsymbol{\eta}(s) = \frac{1}{s-1}\begin{bmatrix} \frac{1}{2} & \frac{1}{2} \\ \frac{1}{2} & \frac{1}{2} \end{bmatrix} + \frac{1}{s+1}\begin{bmatrix} \frac{1}{2} & -\frac{1}{2} \\ -\frac{1}{2} & \frac{1}{2} \end{bmatrix}$$

and then invert to find the form on the RHS of (2) in Example 9.3.

The matrix $\mathbf{A}$ in (10) from Example 9.3 has resolvent matrix

$$\frac{1}{(s-1)^2}\begin{bmatrix} s-1 & 0 \\ 2 & s-1 \end{bmatrix} = \frac{1}{(s-1)^2}\begin{bmatrix} 0 & 0 \\ 2 & 0 \end{bmatrix} + \frac{1}{s-1}\begin{bmatrix} 1 & 0 \\ 0 & 1 \end{bmatrix}$$

Inverting the matrix partial fractions on the RHS yields $\exp t\mathbf{A}$ as given in (12).

In summary, we see that the main effort in using resolvent matrices (once we find them) and Laplace transforms to solve (9.10) consists of finding the partial fraction expansions. This effort replaces the integrations that serve to find particular solutions in the direct method where matrix exponentials are written from JCFs or by other means.

PROBLEMS FOR SOLUTION

9.5.1 Laplace Transforms. Establish:

(1) Result (9.60).

(2) Entries in Table 9.2.

(3) Formal verification of entries in Table 9.3 without appeal to theorems for differentiating under the integral sign in an improper integral, and so on.

(4) A function that does not have a Laplace transform.

(5) A function that cannot be a Laplace transform.

(6) The other member of the pair by using Tables 9.2 and 9.3.

- (i) $t\cos bt$.
- (ii) $s^{-1}(1-e^{-bs})$.
- (iii) $(s^2-a^2)^{-1}$.
- (iv) $s(s^2-a^2)^{-1}$.
- (v) *Periodic function* with period $p: y(t+p)=y(t)$.
- (vi) t on $[0,1]$, 1 on $[1,2]$, $3-t$ on $[2,3]$, and 0 on $[3,\infty)$.
- (vii) $s^{-2}(s-1)^{-1}$.
- (viii) $(s^5+2s^3+1)/[s^4(s-1)^4]$.
- (ix) $(s^4-s^3+s^2-s+1)/[s^4(s-1)^4]$.

(7) (Optional) Repeat using the residue formula (9.64).

9.5.2 Exercises. Establish:

(1) Solutions for 9.3.6 using Laplace transforms for second-order equations.

(2) Repeat to find particular solutions $p(t)$ where $p(0)=p'(0)=0$ for parts (1), (2), (4), (7).

(3) Solutions for 9.4.4 using Laplace transforms for the 2×2 systems.

9.5.3 Problem (9.6). Use Laplace transforms to write a formula for the solution.

9.5.4 Two-Point Problems. Can we use Laplace transforms to solve the problem in 9.3.9?

10
LINEAR DIFFERENCE EQUATIONS

Most of the results in this chapter apply to situations where "time" advances in discrete steps or there are concerns for other phenomena, such as counts of populations, that correspond to integers. Most in the last section concern such counts in continuous time. Chapter 9 is a prerequisite because it serves as a prototype for nearly all of this chapter. Chapter 8 remains optional but its potential contribution—through (10.89) as illustrated in Example 10.10—is far greater than in Chapter 9.

10.1 DISCRETE CALCULUS

Let us start by introducing the material that replaces differential and integral calculus in most of this chapter.

A *discrete domain* is a finite or countably infinite set of real numbers

$$D=\{x_k:\cdots<x_{-2}<x_{-1}<x_0<x_1<x_2<\cdots\}$$

and *discrete calculus* is the study of functions $f:D\to\mathfrak{R}$. We further require

$$x_{k+1}-x_k=h \tag{10.1}$$

for all $k=0,\pm 1,\pm 2,\ldots$ such that both elements on the LHS belong to D; these domains are called *equally spaced*. There is the associated

Definition. The value of the first *difference* of f at $x_k\in D$ is

$$\Delta f(x_k)=f(x_{k+1})-f(x_k)$$

provided $x_{k+1}\in D$. The $(m+1)$st *difference* of f at $x_k\in D$ for $m=1,2,\ldots$ is

$$\Delta^{m+1}f(x_k)=\Delta^m f(x_{k+1})-\Delta^m f(x_k)$$

provided $x_{k+m+1}\in D$. (10.2)

Another name is *forward difference*. Table 10.1 illustrates all possible cases when D has four elements and, in general, such tables are called *difference tables*. Another set of functions used along with f is given in the

Definition. The value of the first *displacement* of f at $x_k \in D$ is

$$Ef(x_k) = f(x_{k+1})$$

provided $x_{k+1} \in D$. The $(m+1)$st *displacement* of f at $x_k \in D$ for $m = 1, 2, \ldots$ is

$$E^{m+1}f(x_k) = E\left(E^m f(x_k)\right)$$

provided $x_{k+m+1} \in D$. (10.3)

Either of $\Delta^m f$ and $E^m f$ can be defined in terms of the other.

One of the most significant properties of differences is that the higher orders eventually become zero, in the same manner as the higher derivatives, for polynomials. That is,

The mth differences of a polynomial $p(x)$ of degree m are constant and all higher order differences are zero. (10.4)

PROOF. Clearly the value for each x and each $r = m, m-1, \ldots, 2, 1$

$$\Delta x^r = (x+h)^r - x^r$$

is a polynomial of degree $r-1$, and the difference of a constant function is zero. Thus $\Delta p(x)$ is a polynomial of degree $m-1$ and $\Delta^m p(x)$ is a constant, namely $m!h^m$ given that the leading coefficient of $p(x)$ is unity. ■

In this chapter it is convenient to use the RHS of $f(x_k) = f(k)$ as simplified notation, even in preference to f_k when f is a sequence which is whenever D is infinite. In this way we gain the effect of using subsets of

TABLE 10.1 A DIFFERENCE TABLE

x_0	$f(x_0)$			
		$\Delta f(x_0)$		
x_1	$f(x_1)$		$\Delta^2 f(x_0)$	
		$\Delta f(x_1)$		$\Delta^3 f(x_0)$
x_2	$f(x_2)$		$\Delta^2 f(x_1)$	
		$\Delta f(x_2)$		
x_3	$f(x_3)$			

consecutive integers from

$$\{\ldots,-2,-1,0,1,2,\ldots\} \tag{10.5}$$

as domains. Most frequently we use

$$\{0,1,2,\ldots\} \tag{10.6}$$

although at times we are interested in finite domains

$$\{0,1,2,\ldots,n\} \tag{10.7}$$

These latter are often obtained by inserting $n-1$ equally spaced points in an interval $[a,b]$. Then $x_k = a + kh$ where $h = (b-a)n^{-1}$ and we see that using (10.5) is the same as using

$$k = \frac{x_k - a}{h} \tag{10.8}$$

which measures the distance from $a = x_0$ to x_k in units of the common spacing h. From this we also see that we lose no generality by using (10.6) in place of any infinite domain of the type

$$\{a, a+h, a+2h, \ldots\} \tag{10.9}$$

or by using (10.5) when D has neither a largest nor a smallest element. All possibilities are therefore covered by the

Definition. The appropriate one of (10.5), (10.6), (10.7) serves as the *standard domain* $\mathfrak{D}$ that replaces any given equally spaced discrete domain D. (10.10)

Formulas become relatively simple when domains $\mathfrak{D}$ are used in place of domains such as (10.9).

If r is a positive integer and x is any real number, the value of the rth *factorial power* of x is the product of r factors

$$x^{(r)} = x(x-1)\cdots(x-r+1) \tag{10.11}$$

This generalizes $r! = r^{(r)}$ which suggests the definition $x^{(0)} = 1$. There is also a rationale for allowing r to be negative. If p and q are integers and $p \geqq q \geqq 1$ then $p = q + (p-q)$ and using (10.11) we can write

$$x^{(p)} = \left[x(x-1)\cdots(x-q+1)\right]\left[(x-q)\cdots(x-q-p+q+1)\right]$$

where the first term within $[\cdot]$ has q factors and the second has $p-q$. This states

$$x^{(p)}=x^{(q)}(x-q)^{(p-q)}$$

and if we formally set $p=0$ we obtain

$$x^{(0)}=x^{(q)}(x-q)^{(-q)}$$

Let us now return to using r for the power. Writing r for q in the last display we obtain

$$\frac{1}{x^{(r)}}=(x-r)^{(-r)} \tag{10.12}$$

If we change variables one more time, via $x-r=y$ we obtain

$$\frac{1}{(y+r)^{(r)}}=y^{(-r)} \tag{10.13}$$

Expressions (10.12) and (10.13) are equivalent, and using them successively for $r=2$

$$\frac{1}{x^{(2)}}=\frac{1}{x(x-1)}=(x-2)^{(-2)}, \qquad x^{(-2)}=\frac{1}{(x+2)^{(2)}}=\frac{1}{(x+2)(x+1)}$$

we see that unless $r=0$

$$x^{(r)}x^{(-r)}\neq 1, \quad \text{that is} \quad x^{(-r)}\neq\frac{1}{x^{(r)}}$$

This motivates the

Definition. For $x\in\mathfrak{R}$ the *factorial power function* has values

$$\begin{aligned} x^{(r)}&=x(x-1)\cdots(x-r+1), && r=1,2,\ldots \\ &=1, && r=0 \\ &=\frac{1}{(x-r)^{(-r)}} && r=-1,-2,\ldots \end{aligned} \tag{10.14}$$

Factorial powers appear in the following generalization of the familiar expression for the number of combinations of x objects taken r at a time.

Definition. For $x \in \mathcal{R}$ the *binomial coefficients* are

$$C(x,r)=\frac{x^{(r)}}{r!}, \qquad r=1,2,\ldots$$

$$=1, \qquad r=0$$

$$=0, \qquad r=-1,-2,\ldots \tag{10.15}$$

Notice that x can be any real number.

One of the most important properties of factorial powers is readily verified by calculation, namely

$$\Delta x^{(r)} = r x^{(r-1)} \tag{10.16}$$

This exhibits $x^{(r)}$ as a discrete counterpart of the ordinary power x^n which satisfies

$$\frac{d}{dx}x^n = nx^{n-1}$$

If we expand the expression for $\Delta^n f(0)$ we obtain *Newton's interpolation formula*, briefly *Newton's formula*

$$f(n)=C(n,0)f(0)+C(n,1)\Delta f(0)+\cdots+C(n,n)\Delta^n f(0) \tag{10.17}$$

which can also be written

$$f(n)=f(0)+\Delta f(0)n+\frac{\Delta^2 f(0)}{2!}n^{(2)}+\cdots+\frac{\Delta^n f(0)}{n!}n^{(n)}$$

as a discrete counterpart of Taylor's formula (1.13).

Table 10.2 illustrates the use of a difference table for writing the formula (10.17). Starting with $C(n,0)$, all coefficients are entered immediately above their associated factors. Then the resulting formula with five terms

$$f(n)=C(n,0)f(0)+C(n,1)\Delta f(0)+C(n,2)\Delta^2 f(0)$$

$$+C(n,3)\Delta^3 f(0)+C(n,4)\Delta^4 f(0) \tag{10.18}$$

gives the correct table entry for $n=0,1,2,3,4$. Formula (10.18) also determines the unique sequence that coincides with the values in Table 10.2 and further satisfies

$$\Delta^m f(k) = \text{constant} \tag{10.19}$$

TABLE 10.2 TABLEAU FOR NEWTON'S FORMULA (10.17)

	$C(n,0)$				
0	$f(0)$				
		$C(n,1)$			
		$\Delta f(0)$	$C(n,2)$		
1	$f(1)$		$\Delta^2 f(0)$		
				$C(n,3)$	
		$\Delta f(1)$		$\Delta^3 f(0)$	$C(n,4)$
2	$f(2)$		$\Delta^2 f(1)$		$\Delta^4 f(0)$
		$\Delta f(2)$		$\Delta^3 f(1)$	
3	$f(3)$		$\Delta^2 f(2)$		
		$\Delta f(3)$			
4	$f(4)$				

for $m=4$ and $k=0,1,2,\ldots$. More generally, if m denotes the lowest order of differences that are constant in any difference table, then the RHS of (10.17) is the polynomial of degree m and it determines a unique sequence in the corresponding sense.

Let us now consider properties of a *finite sum*

$$\sum_{u=0}^{n} f(u)=f(0)+f(1)+\cdots+f(n) \tag{10.20}$$

We adopt the convention that the lower limit never exceeds the upper which is equivalent to using summation signs as mere notation for finite sums. This is a departure from procedures for definite integrals where there is the rule

$$\int_0^b f(t)\,dt=-\int_b^0 f(t)\,dt$$

Just as integration is more difficult to perform than differentiation, it is harder to evaluate sums than it is to find differences. However, if we can identify f as a difference, say

$$f(u)=\Delta y(u)$$

then the RH members telescope in summing

$$\Delta y(0)=y(1)-y(0)$$
$$\Delta y(1)=y(2)-y(1)$$
$$\vdots$$
$$\Delta y(n)=y(n+1)-y(n)$$

and we obtain the formula

$$\sum_{u=0}^{n} \Delta y(u)=y(n+1)-y(0)$$

It is helpful to write the preceeding as

$$\sum_{u=0}^{n} \Delta y(u)=y(u)\Big|_0^{n+1} \tag{10.21}$$

for two reasons. First, it recalls a familiar integration formula. Second, it signals that $n+1$ and not n appears on the RHS. The deceptively simple result (10.21) provides for most of what we can use to evaluate finite sums in closed form.

Table 10.3 lists differences in forms arranged for use in (10.21). With entry (4) we can write the familiar formula for geometric series

$$\begin{aligned}\sum_{u=0}^{n} a^u &= \frac{a^{n+1}-1}{a-1}, \qquad a\neq 1\\ &= n+1, \qquad a=1\end{aligned} \tag{10.22}$$

Entry (2) in Table 10.3 can be rewritten

$$y_1(u)\Delta y_2(u)=\Delta[y_1(u)y_2(u)]-y_2(u+1)\Delta y_1(u)$$

From (10.21) we then obtain the formula for *summation by parts*

$$\sum_{u=0}^{n} y_1(u)\Delta y_2(u)=y_1(u)y_2(u)\Big|_0^{n+1}-\sum_{u=0}^{n} y_2(u+1)\Delta y_1(u) \tag{10.23}$$

Now we can evaluate

$$S=\sum_{u=0}^{n} ua^u, \qquad n=0,1,2,\ldots \tag{10.24}$$

TABLE 10.3 PAIRS FOR EVALUATING $\sum_{u=0}^{n} \Delta g(u)$

	$\Delta g(u)=g(u+1)-g(u)$	$g(u)$	
(1)	$c_1\Delta y_1(u)+c_2\Delta y_2(u)$	$c_1 y_1(u)+c_2 y_2(u)$	
(2)	$y_1(u)\Delta y_2(u)+y_2(u+1)\Delta y_1(u)$	$y_1(u)y_2(u)$	
(3)	$\dfrac{y_2(u)\Delta y_1(u)-y_1(u)\Delta y_2(u)}{y_2(u)y_2(u+1)}$	$\dfrac{y_1(u)}{y_2(u)}$,	$y_2(\cdot)\neq 0$
(4)	a^u	$\dfrac{a^u}{a-1}$,	$a\neq 1$
(5)	$u^{(r)}$	$\dfrac{u^{(r+1)}}{r+1}$,	$r\neq -1$
(6)	$\dfrac{-r}{(u+r+1)^{(r+1)}}$	$\dfrac{1}{(u+r)^{(r)}}$	$r\geqq 1$
(7)	$C(u,r)$	$C(u,r+1)$	
(8)	$y(u+1)$	$\sum_{v=0}^{u} y(v)$	
(9)	$-y(u)$	$\sum_{v=u}^{\infty} y(v)<\infty$	

as follows. For $a\neq 1$,

$$S=\left.\frac{ua^u}{a-1}\right|_0^{n+1}-\sum_{u=0}^{n}\frac{a^{u+1}}{a-1}$$

and then with (10.22) we obtain

$$\begin{aligned}\sum_{u=0}^{n} ua^u &= \frac{a}{(a-1)^2}\left[na^{n+1}-(n+1)a^n+1\right], \qquad a\neq 1\\ &= \frac{n(n+1)}{2}, \qquad a=1 \qquad (10.25)\end{aligned}$$

With the following in (10.23)

$$y_1(u)=g_1(u), \qquad \Delta y_2(u)=g_2(u), \qquad y_2(u)=\sum_{v=0}^{u-1} g_2(v)$$

we obtain the result called *Abel's identity*

$$\sum_{u=0}^{n} g_1(u)\,g_2(u)=g_1(u)\sum_{v=0}^{u-1} g_2(v)\Big|_{u=0}^{u=n+1}-\sum_{u=0}^{n}\sum_{v=0}^{u} g_2(v)\Delta g_1(u) \qquad (10.26)$$

The preceding can be written

$$\sum_{u=0}^{n} uy(u) = n\sum_{u=0}^{n} y(u) - \sum_{u=0}^{n-1}\sum_{v=0}^{u} y(v) \tag{10.27}$$

for working with discrete probability distributions where the LHS equals the mean of the distribution.

Entry (5) in Table 10.3 enables us to sum any LC of factorial powers. The complete result is

$$\begin{aligned} \sum_{u=0}^{n} u^{(r)} &= \left.\frac{u^{(r+1)}}{r+1}\right|_{u=0}^{u=n+1}, \qquad && r \neq -1 \\ &= 1 + \frac{1}{2} + \cdots + \frac{1}{n+1}, \qquad && r = -1 \end{aligned} \tag{10.28}$$

Any polynomial in u can be summed using the first line just above and the expressions

$$u^n = \sum_{m=0}^{n} \mathfrak{S}_n^{(m)} u^{(m)} \tag{10.29}$$

where the coefficients are *Stirling numbers of the second kind.* (Those of the *first kind* are the corresponding coefficients in expressions for factorial powers.)

Reciprocals of polynomials can also be summed with (10.28) and (10.13). Also using the elementary result

If $f(k) = \Delta y(k)$ for $k = 0, 1, 2, \ldots$ where $y(\cdot)$ is a convergent sequence with limit b, then $\sum_{u=0}^{\infty} f(u)$ is a convergent series and

$$y(k) = b - \sum_{u=k}^{\infty} f(u)$$

Conversely, if the above equation holds for all k then so also does $\Delta y(k) = f(k)$. (10.30)

constitutes a way of obtaining sums of infinite series (10.1.5).

Entry (7) in Table 10.3 implies

$$\sum_{u=k}^{n} C(u,k) = C(n+1, k+1), \qquad k = 0, 1, 2, \ldots, n \tag{10.31}$$

This is one of the many summations involving binomial coefficients that can be evaluated in closed form. See Feller (1968) for a number of derivations based on the following result from elementary calculus.

NEWTON'S BINOMIAL SERIES. For any $x \in \Re$ and $t \in (-1, 1)$,

$$(1+t)^x = \sum_{u=0}^{\infty} C(x,u)t^u$$

where the binomial coefficients are defined in (10.15). (10.32)

A large collection of sums involving binomial coefficients is given in Gould (1972). Let us next establish *Newton's summation*

$$\sum_{u=0}^{n} f(u) = \sum_{u=1}^{n+1} C(n+1,u)\Delta^{u-1}f(0) \qquad (10.33)$$

(Notice that both LHS limits have been increased by unity on the RHS.)

PROOF. For a computational proof use (10.17) to expand each summand on the LHS and then use (10.31) to sum the coefficients of $f(0), \Delta f(0), \ldots, \Delta^n f(0)$. ■

Newton's summation (10.33) is much the same kind of expression as Newton's interpolation formula (10.17). Most importantly, if (10.19) holds for some m then

$$\Delta^n f(0) = 0$$

for $n = m+1, m+2, \ldots$ and (10.33) has only $m+1$ terms on the RHS for

TABLE 10.4 TABLEAU FOR NEWTON'S SUMMATION (10.33)

	$C(n+1,1)$				
0	$f(0)$				
		$C(n+1,2)$			
		$\Delta f(0)$	$C(n+1,3)$		
1	$f(1)$		$\Delta^2 f(0)$	$C(n+1,4)$	
		$\Delta f(1)$		$\Delta^3 f(0)$	$C(n+1,5)$
2	$f(2)$		$\Delta^2 f(1)$		$\Delta^4 f(0)$
		$\Delta f(2)$		$\Delta^3 f(1)$	
3	$f(3)$		$\Delta^2 f(2)$		
		$\Delta f(3)$			
4	$f(4)$				

each such n. Consequently, we can sum any polynomial by using a total of $m+1$ lines in Table 10.4 where m is the degree of the polynomial. Actually $m=4$ is illustrated in Table 10.4 where it is to be noticed that the change from Table 10.2 is that both arguments of all factors $C(\cdot,\cdot)$ are increased by unity. If Newton's summation (10.33) is based on Table 10.4 and used for all $n>4$ then we are using the sequence discussed in connection with (10.19). Also see 10.1.4.

EXAMPLE 10.1. Optimization Theory for Sequences. Let us use the notation of the present chapter where $f(k)$ rather that f_k denotes the general term of the infinite sequence of real numbers $\{f_k\}$.

Clearly f can properly be called *monotone nondecreasing* if

$$f(k) \leqq f(k+1)$$

for all k. We are free to disregard any finite initial collection of values of an infinite sequence but let us agree to make adjustments through specification of the standard domain (10.10). For example, if we wish to disregard any finite set of values we define a new appropriate sequence that does commence with $k=0$. Consequently we can say "for all k" in

(1) The sequence $f(k)$ is monotone nondecreasing iff $\Delta f(k) \geqq 0$ for all k.

We similarly find that *monotone nonincreasing* should mean

$$f(k+1) \leqq f(k)$$

for all k and evidently

(2) The sequence is monotone nonincreasing iff $\Delta f(k) \leqq 0$ for all k.

The concept of a convex sequence is a natural extension of a (mid)convex function, namely

(3) The sequence $f(k)$ is a *convex sequence* if

$$f(k+1) \leqq \frac{f(k)+f(k+2)}{2}$$

holds for all k.

Convenient equivalent properties are readily established. Each of the following is necessary and sufficient for $f(k)$ to be a convex sequence

(4) $$\Delta^2 f(k) \geqq 0 \quad \text{for all } k$$

(5) $\Delta f(k)$ is monotone nondecreasing.

Just as for real functions generally, we define $f(k)$ to be *concave* if $-f(k)$

is convex. We define *strict* properties for monotonicity, convexity, and concavity by omitting equality conditions from the corresponding $\leqq$ or $\geqq$ specifications.

Extrema are also defined in straightforward manner. The sequence $f(k)$ has a *strict local minimum* at k_0 if

$$f(k_0-1)>f(k_0) \quad \text{and} \quad f(k_0)<f(k_0+1)$$

An equivalent pair of conditions can be written

$$\Delta f(k_0-1)<0 \quad \text{and} \quad \Delta f(k_0)>0 \tag{6}$$

These mean that $\Delta f(\cdot)$ changes sign from negative to positive at $k=k_0$. At $k=0$ we agree that a strict local minimum means

$$f(0)<f(1)$$

and we similarly use $f(k_\alpha-1)>f(k_\alpha)$ in case the standard domain $\mathfrak{D}$ has a largest element k_α. We summarize for the domain (10.6) as follows. The sequence $f(k)$ has a strict local minimum

(7) At $k_0=0$ iff $\Delta f(0)>0$.

(8) At $k_0>0$ iff $\Delta f(\cdot)$ changes sign from $-$ to $+$ at k_0.

Ordinary—that is, nonstrict—minima are important in practice. Usually we are interested in determining the smallest value of k at which the minimum occurs, but perhaps more significance is attached to the largest such value in the set of successive values $f(k)$. Finally, it is easy to change (7) and (8) to apply to local minima and we can also define global minima in straightforward manner.

If $f(k)$ is a convex sequence then it is also easy to extend the results of Section 7.5. For example, we have the

(9) **THEOREM.** A convex sequence $f(k)$ has a global minimum on (10.6) iff the following condition holds for at least one $k_0 \geqq 0$.

$$\Delta f(k_0) \geqq 0$$

If the condition holds, then either $k_0=0$ or there is a unique smallest $k_0>0$ at which it holds.

PROOF. The condition is necessary because $\Delta f(k)<0$ for all k means that $f(k)$ is strictly decreasing for all k and so it cannot have a minimum. The sufficiency is equally obvious and the final statement follows immediately from the definition of convexity. ■

We can use Theorem (9) to solve the classic newsboy problem of inventory theory. This is the problem of the boy who buys his papers at 2 cents, sells them at 3 cents, and loses 2 cents for each unsold paper because he cannot return any. He supposes that a known probability distribution applies to the number of *demands* by customers and he seeks the number n of papers that he should *stock*—that is, buy—each day to minimize his expected loss for the day.

Although this problem formulation fits a number of different resource allocation problems where there is concern for a single time period, let us continue to deal with matters in terms of the newsboy. However, let us use symbols for unit losses,

(10) A = penalty per unit stocked in excess of number demanded during the day

B = penalty per unit demanded in excess of number stocked for the day

and for simplicity let us suppose A and B are positive. Thus $A=2$ cents and $B=1$ cent for the newsboy but, in general, A and B measure units of utility that need not necessarily be expressed in monetary units. The number A is the *surplus penalty* and B is the *shortage penalty*.

We suppose we have a *discrete probability distribution*

$$\sum_{u=0}^{s} p(u), \qquad s=0,1,2\ldots$$

with finite *mean* m and with application to customers as follows.

(11) $p(k)$ = probability of exactly k units being demanded during one day: $k=0,1,2,\ldots$

By definition, this requires $p(k)\in[0,1]$ for $k=0,1,2,\ldots$, and

$$\sum_{u=0}^{\infty} p(u)=1, \qquad \sum_{u=0}^{\infty} up(u)=m$$

A common assumption is that the *Poisson distribution* with mean m of Section 10.7 applies.

In order to proceed we need to find the expected value of the loss if s units are stocked for the day. First, we have the expected number of units overstocked

$$a(s)=\sum_{u=0}^{s}(s-u)p(u), \qquad s=0,1,2,\ldots \tag{12}$$

given by the *surplus function* a. Second, the *shortage function* b has values

$$b(s)=\sum_{u=s+1}^{\infty}(u-s)p(u), \qquad s=0,1,2,\ldots \tag{13}$$

equaling the expected number of units understocked if s are stocked for the day. The newsboy *loss function* is

$$L(s)=Aa(s)+Bb(s), \qquad s=0,1,2,\ldots \tag{14}$$

and the *newsboy problem* is

$$\begin{aligned} &\text{minimize } L(s) \\ &\quad \text{s.t. } s\in\{0,1,2,\ldots\} \end{aligned} \tag{15}$$

As noted below (8) we are interested in finding the smallest s at which the minimum occurs in cases where the minimum occurs in more than one place.

The loss function (14) is readily shown to be convex using (4) so that Theorem (9) applies. Then the condition on k_0 is

$$A\sum_{u=0}^{k_0} p(u)-B\sum_{u=k_0+1}^{\infty} p(u)\geqq 0$$

and rewriting the second summation as unity minus the first yields

$$(A+B)\sum_{u=0}^{k_0} p(u)\geqq B \tag{16}$$

Since the summation equals $p(0)\geqq 0$ for $k_0=0$ and converges to unity and since $B/(A+B)$ is a proper fraction, there must be a smallest k_0 for which (16) holds. From Theorem (9) in this way we obtain the

(17) Solution to the Newsboy Problem. There exists a unique minimum and it first occurs at

$$n=\min\left\{ s:\ \sum_{u=0}^{s} p(u)\geqq\frac{B}{A+B}\right\}$$

In words, the *optimal number* of units to stock is the smallest integer n for which the associated *protection level*

$$\sum_{u=0}^{n} p(u)$$

is not less than $B/(A+B)$. Notice that the solution depends on the ratio B/A and not on the individual magnitudes of the unit penalties A and B.

Let us also observe that if

$$p(0) \geqq \frac{B}{A+B}$$

that is, if the probability of zero demand is greater than or equal to $B/(A+B)$, then the newsboy should not try to sell any papers at all at the place where this value $p(0)$ applies. Finally, we note that this last statement can be made independent of the particular form of the distribution (11) that may apply to the number of customers: it could be a Poisson distribution or any other type with finite mean.

PROBLEMS FOR SOLUTION

10.1.1 Differences. Establish:

(1) $\Delta f(k) = f(k)$ for $f(k) = ?$
(2) Property (10.16).
(3) If n is a positive integer, $C(n,r) = 0$ if either $r > n$ or $r < 0$.
(4) $C(x,r) = C(x, x-r)$.
(5) $C(-x,r) = (-1)^r C(x+r-1, r)$.
(6) $C(x+1,r) - C(x,r) = C(x,r-1)$.
(7) Newton's formula (10.17).
(8) Lagrange's formulas for $\Delta^m f(k)$.

   (i) $\sum_{u=0}^{m} (-1)^u C(m,u) f(k+m-u)$.

   (ii) $\sum_{u=0}^{m} (-1)^{m-u} C(m,u) f(k+u)$.

(9) Sum of all entries $\Delta^k f(\cdot)$ in the generalization of Table 10.1 for $x_0, x_1, \ldots, x_m$ where $k = 1, 2, \ldots, m-1$.
(10) Entries in Table 10.3.

10.1.2 Numerical Exercises. Establish:

(1) Newton's formula (10.17) for the following and check results using synthetic division.

   (i) $(k, f(k)) : (0,11), (1,3), (2,1), (3,11), (4,39), (5,91)$.
   (ii) $(k, \exp(0.1)k)$ for $k = 0, 1, 2, 3$.
   (iii) $(k, f(k)) : (0,0), (1,-2), (2,0), (3,12), (4,40)$.
   (iv) $(k, f(k)) : (0,1), (1,2), (2,5), (3,16), (4,65), (5,205), (6,517)$.
   (v) $(k, f(k)) : (4,15), (5,41), (6,88), (7,162)$.

(2) Original data for preceding cases starting at the constant difference and "working back" through the table.
(3) Extrapolated data for $f(-1)$ and $f(m+1)$ for the preceding cases where $f(m)$ is the final data element in the original table.

10.1.3 Finite Sums. Establish:

(1) $\sum_{u=0}^{n} a=(n+1)a.$

(2) $\sum_{u=0}^{n} f(u)=\sum_{v=0}^{n} f(v).$

(3) $\sum_{u=0}^{n} [af(u)+bg(u)]=a\sum_{u=0}^{n} f(u)+b\sum_{u=0}^{n} g(u)$

(4) If $f:\mathfrak{D}_1\times\mathfrak{D}_2\to\mathfrak{R}$ then

$$\sum_{t=0}^{m}\sum_{u=0}^{n} f(t,u)=\sum_{u=0}^{n}\sum_{t=0}^{m} f(t,u).$$

(5) Formula (10.25) using Table 10.3 (4).
(6) Standard expression for second line in (10.28)?
(7) Value of (10.20) for $f(u)$ as follows:
   (i) u^2a^u.
   (ii) $\exp(au+b)$.
   (iii) $\sinh(au+b)$.
   (iv) $\cosh(au+b)$.
   (v) $\sin(au+b)$, $\sin(a/2)\neq 0$.
   (vi) $\cos(au+b)$, $\sin(a/2)\neq 0$.
   (vii) $a^u\sin(au+b)$.
   (viii) $a^u\cos(au+b)$.
   (ix) $\sin^2(au+b)$, $\sin(a/2)\neq 0$.
   (x) $\cos^2(au+b)$, $\sin(a/2)\neq 0$.
(8) Values of first four sets of coefficients in (10.29).
(9) Values of (10.20) for polynomials as follows by using Stirling numbers and also by Newton's summation (10.33).
   (i) u.
   (ii) u^2.
   (iii) u^3.
   (iv) u^4.
(10) Values of sums from $u=p$ to $u=q$ where p and q are positive integers, $p<q$, for preceding (i)–(iv).
(11) Formula (10.31).
(12) $\sum_{u=0}^{n} C(u,r-1)=C(n+1,r).$

10.1.4 More on Sums of Polynomials. Establish for

$$S(n)=\sum_{u=0}^{n} f(u), \qquad n=0,1,2,\ldots$$

where $f(u)$ is a polynomial in u whose representation based on (10.29) is

$$f(u)=a_0u^{(0)}+a_1u^{(1)}+\cdots+a_{m-1}u^{(m-1)}+u^{(m)}$$

(1) $S(n)=a_0(n+1)^{(1)}+\dfrac{a_1(n+1)^{(2)}}{2}+\cdots+\dfrac{(n+1)^{(n+1)}}{n+1}.$

(2) $\sum_{u=-n+1}^{-1} f(u)=?$ for $n=1,2,\ldots.$

(3) $\sum_{u=p+1}^{q} f(u)=?$ for $p=\ldots,-2,-1,0,1,2,\ldots$ and $p+1\leqq q$.

(4) The result in (2) is given by replacing n by $-n$ in (1) and prefixing a minus sign to the resulting expression.

10.1.5 Sums of Reciprocals of Polynomials. Establish:

(1) Result (10.30).

(2) $\sum_{u=0}^{\infty}\dfrac{1}{(u+1)(u+2)}=1$

(3) $\sum_{u=0}^{\infty}\dfrac{1}{(u+m)^{(n)}}=\dfrac{1}{(n-1)(m-1)^{(n-1)}}$ where $n=2,3,\ldots$ and $m=n, n+1,\ldots.$

(4) If $m>n-1$ then (3) remains valid even if m is not an integer. Furthermore, in this case (3) becomes a "most general" formulation based on factors linear in u.

10.1.6 Example 10.1. Establish:

(1) Conditions (4) and (5) for convexity.

(2) Counterpart of (9) for maxima.

(3) Initial values, monotonicity, and convexity properties of a, b, and L.

(4) $a(s)-b(s)=s-m$.

(5) Optimum number n for Poisson distribution with $m=10$ when (A,B) equals (i),(1,2), (ii) (2,2), (iii) (2,3), (iv) (4,3), and (v) (20,3).

(6) Properties for nine cases of (A,B) where "penalties" are unrestricted in sign.

10.2 DIFFERENCE EQUATIONS

Let us define what we mean by a difference equation and introduce problems that we solve in this chapter.

What we call difference equations are often called recurrences. Other authors give the name difference equations to what we call functional equations. In its simplest terms, when we work with difference equations we work on domains of consecutive integers, and have nothing to do with intermediate real numbers. When we also deal with the intermediate intervals we say we work with functional equations. (10.2.7).

The general notion of a difference equation is that it may be considered to connect values and differences (10.2) on some equally spaced domain. We say "may be considered to connect" because we wish to identify equivalent pairs of equations, one of which involves only functional values, and the other of which involves differences as well as values. For example, the two equations in

$$y(k+1)=ay(k)+f(k)$$

$$\Delta y(k)+(1-a)y(k)=f(k) \tag{10.34}$$

are considered below to be the same difference equation because each is a simple rewriting of the other based on (10.2). But the equation

$$\Delta y(k)=-y(k)+1$$

is equivalent to $y\,(k+1)=1$ and we prefer the latter form because it immediately reveals the trivial nature of the relationship. The equation

$$y(k+1)=ky(k) \tag{10.35}$$

can be regarded as a difference equation holding only for $k=1,2,3,\ldots$ or it can be considered to be a more general functional equation holding for $k\in(0,\infty)$. The solutions are quite different: In the first case we have the simple factorial

$$y(k)=(k-1)!, \qquad k=1,2,3,\ldots$$

while in the second case there is the much more complicated *gamma function* (10.2.7).

Let us now define a *difference equation* as an equation that can be written as

$$\phi(k,y(k),y(k+1),\ldots,y(k+n))=0 \tag{10.36}$$

where $k\in\mathfrak{D}$ of (10.10) and n is a positive integer. A *solution* of a difference equation is a function

$$y:\mathfrak{D}\rightarrow\mathfrak{R}$$

that satisfies (10.36) for all $k \in \mathfrak{D}$ for which the LHS is defined. For brevity we generally omit qualifications of this latter type, particularly for finite domains where obvious adjustments must be made at or near the largest integer; this amounts to working on domain (10.6) and leaving adjustments for (10.7), and often for (10.5), to the reader (10.2.1). In order that there be no difficulties at points of discontinuity of piecewise continuous functions we agree to define

$$y(k) = \lim_{x \to k+} y(x)$$

whenever we work with difference equations for functions on $\mathfrak{R}$ to $\mathfrak{R}$.

An equation is a *linear difference equation* if it can be written

$$y(k+n) + \alpha_{n-1}(k)y(k+n-1) + \cdots + \alpha_1(k)y(k+1) + \alpha_0(k)y(k) = b(k) \tag{10.37}$$

where $\alpha_{n-1}, \ldots, \alpha_1, \alpha_0, b$ are functions on $\mathfrak{D}$ to $\mathfrak{R}$. The *order* of (10.37) is

$$\max\{n-j : \alpha_j(k) \neq 0 \quad \text{for some } k \in \mathfrak{D}\}$$

In words, the order is the maximum difference between arguments of $y(\cdot)$ that actually appear in (10.37). If, for example, $\alpha_0(k) = 0$ for all k then the equation is at most of order $n-1$; notice that the argument of the known $b(\cdot)$ does not enter into definition.

From the preceding definitions we see that proceeding from Section 9.1 to the present we have the correspondences

$$t \in \mathfrak{I} \text{ to } k \in \mathfrak{D}$$

$$y(t) \text{ to } y(k)$$

$$y'(t) \text{ to } y(k+1)$$

$$\vdots$$

$$y^{(n)}(t) \text{ to } y(k+n)$$

Notice that displacements (9.3), and not differences (9.2), are the counterparts to derivatives.

Again as for the linear differential equation (9.3), (10.37) is referred to as having *variable coefficients* if not all of the α's are constant for all $k \in \mathfrak{D}$. Otherwise, it has *constant coefficients*. If $b(k) = 0$ for all $k \in \mathfrak{D}$ then (10.37) is called *homogeneous*, and otherwise it is *nonhomogeneous*.

Changing the name from Chapter 9 slightly, we are interested in *initial value difference problems*. Here we seek solutions to (10.37) subject to

$$\begin{aligned} y(0)&=y_0\\ y(1)&=y_1\\ &\vdots\\ y(n-1)&=y_{n-1} \end{aligned} \tag{10.38}$$

where the n real numbers $y_0,y_1,\ldots,y_{n-1}$ are prescribed *initial values* just as in (9.4). If we think of $k\in\mathfrak{D}$ as "time," the problem is to find $y(k)$ for $k=n,n+1,n+2,\ldots$ given the prescribed values for $k=0,1,\ldots,n-1$.

The only variable coefficient equation we consider in detail is the counterpart of (9.5), namely

$$\begin{aligned} y(k+1)&=a(k)y(k)+f(k)\\ y(0)&=y_0 \end{aligned} \tag{10.39}$$

The counterpart of (9.6) is

$$\begin{aligned} y(k+2)+\alpha_1 y(k+1)+\alpha_0 y(k)&=b(k)\\ y(0)=y_0,\qquad y(1)&=y_1 \end{aligned} \tag{10.40}$$

Using the notation

$$\begin{aligned} y(k)&=y_1(k)\\ y(k+1)&=y_2(k)\\ &\vdots\\ y(k+n-1)&=y_n(k) \end{aligned} \tag{10.41}$$

we obtain from (10.37) and (10.38) the special case of the *linear difference systems problem*

$$\begin{aligned} \mathbf{y}(k+1)&=\mathbf{A}\mathbf{y}(k)+\mathbf{f}(k)\\ \mathbf{y}(0)&=\mathbf{y}_0 \end{aligned} \tag{10.42}$$

where $\mathbf{A}$ is a constant matrix of order n, $f:\mathfrak{D}\rightarrow\mathfrak{R}^n$ is given, and $\mathbf{y}_0$ is the *initial value vector.* This is the counterpart to (9.10). It is based on *data*

$$\mathbf{A}=\begin{bmatrix} 0 & 1 & 0 & \cdots & 0 \\ 0 & 0 & 1 & \cdots & 0 \\ \vdots & \vdots & \vdots & & \vdots \\ 0 & 0 & 0 & \cdots & 1 \\ -\alpha_0 & -\alpha_1 & -\alpha_2 & \cdots & -\alpha_{n-1} \end{bmatrix}, \qquad \mathbf{y}(k)=\begin{bmatrix} y_1(k) \\ y_2(k) \\ \vdots \\ y_{n-1}(k) \\ y_n(k) \end{bmatrix}$$

$$\mathbf{y}_0=\begin{bmatrix} y_0 \\ y_1 \\ \vdots \\ y_{n-1} \end{bmatrix}, \qquad \mathbf{f}(k)=\begin{bmatrix} 0 \\ 0 \\ \vdots \\ b(k) \end{bmatrix} \tag{10.43}$$

Notice that there are only minor changes in notation from (9.8), we continue to use boldface letters for values of transformations such as y and f, and, most importantly, $\mathbf{A}$ is unchanged. We naturally call (9.9) the *characteristic equation* of (10.37). Given any system (10.42) it is called *homogeneous* if $\mathbf{f}(k)=\mathbf{0}$ for all $k\in\mathfrak{D}$ and otherwise it is *nonhomogeneous*, and

The problem of solving the constant coefficient case of the nth-order equation (10.37) subject to (10.38) is equivalent to solving the linear system (10.42) with data (10.43). (10.44)

PROOF. We repeat the procedure used to prove (9.11). Starting at (10.37) we reach (10.42), and the process is reversible. ■

To obtain a solution procedure we simply change the title in (9.12) to "Solving Initial Value *Difference* Problems" and repeat the statement word for word. But there is something fundamentally different that is not present in the preceding chapter. If the general difference equation (10.36) can be rewritten in equivalent form

$$y(k+n)=\psi(k,y(k),\ldots,y(k+n-1)) \tag{10.45}$$

we say that (10.45) serves as a *recurrence relation* for solutions of (10.36). When this happens we are able at once to assert existence and uniqueness of solutions given a set of initial values (10.38). This is true because (10.38) can be regarded as a set of arbitrary constants so that for any set of n

numbers we can use (10.45) to calculate as many values $y(\cdot)$ as we wish. Linear difference equations (10.37) can always be written in the form (10.45) so the preceding remarks apply to such equations. It is furthermore true that solutions for variable coefficient linear difference equations can be found with (10.45) as easily as for constant coefficient equations.

Despite the fact that (10.45) plays the role of a formula for the solution of (10.36), we remain interested in finding solutions as explicit functions of k. These latter are called *solutions in closed form* and for theoretical purposes they reveal more insights into the nature of solutions than do sequences generated by (10.45). The point at issue is the contrast between having a sequence and having a formula for its general term.

Let us consider two examples. The problem

$$y(k+1)=y(k)+2k+1$$
$$y(0)=0 \tag{10.46}$$

is an instance of (10.39) for which we readily generate the solution on (10.6), namely

$$0,1,4,9,25,\ldots,k^2,\ldots$$

Here the general term $y(k)=k^2$ is easily verified as the solution. More generally, whenever $y(k)$ is a polynomial in k we can generate solutions using (10.45) and then use Newton's formula (10.17) to find the solution in a closed form. The problem

$$y(k+2)=y(k+1)+y(k)$$
$$y(0)=0, \qquad y(1)=1 \tag{10.47}$$

is an instance of (10.40) whose solution is the sequence of *Fibonacci numbers*

$$0,1,1,2,3,5,8,13,\ldots \tag{10.48}$$

Newton's formula does not apply because $y(k)$ is not a polynomial in k. Following is the formula (10.4.7).

$$y(k)=\frac{1}{\sqrt{5}}\left[\left(\frac{1+\sqrt{5}}{2}\right)^k-\left(\frac{1-\sqrt{5}}{2}\right)^k\right] \tag{10.49}$$

and the contrast between (10.48) and (10.49) is striking. On the one hand, it is not obvious that (10.49) is a nonnegative integer for $k=0,1,2,\ldots$; and surely it is easier to generate the sequence using (10.47) in place of (10.49).

On the other hand, (10.49) can be used to show quite easily

$$\frac{y(k)}{y(k+1)} \to \mu \quad \text{as} \quad k \to \infty \tag{10.50}$$

where

$$\mu = \frac{\sqrt{5} - 1}{2}$$

is the *golden mean* 0.61803... of antiquity. The value of μ is of interest, for example, when the difference equation is used to represent a law of population growth. In general, we can say we are interested in finding solutions in closed form and this is our purpose in the remainder of this chapter.

PROBLEMS FOR SOLUTION

10.2.1 Changing Domains. Suppose we are able to solve (10.40) on $\{0, 1, 2, \ldots\}$.

(1) What are the (minor) complications of definitions of domains for y and b in solving for y on $\mathfrak{D} = \{0, 1, \ldots, n\}$?

(2) Show how to solve for y on $\{, \ldots, , -2, -1, 0, 1, 2, \ldots\}$.

10.2.2 Transforming Equations. Exhibit the following as instances of (10.42) by identifying $\mathbf{A}$, $\mathbf{f}$, and $\mathbf{y}_0$.

(1) $y(k+2) - 4y(k) = k^2 - 1$.

(2) $y(k+2) - 2y(k+1) + y(k) = 1$.

(3) $y(k+2) + y(k+1) + y(k) = k^2$
$y(0) = y(1) = 1$.

(4) $y(k+2) + 6y(k+1) + 8y(k) = 0$
$y(0) = 0,\ y(4) = 1$.

(5) $y(k+2) = y(k+1) + 2y(k)$
$y(0) = 0,\ y(1) = 1$.

(6) $y(k+2) + y(k+1) - 2y(k) = k^2$
$y(0) = y(1) = 0$.

10.2.3 An Extension. Generalize the preceding in order to write the following as an $n = 4$ instance of (10.42).

$$y(k+2) + z(k+1) + 2y(k) + z(k) = k$$

$$2y(k+1) - z(k+2) + y(k) + z(k) = 1 + k^2$$

10.2.4 Transforming Systems. Establish the counterpart of 9.1.4 and decide whether or not one of (10.37) or (10.42) is the "more general."

10.2.5 Finding Difference Equations. Find homogeneous linear difference equations of minimal order satisfied by the following.

(1) e^{ck}.
(2) $ck/(k+1)$.
(3) $aC(k,r)$, $k=r,r+1,\dots$.
(4) $a+bk+ck^2$.

10.2.6 Fibonacci Numbers. Establish:

(1) Property (10.50).
(2) How many terms in (10.48) are needed to verify (10.50) to three decimal places?
(3) Consider a rectangle R of width w and length l where $w<l$. Suppose that R is "ideal" or "perfectly proportioned" if upon removing the largest possible square from R there remains a rectangle whose width remains in the original proportion w/l to its length. Then necessarily $\mu=w/l$.
(4) A *golden section* μ can be defined in terms of an "ideal" cut in a line of length l.
(5) A simple model of reproduction and growth in a population of pairs at time k and the difference equation in (10.47) holds.

10.2.7 Functional Equations. There are difficulties in defining general functional equations (see, for example, Acźel (1966) and (1969), Kuczma (1968), or Milne-Thomson (1933) where they are called difference equations). Let us agree to define a functional equation as a difference equation written with x in place of k and holding for all $x\in I$ where I is some real interval containing [0, 1]. Whereas the solution of (10.37) is determined by the values of $y(k)$ at n consecutive integers in $\mathfrak{D}$ we see that the solution of the corresponding linear functional equation is determined by the values of

$$y(x), y(x+1), \dots, y(x+n-1)$$

for $x\in[0,1]$. From among the many areas of interest let us address three.

(1) *Unit periodics* are functions $\omega: \mathfrak{R}\to\mathfrak{R}$ that are periodic with periods equal to unity, that is

$$\omega(x+1)=\omega(x)$$

for all $x\in\mathfrak{R}$. Deduce their significance for functional equations.
(2) A principal concern is to find solutions that are continuous on $\mathfrak{R}$ (or

analytic in $\mathcal{C}$). Verify the following cases where $c \in \mathcal{R}$ is arbitrary.

(i) $f(x_1+x_2)=f(x_1)+f(x_2)$ has $f(x)=cx$.

(ii) $f(x_1+x_2)=f(x_1)f(x_2)$ has $f(x)=\exp(cx)$.

(iii) $f(x_1x_2)=f(x_1)+f(x_2)$ has $f(x)=c\log x$.

(iv) $f(x_1x_2)=f(x_1)f(x_2)$ has $f(x)=x^c$.

(v) $f(x_1+x_2)=f(x_1)+f(x_2)+f(x_1)f(x_2)$ has $f(x)=-1$ and $f(x)=e^{cx}-1$.

(3) Many special functions are defined by functional equations. Verify the following (from the literature) where p and q are polynomials.

(i) The functional equation version of (10.35) with $y(1)=1$ has the gamma function as solution among whose many closed form expressions is

$$\Gamma(x)=\int_0^\infty e^{-u}u^{x-1}du, \qquad x>0$$

where, in particular $\Gamma(1/2)=\sqrt{\pi}$.

(ii) More generally, the special cases of functional equation problem (10.39)

$$y(x+1)=y(x)+p(x)$$

$$y(x+1)=-y(x)+p(x)$$

$$y(x+1)=\frac{p(x)}{q(x)}y(x)$$

have solutions which can be written in terms of *Bernoulli polynomials*, *Euler polynomials*, and gamma functions, respectively.

10.3 FIRST-ORDER EQUATIONS

Relatively few nonlinear difference equations (10.36) can be solved in closed form despite the fact that (10.45), whenever it can be found, works as well for nonlinear cases as well as for linear. In this section we find the solution to the general variable coefficient first order linear difference problem (10.39).

Let us start with the special case for domain (10.5)

$$y(k+1)=y(k)+f(k)$$

$$y(0)=y_0 \tag{10.51}$$

which is an elementary version of what is called the *summation problem* in Milne-Thomson (1933). The difference equation is equivalent to

$$\Delta y(k)=f(k) \tag{10.52}$$

and we can consider the initial value difference problem to be the counterpart to (9.13). Using the difference equation in (10.51) as a recurrence relation for $k=1,2,3,\ldots$ in the equivalent form

$$y(k)=y(k+1)-f(k)$$

for $k=-1,-2,-3,\ldots$ we find

$$\begin{aligned} y(k)&=y_0+\sum_{u=0}^{k-1} f(u), && k=1,2,3,\ldots \\ &=y_0, && k=0 \\ &=y_0-\sum_{u=k}^{-1} f(u), && k=-1,-2,-3,\ldots \end{aligned} \tag{10.53}$$

which is the one and only one solution according to

There exists a unique solution to Problem (10.51) and it is given by (10.53). (10.54)

PROOF. Clearly (10.53) is a solution. If there were a second solution z then $w=y-z$ would be a solution to

$$w(k+1)=w(k)$$
$$w(0)=0$$

which requires $w(k)=0$ for all $k\in\mathfrak{D}$. ■

EXAMPLE 10.2. Handling Zero Coefficients. The first zero coefficient $a(k')$ in any case of Problem (10.39) introduces a special equation for $k=k'$, namely

$$y(k'+1)=f(k')$$

The effect is that $f(k')$ then begins to play the role previously played by y_0 as we determine the solution for $k'+2,k'+3,\ldots$ until we encounter the next vanishing $a(k)$, or reach the end.

Suppose $\mathfrak{D}=\{0,1,2,\ldots\}$ and we have a single exceptional case for Problem (10.51) at $k=4$, namely

(1) $$y(5)=0+f(4)$$

First, we solve (10.51) for $\mathfrak{D}'=\{0,1,2,4\}$ using (10.53) obtaining

$$(2)\qquad y(k)=y_0,\qquad k=0$$

$$=y_0+\sum_{u=0}^{k-1} f(u),\qquad k=1,2,3,4$$

Second, we solve (10.51) for $\mathfrak{D}''=\{5,6,7,\dots\}$ using $f(4)$ as initial value according to (1), and obtain

$$(3)\qquad y(k)=\sum_{u=4}^{k-1} f(u),\qquad k=5,6,7,\dots$$

The solution on $\mathfrak{D}$ is given by (2) and (3).

By adapting the procedure used for (9.19), or by using the difference equation as a recurrence, we find

If $a(k)\neq 0$ for all $k\in\mathfrak{D}$ then Problem (10.39) has solution $y(k)$ given by

$$m(k)y(k)=y_0+\sum_{u=0}^{k-1} m(u+1)f(u),\qquad k=1,2,3,\dots$$

$$=y_0,\qquad k=0$$

$$=y_0-\sum_{u=k}^{-1} m(u+1)f(u),\qquad k=-1,-2,-3,\dots$$

where

$$m(k)=\frac{1}{a(0)a(1)\cdots a(k-1)},\qquad k=1,2,3,\dots$$

$$=1,\qquad k=0$$

$$=a(-1)a(-2)\cdots a(k),\qquad k=-1,-2,-3,\dots \tag{10.55}$$

As should be expected, it is the only solution according to the

EXISTENCE AND UNIQUENESS THEOREM. There exists a unique solution $y(k)$ to Problem (10.39) and it is given by (10.55). (10.56)

PROOF. If $a(k)\neq 0$ on $\mathfrak{D}$ then existence follows at once from (10.55) and uniqueness results from the argument used for (10.53). Whenever

$a(k')=0$ there is the unique value $y(k'+1)=f(k')$ and we can adapt (10.53) to apply on consecutive integers between zero values using the method illustrated in Example 10.2. ■

The formula for $y(k)$ in (10.55) can be rewritten in equivalent forms by dividing both sides by $m(k)$ (10.3.2). It is in general also true that solutions to particular cases can often be transformed into numerous equivalent forms; in fact, such variety is a characteristic of many aspects of discrete calculus.

The (scalar) problem with $a \neq 0$ is again the prototype for the (vector) linear systems problem with matrix **A**. Here we have

When $a(k)=a$ is nonzero for all $k \in \mathfrak{D}$ the unique solution to Problem (10.39) is given by either of the following two equivalent forms for $k=1,2,3,\ldots$

$$y(k)=a^k y_0+\sum_{u=0}^{k-1} a^{k-1-u} f(u)$$

$$y(k)=a^k y_0+\sum_{v=0}^{k-1} a^{v} f(k-1-v) \tag{10.57}$$

(See 10.3.4 for formulas covering $k=-1,-2,-3,\ldots$.)

PROOF. We specialize (10.55) to obtain the first form. Changing summation variable via $k-1-u=v$ yields the second form. ■

When $f(k)=$constant for all k the second form is somewhat more convenient than the first because all $f(k-1-v)$ equal that same constant.

EXAMPLE 10.3. Working with Finite Sums. Let us use (10.53) to define a right inverse to the forward difference operation (10.3). Just as the latter is defined for all integers when f is so defined, so also is the *summation operation*

(1)
$$\begin{aligned} \Sigma f(k) &= y_0+\sum_{u=0}^{k-1} f(u), && k=1,2,3,\ldots \\ &= 0, && k=0 \\ &= y_0-\sum_{u=k}^{-1} f(u), && k=-1,-2,-3,\ldots \end{aligned}$$

where y_0 now represents an arbitrary constant. Clearly $\Delta\Sigma f(k)=f(k)$ for all k so Σ is a right inverse. However, by (10.21)

$$\Sigma\Delta f(k)=f(k)+y_0-f(0)$$

where $y_0 - f(0)$ is arbitrary, and so Σ is not a left inverse. In this respect the pair Δ and Σ are related in the same way as differentiation and integration where

$$\frac{d}{dt}\int_0^t f(u)\,du = f(t) \quad \text{but} \quad \int_0^t f'(u)\,du = f(t) - f(0)$$

so neither is the inverse of the other.

We noted that (10.51) is an elementary version of the summation problem treated in Milne-Thomson (1933). Let us make two remarks on this situation. First, it is instructive to derive (1) using a difference table to find unknown values of $\Sigma f(k)$ for $k=0,-1,-2,\ldots$ based on $\Delta\Sigma f(k)= f(k)$ for all k and the obvious values of $\Sigma f(k)$ for $k=1,2,3,\ldots$(10.3.5). Second, it is worth noting that (1) can be used to evaluate summations over any set of consecutive integers

(2) $$\{p, p+1, p+2, \ldots, q\}, \qquad p \leqq q$$

where $f(\cdot)$ is defined. The formula is

(3) $$\sum_{u=p}^{q} f(u) = \Sigma f(u)\Big|_p^{q+1}$$

where $\Sigma f(k)$ is defined using $y_0 = 0$ in (1). We observe that the result is particularly simple when f is a polynomial.

Finally, it is possible to regard (1) as a definition of a counterpart to a definite integral. Since we have agreed that the lower summation limit never exceeds the upper, we introduce a new symbol in the

(4) Definition. Let p and q be any integers and let f be defined for

$$\min\{p, q+1\} \leqq k \leqq \max\{p-1, q\}$$

Then the *definite summation* of f from p to q is

$$\underset{u=p}{\overset{q}{\mathrm{S}}} f(u) = \sum_{u=p}^{q} f(u), \qquad p \leqq q$$

$$= 0, \qquad p = q+1$$

$$= -\sum_{u=q+1}^{p-1} f(u), \qquad p \geqq q+2$$

This can be motivated by setting $y_0 = 0$ and considering $p=0$, $q=0$,

$q=k-1$ in (1). Alternatively, (4) represents the solution of

(5) $$\Delta y(q)=f(q+1)$$

$$y(p)=f(p)$$

on the set of integers specified in (4). Two properties of definite summations

(6) $$\underset{u=p}{\overset{q}{\mathrm{S}}} f(u)=f(p)-\underset{u=q}{\overset{p}{\mathrm{S}}} f(u)+f(q)$$

(7) $$\underset{u=a}{\overset{b-1}{\mathrm{S}}} f(u)+\underset{u=b}{\overset{c}{\mathrm{S}}} f(u)=\underset{u=a}{\overset{c}{\mathrm{S}}} f(u)$$

are counterparts of familiar properties of definite integrals. Property (6) causes definite summations to be less convenient than Σ-summations where variables can be changed between "uphill" and "downhill," as in (10.57), without corresponding complications.

PROBLEMS FOR SOLUTION

10.3.1 Summation Problem. Supply for (10.53):

(1) Derivation.

(2) Verification that it satisfies (10.51) for all k.

10.3.2 General First-Order Problem. Supply for (10.55):

(1) Derivation.

(2) Verification that it satisfies (10.39) for all k.

(3) Equivalent version where $y(k)$ alone appears on LHS.

10.3.3 Exercises. Solve and check the solutions.

(1) Problem (10.39) on $\{r, r+1, \ldots\}$.

(2) $y(k+1)=(k+1)y(k)+1$, $y(0)=1$.

(3) $y(k+1)=y(k)+e^k$, $y(0)=0$

(4) $y(k+1)=ay(k)-1$, $y(0)=y_0$, $a\neq 1$.

(5) Repeat for $a=1$.

(6) $y(k+1)=2y(k)+k$, $y(0)=1$, on $\{0,1,\ldots,7\}$.

(7) Repeat given $y(4)=27$.

10.3.4 Constant Coefficient Case. Establish:

(1) Extension of (10.57) for $k=-1,-2,-3,\ldots$.

(2) Solution of (10.39) for $a(k)=1$ for all k.

10.3.5 Example 10.3. Establish:
(1) Result (1) by solving (10.51).
(2) Result (3).
(3) Specialization of (3) for polynomials, using 10.1.4.
(4) Result (4) by solving (10.51).
(5) Results (6) and (7).

10.4 SECOND-ORDER EQUATIONS

In this section we confine our attention to nonnegative integers. Solutions for negative integers are covered by (10.85).

Let us start with the homogeneous case of Problem (10.40)

$$y(k+2)+\alpha_1 y(k+1)+\alpha_0 y(k)=0$$
$$y(0)=y_0, \qquad y(1)=y_1 \tag{10.58}$$

We seek constants λ_1 and λ_2 such that under $z(k)=y(k+1)-\lambda_1 y(k)$ the homogeneous problem in y becomes the counterpart of (9.21), namely

$$z(k+1)=\lambda_2 z(k)$$
$$z(0)=y_1-\lambda_1 y_0$$

This again leads to

$$\lambda^2+\alpha_1\lambda+\alpha_0=0 \tag{10.59}$$

If the roots λ_1,λ_2 are distinct then the solution of (10.58) is

$$y(k)=c_1\lambda_1^k+c_2\lambda_2^k \tag{10.60}$$

where c_1 and c_2 are the unique solutions of

$$c_1 \quad +c_2=y_0$$
$$c_1\lambda_1+c_2\lambda_2=y_1$$

If $\lambda_1=\lambda_2=\lambda$ then the solution of (10.58) is

$$y(k)=c_1\lambda^k+c_2 k\lambda^k \tag{10.61}$$

where c_1 and c_2 are the unique solutions of

$$c_1 \qquad =y_0$$
$$c_1\lambda+c_2\lambda=y_1$$

The counterpart to (9.26) is

There exists a unique solution to Problem (10.58) and it is given by (10.60) when the roots of (10.59) are distinct, and by (10.61) when the roots coincide. (10.62)

When the roots are not real, say

$$\lambda = \rho e^{i\theta}, \qquad \bar{\lambda} = \rho e^{-i\theta}$$

then we can use the real form of (10.60)

$$y(k) = c_1 \rho^k \cos k\theta + c_2 \rho^k \sin k\theta \tag{10.63}$$

where c_1 and c_2 are the unique solutions of

$$\begin{aligned} c_1 \qquad\qquad\qquad &= y_0 \\ c_1 \rho \cos\theta + c_2 \rho \sin\theta &= y_1 \end{aligned}$$

Continuing, we find that we are entitled to call the appropriate expression (10.60), (10.61), or (10.63) the *general solution* of the homogeneous difference equation

$$y(k+2) + \alpha_1 y(k+1) + \alpha_0 y(k) = 0 \tag{10.64}$$

and the functions λ_1^k, λ_2^k, λ^k, and $k\lambda^k$ are *fundamental solutions* of (10.64). There is the

FUNDAMENTAL SOLUTIONS THEOREM. Any solution of (10.64) can be written as a LC of two fundamental solutions. (10.65)

Illustrations of vector space aspects that parallel those in Example 9.2 are given in the next example.

EXAMPLE 10.4. More on Solution Spaces. The set of all functions on a standard domain $\mathfrak{D}$ to $\mathfrak{R}$ forms a vector space. The set Σ of all solutions to

(1) $$y(k+2) + \alpha_1 y(k+1) + \alpha_0 y(k) = 0$$

cited in (11) of Example 2.1 forms a subspace of that space. Essentially all of the material for solution spaces S in Example 9.2 translates directly into results for Σ. It is again true that study of solutions for variable coefficient cases of (1), and linear difference equations in general, benefits considerably from interpretation of solutions as vectors in a vector space.

As in (9.29) we have the

(2) **THEOREM.** The solution space Σ of (1) has dimension $n=2$ and pairs of fundamental solutions provide bases.

The superposition principle, properties of LD in Σ, and so on, can also be copied from those in Example 9.2 almost without change. In place of Wronskians there now appear

$$c(y_1(k),y_2(k))=\begin{vmatrix} y_1(k) & y_2(k) \\ y_1(k+1) & y_2(k+1) \end{vmatrix}$$

called *Casorati determinants* and the counterpart of (4) in Example 9.2 holds in Σ.

Any solution of the nonhomogeneous equation

$$y(k+2)+\alpha_1 y(k+1)+\alpha_0 y(k)=b(k) \tag{10.66}$$

is called a *particular solution* of that equation and (recall (3.20)) there is the

THEOREM. The general solution of (10.66) is given by the sum of the general solution of (10.64) and any particular solution of (10.66). (10.67)

Proof results from the approach used for (9.31). Next we have the

EXISTENCE AND UNIQUENESS THEOREM. There exists a unique solution to Problem (10.40) and it can be found by finding the unique pair of coefficients that makes the general solution in (10.67) satisfy the initial values. (10.68)

We express the results of this section as follows.

DIRECT PROCEDURE FOR SOLVING PROBLEM (10.40).

(1) Find the roots of (10.59) and write the general solution of (10.58) using (10.60), (10.61), or (10.63).
(2) Find a particular solution of (10.66).
(3) Form the sum of the two solutions and determine the coefficients that satisfy the initial values.
(4) Check that the solution satisfies both the difference equation and the initial values. (10.69)

PROBLEMS FOR SOLUTION

10.4.1 Casorati Determinants. State and prove the counterpart of (4) in Example 9.2.

10.4.2 Reduction to First-Order Equation. Establish counterparts for results in 9.3.2.

10.4.3 Euler's Rationale. Repeat for 9.3.3.

10.4.4 Undetermined Coefficients. Repeat for 9.3.4.

10.4.5 Variation of Parameters. Repeat for 9.3.5.

10.4.6 Details. Establish:
(1) Results (10.59)–(10.63).
(2) Results (10.67) and (10.68).

10.4.7 Exercises. Solve and check:
(1) Problems in 10.2.2.
(2) Fibonacci problem (10.47).

10.4.8 Strict Stability. Establish counterparts for results in 9.3.7.

10.4.9 Vector Space for $n=1$. Repeat for 9.3.8.

10.4.10 Two-Point Problems. Repeat for 9.3.9.

10.5 SYSTEMS OF EQUATIONS

This section is the counterpart to Section 9.4.

Finding the solution of a system of first-order linear difference equations (10.42) is equivalent to finding the kth power of the coefficient matrix $\mathbf{A}$. To avoid citing the trivial exceptional case where $y(k)=f(k-1)$, we suppose $\mathbf{A}\neq\mathbf{O}$. We have no major tasks of developing properties for $\mathbf{A}^k$ such as we had for $\exp t\mathbf{A}$, but if we are interested in solutions on all the integers, rather than merely on $\{0,1,2,\dots\}$, then we need to be concerned as to whether or not $\mathbf{A}$ is singular.

Following earlier patterns we start with the homogeneous case

$$\mathbf{y}(k+1)=\mathbf{A}\mathbf{y}(k)$$

$$\mathbf{y}(0)=\mathbf{y}_0 \tag{10.70}$$

for $k=0,1,2,\dots$. Here we have a recurrence relation that we can use as follows

$$\mathbf{y}(1)=\mathbf{A}\mathbf{y}(0), \qquad \mathbf{y}(2)=\mathbf{A}\mathbf{y}(1)=\mathbf{A}^2\mathbf{y}(0),\dots$$

to obtain the result

There exists a unique solution to Problem (10.70) and it is given by

$$\mathbf{y}(k)=\mathbf{A}^k\mathbf{y}_0$$

for $k=0,1,2,\dots$ where $\mathbf{A}^0=\mathbf{I}$. (10.71)

PROOF. By induction, the given $\mathbf{y}(k)$ is a solution as stated. If there were a second solution $\mathbf{z}$ then $\mathbf{w}=\mathbf{y}-\mathbf{z}$ would be zero for all k because the initial value would be $\mathbf{0}$ in (10.70). ■

If $\mathbf{A}$ is singular there is no unique solution $y(-1)$ satisfying

$$\mathbf{A}\mathbf{y}(-1)=\mathbf{y}_0$$

and so we cannot find a unique solution to (10.70) for negative k. When $\mathbf{A}$ is nonsingular we have the recurrence relation

$$\mathbf{y}(k)=\mathbf{A}^{-1}\mathbf{y}(k+1)$$

that we can use to obtain

$$\mathbf{y}(-1)=\mathbf{A}^{-1}\mathbf{y}_0, \qquad \mathbf{y}(-2)=\mathbf{A}^{-1}\mathbf{y}(-1)=\mathbf{A}^{-2}\mathbf{y}_0,\ldots$$

where $\mathbf{A}^{-2}=\mathbf{A}^{-1}\mathbf{A}^{-1}$ and in general

$$\mathbf{A}^{-k}=\mathbf{A}^{-1}\mathbf{A}^{-k+1}=(\mathbf{A}^{-1})^k$$

for $k=2,3,4,\ldots$. In this way we obtain the following where we write $\mathbf{PJP}^{-1}$ for $\mathbf{A}$ and consequently $\mathbf{PJ}^k\mathbf{P}^{-1}$ for $\mathbf{A}^k$ by the familiar "telescoping" process.

If $\mathbf{A}$ is nonsingular then $y(k)$ in (10.71) is also the unique solution for $k=-1,-2,-3,\ldots$. (10.72)

PROOF. The result follows from (2.23) and (10.71). ■

In order to see how to find the solution, let us use the transformation that served for (9.34), namely

$$\mathbf{y}(k)=\mathbf{P}\mathbf{z}(k)$$

where $\mathbf{P}$ is the nonsingular matrix appearing as $\mathbf{T}$ in (5.38). We obtain

$$\mathbf{z}(k+1)=\mathbf{J}\mathbf{z}(k)$$

$$\mathbf{z}(0)=\mathbf{P}^{-1}\mathbf{y}_0 \tag{10.73}$$

where $\mathbf{J}=\mathbf{P}^{-1}\mathbf{AP}$ is the JCF. Consequently, we have the

EXISTENCE AND UNIQUENESS THEOREM. There exists a unique solution to Problem (10.70) and it is given by

$$\mathbf{y}(k)=\mathbf{PJ}^k\mathbf{P}^{-1}\mathbf{y}_0$$

where $k=0,1,2,\ldots$. If and only if $\mathbf{A}$ is nonsingular there also exists a unique solution for $k=-1,-2,-3,\ldots$ and it is given by the same expression. (10.74)

PROOF. This combines (10.71) and (10.72) for the equivalent Problem (10.73). ■

On the strength of this theorem we are entitled to call

$$\mathbf{y}(k)=\mathbf{A}^k\mathbf{c} \tag{10.75}$$

where $\mathbf{c}\in\mathfrak{R}^n$ is arbitrary, the *general solution* of

$$\mathbf{y}(k+1)=\mathbf{A}\mathbf{y}(k) \tag{10.76}$$

so (10.75) is the counterpart to (9.48). The n columns of $\mathbf{A}^k$ comprise a set of *fundamental solutions* and there is the

FUNDAMENTAL SOLUTIONS THEOREM. Any solution of (10.76) can be written as a LC of n fundamental solutions. (10.77)

Any solution of the nonhomogeneous equation

$$\mathbf{y}(k+1)=\mathbf{A}\mathbf{y}(k)+\mathbf{f}(k) \tag{10.78}$$

is called a *particular solution* of that equation. The counterpart to (9.52) is the readily established

THEOREM. The general solution of (10.78) is given by the sum of the general solution of (10.76) and any particular solution of (10.78). (10.79)

If $k\mapsto\mathbf{h}(k)\in\mathfrak{R}^n$ is any transformation on a standard domain $\mathfrak{D}$ we define

$$\sum_{u\in\mathfrak{D}'}\mathbf{h}(u)=\begin{bmatrix}\sum_{u\in\mathfrak{D}'}h_1(u)\\ \vdots\\ \sum_{u\in\mathfrak{D}'}h_n(u)\end{bmatrix}$$

for any set of consecutive integers $\mathfrak{D}'\subset\mathfrak{D}$. This is the counterpart to (9.53) and if

$$\mathfrak{D}'=\{0,1,\ldots,n\}$$

then

$$\Delta\sum_{u=0}^{n}\mathbf{h}(u)=\mathbf{h}(n+1)$$

because the corresponding relation holds for each component on the RHS by Table 10.3(8). We can write a useful general formula for a particular solution of the nonhomogeneous equation by collecting together results for individual components. If we use the second terms in (10.57) we are led to the following important result which generalizes both (10.56) and (10.68).

A particular solution of (10.78) for $k=1,2,3,\ldots$ is given by the following equivalent expressions

$$\mathbf{p}(k)=\sum_{u=0}^{k-1}\mathbf{A}^{u}\mathbf{f}(k-1-u)$$

$$\mathbf{p}(k)=\sum_{v=0}^{k-1}\mathbf{A}^{k-1-v}\mathbf{f}(v) \tag{10.80}$$

PROOF. Multipy the first RHS by $\mathbf{A}$ and verify that it coincides with $\mathbf{p}(k+1)-\mathbf{f}(k)$. Change summation variable to obtain the second expression for $\mathbf{p}(k)$. ■

We also have

If $\mathbf{A}$ is nonsingular, a particular solution of (10.78) for $k=-1,-2,-3,\ldots$ is given by the following equivalent expressions

$$\mathbf{p}(k)=-\sum_{u=k}^{-1}\mathbf{A}^{u}\mathbf{f}(k-1-u)$$

$$\mathbf{p}(k)=-\sum_{v=k}^{-1}\mathbf{A}^{k-1-v}\mathbf{f}(v) \tag{10.81}$$

The particular solutions in (10.80) and (10.81) are readily verified as being the unique solutions to

$$\mathbf{p}(k+1)=\mathbf{A}\mathbf{p}(k)+\mathbf{f}(k)$$

$$\mathbf{p}(0)=\mathbf{0} \tag{10.82}$$

and we have all we need to establish the

EXISTENCE AND UNIQUENESS THEOREM. There exists a unique solution to Problem (10.42) and it is given by

$$\mathbf{y}(k)=\mathbf{A}^{k}\mathbf{y}_0+\mathbf{p}(k)$$

where $\mathbf{p}(k)$ is given by (10.80) for $k=0,1,2,\ldots$. If and only if $\mathbf{A}$ is nonsingular there also exists a unique solution for $k=-1,-2,-3,\ldots$ and it is given by the same expression where $\mathbf{p}(k)$ is given by (10.81). (10.83)

Let us state the principal

DIRECT PROCEDURE FOR SOLVING PROBLEM (10.42).

(1) Find the JCF of **A**.
(2) Evaluate the formula in (10.83).
(3) Check the result. (10.84)

We can solve any constant coefficient linear difference equation of order n (10.37) with prescribed initial values (10.38) using the

DIRECT PROCEDURE FOR SOLVING (10.37) SUBJECT TO (10.38).

(1) Solve Problem (10.42) for data (10.43).
(2) Extract the solution $y(k)=y_1(k)$.
(3) Check the result. (10.85)

Notice that the companion matrix **A** will be singular in (1) iff $\alpha_0=0$ so this is the only case where we cannot apply (10.85) for $k=-1,-2,-3,\ldots$.

EXAMPLE 10.5. Direct Solutions of Linear Difference Systems. The problem is to find a closed form expression for $\mathbf{A}^k$ where k is any positive integer, and if **A** is nonsingular, for any integer.

The following instance of Problem (10.42)

$$\text{(1)} \qquad \begin{aligned} y_1(k+1)&=y_2(k)+1, \qquad & y_1(0)&=3\\ y_2(k+1)&=y_1(k)+k, \qquad & y_2(0)&=1 \end{aligned}$$

where $k=0,1,2,\ldots$, is the counterpart of (1) in Example 9.3 in the sense that it is defined by the counterpart data **A**, **f**, and $\mathbf{y}_0$. Using **P** and **J** from the latter example

$$\mathbf{A}^k=\frac{1}{2}\begin{bmatrix} 1+(-1)^k & 1-(-1)^k \\ 1-(-1)^k & 1+(-1)^k \end{bmatrix}$$

and the solution to the homogeneous case that assumes the given initial values is

$$\text{(2)} \qquad \mathbf{A}^k\mathbf{y}_0=\begin{bmatrix} 2+(-1)^k \\ 2-(-1)^k \end{bmatrix}$$

In order to find a particular solution to (1) we solve (10.82) using the first formula in (10.80)

$$(3)\qquad \mathbf{p}(k)=\frac{1}{2}\sum_{u=0}^{k-1}\begin{bmatrix} k-u+(-1)^u(2-k+u) \\ k-u-(-1)^u(2-k+u) \end{bmatrix}$$

The RHS can be evaluated by using elementary results from Section 10.1 to calculate individual sums as follows.

$$\frac{1}{2}\sum_{u=0}^{k-1}(k-u)=\frac{1}{2}\,\frac{k(k+1)}{2}$$

$$\frac{1}{2}(2-k)\sum_{u=0}^{k-1}(-1)^u+\frac{1}{2}\sum_{u=0}^{k-1}u(-1)^u=\frac{3}{8}-\frac{3}{8}(-1)^k-\frac{k}{2}$$

Using these to calculate the summations of the two components in (3) we obtain

$$(4)\qquad \mathbf{p}(k)=\frac{1}{8}\begin{bmatrix} 2k^2-3(-1)^k+3 \\ 2k^2+4k+3(-1)^k-3 \end{bmatrix}$$

The solution to (1) is given according to (10.83) by the sum of the general solution (2) and the particular solution (4), namely

$$(5)\qquad \mathbf{y}(k)=k^2\begin{bmatrix} \frac{1}{4} \\ \frac{1}{4} \end{bmatrix}+k\begin{bmatrix} 0 \\ \frac{1}{2} \end{bmatrix}+(-1)^k\begin{bmatrix} \frac{5}{8} \\ -\frac{5}{8} \end{bmatrix}+\begin{bmatrix} \frac{19}{8} \\ \frac{13}{8} \end{bmatrix}$$

It is readily verified that (5) satisfies (1)

Two features of Problem (1) deserve comment. First, it is quite special because $\mathbf{A}=\mathbf{A}^{-1}$ and this makes it easy to solve for $k=-1,-2,-3,\ldots$. Second, we can find a particular solution $\mathbf{p}(k)$ using undetermined coefficients but there are complications including some that are commonly encountered when the (scalar) procedures of 10.4.4 are used for a (vector) problem such as (1) (10.5.2(3)).

Using (10.85) we can solve

$$(6)\qquad y(k+2)-y(k)=1$$

$$y(0)=3,\qquad y(1)=1$$

With the transformation (10.41) we obtain the following instance of

Problem (10.42)

$$\mathbf{y}(k+1)=\begin{bmatrix} 0 & 1 \\ 1 & 0 \end{bmatrix}\mathbf{y}(k)+\begin{bmatrix} 0 \\ 1 \end{bmatrix}, \qquad \mathbf{y}_0=\begin{bmatrix} 3 \\ 1 \end{bmatrix} \tag{7}$$

where the matrix happens to coincide with **A** in (1). In this case the general solution of the homogeneous case is also given by (2), and the particular solution from (10.80) is

$$\frac{1}{2}\sum_{u=0}^{k-1}\begin{bmatrix} 1-(-1)^u \\ 1+(-1)^u \end{bmatrix}=\frac{1}{4}\begin{bmatrix} 2k+(-1)^k-1 \\ 2k-(-1)^k+1 \end{bmatrix} \tag{8}$$

Consequently, the solution to (7) is the sum of (2) and (8)

$$\mathbf{y}(k)=\begin{bmatrix} \frac{7}{4}+\frac{5}{4}(-1)^k+\frac{1}{2}k \\ \frac{9}{4}-\frac{5}{4}(-1)^k+\frac{1}{2}k \end{bmatrix} \tag{9}$$

and the solution to (6) is the first component

$$y(k)=\frac{7}{4}+\frac{5}{4}(-1)^k+\frac{1}{2}k \tag{10}$$

As we remarked for (9) in Example 9.3, there are easier ways to find the solution we have just obtained but the preceding illustrates how to solve an nth-order problem by solving a system of n first-order equations.

If we use the nondiagonalizable matrix **A** and initial vector $\mathbf{y}_0$ in (10) of Example 9.3 we obtain the problem

$$\mathbf{y}(k+1)=\begin{bmatrix} 1 & 0 \\ 2 & 1 \end{bmatrix}\mathbf{y}(k), \qquad \mathbf{y}(0)=\begin{bmatrix} 1 \\ 1 \end{bmatrix} \tag{11}$$

Direct Procedure (10.84) yields

$$\mathbf{y}(k)=\begin{bmatrix} 1 \\ 2k+1 \end{bmatrix} \tag{12}$$

which is readily verified as a solution for both negative and positive k.

It is also readily verified that the counterpart of (15) in Example 9.3 is the formula

$$\mathbf{A}^k=\frac{\lambda_1\lambda_2^k-\lambda_1^k\lambda_2}{\lambda_1-\lambda_2}\mathbf{I}+\frac{\lambda_1^k-\lambda_2^k}{\lambda_1-\lambda_2}\mathbf{A} \tag{13}$$

which is correct for $n=2$ when $\lambda_1 \neq \lambda_2$. The same can be said for the counterpart of (17) in Example 9.3, namely

$$\mathbf{A}^k = (1-k)\lambda^k \mathbf{I} + k\lambda^{k-1}\mathbf{A} \tag{14}$$

which is correct for $n=2$ when $\lambda_1 = \lambda_2 = \lambda$.

PROBLEMS FOR SOLUTION

10.5.1 Details. Establish:

(1) Equivalence of (10.70) and (10.73).
(2) Results (10.79), (10.81), and (10.83).

10.5.2 Example 10.5. Establish:

(1) Summation (4).
(2) Formula (5) furnishes a solution to (1).
(3) A solution to (1) using undetermined coefficients.
(4) Solution of (1) for $k = -1, -2, -3, \ldots$.
(5) A particular solution to (1) using 10.4.4.
(6) Formulas (10) and (12) furnish solutions to Problems (6) and (11), respectively.
(7) Formulas (13) and (14).

10.5.3 Exercises. Solve and check:

(1) Problems in 10.2.2 as 2×2 systems.
(2) Problem (10.42) given $\mathbf{A}$, $\mathbf{f}(k)$, and $\mathbf{y}_0$ as follows

(i) $\begin{bmatrix} -3 & -2 \\ 4 & 3 \end{bmatrix}, \begin{bmatrix} e^k \\ -3k \end{bmatrix}, \begin{bmatrix} 1 \\ 0 \end{bmatrix}$

(ii) $\begin{bmatrix} 1 & 2 \\ 0 & 1 \end{bmatrix}, \begin{bmatrix} 0 \\ k \end{bmatrix}, \begin{bmatrix} 1 \\ 1 \end{bmatrix}$

(iii) $\begin{bmatrix} 4 & -2 \\ 3 & -1 \end{bmatrix}, \begin{bmatrix} 1 \\ 0 \end{bmatrix}, \begin{bmatrix} 1 \\ 2 \end{bmatrix}$

(3) Periodic Sequences. Find the general term of

$$1, 2, 3, 1, 2, 3, \ldots, 1, 2, 3, \ldots$$

by solving an appropriate

(i) Second-order nonhomogeneous difference equation.
(ii) Third-order homogeneous difference equation.
(iii) 2×2 nonhomogeneous system.
(iv) 3×3 homogeneous system.

(4) Repeat for $1, 2, 1, 1, 2, 1, \ldots$.

(5) Gross and Harris (1974, 499). The system

$$y(k+1)-2y(k)+2z(k)=2$$

$$2y(k+1)-3y(k)+3z(k+1)-z(k)=6(4)^k$$

10.5.4 Strict Stability. Establish counterparts for results in 9.4.5.

10.6 z-TRANSFORMS

Here we have the indirect methods that are counterparts to those in Section 9.5.

The subject transforms are widely used in engineering and applied science as counterparts to Laplace transforms. They are completely interchangeable with generating functions commonly used in mathematics, probability, and statistics; in Example 10.9 we show how anything done with one can also be done with the other. We feature z-transforms because

1. The algebraic manipulations associated with z-transforms match those for Laplace transforms exactly (rather than through a reciprocal relationship).
2. The inversion formula is simply the formula for the $(-k)$th coefficient in Laurent's series (8.10) for the z-transform and, as in cases for the Laplace transform, it holds for sufficiently large (rather than small) values of the parameter.

Chapter 8 remains optional despite (2). Wherever we use or refer to advanced results for complex variables, it is for an alternative method or an insight that can be omitted. We do define z-transforms for $z \in \mathcal{C}$ and we do use calculus for $\mathcal{C}$—for example, series of complex terms converge iff the series of real and imaginary parts converge separately—but only in elementary ways that are comparable to the reliance on algebra for $\mathcal{C}$ in Section 3.4 and Chapter 5.

The basic formula for the *z-transformation* y to η is

$$\eta(z)=\sum_{u=0}^{\infty}\frac{y(u)}{z^u} \tag{10.86}$$

where $z \in \mathcal{C}$. We call the function η the *z-transform* of y if there is some $r \in \mathcal{R}$ such that (10.86) converges whenever $|z|>r$; in such cases, y is the *inverse z-transform* of η. We write

$$\eta(z)=\mathcal{Z}\{y(t)\}, \qquad y(t)=\mathcal{Z}^{-1}\{\eta(s)\}$$

and we denote z-transforms of functions $b, f, z, \ldots$ by the generally corresponding Greek letters β, ϕ, $\zeta, \ldots$ used for Laplace transforms. We have the following criteria for

EXISTENCE OF z-TRANSFORMS. If y satisfies
(1) $y(k)=0$ for $k=-1,-2,-3,\ldots$
(2) $y(k)=O(k^n)$ for some positive integer n
then y has a z-transform. (10.87)

PROOF. By (2) there exists $M \geqq 0$ such that

$$|y(k)| \leqq Mk^n$$

as $k \to \infty$ so the conclusion will follow if $\mathfrak{Z}\{k^n\}$ exists. By calculation, $k^{(m)}$ has a z-transform for every positive integer m and the same can be said for k^m because of (10.29). ■

As an example, $y(k)=k^k$ has no z-transform. Condition (1) can be omitted but we include it, and impose it whenever we use (10.86), to provide for unique inverse z-transforms. If we need to work on negative integers, or on all the integers, we work separately with functions such as

$$\begin{aligned} y_1(k) &= y(k), \qquad & k &= 0,1,2,\ldots \\ &= 0, & k &= -1,-2,-3,\ldots \end{aligned}$$

and

$$\begin{aligned} y_2(k) &= y(-k), \qquad & k &= 0,1,2,\ldots \\ &= 0, & k &= -1,-2,-3,\ldots \end{aligned}$$

as appropriate. We continue to use $y(k)=\lim_{x\to k+} y(x)$ whenever we work with difference equations for functions on $\mathfrak{R}$ to $\mathfrak{R}$.

Clearly we must also restrict $\eta(z)$; for example, z could never be the LHS of (10.86) and so $\eta(z)=z$ is not a z-transform. But before we consider sufficient conditions, let us state that in practice—where usually there is reason to believe that a given $\eta(z)$ is a z-transform—the following (optional) result can be used to find $y(k)$.

If $\eta(z)$ is the z-transform of some function for $|z|>r$, then that function is

$$\begin{aligned} y(k) &= \frac{1}{2\pi i}\int_C z^{k-1}\eta(z)\,dz, \qquad & k &= 0,1,2,\ldots \\ &= 0, & k &= -1,-2,-3,\ldots \end{aligned}$$

where C is any positively oriented circle of radius $r' > r$ and center $z = 0$. (10.88)

PROOF. We can be sure that the RHS of (10.86) is the Laurent's series in the annular region $A = \{z : |z| > r\}$. Theorem (8.10) also specifies that the coefficients (changing n to $-k$) have the stated form. ■
Special results for complex variables are used to evaluate the integral.

By restricting $\eta(z)$ further we obtain the (optional)

INVERSION FORMULA. Suppose $\eta(z)$ is a z-transform that is furthermore analytic except for a finite number of isolated singu'arities $z_1, z_2, \ldots, z_m$. Then its inverse z-transform is given by

$$y(k) = \sum \operatorname{res}(z^{k-1}\eta(z), p), \qquad k = 0, 1, 2, \ldots$$

$$= 0, \qquad k = -1, -2, -3, \ldots$$

where the summation is taken over all poles p of $z^{k-1}\eta(z)$ including, perhaps, a pole at $p = 0$ for $k = 0$. (10.89)

PROOF. We use the residue theorem (8.13) to evaluate the integral in (10.88) and, as stated, we must provide for a possible pole at the origin (even when $\eta(z)$ has no pole there). ■
As already mentioned, Example 10.10 illustrates the value of this optional formula.

We call the preceding two results optional on account of the following. Table 10.5, and larger tables in works such as Cadzow (1973), can often be used together with Table 10.6 to find $y(t)$ when a z-transform $\eta(z)$ is given. In the remaining cases of interest in this book, a division algorithm, in the manner of Example 8.1, can be used (as illustrated in Example 10.10).

Each formula in Table 10.5 is valid outside a circle of radius r and center $z_0 = 0$; ordinarily we are not concerned with the actual value of r as long as we are certain that a value does exist. Entry (1) is the familiar result for a geometric series, and the remaining entries are similarly established by elementary methods consisting, in most cases, of division algorithms.

Entries (1)–(12) in Table 10.6 can be established as properties of formal series that can be manipulated in certain ways without being concerned with convergence. However, there is a specific region of validity for each of these entries; ordinarily it is outside the larger of two circles when $y_1(k)$ and $y_2(k)$ are involved, or an expression such as

$$|z| > |a|$$

TABLE 10.5 A BRIEF TABLE OF z-TRANSFORMS

	Inverse $y(k)$	z-Transform $\eta(z)$	Valid $\lvert z\rvert > r$ r
(1)	1	$\dfrac{z}{z-1}$	1
(2)	k	$\dfrac{z}{(z-1)^2}$	1
(3)	k^2	$\dfrac{z(z+1)}{(z-1)^3}$	1
(4)	k^3	$\dfrac{z(z^2+4z+1)}{(z-1)^4}$	1
(5)	$k^{(m)},\ m=0,1,2,\ldots$	$\dfrac{m!z}{(z-1)^{m+1}}$	1
(6)	a^k	$\dfrac{z}{z-a}$	$\lvert a\rvert$
(7)	ka^k	$\dfrac{az}{(z-a)^2}$	$\lvert a\rvert$
(8)	$\dfrac{a^k}{k!}$	$\exp\left(\dfrac{a}{z}\right)$	0
(9)	e^{-ak}	$\dfrac{z}{z-e^{-a}}$	e^{-a}
(10)	$\sin bk$	$\dfrac{z\sin b}{z^2-2z\cos b+1}$	1
(11)	$\cos bk$	$\dfrac{z(z-\cos b)}{z^2-2z\cos b+1}$	1
(12)	$e^{-ak}\sin bk$	$\dfrac{ze^{-a}\sin b}{z^2-2ze^{-a}\cos b+e^{-2a}}$	e^{-a}
(13)	$e^{-ak}\cos bk$	$\dfrac{z(z-e^{-a}\cos b)}{z^2-2ze^{-a}\cos b+e^{-2a}}$	e^{-a}
(14)	$C(k,m),\ m=0,1,2,\ldots$	$\dfrac{z}{(z-1)^{m+1}}$	1
(15)	$C(a,k)$	$\left(1+\dfrac{1}{z}\right)^a$	1

which applies to (6). Series (10.86) are then absolutely and uniformly convergent outside sufficiently large circles about the origin so that Entries (1)–(12) follow by changes of variable, differentiations, multiplications, and other such operations. The transformation is linear as stated in (2) so that with (4) we can be sure that a linear algebraic equation for $\eta(z)$ will result whenever we transform both members of any constant coefficient

TABLE 10.6 PROPERTIES OF z-TRANSFORMS

	Inverse	z-Transform
(1)	$y(k)$	$\eta(z)$
(2)	$ay_1(k)+by_2(k)$	$a\eta_1(z)+b\eta_2(z)$
(3)	$y(k+1)$	$z\eta(z)-zy(0)$
(4)	$y(k+n)$	$z^n\eta(z)-z^ny(0)$ $-z^{n-1}y(1)-\cdots-zy(n-1)$
(5)	$y(k-c),\ c\geqq 0$	$z^{-c}\eta(z)$
(6)	$a^ky(k)$	$\eta\left(\frac{z}{a}\right)$
(7)	$ky(k)$	$-z\frac{d\eta(z)}{dz}$
(8)	$k^2y(k)$	$-z\frac{d}{dz}[-z\eta'(z)]$
(9)	$k^my(k),\ m=0,1,2,\ldots$	$\left(-z\frac{d}{dz}\right)^m\eta(z)$
(10)	$\sum_{u=0}^{k} y_1(k-u)y_2(u)$	$\eta_1(z)\eta_2(z)$
(11)	$y_1(k)y_2(k)$	$\frac{1}{2\pi i}\int_C p^{-1}\eta_1(p)\eta_2(p^{-1}z)\,dp$
(12)	$\sum_{u=0}^{k} y(u)$	$\frac{z}{z-1}\eta(z)$
	Special Values	
(13)	Initial value	$y(0)=\lim_{z\to\infty}\eta(z)$
(14)	Intermediate values	$y(k)=\lim_{z\to\infty}\frac{1}{k!}\left(-z^2\frac{d}{dz}\right)^k\eta(z)$
(15)	Final value	$\lim_{k\to\infty}y(k)=\lim_{z\to 1}(z-1)\eta(z)$
(16)	First moment	$\lim_{k\to\infty}\sum_{u=0}^{k}uy(u)=\lim_{z\to 1}[-\eta'(z)]$
(17)	Factorial moments $m=1,2,\ldots$	$\lim_{k\to\infty}\sum_{u=0}^{k}(u+m-1)^{(m)}y(u)$ $=\lim_{z\to 1}[(-1)^m\eta^{(m)}(z)]$

case of (10.37). Entry (3) is readily verified by forming

$$\mathfrak{Z}\{y(k+1)\}=y(1)+\frac{y(2)}{z}+\frac{y(3)}{z^2}+\cdots$$

and showing that it equals

$$z\eta(z)-zy(0)$$

Notice that the preceding is the counterpart of the Laplace transform of $y'(t)$, namely

$$s\eta(s)-y(0)$$

Presence of the multiplier z in $zy(0)$ causes no essential change in the algebraic procedure. Entry (5) covers "backward displacement" counterparts of translations $f_c(t)$ illustrated in Figure 9.2. Since $y(k)=0$ for $k=-1,-2,-3,\ldots$, we actually use $y(k-c)$ in (5) as notation for

$$\begin{aligned} y_c(k)&=0, && k=0,1,\ldots,c-1\\ &=y(k-c), && k=c,c+1,c+2,\ldots \end{aligned}$$

Entry (9) is the general mth case where $m=1$ in (7) and $m=2$ in (8). The entry for $\eta(z)$ in (9) means "apply the operation m times in succession, first differentiating with respect to z and then multiplying by $-z$." Entry (10) is the important *convolution* property of z-transforms. It also represents the formula for the elementary (Cauchy) product of the two series η_1 and η_2. The optional Entry (11) has no counterpart in Chapter 9; in practice the RHS is evaluated by summing all the residues of the integrand inside the circle of convergence. Entry (12) can be used to write expressions for finite sums using a table of z-transforms. It represents a special case of (10) where $y_1(k)=1$ is specified. Entries (13) and (14) are often important in practice where all $y(k)$ are not found; also see Approach 1 in Example 10.10. Entry (15) is of interest in connection with strict stability (10.5.4) of $y(k)$. The final entries (16) and (17) are used when working with discrete probability distributions; in some cases, the "moments" can be found where it is quite difficult and perhaps impractical to find the entire set of values of $y(k)$.

Let us now return to Table 10.5 and notice that, with few exceptions, the factor z appears explicitly in $\eta(z)$. It is consequently effective to proceed as follows in

INVERTING RATIONAL FUNCTIONS USING PARTIAL FRACTIONS. To find the inverse z-transform of

$$\eta(z)=\frac{p(z)}{q(z)}$$

where p/q has the form given in Example 8.2(8), "withhold" a factor z and expand $z^{-1}\eta(z)$ so that in place of (11) there is obtained in the end

$$\frac{p(z)}{q(z)}=zA_1(z)+\cdots+zA_m(z)$$

which can always be inverted using Tables 10.5(6) and 10.6(9). (10.90)

PROOF. The final statement can be made because, as readily verified, the special form of the rational function leads only to terms that can be inverted using the stated (6) and results of first differentiating with respect to z and then multiplying by $-z$ according to the stated (9). ■

All we need do to specify the general procedure for solving linear difference equations is to copy (9.61) almost word for word as follows.

USING z-TRANSFORMS. Proceed as follows to solve an initial value difference problem, based on a linear difference equation with constant coefficients, to which an existence and uniqueness theorem applies.

(1) Transform the difference equation in $y(k)$ into an equation in $\eta(z)$ using Tables 10.5 and 10.6.
(2) Solve the resulting linear algebraic equation for the z-transform $\eta(z)=\mathfrak{Z}\{y(k)\}$.
(3) Use Tables 10.5 and 10.6, (or (10.89) and (10.90)) to find the inverse z-transform $y(k)=\mathfrak{Z}^{-1}\{\eta(z)\}$.
(4) Verify that $y(k)$ is a solution, and hence that it must be the unique solution. (10.91)

We omit the modifier "formal," even though the preceding is a procedure where all steps along the way are not justified, because the one-to-one nature of the transformation and the validity of Tables 10.5 and 10.6 can be established using results stated in Chapter 8. Advantages from using (10.91) are that it can effectively handle

1. Numerical (for example, empirical) RHSs almost as easily as polynomials, powers, and so forth
2. Nonhomogeneous equations in the same manner as homogeneous ones

3. Prescribed initial values directly

as counterparts to advantages from using (9.61).

EXAMPLE 10.6. Solving Low-Order Difference Equations. Let us illustrate (10.91) by solving the following instance of Problem (10.39)

$$(1) \qquad y(k+1)=ay(k)+f(k), \qquad a\neq 0$$
$$y(0)=y_0$$

where $k=0,1,2,\ldots$ and $y(k)=0$ for $k=-1,-2,-3,\ldots$ is understood everywhere in this example. Taking z-transforms of both members we obtain

$$z\eta(z)-zy_0=a\eta(z)+\phi(z)$$

so the z-transform of the solution $y(k)$ is

$$(2) \qquad \eta(z)=\frac{z}{z-a}y_0+\frac{1}{z-a}\phi(z)$$

The first term on the RHS can be inverted using Table 10.5. Then if we rewrite the second ("withholding" z as specified in (10.90)) in the form

$$\frac{z}{z-a}\,\frac{\phi(z)}{z}$$

we recognize it as the product of two transforms that we can invert by the convolution formula Table 10.6(10). While we can use either factor as $\eta_1(z)$ in the convolution, let us specify

$$\eta_1(z)=\frac{\phi(z)}{z}$$

so $y_1(k)=f(k-1)$ by Table 10.6(5) for $c=1$. Then by convolution the inverse z-transform of (2) is

$$(3) \qquad y(k)=a^ky_0+\sum_{u=0}^{k} f(k-1-u)a^u$$

This coincides with the second form in (10.57) because we can change the upper limit in (3) to $k-1$ without affecting its value.

Next we solve

$$(4) \qquad y(k+1)=y(k)+2e^k$$
$$y(0)=1$$

to illustrate (10.90). The transformed equation is

$$z\eta(z)-z=\eta(z)+\frac{2z}{z-e}$$

and the z-transform is

$$\eta(z)=\frac{z}{z-1}+\frac{2z}{(z-1)(z-e)} \tag{5}$$

We "withhold" z from the second term on the RHS as specified in (10.90) and expand by Heaviside coverup

$$\frac{2}{(z-1)(z-e)}=\frac{2}{1-e}\,\frac{1}{z-1}+\frac{2}{e-1}\,\frac{1}{z-e}$$

Using this in (5) and going to Table 10.5 yields

$$y(k)=1+\frac{2}{e-1}(e^k-1) \tag{6}$$

If we had not "withheld" z we would have obtained

$$\eta(z)=\frac{z}{z-1}+\frac{2}{1-e}\,\frac{1}{z-1}+\frac{2e}{e-1}\,\frac{1}{z-e} \tag{7}$$

which, though correct, is awkward, as we now show. The second and third terms on the RHS do not appear in Table 10.5 but we can use its (1) and (6) with Table 10.6(5) to write

$$\begin{aligned} y(k)&=1, && k=0 \\ &=1+\frac{2}{1-e}+\frac{2e}{e-1}e^{k-1}, && k=1,2,3,\ldots \end{aligned} \tag{8}$$

This is equivalent to (6) and so the effect of "withholding" z is to obtain the one-line formula (6) directly.

The next example illustrates the feature, common to Laplace and z-transform methods, that the same approach is used for homogeneous and nonhomogeneous cases. If we transform

$$y(k+2)-2y(k+1)-3y(k)=1+k^2 \tag{9}$$

$$y(0)=0, \qquad y(1)=1$$

and solve for $\eta(z)$ we obtain

$$(10)\qquad \eta(z)=\frac{z}{(z-3)(z+1)}+\frac{z}{(z-3)(z-1)(z+1)}+\frac{z}{(z-3)(z-1)^3}$$

The second and third terms on the RHS are absent in the homogeneous case and this is typical. The additional effort in nonhomogeneous cases is that of finding additional partial fraction expansions. These can readily be found using a computer and, although they are not strictly simpler, finding them may be more straightforward than evaluating summations from (10.80), or using complicated undetermined coefficient schemes.

Returning to (10), the result of finding the partial fraction expansions is

$$(11)\qquad y(k)=\frac{1}{2}3^k-\frac{1}{8}(-1)^k-\frac{1}{4}k^{(2)}-\frac{1}{4}k-\frac{3}{8}$$

Is this the correct solution? First, we verify the initial conditions $y(0)=0$ and $y(1)=1$. Second, we need not substitute the first and second terms into the difference equation in (9) because the characteristic roots are clearly 3 and -1. Third, we complete the check by verifying that

$$(12)\qquad -\frac{1}{4}k^{(2)}-\frac{1}{4}k-\frac{3}{8}=-\frac{1}{4}k^2-\frac{3}{8}$$

is indeed a particular solution, and so the answer is "yes."

Would it be less work to invert (10) using residues in place of partial fractions? The answer is that there is no appreciable difference. If we use (10.90) without any combining of terms in (10) we need to find seven individual residues on the RHS of

$$z^{k-1}\eta(z)=\frac{z^k}{(z-3)(z-1)}+\frac{z^k}{(z-3)(z-1)(z+1)}+\frac{z^k}{(z-3)(z-1)^3}$$

The first six, proceeding with $3,1,3,1,-1,3$, result from Heaviside coverups and the new phenomenon is that we do not have to consult Table 10.5 but we are instead finished once we finish a coverup. For the pole at $z=1$ of order 3 we go through about the same effort as required for finding the partial fraction expansion. Some may prefer using (10.90) to consulting Tables 10.5 and 10.6, but it seems fair to say that significant advantage comes in more complicated cases where, at least initially, it appears that we simply must use (10.90). We qualify this last statement because once we find a solution we can usually devise other methods, at the least a division algorithm, that would produce the solution.

Let us illustrate that we can effectively use z-transforms to find particular solutions to accompany solutions of the homogeneous equation that satisfy the initial values. For (9) we can immediately write the general solution for the homogeneous case

$$y(k)=c_1 3^k+c_2(-1)^k$$

and quickly determine c_1 and c_2 so we merely seek $p(k)$ in

$$y(k)=\frac{1}{4}3^k-\frac{1}{4}(-1)^k+p(k) \tag{13}$$

Thus we use z-transforms to solve

$$p(k+2)-2p(k+1)-3p(k)=1+k^2$$

$$p(0)=0, \qquad p(1)=0$$

and $\pi(z)$ is given by (10) with the first term on the RHS omitted. What happens is that there are two less partial fraction coefficients (or residues) to find in working with π in place of η. This is not significantly less effort because we had to find c_1 and c_2. The point is that if we wish to work with different particular solutions, for different RHSs, then the approach represented by (13) is effective, and markedly so in more complicated cases than the simple example (9). Finally, let us observe that $p(k)$ is a different particular solution than (11) but this is entirely inconsequential.

In order to solve the linear difference systems Problem (10.42) we find the z-transforms of the components of $\mathbf{y}(k)$ and then write the results in vector notation as

$$(z\mathbf{I}-\mathbf{A})\boldsymbol{\eta}(z)=z\mathbf{y}_0+\boldsymbol{\phi}(z) \tag{10.92}$$

where $\boldsymbol{\eta}(x)=[\eta_1(z),\ldots,\eta_n(z)]^T$ is the vector of z-transforms of the components of $\mathbf{y}$, and we similarly use the corresponding notation for the z-transform of $\mathbf{f}(k)$. If z is not an eigenvalue of $\mathbf{A}$, the coefficient matrix in (10.92) is nonsingular, and for $|z|$ sufficiently large, (10.92) has the unique solution

$$\boldsymbol{\eta}(z)=z(z\mathbf{I}-\mathbf{A})^{-1}\mathbf{y}_0+(z\mathbf{I}-\mathbf{A})^{-1}\boldsymbol{\phi}(z) \tag{10.93}$$

This is the counterpart of (9.68) distinguished merely by the factor z (that we are in general pleased to see for the reasons explained in (10.90)) in the first term on the RHS. The counterpart of (9.69) is

$$\mathfrak{Z}\{\mathbf{A}^k\}=z(z\mathbf{I}-\mathbf{A})^{-1} \tag{10.94}$$

so we have the same algebraic problem as in working with Laplace transforms: we must express the resolvent matrix $(z\mathbf{I}-\mathbf{A})^{-1}$ as a function of z, or else we must solve (10.92) for $\boldsymbol{\eta}(z)$ by some other means. In summary, we have the

z-TRANSFORM PROCEDURE FOR SOLVING PROBLEM (10.42). Applying (10.91) for component functions yields (10.92) which must be solved for $\boldsymbol{\eta}(z)$ as an explicit function of z. The remaining steps (3) and (4) of (10.91) are carried out for the components of $\boldsymbol{\eta}(z)$ using appropriate vector-matrix methods.

In order to find a particular solution (say, to avoid using (10.80)) solve

$$(z\mathbf{I}-\mathbf{A})\,\mathbf{p}(z)=\boldsymbol{\phi}(z)$$

for $\mathbf{p}(z)$ and find its inverse z-transform. (10.95)

EXAMPLE 10.7. z-Transform Solutions of Linear Difference Systems. Let us again solve the first problem in Example 10.5

(1)
$$y_1(k+1)=y_2(k)+1, \qquad y_1(0)=3$$
$$y_2(k+1)=y_1(k)+k, \qquad y_2(0)=1$$

Using Example 9.5(2) and (10.93)

$$\boldsymbol{\eta}(z)=\frac{z}{(z-1)(z+1)}\begin{bmatrix} z & 1 \\ 1 & z \end{bmatrix}\begin{bmatrix} 3 \\ 1 \end{bmatrix}+\frac{1}{(z-1)(z+1)}\begin{bmatrix} z & 1 \\ 1 & z \end{bmatrix}\begin{bmatrix} \dfrac{z}{z-1} \\ \dfrac{z}{(z-1)^2} \end{bmatrix}$$

(2)

is the z-transform of the solution. Multiplication and determination of partial fractions in the first member on the RHS yields

(3)
$$\mathcal{Z}\{\mathbf{A}^k\mathbf{y}_0\}=\frac{z}{z-1}\begin{bmatrix} 2 \\ 2 \end{bmatrix}+\frac{z}{z+1}\begin{bmatrix} 1 \\ -1 \end{bmatrix}$$

From the second member we similarly obtain the z-transform of a particular solution to the homogeneous equation

(4)
$$\boldsymbol{\pi}(z)=\frac{z}{(z-1)^3}\begin{bmatrix} \frac{1}{2} \\ \frac{1}{2} \end{bmatrix}+\frac{z}{(z-1)^2}\begin{bmatrix} \frac{1}{4} \\ \frac{3}{4} \end{bmatrix}+\frac{z}{z-1}\begin{bmatrix} \frac{3}{8} \\ -\frac{3}{8} \end{bmatrix}+\frac{z}{z+1}\begin{bmatrix} -\frac{3}{8} \\ \frac{3}{8} \end{bmatrix}$$

Taking the sum of the inverse z-transforms of (3) and (4) we obtain the solution (5) in Example 10.5.

From (2) we can write (10.94) and expand

$$\mathfrak{Z}\{\mathbf{A}^k\}=\frac{z}{z-1}\begin{bmatrix}\frac{1}{2} & \frac{1}{2}\\ \frac{1}{2} & \frac{1}{2}\end{bmatrix}+\frac{z}{z+1}\begin{bmatrix}\frac{1}{2} & -\frac{1}{2}\\ -\frac{1}{2} & \frac{1}{2}\end{bmatrix}$$

in case we wish to find $\mathbf{A}^k$.

Problem (6) of Example 10.5

(5) $$y(k+2)-y(k)=1$$

$$y(0)=3, \qquad y(1)=1$$

can be solved by using z-transforms for the equivalent 2×2 problem. The only change from (1) is a new

$$\boldsymbol{\phi}(z)=\begin{bmatrix}0\\ \frac{z}{z-1}\end{bmatrix}$$

We solve (5) by finding the inverse z-transform of the first component of

(6) $$\boldsymbol{\eta}(z)=\frac{z}{z-1}\begin{bmatrix}\frac{7}{4}\\ \frac{9}{4}\end{bmatrix}+\frac{z}{z+1}\begin{bmatrix}\frac{5}{4}\\ -\frac{5}{4}\end{bmatrix}+\frac{z}{(z-1)^2}\begin{bmatrix}\frac{1}{2}\\ \frac{1}{2}\end{bmatrix}$$

Verifications are requested in 10.6.1.

EXAMPLE 10.8. Solving a Partial Difference Equation. Where one application of the z-transformation replaces a linear difference equation in one variable by an algebraic equation, each application on certain partial difference equations results in a difference equation with one less variable. In this way we can solve

(1) $$y(k+1,t+1)=ay(k+1,t)+by(k,t)$$

$$y(0,t+1)=ay(0,t)$$

$$y(k,0)=1$$

where $k=0,1,2,\ldots$ and $t=0,1,2,\ldots$. We first transform (1), k to z, via

z-transforms to obtain

$$z[\eta(z,t+1)-y(0,t+1)]=az[\eta(z,t)-y(0,t)]+b\eta(z,t)$$

$$y(0,t+1)=ay(0,t)$$

$$\eta(z,0)=\frac{z}{z-1}$$

where we have held t fast as a parameter. Combining the first two equations produces

$$\eta(z,t+1)=\frac{az+b}{z}\eta(z,t) \tag{2}$$

$$\eta(z,0)=\frac{z}{z-1}$$

which is an instance of the homogeneous case of (10.39). Moreover, holding z fast as a parameter, (2) is a constant coefficient problem whose solution is

$$\eta(z,t)=\left(\frac{az+b}{z}\right)^t\frac{z}{z-1} \tag{3}$$

Now (3) is the z-transform of the sought-for solution $y(k,t)$ and we rewrite it as follows so we can use tabulated results.

$$\eta(z,t)=a^t\left(1+\frac{b}{a}\frac{1}{z}\right)^t\frac{z}{z-1}$$

Using Table 10.5(15) and 10.6(6),

$$\left(1+\frac{b}{a}\frac{1}{z}\right)^t=\mathfrak{Z}\left\{\left(\frac{b}{a}\right)^k C(t,k)\right\} \tag{4}$$

so that with convolution

$$y(k,t)=\sum_{u=0}^{k}C(t,u)a^{t-u}b^u \tag{5}$$

which is readily verified as a solution to (1). More than this, it can be shown to be the only solution (10.6.1).

EXAMPLE 10.9. Using Generating Functions. As remarked at the start of the present section, generating functions are commonly used in mathematics, probability, and statistics in place of z-transforms. Here we show how either can be used in place of the other with change of variable

$$(1) \qquad z=\frac{1}{s} \quad \text{or} \quad s=\frac{1}{z}$$

where s is the variable used for generating functions.

The counterpart of (10.86) is

$$(2) \qquad Y(s)=\sum_{k=0}^{\infty} s^k y(k)$$

where $s \in \mathcal{C}$. We call the function Y the *generating function* of y if there is some $\sigma \in \mathcal{R}$ such that (2) converges whenever $|s|<\sigma$. We use upper case symbols for generating functions as an exception to our customary practice of denoting functions by lowercase letters. There is the (optional) *inversion integral*

$$(3) \qquad y(k)=\frac{1}{2\pi i}\int_{\Gamma} s^{-k-1} Y(s)\,ds$$

where Γ is any circle with center at the origin and radius *less* than the critical radius σ for (2).

Evidently we have counterparts for all the results established for z-transforms because, roughly speaking, the only effect of the choice between s and z in (1) is to work inside, or outside, of circles about the origin. Using $\mathcal{G}\{\cdot\}$ for generating functions, the counterpart of

$$(4) \qquad \mathfrak{Z}\{y(k+1)\}=z\mathfrak{Z}\{y(k)\}-z\{y(0)\}$$

is the relation

$$(5) \qquad \mathcal{G}\{y(k+1)\}=\frac{1}{s}\mathcal{G}\{y(k)\}-\frac{1}{s}\{y(0)\}$$

These explain our opening remark noting that algebraic manipulations for z-transforms more directly match those for Laplace transforms than do those for generating functions.

There is one caution in connection with interchanges between z-transforms and generating functions in work with differential equations. In the present section we have worked entirely with difference equations and in these cases we can freely interchange the two by merely using the substitutions in (1). Such freedom also exists in the following section except that we must be sure to transform correctly derivatives with respect to s and z.

All that is required is use of the simplest case of the chain rule (6.64). The first set of procedures is the following for

(6) **GOING FROM z-TRANSFORMS TO GENERATING FUNCTIONS.** The replacements involving $\eta(z,t)$ for $t \in \mathfrak{R}$ are

$$z \text{ to } \frac{1}{s}, \qquad \eta \text{ to } Y$$

$$\frac{\partial \eta}{\partial t} \text{ to } \frac{\partial Y}{\partial t}, \qquad \frac{\partial \eta}{\partial z} \text{ to } -s^2 \frac{\partial Y}{\partial s}$$

For example, if the z-transform $\eta(z,t)$ satisfies

$$\frac{\partial \eta(z,t)}{\partial t} = -\left[\mu z^2 - (\lambda+\mu)z + \lambda\right] \frac{\partial \eta(z,t)}{\partial z}$$

then the equivalent condition for the generating function is

$$\frac{\partial Y(s,t)}{\partial t} = \left[\lambda s^2 - (\lambda+\mu)s + \mu\right] \frac{\partial Y(s,t)}{\partial s}$$

The companion procedure covers

(7) **GOING FROM GENERATING FUNCTIONS TO z-TRANSFORMS.** The replacements involving $Y(s,t)$ for $t \in \mathfrak{R}$ are

$$s \text{ to } \frac{1}{z}, \qquad Y \text{ to } \eta$$

$$\frac{\partial Y}{\partial t} \text{ to } \frac{\partial \eta}{\partial t}, \qquad \frac{\partial Y}{\partial s} \text{ to } -z^2 \frac{\partial \eta}{\partial z}$$

Both (6) and (7) are of importance for the material in Section 10.7 where partial derivatives occur and where both z-transforms and generating functions are commonly used in the literature.

PROBLEMS FOR SOLUTION

10.6.1 z-Transforms. Establish:

(1) Details omitted in text for Table 10.5.

(2) Repeat for Table 10.6.

(3) The other member of the pair by using Tables 10.5 and 10.6

(i) $k\cos bk$.
(ii) $k\sin bk$.
(iii) ke^{-ak}.
(iv) k^4.
(v) k^2a^k, several different ways.
(vi) $z^2(z-1)^{-3}$.
(vii) $z(z^2+1)^{-1}$.
(viii) z^{-k}.
(ix) $(z^2+1)^{-1}$.

(4) Repeat (vi)–(ix) using (10.90).
(5) Repeat using division algorithms.
(6) Details omitted in Example 10.6.
(7) Repeat for Example 10.7.
(8) Repeat for Example 10.8.
(9) Formal correspondences between Laplace and z-transforms via t to k, s to $z=e^s$, and so on.

10.6.2 Exercises. Solve using z-transforms and check solutions.

(1) Problems in 10.3.3.
(2) Problems in 10.2.2.
(3) Problems in 10.5.3.

10.6.3 Generating Functions. Establish:

(1) Solution of (10.47) using generating functions.
(2) Equations for generating function if the z-transform satisfies

(i) $$\frac{\partial\eta(z,t)}{\partial t}=\lambda\left(\frac{1}{z}-1\right)\eta(z,t).$$

(ii) $$\frac{\partial\eta(z,t)}{\partial t}+\lambda(z-1)^2\frac{\partial\eta(z,t)}{\partial z}=0.$$

10.7 DIFFERENTIAL-DIFFERENCE EQUATIONS

In this section we solve equations that are both differential equations and difference equations. There is no uniform terminology and at least three terms

differential–difference equations
difference–differential equations
equations of mixed type

are used with various meanings by authors. We favor the first, avoid the second, and reserve the third for what are also called differential equations with deviating arguments.

The subject has a long history and is generally conceded to be difficult for reasons including those that make functional equations difficult. Standard works include Pinney (1958), Bellman and Cooke (1963), and El'sgol'ts and Norkin (1973). See both of the latter references for terminology *retarded*, *advanced*, and *neutral* used to indicate how rates of change depend on past, future, and present values of "time."

We treat four sets of problems in detail with considerable attention to matters of technique. Representative equations are as follows, where throughout this section we continue the practice of the preceding section by agreeing

$$y(k,t)=0 \quad \text{for } k=-1,-2,\ldots, t<0$$

and the same for all functions of either k or t. In (10.96) we have

$$\frac{\partial y(k+1,t)}{\partial t}=ay(k,t)+f(k,t)$$

where, as elsewhere, we write $\partial/\partial t$ as a matter of personal preference, given that we follow the practice of writing two arguments rather than, for example, $y_k(t)$. The Poisson process is approached in (10.101) through

$$\frac{\partial y(k+1,t)}{\partial t}=-\lambda y(k+1,t)+\lambda y(k,t)$$

Example 10.10 is devoted to a model of linear growth based on

$$\frac{\partial y(k+1,t)}{\partial t}=(k+2)\lambda y(k+2,t)-2k\lambda y(k+1,t)+ky(\lambda,t)$$

Problem (10.112) is based on

$$\frac{dy(t+1)}{dt}=ay(t)+f(t+1)$$

Equations such as the first three figure importantly in probability theory and the fourth represents a class that has found wide application in automatic control theory, particularly, and in engineering and applied science, generally.

Alternative methods for finding solutions are demonstrated for all four sets of problems; in more complicated cases this could not be expected to be possible. Three different approaches are given for (10.96) but, in general, the choice is between direct solution methods and indirect ones based on transforms. Sometimes, as for (10.96), there turn out to be distinct advantages in one particular approach.

The first problem can be viewed as a generalization of both a first-order linear differential equation and a difference equation. Roughly speaking, we have a kind of differential equation in t and something like a difference equation in k. Precisely, the Laplace transform of $y(k,t)$ satisfies (10.39) while the z-transform satisfies (9.5) in the problem

$$\frac{\partial y(k+1,t)}{\partial t} = ay(k,t)+f(k,t), \qquad k=0,1,2,\ldots,t>0$$

$$y(k,0)=g(k)$$

$$y(0,t)=h(t) \tag{10.96}$$

where h and all f are continuous functions of t, and where $g(0)=h(0)$ so $y(0,0)$ is defined. We call the first equation a *first-order linear differential–difference equation with constant coefficient a*. When $f(k,t)=0$ for all k and t of interest the problem is called *homogeneous* and otherwise it is *nonhomogeneous*. If t represents "time" while k identifies "state," then $g(k)$ can be said to prescribe *initial conditions* while $h(t)$ prescribes *boundary values*. If k represents "time" then we would associate $h(t)$ with initial conditions but, for brevity, we use only the preceding names.

An existence and uniqueness theorem can be established for Problem (10.96) in straightforward fashion using the corresponding theorems for (9.5) and (10.39); see 10.7.1. There are three methods of solution that we can immediately identify.

1. The direct method of using (9.20) to solve successive differential equations for $k=0,1,2,\ldots$.
2. The z-transform method using (9.20) to solve for $\eta(z,t)$.
3. The Laplace transform method using (10.57) to solve a difference equation for $\eta(k,s)$.

In 1 we proceed as though we were using a recurrence relation (10.45) and it may be difficult to find a closed form solution. In 2 it appears initially that $h(t)$ needs to be differentiable but it later turns out that such is not required. We will demonstrate 3 which is perhaps the best approach; see 10.7.1 for solutions using 1 and 2.

If we take Laplace transforms, t to s, in (10.96) then we obtain

$$\eta(k+1,s)=\frac{a}{s}\eta(k,s)+\frac{\phi(k,s)}{s}+\frac{g(k+1)}{s}$$

$$\eta(0,s)=\chi(s) \tag{10.97}$$

where we have adapted the notation of Section 9.5 so that $\chi(s)=\mathfrak{L}\{h(t)\}$. It should also be noted that $g(k+1)=y(k+1,0)$ appears as the initial value and not $g(k)$. Using (10.57) we write the unique solution

$$\eta(k,s)=a^k\frac{1}{s^k}\chi(s)+\sum_{u=0}^{k-1}a^u\frac{1}{s^{u+1}}\left[\phi(k-1-u,s)+g(k-u)\right]$$

From Table 9.2(3) and convolution (9) in Table 9.3 we obtain the general expression

$$y(k,t)=\sum_{u=0}^{k-1}a^u\int_0^t\frac{(t-x)^u}{u!}f(k-1-u,x)\,dx+\sum_{u=0}^{k-1}a^u\frac{t^u}{u!}g(k-u)$$

$$+\frac{a^k}{(k-1)!}\int_0^t(t-x)^{k-1}h(x)\,dx \tag{10.98}$$

which is verifiable as a solution, and hence the unique solution, to Problem (10.96). See 10.7.1 where some special cases are considered. Here we note that the expression involving f is a particular solution to the nonhomogeneous equation. Finally, if

$$h(t)=\text{constant}$$

then (as is apparent from the preliminary form of the solution obtained by first taking z-transforms in place of Laplace transforms (10.7.1)) there are two significant changes in (10.98). First, the integral involving h drops out. Second, $h(0)=g(0)$ appears in the second summation where now the upper limit is k in place of $k-1$.

What is the vector-matrix problem that generalizes (10.96) in the way that the systems problems in Sections 9.4 and 10.5 generalize their scalar predecessors? It can be written at once, and since we can solve the individual problems for components, we can also solve the problem in $y(k,t)$. In place of pursuing these matters here, let us note that the accompanying $n=2$ generalization, that is, the $n=2$ scalar equation that is equivalent to the 2×2 system equation is

$$\frac{\partial^2 y(k+2,t)}{\partial t^2}+\alpha_1\frac{\partial y(k+1,t)}{\partial t}+\alpha_0 y(k,t)=b(k,t) \qquad (10.99)$$

For proofs of the preceding assertions, examples, and the initial and boundary values that must accompany (10.99), see 10.7.2.

Next we consider a special case of an important differential–difference problem from probability theory.

$$\frac{\partial y(k+1,t)}{\partial t}=\mu y(k+2,t)-(\lambda+\mu)y(k+1,t)+\lambda y(k,t)$$

$$\frac{\partial y(0,t)}{\partial t}=\mu y(1,t)-\lambda y(0,t)$$

$$y(k,0)=1, \qquad k=i$$

$$=0, \qquad k\neq i \qquad (10.100)$$

where $k=0,1,2,\ldots,t>0$, and $k=i$ is the prescribed *initial state*. This is the constant coefficient (that is, not state dependent) case of a *birth-and-death process* where λ turns out to be the *mean birth rate* and μ the *mean death rate*; see Feller (1968, XVII.5). Also see Gross and Harris (1974, Chapter 2) where (10.100) is developed and solved in the context of a *single-channel queueing model* with *exponential interarrival times* and *exponential service times*. In general, $y(k,t)$ is the probability of finding a "system" in "state" k at "time" t. *State* k identifies the situation where there are exactly k members of a "population" possessing some specified property. For example, in queueing theory

$y(k,t)=$ probability that there are exactly k units calling for service at time t

$\lambda=$ mean arrival rate

$\mu=$ mean service rate

$i=$ initial queue length

and (10.100) describes what is called the *transient behavior*. Equating the RHSs of the two differential–difference equations to zero and solving yields the *steady-state solution* when $y(k)$ is independent of time t; see 10.7.3.

The special case of (10.100) we consider describes the *Poisson process*. Here $\mu=0$ which means units cannot die so we have only a birth process. The initial state is $i=0$ and the complete set of equations for $k=0,1,2,\ldots,$ $t>0$ is

$$\begin{aligned}
\frac{\partial y(k+1,t)}{\partial t} &= -\lambda y(k+1,t)+\lambda y(k,t)\\
\frac{\partial y(0,t)}{\partial t} &= -\lambda y(0,t)\\
y(0,0)&=1,\\
y(k,0)&=0, \qquad k\neq 0
\end{aligned} \tag{10.101}$$

These equations can be solved using (9.20) for $k=0,1,2,\ldots$ and we can write

$$\sum_{u=0}^{k} y(u,t)=\sum_{u=0}^{k}\frac{(\lambda t)^u}{u!}e^{-\lambda t} \tag{10.102}$$

which is the distribution function of the *Poisson probability distribution with mean* λt. See 10.7.4 for the direct solution.

Let us solve (10.101) using z-transforms to illustrate how to find an equation satisfied by the transform. We cannot proceed exactly as in earlier cases where we had a single equation holding for all k because the second equation in (10.101) is not a special case of the first. Such exceptions are commonly encountered in working with particular differential–difference equations where they are natural consequences of initial or boundary "discontinuities." For example, there is no $y(-1,t)$ term in the second equation in (10.101) because there is no state $k=-1$ from which one "birth" could have led to the subject state $k=0$.

Since we do not have a single equation holding for all k and t, we use a tabular arrangement to make sure we obtain a correct single equation for the z-transform. The procedure is the simple one shown in Table 10.7. We merely multiply each equation by its indicated multiplier and then sum the resulting equations. In Table 10.7 we first obtain

$$\frac{\partial\eta(z,t)}{\partial t}=-\lambda\eta(z,t)+\lambda\sum_{k=1}^{\infty}\frac{y(k-1,t)}{z^k}$$

TABLE 10.7 TABLE FOR POISSON PROCESS

Index	Multiplier	Equation
$k=0$	1	$\frac{\partial y(0,t)}{\partial t}=-\lambda y(0,t)$
$k=1$	z^{-1}	$\frac{\partial y(1,t)}{\partial t}=-\lambda y(1,t)+\lambda y(0,t)$
$k=2$	z^{-2}	$\frac{\partial y(2,t)}{\partial t}=-\lambda y(2,t)+\lambda y(1,t)$
$\vdots$	$\vdots$	$\vdots$
k	z^{-k}	$\frac{\partial y(k,t)}{\partial t}=-\lambda y(k,t)+\lambda y(k-1,t)$

Either by rewriting this last sum as follows with $k-1=u$

$$\frac{\lambda}{z}\sum_{u=0}^{\infty}\frac{y(u,t)}{z^{u}}$$

or by (meticulous) use of Table 10.6(5), we recognize the sum as

$$\frac{\lambda}{z}\eta(z,t)$$

Thus the new problem obtained from transforming (10.101) is

$$\frac{\partial\eta(z,t)}{\partial t}=\left(-\lambda+\frac{\lambda}{z}\right)\eta(z,t)$$

$$\eta(z,0)=1 \tag{10.103}$$

where the second equation results from the fact that the only nonzero term in (10.86) for $t=0$ is

$$y(0,0)=1$$

Again by (9.20), the unique solution to (10.103) is

$$\eta(z,t)=e^{-\lambda t}e^{\frac{\lambda t}{z}} \tag{10.104}$$

and the inverse transform from Table 10.5(8) coincides with the summand in (10.102) found by the direct method.

Once we have found the z-transform of $y(k,t)$ in any case where

$$Y(k,t)=\sum_{u=0}^{k} y(u,t) \tag{10.105}$$

is a discrete probability distribution function, we can use its derivatives with respect to z to study the moments of the distribution. This can be done without the need to find $y(u,t)$, that is, without inverting the z-transform, because we can use Table 10.6(17). The rth moment of (10.105) is the expected value of k^r which is defined for $r=1,2,3,\ldots$ as

$$\sum_{u=0}^{\infty} u^r y(u,t) \tag{10.106}$$

whenever this series converges. Thus, the *mean* $m(t)$, which is given by (10.106) for $r=1$, can be found from

$$m(t)=\lim_{z\to 1} -\frac{\partial\eta(z,t)}{\partial z} \tag{10.107}$$

If we define

$$n(t)=\lim_{z\to 1} \frac{\partial^2\eta(z,t)}{\partial z^2} \tag{10.108}$$

then the *variance* $v(t)$, which is the expected value of $[k-m(t)]^2$, that is,

$$v(t)=\sum_{u=0}^{\infty} \left[u-m(t)\right]^2 y(u,t)$$

can be found as

$$v(t)=n(t)-m(t)-\left[m(t)\right]^2 \tag{10.109}$$

Using these results with (10.104) we find that both the mean and variance exist for the Poisson distribution and have the common value λt (10.7.4).

There is another method that can perhaps be used to find moments of a probability distribution when we have the associated differential–difference equations. The method consists of using the latter equations to find differential equations that can be solved to yield $m(t)$, $v(t)$, and so on. Here the first task is to find single equations for the moments given that there usually will not be a single differential–difference equation covering all $k=0,1,2,\ldots$. In this latter respect we have the same concern that prompted use of Table 10.7 for the Poisson distribution. Indeed, if we proceed for that distribution using multipliers k in (10.101) and summing the equations (compare the equations in Table 10.7) we obtain

$$\sum_{k=0}^{\infty} k\frac{\partial y(k,t)}{\partial t} = -\lambda \sum_{k=0}^{\infty} ky(k,t)+\lambda \sum_{k=0}^{\infty} (k+1)y(k,t)$$

With this, and the fact that $y(0,0)=1$, we see that the mean $m(t)$ must satisfy

$$m'(t)=\lambda$$

$$m(0)=0$$

and consequently $m(t)=\lambda t$. Working with k^2 in place of k and with notation

$$q(t)=\sum_{u=0}^{\infty} u^2y(u,t) \tag{10.110}$$

we obtain the problem

$$q'(t)=2\lambda m(t)+\lambda$$

$$q(0)=0$$

with solution $q(t)=\lambda^2t^2+\lambda t$. Since the variance satisfies

$$v(t)=q(t)-[m(t)]^2 \tag{10.111}$$

we have another method of showing $v(t)=m(t)$ for the Poisson probability distribution.

EXAMPLE 10.10. A Linear Growth Process. The subject process is studied in Feller (1968, XVII.5). We are interested in solving the equations using z-transforms and showing that three quite different approaches can be used.

We consider a population of elements that can split or die. During any short interval of length h the probability for any living element to split into two is

$$\lambda h+o(h)$$

and the probability that the unit will die equals this same quantity; what remains of unity of course gives the probability of no change. The differential–difference system for this special birth-and-death process for $k=0,1,2,\ldots$ and $t>0$ is

$$(1) \qquad \frac{\partial y(k+1,t)}{\partial t}=(k+2)\lambda y(k+2,t)-2k\lambda y(k+1,t)+k\lambda y(k,t)$$

$$\frac{\partial y(0,t)}{\partial t}=\lambda y(1,t)$$

$$y(1,0)=1$$

$$y(k,0)=0, \qquad k\neq 1$$

The final two equations specify that the initial state is $i=1$; in other words, we start at $t=0$ with a population consisting of a single element.

Table 10.8 displays the data used to find the equation satisfied by the z-transform. Just as in Table 10.7 for (10.101) we use this method because we do not have a single equation for $y(k,t)$. Let us denote the result of summing the products of multipliers and equations in Table 10.8 by

$$(2) \qquad \frac{\partial \eta(z,t)}{\partial t}=Q+R+S$$

The first summation Q equals

$$\lambda \sum_{k=0}^{\infty} \frac{(k+1)y(k+1,t)}{z^k}=\lambda z \sum_{u=0}^{\infty} \frac{uy(u,t)}{z^u}$$

where we have changed summation variable via $k=u-1$ and then altered the lower limit. Thus Q equals

$$(3) \qquad \lambda z \mathfrak{Z}\{ky(k,t)\}=-\lambda z^2 \frac{\partial \eta(z,t)}{\partial z}$$

by Table 10.6(7). In similar fashion R equals

$$(4) \qquad \lambda \sum_{k=2}^{\infty} \frac{(k-1)y(k-1,t)}{z^k}=-\lambda \frac{\partial \eta(z,t)}{\partial z}$$

and finally

$$(5) \qquad S=2\lambda z \frac{\partial \eta(z,t)}{\partial z}$$

TABLE 10.8 TABLE FOR LINEAR GROWTH PROCESS

Index	Multiplier	Equation
$k=0$	1	$\frac{\partial y(0,t)}{\partial t}=\lambda y(1,t)$
$k=1$	z^{-1}	$\frac{\partial y(1,t)}{\partial t}=2\lambda y(2,t)-2\cdot 1\lambda y(1,t)$
$k=2$	z^{-2}	$\frac{\partial y(2,t)}{\partial t}=3\lambda y(3,t)-2\cdot 2\lambda y(2,t)+\lambda y(1,t)$
$k=3$	z^{-3}	$\frac{\partial y(3,t)}{\partial t}=4\lambda y(4,t)-2\cdot 3\lambda y(3,t)+2\lambda y(2,t)$
$\vdots$	$\vdots$	$\vdots$
k	z^{-k}	$\frac{\partial y(k,t)}{\partial t}=(k+1)\lambda y(k+1,t)-2k\lambda y(k,t)$ $+(k-1)\lambda y(k-1,t)$

Using (3), (4), (5) in (2) we obtain the transformed problem

$$\frac{\partial\eta(z,t)}{\partial t}+\lambda(z-1)^2\frac{\partial\eta(z,t)}{\partial z}=0 \tag{6}$$

$$\eta(z,0)=\frac{1}{z}$$

This problem can be solved using an elementary solution method for partial differential equations (10.7.7). We integrate both sides of

$$\frac{dt}{1}=\frac{dz}{\lambda(z-1)^2}$$

and obtain

$$t=\frac{-1}{\lambda(z-1)}+c$$

which we rewrite as

$$\lambda t+\frac{1}{z-1}=\lambda c$$

Since the RHS is a constant, according to 10.7.7 the solution of (6) is

$$\eta(z,t)=\psi\left(\lambda t+\frac{1}{z-1}\right)$$

where $\psi(\cdot)$ is a function that is to be determined. From the second line in (6), $\eta(z,0)$ equals

$$\frac{1}{z}=\psi\left(\frac{1}{z-1}\right)$$

Finally, we use the elementary technique of 1.1.8 and find (10.7.8)

$$\eta(z,t)=\frac{\lambda t+1/(z-1)}{1+\lambda t+1/(z-1)} \tag{7}$$

is the solution to (6).

Now Table 10.5 does not list the inverse z-transform of (7). However, we can invert (7), and consequently solve (1), in three distinct ways. The first is not applicable except in special cases.

APPROACH 1. Use the "initial value" result, Table 10.6(13) to find $y(0,t)$. Then find all $y(k,t)$ by differentiation (without further use of $\eta(z,t)$).

The limiting value of (7) is

$$y(0,t)=\frac{\lambda t}{1+\lambda t} \tag{8}$$

Using the equation for $k=0$ in Table 10.8

$$y(1,t)=\frac{1}{(1+\lambda t)^2}$$

Continuing in this way we find

$$y(k,t)=\frac{(\lambda t)^{k-1}}{(1+\lambda t)^{k+1}}, \qquad k=1,2,3,\ldots \tag{9}$$

which, together with (8), gives the solution to (1). Notice that as soon as we find (8) we can verify $y(0,t)\to 1$ as $t\to\infty$ so the population ultimately becomes extinct. There is the optional

APPROACH 2. Use the Inversion formula (10.89) to find the inverse z-transform of (7).

We must find the residues of

$$z^{k-1}\eta(z,t)=\frac{z^{k-1}}{1+\lambda t}\,\frac{\lambda t z-\lambda t+1}{z-\lambda t/(1+\lambda t)} \tag{10}$$

and we notice first that (10) has a pole of order one at $z=0$ although $\eta(z,t)$ has none there. Using (8.15), we evaluate the RHS of (10) for $k=0$, multiply by z, and let $z \to 0$ to obtain

$$\text{(11)} \qquad \operatorname{res}\left(z^{-1}\eta(z,t),0\right)=\frac{\lambda t-1}{\lambda t}$$

From the same formula, and for $k=0,1,2,\ldots$

$$\text{(12)} \qquad \operatorname{res}\left(z^{k-1}\eta(z,t),\frac{\lambda t}{1+\lambda t}\right)=\frac{(\lambda t)^{k-1}}{(1+\lambda t)^{k+1}}$$

The sum of (11) and (12) for $k=0$ yields $y(0,t)$ as given in (8). For $k=1,2,3,\ldots$, (12) coincides with (9) so again we have found the solution. There is also

APPROACH 3. Use division algorithms as in Example 8.2 to find the inverse z-transform of (7).

First, we can expand (7) in powers of z^{-1} through "long division" arranged as follows

$$\text{(13)} \qquad \left\{\lambda t+\frac{1}{z-1}\right\} \div \left\{(1+\lambda t)+\frac{1}{z-1}\right\}$$

where the geometric series

$$\frac{-1}{z-1}=1+z+z^2+\cdots$$

is used in both dividend and divisor. From such a division we obtain (8) and (9). As an equivalent process, we can seek to rewrite (7) in a form

$$\frac{1}{1-\alpha(\lambda,t)z^{-1}}$$

possibly multiplied by a power of z^{-1}. Then $y(k,t)$ can be found directly from the formula for a geometric series in z^{-1}. In the present case, we can rewrite (7) as

$$\text{(14)} \qquad \eta(z,t)=\frac{\lambda t}{1+\lambda t}+\frac{1}{(1+\lambda t)^2}\cdot\frac{1}{z}\cdot\frac{1}{1-\dfrac{\lambda t}{1+\lambda t}\cdot\dfrac{1}{z}}$$

from which (8) and (9) result at once.

Approach 1 works well and requires no knowledge of complex variables; however, it is not a generally applicable method. Whenever we can succeed in finding $\eta(z,t)$, Approach 3, which also does not use complex variables, succeeds; however, in complicated cases this requires something more like "art" than "science." Approach 2 is applicable whenever we can determine the z-transform; however, its greatest advantage is that it is markedly straightforward and systematic.

We call the problem

$$\frac{dy(t+1)}{dt} = ay(t) + f(t+1), \qquad t>0$$

$$y(t) = g(t), \qquad 0<t<1 \tag{10.112}$$

where f and g are continuous, a (differential–difference) *problem of mixed type* because the single variable t is involved in both differentiation and displacement. The terms *constant coefficient*, *nonhomogeneous*, *homogeneous*, and *initial value* naturally refer to a, $f \neq 0$, $f=0$, and $g(t)$, respectively, where the *initial period* corresponds to the interval $(0,1)$. Problems such as (10.112) are also called differential equations with *deviating arguments*.

Problem (10.112) is closer to a functional equation problem than to any of the preceding problems of this section because the displaced argument t is defined on a real halfline. We find solutions neglecting arbitrary unit periodics (10.2.7).

If we impose the requirement in (10.112) that y is continuous then there exists a unique solution. This can be shown as follows where we use the direct solution method called *solution by continuation* which consists of integrating from one interval to the next. To this end let us rewrite the first line as

$$y'(t) = ay(t-1) + f(t), \qquad t \geqq 1 \tag{10.113}$$

and commence. On the first interval we impose the given initial condition

$$y(t) = g(t), \qquad 0 \leqq t \leqq 1$$

so that on the second interval we have an instance of Problem (9.13), namely

$$y'(t) = ag(t-1) + f(t), \qquad 1 \leqq t \leqq 2$$

$$y(1) = g(1)$$

The second portion of the solution is therefore

$$y(t) = g(1) + \int_1^t \left[ag(x-1) + f(x) \right] dx, \qquad 1 \leqq t \leqq 2$$

Clearly this process yields the solution in the form of a recurrence relation

$$y(t)=g(t), \qquad 0\leqq t\leqq 1$$
$$=y(k)+\int_k^t [ay(x-1)+f(x)]\,dx, \quad k\leqq t\leqq k+1, t\geqq 1 \qquad (10.114)$$

where $k=[t]$ denotes the greatest integer in t, and where, as noted previously, we neglect arbitrary unit periodics.

Given particular continuous functions f and g, we can use (10.114), or go back to (10.113), to write the solution, interval by interval (generally hoping that we can find a closed form expression, or that we can make do without the need to find one). For examples, see 10.7.9.

Before considering the use of Laplace transforms to solve (10.112) let us consider the simpler problem

$$y'(t+1)=ay(t)+f(t+1), \qquad t>0$$
$$y(t)=\int_0^t f(u)\,du, \qquad 0\leqq t\leqq 1 \qquad (10.115)$$

where again f is continuous and we seek a continuous solution y. We say (10.115) is simpler because here we can write a simple equation for the Laplace transform. To see this, and to be sure that we obtain a correct equation, we rewrite (10.115) in an equivalent form (recall $y(t)=0$ for $t<0$ throughout the present section)

$$y'(t)=ay(t-1)+f(t), \qquad t>0$$
$$y(0)=0 \qquad (10.116)$$

Now it is clear from (3) and (5) in Table 9.3 that the Laplace transform satisfies

$$s\eta(s)=ae^{-s}\eta(s)+\phi(s)$$

Thus we can express

$$\eta(s)=\frac{\phi(s)}{s}\,\frac{1}{1-ae^{-s}/s}$$

and below we show that we can use the series

$$\gamma(s)=\frac{1}{s}+\frac{a}{s^2}e^{-s}+\frac{a^2}{s^3}e^{-2s}+\cdots \qquad (10.117)$$

to write

$$\eta(s)=\phi(s)\gamma(s)$$

If we use convolution, Table 9.3(9), we can write $y(t)$ in either of the equivalent forms

$$y(t)=\int_0^t f(t-u)\,g(u)\,du \tag{10.118}$$

$$y(t)=\int_0^t f(u)\,g(t-u)\,du \tag{10.119}$$

where

$$g(t)=\mathcal{L}^{-1}\{\gamma(s)\} \tag{10.120}$$

is the inverse Laplace transform of (10.117). Are we correct in using the series (10.117) in (10.120)? The answer is "yes," and not for the reason that we can use an infinite series of inverse transforms. We are correct in (10.120) because the inverse transform of the infinite series (10.117) is a finite sum on any finite subinterval of the real halfline. First, we have the individual inverse transforms

$$\begin{aligned}
\mathcal{L}^{-1}\left\{\frac{1}{s}\right\} &= 1, && 0<t\\
\mathcal{L}^{-1}\left\{\frac{1}{s^2}e^{-s}\right\} &= 0, && 0<t\leqq 1\\
&= t-1, && 1\leqq t\\
\mathcal{L}^{-1}\left\{\frac{1}{s^3}e^{-2s}\right\} &= 0, && 0<t\leqq 2\\
&= \frac{(t-2)^2}{2!}, && 2\leqq t
\end{aligned} \tag{10.121}$$

and so on. Second, we use these to write finite expressions for (10.120)

$$\begin{aligned}
g(t)&=1, && 0\leqq t\leqq 1\\
&=1+a(t-1), && 1\leqq t\leqq 2\\
&=1+a(t-1)+\frac{a^2}{2!}(t-2)^2, && 2\leqq t\leqq 3
\end{aligned} \tag{10.122}$$

and so on, for $t>0$. Using (10.118) we have

$$\begin{aligned} y(t) &= \int_0^t f(t-u)\,du, && 0 \leqq t \leqq 1 \\ &= \int_0^t f(t-u)\,du + a\int_1^t (u-1)f(t-u)\,du, && 1 \leqq t \leqq 2 \\ &= \int_0^t f(t-u)\,du + a\int_1^t (u-1)f(t-u)\,du \\ &\quad + \frac{a^2}{2!}\int_2^t (u-2)^2 f(t-u)\,du, && 2 \leqq t \leqq 3 \end{aligned} \tag{10.123}$$

and so on, as an explicit solution that can be rewritten in a number of ways (10.7.9).

Let us now consider the use of Laplace transforms for our original problem of mixed type (10.116). If we suppose g is differentiable and express (10.116) in the equivalent form

$$\begin{aligned} y'(t) &= g'(t), && 0 < t \leqq 1 \\ &= ay(t-1) + f(t), && 1 \leqq t \\ y(0) &= g(0) \end{aligned} \tag{10.124}$$

we soon see exactly what is involved in finding an equation for $\eta(s)$. Forming $\mathfrak{L}\{y'(t)\}$, we have

$$s\eta(s) - g(0) = \int_0^1 e^{-st}g'(t)\,dt + a\int_1^\infty e^{-st}y(t-1)\,dt + \int_1^\infty e^{-st}f(t)\,dt \tag{10.125}$$

The second term on the RHS in another instance of Table 9.3(5), namely

$$ae^{-s}\eta(s)$$

so we must find expressions for the first and third terms. Given particular functions f and g we can perhaps evaluate the RHS of (10.125) using tables of integrals. Alternatively we can proceed as follows.

If we transform t to $u+1$ in the extreme RHS member we obtain, formally

$$e^{-s}\mathfrak{L}\{f(t+1)\}$$

which we may be able to evaluate using a table of transforms (10.7.9). Similarly, the first RHS member in (10.125) is, again formally

$$\mathfrak{L}\{g'(t)\} - \int_1^{\infty} e^{-st} g'(t)\,dt = s\gamma(s) - g(0) - e^{-s}\mathfrak{L}\{g'(t+1)\}$$

In this way we are led to the following expression for the formal Laplace transform

$$\eta(s) = \gamma(s)\frac{s}{s-ae^{-s}} + \mathfrak{L}\{f(t+1) - g'(t+1)\}\frac{e^{-s}}{s-ae^{-s}} \qquad (10.126)$$

for the solution of (10.116). This expression could be manipulated further but its usefulness clearly depends on particular properties of f and g. Given particular f and g, we can try to evaluate the RHS.

PROBLEMS FOR SOLUTION

10.7.1 Problem (10.96). Establish:

(1) A general existence and uniqueness theorem.

(2) The solution for the homogeneous case given $g(k) = 1$ and $h(t) = 1$, using the direct method.

(3) Repeat using Laplace transforms.

(4) Repeat using z-transforms.

(5) The solution $y(k,t)$ is a k-stage truncation of the solution of

$$y'(t) = ay(t)$$

$$y(0) = 1$$

(6) Formula (10.98) is a solution.

(7) Solutions for the following special cases (and check).

  (i) $f(k,t) = f(k)$ is independent of t.

  (ii) $f(k,t) = b$ where $b \in \mathfrak{R}$.

  (iii) $g(k) = k$ and $h(t) = te^t$.

(8) The solution can be found (in the form where $h'(x)$ appears under the integral sign) by using z-transforms.

(9) The preceding can be used to write the solution for $h(t) = c \in \mathfrak{R}$.

(10) The result in (8) just above coincides with that in (10.98).

10.7.2 A First-Order System. Establish:

(1) The nth order generalization of (10.96) and the accompanying $n \times n$ problem for $n = 2$.

(2) An existence and uniqueness theorem.

(3) The solution for the homogeneous case where

$$\mathbf{A}=\begin{bmatrix} 4 & -2 \\ 3 & -1 \end{bmatrix}, \qquad \mathbf{y}(k,0)=\mathbf{y}(0,t)=\begin{bmatrix} 1 \\ 2 \end{bmatrix}$$

(4) Solution for (10.99) where

(i) $\alpha_1=-3$, $\alpha_0=2$, $y(0,t)=1$, $\dfrac{\partial y(1,t)}{\partial t}=2.$

(ii) Repeat for $-2,1,1,1$, respectively.

10.7.3 Steady-State Solution. Find the subject solution $y(k)$ of (10.100) subject to $\Sigma_{u=0}^{\infty} y(u)=1$ and check.

10.7.4 Poisson Process. Establish:

(1) Solution of (10.101) by direct integration.
(2) The mean and variance from the definitions.
(3) Repeat using the z-transform.
(4) Use of "backward recursion" to solve (10.100) for the "pure death" case $\lambda=0$ where initially there are n units and

$$\sum_{u=0}^{n} y(u,t)=1, \qquad t\geqq 0$$

(5) How does the preceding relate to a Poisson distribution?

10.7.5 Finding Equation for z-Transform. Use a counterpart to Tables 10.7, 10.8 to find a partial differential equation for $\pi(z,t)$ given

$$\frac{\partial p(0,t)}{\partial t}=-\lambda p(0,t)+\mu p(1,t)$$

$$\frac{\partial p(k,t)}{\partial t}=-(\lambda+k\mu)p(k,t)+\lambda p(k-1,t)+(k+1)\mu p(k+1,t),$$

$$k=1,2,3,\ldots$$

10.7.6 Yule–Furry Distribution. Find:

(1) Solution using a recurrence relation.

$$\frac{\partial p(n,t)}{\partial t}=-n\lambda p(n,t)+(n-1)\lambda p(n-1,t)$$

$$\text{for } n=1,2,3,\ldots, t\geqq 0$$

$$p(u,t)=e^{-u\lambda t}$$

$$\begin{aligned} p(n,0)&=1, \qquad n=u \\ &=0, \qquad n\neq u \end{aligned}$$

(2) Mean.
(3) Variance.
(4) Probability of extinction.

10.7.7 A Partial Differential Equation. Show that we can solve

$$u(x,y)\frac{\partial w}{\partial x}+v(x,y)\frac{\partial w}{\partial y}=0$$

as follows. First obtain the solution $\theta(x,y)=\text{constant}$ by integrating

$$\frac{dx}{u(x,y)}=\frac{dy}{v(x,y)}$$

Second, write the solution $w=\psi(\theta(x,y))$ where $\psi(\cdot)$ is an arbitrary differentiable function of a single variable.

10.7.8 Example 10.10. Establish:
(1) Formula (7).
(2) Repeat for (9) using induction.
(3) Use of (13) to obtain the solution.
(4) Repeat for (14).
(5) Mean, variance, and probability of extinction.
(6) Repeat for $\lambda\neq\mu$ and initial population i.

10.7.9 Equations of Mixed Type. Establish for (10.112):
(1) Formula (10.114) determines a solution and it is the unique solution.
(2) Solutions by continuation for the following cases and check.
  (i) Homogeneous case for $a=1$ and $g(t)=0$.
  (ii) Repeat for $g(t)=1$.
  (iii) $f(t)=t$ and $g(t)=1$.
(3) Formula (10.123) provides a solution of (10.115).
(4) Solutions using Laplace transforms for
  (i) $f(t)=b$.
  (ii) $f(t)=t$.
(5) Different versions of (10.123) using
  (i) Formula (10.119).
  (ii) Upper limits $t,t-1,t-2,\ldots$ in the successive integrals.
(6) The counterpart of Table 9.2 with entries
  (i) $y(t+1)$.
  (ii) $y'(t+1)$.
(7) Repeat of (2) just above, using Laplace transforms.

ANSWERS, HINTS, AND COMMENTS

SECTION 1.1

1.1.1 (1) Life exists outside of our solar system. (2) This sentence is false.

1.1.2 (1) There exists an angle that cannot be trisected using ....

1.1.3 (1) p only if q; p is sufficient for q; for p, q is necessary; q is implied by p; q if p; for q, p is sufficient; q is necessary for p. (2) if not q then not p. (3) if q then p. (4) p_1 implies q where p iff (p_1 and p_2). (5) p implies (q and q_1). (6) p is equivalent to q; q is necessary and sufficient for p; theorem and its converse are both true.

1.1.4 (1) Proof not involving assumption "not q," usually amounting to exhaustion of cases. (2) Prove contrapositive. (3) Exhibit one instance where p is true and q false. (4) First, prove that $q(1)$ is true; second, prove that for any positive integer k, $q(k)$ implies $q(k+1)$. (5) First, prove $q(1)$; second, prove that for any positive integer k, $q(m)$ true for $m=1,2,\ldots,k$ implies $q(k+1)$. (6) First, prove $q(2)$. Second, prove $q(2^k)$ for any positive integer by (4) or (5) above. Third, prove that for any n, $q(n)$ implies $q(n-1)$.

1.1.5 (2) Show $C(x)=C(y)$ or else $C(x)$ and $C(y)$ do not intersect.

1.1.6 Drill on terminology.

1.1.7 (1) Suppose first that f^{-1} exists. If $f(x_1)=f(x_2)$, then $(f(x_1),x_1)$ and $(f(x_2),x_2)$ are both in f^{-1}. Conclude f is one-to-one because $x_1=x_2$ results from the fact that f^{-1} is a function. Given any $y \in Y$ it must be the image under f of $f^{-1}(y)$ so that f is onto. Conversely, if f is onto, then any $y \in Y$ must be the second coordinate of an element $(x,y) \in f$. If f is also one-to-one, the elements (y,x) form a function on Y to X because then $x_1=x_2$ whenever (y,x_1) and (y,x_2) are both members. ■ (2) Use an indirect proof.

1.1.8 (1) If $t=(x-1)^{-1}$ then $x=t^{-1}(1+t)$, $x^{-1}=t(1+t)^{-1}$, and $f(t)=t(1+t)^{-1}$. Conclude $f(x)=x(1+x)^{-1}$.

SECTION 1.2

1.2.1 Construct the group table covering all cases for $a+b=?$ and similarly for $ab=?$

1.2.2 Prove both parts.

1.2.3 (1) Since $-1<0$ then $-z<-y<0$. Then add $1-x$ to each member. (2) Since $t^2 \leqq t$, $t^2(b-a)^2 \leqq t(b-a)^2$ and also $ab \leqq ab+t(b-a)^2-t^2(b-a)^2$. Rewrite this last as

$$ab \leqq \left[ta+(1-t)b\right]\left[(1-t)a+tb\right]$$

and then divide both sides by the positive quantity $ab[ta+(1-t)b]$ to obtain the desired inequality. The equality statements are readily verified. (3) $x \in (2,5)$.

1.2.4 Drill on definitions and use of (1.5).

1.2.5 Since $i \neq 0$ necessarily (1) $i<0$ or (2) $i>0$. From (1) there follows $0<1$ and $1<0$; similarly (2) is impossible.

1.2.6 Drill using "xy notation" wherein x and y are manipulated, or "z notation" using relations such as $z\bar{z}=|z|^2$.

1.2.7 Drill using order in $\mathcal{R}$. For example, the triangle inequality (4) results from 1.2.6 via $|z_1+z_2|^2=|z_1|^2+|z_2|^2+2\operatorname{Re} z_1\bar{z}_2$.

1.2.8 (1) $\cos\theta=\cos(\theta+2k\pi)$ for $k=0, \pm 1, \pm 2, \ldots$ and similarly for $\sin\theta$. (2) $\arg\theta$ is single-valued. (3) Whenever z crosses the negative real axis $\arg z$ jumps between π and $-\pi$.

1.2.9 Use the unique factorizations of numerator and denominator.

SECTION 1.3

1.3.1 Note that (6) and (7) do not always determine real-valued functions.

1.3.2 Global and local extrema can occur only at (1) *critical points* which are points x_0 in (1.11), (2) endpoints of I, or (3) points where f is not differentiable.

1.3.3 $f(x_0+h)=f(x_0)+f'(x_0)h+\cdots+\dfrac{f^{(n-1)}(x_0)}{(n-1)!}h^{n-1}+\dfrac{f^{(n)}(u)}{n!}h^n$

1.3.4 The rule may need to be repeated.

1.3.5 Other types 1^∞, ∞^0, 0^0 and $\infty-\infty$ can sometimes be transformed, usually algebraically or with logarithms, into suitable form.

1.3.6 Usually the first method is not available as, for example, in $G(y)=a\int_0^y (y-x)f(x)\,dx+b\int_y^\infty (x-y)f(x)\,dx$ unless some $f(\cdot)$ is specified.

SECTION 2.1

2.1.1 First, identify the field, the vectors, and the two operations. Second, identify the zero vector, additive inverses, establish closure and the remaining group properties. Third, verify the distributive and associative properties.

2.1.2 Direct appeal to definition or specific identification of coefficients as in proof of (2.4).

2.1.3 Use (2.5) as soon as it is itself established.

2.1.4 (6) $\{1+0i\}$ is a basis since $\mathcal{C}$ furnishes scalars.

2.1.5 (4) Form $\{\mathbf{u}_1, \mathbf{u}_2, \ldots, \mathbf{u}_k, \mathbf{a}_1, \mathbf{a}_2, \ldots, \mathbf{a}_n\}$ where the **u**'s are LI and the **a**'s form a basis. Consider the vectors in turn and apply (2.4) to (i) remove no **u**'s and (ii) surely reach a basis.

SECTION 2.2

2.2.1 Use order 2 matrices for counterexamples.

2.2.2 Recall definition of **AB** and also see below (2.14).

2.2.3 Where $\mathbf{A}^T$ appears let $\mathbf{A}^T=\mathbf{C}=[c_{ij}]$, work out the expression in terms of **C**, and then at the end convert to **A** using $c_{ij}=a_{ji}$. (9) Using indices h, i, j

$$\left[u_h=\sum_{i=1}^{n}\sum_{j=1}^{m} a_{hi}a_{ji}y_j\right] \text{ is } m\times 1$$

2.2.4 (2) $[\mathbf{A}_{ij}]^T$.

2.2.5 (2) The necessary and sufficient condition is: There is a partition following row r in **B** iff there is one following column r in **A**. Thus no restrictions are imposed on partitioning rows of **A** and columns of **B**.

2.2.6 (1) $[\mathbf{z}_i^T\mathbf{a}_j]$ is $k\times n$. (2) $[\mathbf{A}^T\mathbf{z}_1, \mathbf{A}^T\mathbf{z}_2, \ldots, \mathbf{A}^T\mathbf{z}_k]^T$.

2.2.7 Use $[\mathbf{a}_1, \mathbf{0}, \ldots, \mathbf{0}]\ \mathbf{B}=\mathbf{a}_1\mathbf{b}_1^T$, and so on. (Compare 5.1.18.)

2.2.8 See, for example, Stoer and Witzgall (1970).

2.2.9 (3) Interchange LH and center members of defining properties to obtain $\mathbf{A}^{-1}\mathbf{A}=\mathbf{A}\mathbf{A}^{-1}=\mathbf{I}$ so $\mathbf{A}=(\mathbf{A}^{-1})^{-1}$. (6) Establish negations of (1) and (2) in (2.22), $\mathbf{A}^T$ is nonsingular, and so on.

2.2.10, 2.2.11, and **2.2.12** Use partitioned matrices and (2.13), (2.17), (2.15), (2.16).

2.2.13 (4) Use LI of columns of rectangular matrices of bases.

2.2.14 The given forms are readily verified by elementary differential calculus. Also see Example 7.1.

SECTION 2.3

2.3.1 (1) Recall (2.5).

2.3.2 (2) Not a vector space unless $\mathbf{b}=\mathbf{0}$. (3) Cases (1), (2), (3) and especially the latter two, are not obviously finite-dimensional but later they are shown to be. (4)–(10), respectively, $mn, mn, n, n^2, n^3, mn, mn^2$.

2.3.3 (1) $\mathbf{a}_j$, $\mathbf{v} \in \mathcal{V}$; $\mathbf{c}_j, D\mathbf{v}, T\mathbf{a}_j, D\mathbf{a}_j \in \mathcal{W}$; $\mathbf{x}, \mathbf{e}_j \in \mathcal{V}^n$; $\mathbf{Dx}, \mathbf{d}_j \in \mathcal{V}^m$.

2.3.4 (4) $\mathbf{U}=\mathbf{Q}^{-1}\mathbf{TP}$.

2.3.5 (2) Use 2.1.5(6).

2.3.6 (4) Establish (i) Any set $\{\mathbf{u}_1, \mathbf{u}_2, \ldots, \mathbf{u}_k\}$ is LI iff $\{T\mathbf{u}_1, T\mathbf{u}_2, \ldots, T\mathbf{u}_k\}$ is LI, (ii) rank $T=\dim \mathcal{V}$, (iii) nullity $T=0$, and so on.

SECTION 3.1

3.1.1 (1) Consider three cases: $m<n$, $m=n$, and $m>n$. In the first case there are two configurations when the first r columns are LI: $r<m$ and $r=m$. There are also two configurations when the first r columns are LD provided we agree to sketch only one "sawtooth" break in the profile of 1*s* forming the LH "boundary" of **S**. (2) Actual zero rows appear if enter from Step 1$(r+1)$ or from Step 4(r) with $r<m$ and $p_r=n$. (3) At any place in a succession of elementary row operations the resulting matrix has a zero column where and only where one occurs in the original matrix.

3.1.2 The first nonzero element in row i is unity and it stands in column p_i for $i=1,2,\ldots,r$; whenever present, rows $r+1, r+2, \ldots, m$ are zero rows.

3.1.3 and **3.1.4** Note essential differences for **A** and $\mathbf{A}^T$.

3.1.5 (2) Use 2.2.10 and (3.7).

3.1.6 If **S** has no zero rows then $\mathbf{D}=\mathbf{I}_m$. Otherwise the upper LH corner of **D** contains $\mathbf{I}_r$, the lower LH corner contains $\mathbf{O}_{m-r,r}$, and the final $m-r$ columns are not uniquely determined.

3.1.7 The *determinant* of a square matrix $\mathbf{A}=[a_{ij}]$ of order n is a single scalar written $\det \mathbf{A}$. If $n=1$, $\det \mathbf{A} = a_{11}$; and, $\det \mathbf{A} = a_{11}a_{22}-a_{12}a_{21}$ for $n=2$. The *minor* m_{ij} of element a_{ij} is the determinant of the submatrix obtained by removing row i and column j. The *cofactor* of a_{ij} is $c_{ij}=(-1)^{i+j}\, m_{ij}$ and the RHS of

$$\det \mathbf{A} = a_{11}c_{11} + a_{12}c_{12} + \cdots + a_{1n}c_{1n}$$

is the *expansion by the first row*. Any row (or column) can be used but whenever $i \neq j$ then

$$a_{i1}c_{j1} + a_{i2}c_{j2} + \cdots + a_{in}c_{jn} = 0$$

Determinants have the following properties where **A**, **B**, and **C** are matrices of the same order. (1) If two rows (or two columns) are interchanged then the determinant changes sign. (2) If each element of a row (or a column) is multiplied by k then the determinant is multiplied by k. (3) If k times the rth row (column) is added to the sth so that the sum becomes the new row (column) s, where $s \neq r$, then the determinant is unchanged. (4) If **A** has a zero row or a zero column, then $\det \mathbf{A}=0$. (5) If two rows (or two columns) coincide, then $\det \mathbf{A}=0$. (6) $\det \mathbf{A}=\det \mathbf{A}^T$. (7) $\det \mathbf{AB}=(\det \mathbf{A})(\det \mathbf{B})$. (8) Let A in (2.13), **B** and **C** differ only in their sth columns $\mathbf{c}_s=\mathbf{a}_s+k\mathbf{b}_s$. Then $\det \mathbf{C}=\det \mathbf{A}+k\cdot\det \mathbf{B}$ and a corresponding result holds for rows.

3.1.8 (1) If $\det \mathbf{A}\neq 0$ then the components x_i of the solution **x** to $\mathbf{Ax}=\mathbf{b}$ satisfy

$$x_i\cdot\det \mathbf{A}=\det\left[\mathbf{a}_1,\ldots,\mathbf{a}_{i-1},\mathbf{b},\mathbf{a}_{i+1},\ldots,\mathbf{a}_n\right]$$

for $i=1,2,\ldots,n$. (2) If $\det \mathbf{A}\neq 0$ then

$$(\det \mathbf{A})\mathbf{A}^{-1}=\begin{bmatrix} a_{22} & -a_{12} \\ -a_{21} & a_{11} \end{bmatrix}$$

3.1.9 (1) Let the determinant rank of **A** be r. If $r=0$ then $\mathbf{A}=\mathbf{O}$ and the conclusion holds. Otherwise there is an rth order submatrix **C** with $\det \mathbf{C}\neq 0$. Type I. If two rows of **A** that do not enter **C** are interchanged, then $\det \mathbf{C}$ clearly remains unchanged. Otherwise two rows of **C** have been interchanged or a new row has been brought in. In the first case $\det \mathbf{C}$ changes sign but remains nonzero. In the second case transfer attention to the rth order submatrix formed by the row that was removed, the $r-1$ rows that remain, and the same set of columns as before. This new submatrix involves the same set of rows as did **C** except that they possibly appear in a different order. If they do so appear, interchange rows, two at a time, to regain the original order so as to exhibit an rth order nonzero determinant. Type II. Multiplying a row of **A** by $k\neq 0$ may again leave **C** unchanged. Otherwise the new submatrix occupying the positions in question has a nonzero determinant equal to $k\cdot\det \mathbf{C}$. Type III. Multiply row i by $k\neq 0$ and add to row $j\neq i$ where at least one of the rows in **C** is involved. If both rows appear in **C** the new determinant is unchanged. It will also be unchanged if row j does not appear in **C**. If row i only appears in **C** it is possible that the new submatrix has zero determinant. In this case, however, rearrange rows in **A** so there are three submatrices differing only in a single row: **C**; $\mathbf{C}_1$ with $\det \mathbf{C}_1=0$ has the new row replacing row j in **C**; and $\mathbf{C}_2$ has the elements from row i replacing row j in **C**. Then $\det \mathbf{C}_1=\det \mathbf{C}+k\cdot\det \mathbf{C}_2$ and $\det \mathbf{C}_1=0$ implies $\det \mathbf{C}_2=-k^{-1}\cdot\det \mathbf{C}\neq 0$. (2) If one did, then its inverse would decrease determinant rank in violation of what is proved in (1).

3.1.10 (3) The columns of **AB** are LCs of the columns of **A** and similarly Range $\mathbf{B}^T\mathbf{A} \subset$ Range $\mathbf{B}^T$. (4) Establish and use: if $S \subset \mathcal{V}^n$ is a subspace of dimension k and nullity $T = c$, then $\dim T(S) \geqq k - c$.

3.1.11 Drill.

SECTION 3.2

3.2.1 All follow immediately from results in text.

3.2.2 (3) rank$[\mathbf{A}, \mathbf{b}] \leqq n+1$. (5) Relate to GJSF.

3.2.3 Same as 3.2.1.

3.2.4 (5) See (3.32) and following.

3.2.5 Full row rank enters here, and existence of right inverses establishes consistency rather than uniqueness.

SECTION 3.3

3.3.1 A good approach is to write a simple computer program that given **A** and **b** produces the GJSFs of $[\mathbf{A}, \mathbf{b}, \mathbf{I}_m]$ and $[\mathbf{A}^T, \mathbf{I}_n]$ for real or complex scalars. (9) Use (3.36) for transpose and then transpose both sides.

3.3.2 (5) The row form (2.17) has successive rows $\mathbf{e}_5^T, \mathbf{e}_1^T, \mathbf{e}_2^T, \mathbf{e}_4^T, \ldots$ because $\mathbf{e}_1$ appears in column 5, $\mathbf{e}_2$ in 1, $\mathbf{e}_3$ in 2, $\mathbf{e}_4$ in 4, and so on, in the given **F**.

3.3.3 Place $\mathbf{I}_n$ below **S**, interchange two columns if necessary to bring $\mathbf{e}_1$ to column 1 in **S**, then interchange as necessary to bring $\mathbf{e}_2$ to column 2, and so on. At the end $\mathbf{I}_n$ has been replaced by **F**.

3.3.4 (2) Find the matrix of the basis for the case of **AF** and then multiply on left by $\mathbf{F}^T$. (Relate to interchanging columns first and rows later.)

3.3.5 Use (3.24) and also $\mathcal{V}^p$.

SECTION 3.4

3.4.1 (1) $\bar{a}\langle \mathbf{u}, \mathbf{w}\rangle + \bar{b}\langle \mathbf{v}, \mathbf{w}\rangle$

3.4.2 (3) Here $\mathbf{I}_n$ in (3.40) is replaced by Hermitian and positive definite **A** as studied in Section 5.3.

3.4.3 Suppose $\dim S = k$. Expand $\langle \mathbf{u}, \mathbf{v}\rangle$ using properties (3.41) to obtain the sum of k^2 terms $\bar{x}_i y_j \langle \mathbf{b}_i, \mathbf{b}_j\rangle$ which reduces to the stated sum of n terms.

3.4.4 For (iii) let $\mathbf{u} = \mathbf{a} - \mathbf{b}$ and $\mathbf{v} = \mathbf{b} - \mathbf{c}$.

3.4.5 (7) Use (2.8) and (3.47). (8) Use $\langle u_i \mathbf{b}_j, \mathbf{u}\rangle$ and (3.45).

3.4.6 (1) Select convenient multiples to avoid fractions in formation of intermediate vectors. For example, $\mathbf{v}_1 = \mathbf{a}_1$, $\mathbf{v}_2 = \langle \mathbf{v}_1, \mathbf{v}_1\rangle \mathbf{a}_2 - \langle \mathbf{v}_1, \mathbf{a}_2\rangle \mathbf{v}_1$, $\mathbf{v}_3 = \langle \mathbf{v}_1, \mathbf{v}_1\rangle\langle \mathbf{v}_2, \mathbf{v}_2\rangle \mathbf{a}_3 - \langle \mathbf{v}_1, \mathbf{a}_3\rangle\langle \mathbf{v}_2, \mathbf{v}_2\rangle \mathbf{v}_1 - \langle \mathbf{v}_2, \mathbf{a}_3\rangle\langle \mathbf{v}_1, \mathbf{v}_1\rangle \mathbf{v}_2$ are mutually orthogonal and span Range $[\mathbf{a}_1, \mathbf{a}_2, \mathbf{a}_3]$. Finally, defer normalization until the end of Step k.

3.4.7 Use (3.23) and (3.47).

3.4.8 (1) Write a nontrivial LC equalling **0**, say with coefficients $c_1, c_2, \ldots, c_k$, and then form successive inner products with **u**'s to obtain $\mathbf{Gc}=\mathbf{0}$. (2) Use the equation in the preceding to show $\|c_1\mathbf{u}_1+\cdots+c_k\mathbf{u}_k\|=0$.

3.4.9 (5) (Note that this does not state that any $\mathbf{v}\in\mathcal{C}^n$ belongs either to S or to $S^\perp$.) If $\mathbf{p}\in S\cup S^\perp$ the representation is obvious. Otherwise consider

$$\mathbf{q}_0=\langle\mathbf{p},\mathbf{b}_1\rangle\mathbf{b}_1+\cdots+\langle\mathbf{p},\mathbf{b}_k\rangle\mathbf{b}_k$$

where the **b**'s form an orthonormal basis for S. (6) Use indirect proof.

SECTION 4.1

4.1.1 The projection of **y** on **0** is **0**. The projection of **y** on $\mathbf{x}\neq\mathbf{0}$ is

$$\mathbf{u}=\frac{\langle\mathbf{y},\mathbf{x}\rangle}{\|\mathbf{x}\|}\cdot\frac{\mathbf{x}}{\|\mathbf{x}\|}$$

Its magnitude is $|\langle\mathbf{y},\mathbf{x}\rangle|\cdot\|\mathbf{x}\|^{-1}$, and if $\langle\mathbf{y},\mathbf{x}\rangle>0$ it has the same direction as **x** while if $\langle\mathbf{y},\mathbf{x}\rangle<0$ it has the opposite.

4.1.2 Let $\{\mathbf{x},\mathbf{y}\}$ be LI in $\mathcal{R}^2$. Then

$$\mathbf{z}=\mathbf{y}-\frac{\langle\mathbf{x},\mathbf{y}\rangle}{\langle\mathbf{x},\mathbf{x}\rangle}\cdot\mathbf{x}$$

is the projection of **y** on the line through **0** orthogonal to **x** in Range $[\mathbf{x},\mathbf{y}]$.

4.1.3 (3) Use (2).

4.1.4 (1) Find $\mathbf{M}_{11}^T$ for Range **A**, and iff GJSF of entire $[\mathbf{A}^T\mathbf{I}_n]$ has been found, $\mathbf{D}_{21}^T$ is the matrix of the standard basis for Null **A**. (3) $\mathbf{B}=\mathbf{E}_{21}$.

4.1.5 Compare (4.5).

4.1.6 (1) The transformation is $P:\mathcal{R}^n\to S$ defined by $\mathbf{x}\mapsto\mathbf{y}$ where **y** is given by $\mathbf{x}=\mathbf{y}+\mathbf{z}$. (3) Let **B** be the matrix of a basis for S and determine **U** such that $\mathbf{P}=\mathbf{UB}^T$ by using (4.5). (4) First use an indirect proof to show that $\mathbf{q}_0$ is uniquely the closest point (using Pythagorean theorem). Second prove that if $\mathbf{q}_0$ is the closest point then $\mathbf{q}_0$ must be the projection (using indirect proof for $\|\mathbf{u}\|=1$ expanding $\|\mathbf{p}-(\mathbf{q}^*+t\mathbf{u})\|^2$). (5) First, the projection is $\mathbf{y}=\mathbf{Bq}$ where $\mathbf{B}^T\mathbf{Bq}=\mathbf{B}^T\mathbf{x}$. Second, use orthogonal basis for S. Third, use orthonormal basis and Fourier expansion from 3.4.9. In the exercise, $5\mathbf{q}_0=[2,1,-1,0,2]^T$. (6) Use transformations P and exploit the unique decomposition $\mathbf{x}=\mathbf{y}+\mathbf{z}$. If P represents a projection then verification is straightforward. Conversely, if $\mathbf{w}\in W=\text{Range}\,P$ is arbitrary then $\mathbf{x}-\mathbf{Px}\in W^\perp$ establishes the unique decomposition. (7) Simply require $\text{Range}\,P\oplus\text{Null}\,P=\mathcal{R}^n$ that is, omit self-adjointness.

4.1.7 (2) Using (4.7) the system of n scalar equations in two unknowns t and t' is

$$\mathbf{p}_1+t(\mathbf{q}_1-\mathbf{p}_1)=\mathbf{p}_2+t'(\mathbf{q}_2-\mathbf{p}_2)$$

In terms of the coefficient matrix $\mathbf{C}$ and the augmented matrix $\mathbf{A}$ the criteria are: identical iff $\operatorname{rank}\mathbf{C}=\operatorname{rank}\mathbf{A}=1$; intersecting iff $\operatorname{rank}\mathbf{C}=\operatorname{rank}\mathbf{A}=2$; parallel iff $\operatorname{rank}\mathbf{C}=1$ and $\operatorname{rank}\mathbf{A}=2$; and skew iff $\operatorname{rank}\mathbf{A}=3$. (6) Construct a table that summarizes entire situation and shows how to go from any one form to any other. Use row labels Given, Condition, Restatement, Interrelations and column labels Two-Point, Parametric, Hyperplane. Entries by rows are: $\mathbf{p}_1,\mathbf{p}_2$; $\mathbf{p}_1,\mathbf{q}_1$; $\mathbf{p}_1,\mathbf{A}$; $\mathbf{p}_1\neq\mathbf{p}_2$; $\mathbf{q}_1\neq\mathbf{0}$; $\operatorname{nullity}\mathbf{A}=1$; $\{\mathbf{p}_2-\mathbf{p}_1\}$LI; $\{\mathbf{q}_1\}$LI; $\operatorname{rank}\mathbf{A}=n-1$; $\mathbf{p}_2-\mathbf{p}_1=\mathbf{q}_1$; $\mathbf{p}_1+\mathbf{q}_1=\mathbf{p}_2$; $\operatorname{Null}\mathbf{A}=\operatorname{Range}[\mathbf{p}_2-\mathbf{p}_1]$.

4.1.8 (4) Iff $\mathbf{a}=t\mathbf{c}$ for some t. (5) Also $b=td$.

4.1.9 (1) (Some hyperplanes are "superfluous" or "redundant.") The system $\mathbf{Cx}=\mathbf{d}$ is consistent by (3.17) and has the same solution set as $\mathbf{Ax}=\mathbf{b}$ where $\mathbf{A}=\mathbf{EC}$ is the $(n-k)\times n$ GJSF of $\mathbf{C}$ and $\mathbf{Ed}=\mathbf{b}\in\mathcal{R}^{n-k}$. (2) With 4.1.4 and using GJSFs to produce bases, find $\mathbf{A}$ such that $\operatorname{Null}\mathbf{A}=\operatorname{Range}\mathbf{D}$, that is, the given columns of $\mathbf{D}$ span $\operatorname{Null}\mathbf{A}$. Here $\mathbf{A}$ can be of full row rank $n-k$ and order $(n-k)\times n$. If also $\mathbf{b}=\mathbf{Ap}$ and $\mathbf{x}\in V^k$ then $\mathbf{Ax}=\mathbf{Ap}+\mathbf{ADt}=\mathbf{b}+\mathbf{0}=\mathbf{b}$. Conversely, reverse the argument to show that if $\mathbf{x}\in M^k$ in (4.10) then $\mathbf{x}\in V^k$ for suitable $\mathbf{D}$ and $\mathbf{p}$. (3) The points are distinct. First show that if $\mathbf{x}\in W^k$ then there exist $\mathbf{A}$ and $\mathbf{b}$ such that $\mathbf{x}\in H^k$ according to (4.10). Rewriting $\mathbf{x}=\mathbf{Pw}$,

$$\mathbf{x}=\mathbf{p}_1+w_2(\mathbf{p}_2-\mathbf{p}_1)+w_3(\mathbf{p}_3-\mathbf{p}_1)+\cdots+w_{k+1}(\mathbf{p}_{k+1}-\mathbf{p}_1)$$

determine $\mathbf{A}$ such that $\operatorname{Null}\mathbf{A}=\operatorname{Range}\mathbf{Z}$ which is possible because for $i\neq1$, $\mathbf{p}_i-\mathbf{p}_1\neq\mathbf{0}$. Then $\mathbf{b}=\mathbf{Ap}_1$ is the required vector. Conversely, consider H^k and find $\mathbf{Q}$ such that $\operatorname{Null}\mathbf{A}=\operatorname{Range}\mathbf{Q}$ where $\mathbf{Q}$ has full column rank k. Because $\operatorname{rank}\mathbf{A}=n-k$ there exists a solution $\mathbf{p}_1$ to $\mathbf{Ap}_1=\mathbf{b}$ and then the $k+1$ points

$$\mathbf{p}_1,\mathbf{p}_2=\mathbf{p}_1+\mathbf{q}_1,\mathbf{p}_3=\mathbf{p}_1+\mathbf{q}_2,\ldots,\mathbf{p}_{k+1}=\mathbf{p}_1+\mathbf{q}_k$$

all belong to H^k. These points also determine W^k and every solution $\mathbf{x}=\mathbf{p}_1+\mathbf{Qt}$, $\mathbf{t}\in\mathcal{R}^k$, has the form $\mathbf{x}=\mathbf{Pw}$ displayed above. ■ (4) (In words, the columns of $\mathbf{P}$ are LI iff W^k is not a subspace.) If the rank is $k+1$ then $\mathbf{0}$ cannot be any kind of a nontrivial LC of the $\mathbf{p}_i$ so $\mathbf{0}\notin W^k$. Prove the converse as: if $\operatorname{rank}\mathbf{P}<k+1$ then $\mathbf{0}\in W^k$. If the rank is k, so also is the rank of $\mathbf{D}=[\mathbf{p}_1,\mathbf{p}_2-\mathbf{p}_1,\ldots,\mathbf{p}_{k+1}-\mathbf{p}_1]$ and there exists a nontrivial LC

$$\mathbf{0}=t_1\mathbf{p}_1+t_2(\mathbf{p}_2-\mathbf{p}_1)+\cdots+t_{k+1}(\mathbf{p}_{k+1}-\mathbf{p}_1)$$

Here $t_1 \neq 0$ because $\{\mathbf{p}_2 - \mathbf{p}_1, \ldots, \mathbf{p}_{k+1} - \mathbf{p}_1\}$ is LI; divide by t_1 to exhibit $\mathbf{0} \in W^k$. If $\operatorname{rank} \mathbf{P} = k-1$ so also is the rank of $\mathbf{D}$ and this is impossible because its final k columns are LI; consequently there are but two possibilities for $\operatorname{rank} \mathbf{P}$; $k+1$ or k. ■ (6) (The idea is that matrices of full rank can be used in the first two forms.) Use the preceding results for (6), (7), (8). (9) This is the criterion for their determining a $(k-1)$-plane.

4.1.10 Drill on definitions.

4.1.11 (5) Any convex set containing A must also contain the convex hull of A so the convex hull must be included in the intersection of all such convex sets. On the other hand, this intersection is convex by (4.18) and must coincide with the convex hull.

4.1.12 (1) Use vertexes $\mathbf{0}$, $\mathbf{x} > \mathbf{0}$ and $k\mathbf{e}_1$ where $k > 0$ and work with barycentric coordinates. (4) $[1 - \langle \mathbf{x}, \mathbf{e} \rangle, x_1, x_2, \ldots, x_n]^T$. (5) Set $t_i = 0$ to obtain *face* of X^k *opposite* vertex $\mathbf{p}_i$. (6) $C(k+1, r+1)$. (7) Use dimension of highest dimension simplex contained in the set.

4.1.13 and **4.1.14** Drill on definitions.

4.1.15 (1) Let $\mathbf{x}$ be a convex combination of $r > n+1$ points $\mathbf{x}_i \in A$, where all coefficients $t_i > 0$, and prove that $\mathbf{x}$ must be a convex combination of $r-1$ points of A. Since any set of more than n points in $\mathfrak{R}^n$ must be LD there must exist $c_2, c_3, \ldots, c_r$ not all 0 such that

$$c_2(\mathbf{x}_2 - \mathbf{x}_1) + c_3(\mathbf{x}_3 - \mathbf{x}_1) + \cdots + c_r(\mathbf{x}_r - \mathbf{x}_1) = \mathbf{0}$$

Then define $c_1 = -(c_2 + c_3 + \cdots + c_r)$ so $c_1\mathbf{x}_1 + c_2\mathbf{x}_2 + \cdots + c_r\mathbf{x}_r = \mathbf{0}$. If for some subscript, say j

$$\frac{1}{a} = \max\left\{ \frac{t_1}{c_1}, \frac{t_2}{c_2}, \ldots, \frac{t_r}{c_r} \right\} = \frac{t_j}{c_j}$$

then the coefficients being sought are $d_i = t_i - ac_i$ for $i = 1, 2, \ldots, r$ for three reasons. First, $a > 0$ because no t_i, and not all c_i, are nonpositive. Second, $d_i \geqq 0$ for $i = 1, 2, \ldots, r$ and $d_j = 0$ by construction. Third, the jth term can be omitted in

$$\mathbf{x} = d_1\mathbf{x}_1 + d_2\mathbf{x}_2 + \cdots + d_r\mathbf{x}_r$$

because it equals $\mathbf{0}$ and the remaining d_i sum to unity as required for a convex combination. This process can be repeated, reducing the number of terms in the convex combination by one each time, until $r-1 = n+1$ is reached. ■

4.1.16 (8) Use an indirect proof.

4.1.17–4.1.20 Drill in supplying details.

4.1.21 (1) All LCs of its points. (2) All lines determined by pairs of its points. (3) All line segments joining pairs of its points. (4) All nonnegative multiples of its points. (5) All LCs of its points with nonnegative coefficients.

SECTION 4.2

4.2.1 (1) Yes. (2) Yes. (3) Yes.

4.2.2 (1)(ii) No. (5) Use GJSFs.

4.2.3 (1) Yes. (2) and (3) Use "dual" in former and "standard" in latter. (6) Use GJSF for Farkas I.

4.2.4 Drill

4.2.5–4.2.14 Drill with processes used in 4.2.2 and 4.2.3.

4.2.15 Successively establish Tucker's lemma, three existence theorems, and Tucker's theorem of the alternative in Mangasarian (1969, 22–30).

4.2.16 There are two typical uses. First, show (or specify) that a particular Problem I has a solution and then use the theorem to guarantee that the corresponding II has no solution. Second, show (or specify) that Problem I has no solution and then use the theorem to conclude that II must have a solution. In actual cases it is at times relatively easy to show that one or the other does or does not have a solution, and the essential contribution of a theorem of the alternative is to provide for all possibilities.

SECTION 4.3

4.3.1 (7) All order r submatrices of $[\mathbf{A}, \mathbf{b}]$ are nonsingular.

4.3.2–4.3.13 Drill in supplying details.

4.3.14 An arduous set of tasks.

SECTION 4.4

4.4.1 (3) Note that (iii) can be expressed as

If there is no positive vector belonging to a given polyhedral convex cone C, then its polar C^* contains a semipositive vector.

(4) Given $K \subset \mathfrak{R}^m$ as in (4.62), consider

$$C = \{\mathbf{z} : [\mathbf{A}^T, -\mathbf{c}], \quad \mathbf{z} \leqq \mathbf{0}\} \subset \mathfrak{R}^{m+1}$$

and establish a one-to-one correspondence between K's and certain C's. Then show that if K is bounded, the preceding can be used, together with (4.62) and other properties, to prove that K is a convex polyhedron.

4.4.2 Use the one-to-one correspondence of (4) just above, together with a finite set of generators for C to express $C=P+Q$ where P consists of LCs of generators having $z_{m+1}=1$, where Q consists of LCs of generators having $z_{m+1}=0$, and where all LCs have nonnegative coefficients. Then use the preceding to complete the proof that any K can be written in the form stated. Finally, show that any sum of a convex polyhedron and a polyhedral convex cone is a polytope.

4.4.3 Drill.

SECTION 5.1

5.1.1 Use (3.19) and (3.13) to show: $\mathbf{A}-\lambda\mathbf{I}$ is singular iff $\mathbf{A}^H-\bar{\lambda}\mathbf{I}$ is singular. For eigenvectors produce counterexamples for real $\mathbf{A}\neq\mathbf{A}^T$ of order 2. (2) Consider $\mathbf{I}_2$ and $\mathbf{J}$ above (5.5). (9) Consider any singular real $\mathbf{A}=\mathbf{A}^T$. (11) If $\mathbf{AB}-\lambda\mathbf{I}$ is singular, multiply on left by $\mathbf{A}^{-1}$ to show $\mathbf{B}-\lambda\mathbf{A}^{-1}$ singular and then multiply on right by $\mathbf{A}$ to show $\mathbf{BA}-\lambda\mathbf{I}$ singular. ■ (14) For order $n=1$ the result holds because there is one (nonzero) eigenvector whether the scalar a is zero or not. Suppose the result holds for $n-1$ but fails for n so that

$$c_1\mathbf{x}_1+c_2\mathbf{x}_2+\cdots+c_n\mathbf{x}_n=\mathbf{0}$$

for, say $c_1\neq 0$. Then multiply both sides by $\mathbf{A}$

$$c_1\lambda_1\mathbf{x}_1+c_2\lambda_2\mathbf{x}_2+\cdots+c_n\lambda_n\mathbf{x}_n=\mathbf{0}$$

and eliminate $\mathbf{x}_n$ between the two equations to find

$$c_1(\lambda_n-\lambda_1)\mathbf{x}_1+c_2(\lambda_n-\lambda_2)\mathbf{x}_2+\cdots+c_{n-1}(\lambda_n-\lambda_{n-1})\mathbf{x}_{n-1}=\mathbf{0}.$$

This last is an LC of $n-1$ LI vectors and each coefficient must equal zero. In particular $c_1(\lambda_n-\lambda_1)=0$ which is impossible because $c_1\neq 0$ and $\lambda_n\neq\lambda_1$. ■ (15) First, $1\leqq d_i$ because eigenvectors are nonzero. Second, for any scalar μ the eigenvalues of $\mathbf{A}$ are those of $\mathbf{A}-\mu\mathbf{I}$ increased by μ. Third,

$$(\mathbf{A}-\mu\mathbf{I})\mathbf{x}=\lambda\mathbf{x} \text{ implies } \mathbf{Ax}-\mu\mathbf{x}=\lambda\mathbf{x} \text{ implies } \mathbf{Ax}=(\lambda+\mu)\mathbf{x}.$$

Fourth, since zero is an eigenvalue of $\mathbf{A}-\lambda_i\mathbf{I}$ of multiplicity d_i, $0+\lambda_i=\lambda_i$ is an eigenvalue of $\mathbf{A}$ of multiplicity at least d_i. ■

5.1.2 It is expected that present approach would be first to find the characteristic equation and second, to find its roots. (Procedures of numerical analysis usually do not require finding the characteristic equation for larger n.)

5.1.3 Use 3.3.4 for hand calculations. (The procedure is mathematically correct but does not allow for efficiency of calculation or effects of rounding off numbers.)

5.1.4 and **5.1.5** Use the preceding based on GJSFs, and 3.3.4 for hand calculations.

5.1.6 For $n=2$ multiply column 2 of $\mathbf{C}-\lambda\mathbf{I}$ by λ and add to column 1. Extend for $n=3$ and then for any n.

5.1.7 (2) Given $\mathbf{A}=\mathbf{A}^H$ and $\mathbf{B}=\mathbf{B}^H$, if $\mathbf{AB}=\mathbf{BA}$ then $(\mathbf{AB})^H=\mathbf{B}^H\mathbf{A}^H=\mathbf{BA}=\mathbf{AB}$. Conversely, if $\mathbf{AB}$ is Hermitian then $\mathbf{AB}=\mathbf{B}^H\mathbf{A}^H=\mathbf{BA}$. ■ (7) Show $\langle\mathbf{x},\mathbf{Ax}\rangle=\langle\mathbf{Ax},\mathbf{x}\rangle$. (8) Use (3.51). (9) Let $\mathbf{x}$ be an eigenvector corresponding to λ. Then $\mathbf{x}\neq\mathbf{0}$ and $\langle\mathbf{x},\mathbf{Ax}\rangle=\lambda\langle\mathbf{x},\mathbf{x}\rangle$ by calculation. The LHS is real by (7) and on the RHS $\langle\mathbf{x},\mathbf{x}\rangle$ is real. ■ (10) Use an indirect proof based on the smallest number of LD eigenvectors: multiply the nontrivial LC by $\mathbf{A}$. (13) Use (5.9). (14) Find a symmetric nonreal matrix with at least one nonreal eigenvalue.

5.1.8 (8) Show $\lambda\bar{\lambda}=1$ for every eigenvalue. (10) Use Gram–Schmidt for "only if." (11) Use (5.9).

5.1.9 (6) Show $\det\mathbf{A}=(-1)^n\cdot\det\mathbf{A}$.

5.1.10 (6) Use (5.9) as $\mathbf{AQ}=\mathbf{Q}\,\mathbf{diag}(\lambda_1,\lambda_2,\ldots,\lambda_n)$. (7) Use $\mathbf{A}$ and $\mathbf{B}$ in their diagonalized forms.

5.1.11 (9) Apply the following to columns $1,2,\ldots,n-2$, and alter as appropriate for column $n-1$.

> Column j. If all elements (p,j) are zero for $p>j$ go to column $(j+1)$. Otherwise, continue. If element (j,j) is nonzero add multiples of row j to lower rows to place zeros in all positions (p,j) for $p>j$; if element (j,j) is zero, add a multiple of some lower row to row j so as to make (j,j) nonzero and then add multiples of new row j to place zeros in column j below row j. Go to column $(j+1)$.

5.1.12 (2) For example, establish for $\mathbf{A}$ nonsingular

$$\mathbf{A}^{-1}=\operatorname{diag}\left(\mathbf{A}_{11}^{-1},\mathbf{A}_{22}^{-1},\ldots,\mathbf{A}_{tt}^{-1}\right)$$

which need not hold for any set of blocks $\mathbf{A}_{ii}$.

5.1.13 (1) There are eight subcases to be resolved in this first case: b,c,e nonzero; $b=0$ only; $c=0$ only; $e=0$ only; $b=c=0$ only; $b=e=0$ only; $c=e=0$ only; and $b=c=e=0$.

5.1.14 Evaluate for $n=1,2,3$ and then use induction.

5.1.15 Use results supplied for 3.1.7.

5.1.16 Pivot in an appropriate partitioned matrix (recall (3.29)).

5.1.17 (1) $\mathbf{A}^2$ and $\mathbf{A}$ have same eigenvalues which implies $\lambda=1$ or $\mathbf{Ax}=\mathbf{0}$.

5.1.18 (1) Find GJSF. (2) Use row operations to put unity in (1, 1) position for proof by induction.

5.1.19 Drill on definitions.

5.1.20 (1) Show that column sums of $\mathbf{A}-\mathbf{I}$ are zero and consequently unity is an eigenvalue of $\mathbf{A}$.

5.1.21 For "if" use $(\det\mathbf{A})(\det\mathbf{A}^{-1})=1$ and for "only if" use 5.1.15.

5.1.22 Use elementary row operations on the matrix (or its transpose) to show $\det\mathbf{V}$ equals the product of all factors (a_i-a_j) for $j<i$.

SECTION 5.2

5.2.1 (5) Provide for two cases: $\mathbf{A}$ singular and $\mathbf{A}$ nonsingular.

5.2.2 Evaluating Determinants. Find the GRF of $\mathbf{A}$ in (5.11) and use: $\det\mathbf{A}=d$.

5.2.3 (1) Once a particular row is added in to produce a pivot row, the resulting row appears unchanged in the final GRF. For example, use

$$\mathbf{B}=\begin{bmatrix}0&1&1\\1&0&1\\1&2&4\end{bmatrix}$$

(2) Use one Type I and one Type II operation.

5.2.4 Drill.

5.2.5 (1) First use (5.11) $[\mathbf{A},\mathbf{I}_n]\overset{R}{\rightarrow}[\mathbf{G},\mathbf{E}]$ where $\mathbf{G}=\mathbf{DU}$ and second use (3.2)

$$[\mathbf{E},\mathbf{I}_n]\overset{R}{\rightarrow}[\mathbf{I}_n,\mathbf{E}^{-1}]$$

where $\mathbf{E}^{-1}=\mathbf{L}$ is the required lower-triangular factor. (2) Consider

$$\begin{bmatrix}0&1\\1&0\end{bmatrix}\overset{?}{=}\begin{bmatrix}a&0\\b&c\end{bmatrix}\begin{bmatrix}d&e\\0&f\end{bmatrix}$$

(3) Use a succession of Type I operations. Repeat the interchanges on a second matrix, initially $\mathbf{I}_n$ and finally $\mathbf{T}$ where $\mathbf{TA}$ is row-equivalent to $\mathbf{A}$ and factorable according to (5.15). Here $\det\mathbf{TA}=\pm\det\mathbf{A}$ and $\mathbf{T}^{-1}=\mathbf{T}^T$. (4) Since $\mathbf{A}$ is Hermitian, $\mathbf{U}^H\mathbf{D}^H\mathbf{L}^H=\mathbf{LDU}$ and this implies

$$(\mathbf{D}^H\mathbf{L}^H)\mathbf{U}^{-1}=(\mathbf{L}^H\mathbf{U}^{-1})^H\mathbf{D}$$

where, since the LHS is upper-triangular and the RHS is lower-triangular, both sides are actually diagonal matrices. Moreover

$$\mathbf{L}^H\mathbf{U}^{-1}=\mathbf{I}$$

because the LHS is diagonal and both factors have unity in all main diagonal positions. Consequently $\mathbf{L}^H=\mathbf{U}$ and the first display simplifies to $\mathbf{D}^H=\mathbf{D}$. ■

5.2.6 (3) $n^3/2+O(n^2)$. (4) $n^3/3+O(n^2)$. (5) $n^3+O(n^2)$.

5.2.7 Drill.

SECTION 5.3

5.3.1 (3) Consider indefinite **A**. (4) True. If $\mathbf{ABx}=\lambda\mathbf{x}$ then $\mathbf{Bx}=\lambda\mathbf{A}^{-1}\mathbf{x}$ and $\mathbf{x}^H\mathbf{Bx}=\lambda\mathbf{x}^H\mathbf{A}^{-1}\mathbf{x}$ shows $\lambda\in\mathcal{R}$. (6) Use 5.1.1(13) and 5.23. (8) (iii) If **A** is positive definite find **B** by replacing λ_i by $(\lambda_i)^{1/2}$ in diagonalized form of **A**; if also $\mathbf{C}^2=\mathbf{A}$ show that **C** can be found by rearranging diagonal matrix used for **B**. Conversely, **B** as specified and $\mathbf{B}^2=\mathbf{A}$ imply all eigenvalues of **A** are positive.

5.3.2 (7) Order 4: $t>0$ and $t^2>(3+\sqrt{5})/2$. Order 5: $t>0$ and $t^2>3$. . . . ?

5.3.3 (1) If $\mathbf{y}\in\mathcal{C}^n$ then $\{\mathbf{Xy},\mathbf{YXy}\rangle$ is positive if $\mathbf{Xy}\neq\mathbf{0}$ and zero if $\mathbf{Xy}=\mathbf{0}$. (4) Use (2.21) and $\langle\mathbf{Xu},\mathbf{Xu}\rangle\geqq 0$ for all $\mathbf{u}\in\mathcal{C}^n$, or let $\mathbf{Y}=\mathbf{I}$ in (2).

5.3.4 (1) Procedure (5.26) preserves determinants and $q_1=\det[a_{11}]$,

$$q_1q_2=\det\begin{bmatrix} a_{11} & \bar{a}_{21} \\ a_{21} & a_{22}\end{bmatrix}$$

and so on for all leading principal submatrices. The criterion follows from (5.28) and it is "iff" because criteria in (5.23) are "iff."(2) A counterexample is the indefinite

$$\begin{bmatrix} 1 & 1 & 1 \\ 1 & 1 & 1 \\ 1 & 1 & 0\end{bmatrix}$$

5.3.5 (1) Since nonsingular transformations do not change rank it suffices to show that positive terms must number p. Suppose there were a second diagonal matrix **G** with $s<p$ positive elements, and suppose both diagonal matrices have positive elements appearing in initial positions, negative elements immediately following, and consequently, zeros in the final r

positions. Suppose also that the diagonal matrices are

$$\mathbf{P}^H\mathbf{A}\mathbf{P}=\mathbf{D} \text{ via } \mathbf{x}=\mathbf{P}\mathbf{y} \quad \text{and} \quad \mathbf{F}^H\mathbf{A}\mathbf{F}=\mathbf{G} \text{ via } \mathbf{x}=\mathbf{F}\mathbf{z}$$

Consider $\mathbf{y}'$ having zeros in components corresponding to negative and zero diagonal elements in $\mathbf{D}$, and $\mathbf{z}'$ having zeros where positive entries appear for $\mathbf{G}$. Then there exists $\mathbf{x}'\neq\mathbf{0}$ satisfying $s+n-p$ homogeneous equations in

$$\mathbf{P}^{-1}\mathbf{x}=\mathbf{y}' \quad \text{and} \quad \mathbf{F}^{-1}\mathbf{x}=\mathbf{z}'$$

where $\mathbf{y}'$ and $\mathbf{z}'$ are nonzero. Now $\mathbf{D}$ shows $\mathbf{x}^H\mathbf{A}\mathbf{x}\geqq 0$ for $\mathbf{x}'$ while $\mathbf{G}$ shows it to be nonpositive so it must be zero. Consequently $\mathbf{y}'=\mathbf{0}$ because of $\mathbf{D}$ and this contradicts $\mathbf{x}'\neq\mathbf{0}$. ■ (2) No.

5.3.6 Use the preceding and (5.25).

5.3.7 (1) Use $\mathbf{Q}=\mathbf{RST}$ where $\mathbf{R}$ is a unitary matrix of eigenvectors of $\mathbf{A}$,

$$\mathbf{S}=\mathbf{diag}\left(\lambda_1^{-\frac{1}{2}},\lambda_2^{-\frac{1}{2}},\ldots,\lambda_n^{-\frac{1}{2}}\right)$$

where the λs are the (positive!) eigenvalues of $\mathbf{A}$, and $\mathbf{T}$ is unitary matrix of eigenvectors of $\mathbf{S}^H\mathbf{R}^H\mathbf{BRS}$. (2) Follows directly from (1).

SECTION 5.4

5.4.1 (6) (i) implied by (ii). (ii) If the set were LD due to

$$c_1\mathbf{x}+c_2\mathbf{Z}\mathbf{x}+\cdots+c_p\mathbf{Z}^{p-1}\mathbf{x}=\mathbf{0}$$

then successive multiplication by $\mathbf{Z}$ would lead to $c_1\mathbf{Z}^{p-1}\mathbf{x}=\mathbf{0}$ whence $c_1=0$. Similarly all $c_i=0$. (iii) $\mathbf{Z}$ is singular due to (4), the inclusions obviously hold, and Null $\mathbf{O}=\mathcal{C}^n$. (iv) Follows from (4).

5.4.2 Use results from 5.4.1 and induction as appropriate.

5.4.3 (1) For example, $n=3$ can be written $1+1+1$, $2+1$, or 3 so possible pairs are (3, 1), (2,2), or (1,3), respectively. (2) False for $n=7$. (3) Tables in (1) can be used for Jordan blocks through $n=6$. For example, there are 15 for $n=4$: to $(\lambda,\lambda,\lambda,\lambda),(\lambda,\lambda,\lambda,\mu),(\lambda,\lambda,\mu,\mu),(\lambda,\lambda,\mu,\nu),(\lambda,\mu,\nu,\xi)$ there correspond 5, 3, 4, 2, 1 configurations, respectively.

5.4.4 (1) Use eigenvalues and eigenvectors from 5.1.4.

SECTION 6.1

6.1.1 (3) (Familiar ε and k techniques suffice.) Specify $\varepsilon/2$ and determine k', k'' in (6.1) for $\{\mathbf{x}_k\}, \{\mathbf{y}_k\}$, respectively. Then $\mathbf{x}_k + \mathbf{y}_k \in N_\varepsilon(\mathbf{a}+\mathbf{b})$ whenever $k > \max\{k', k''\}$.

6.1.2 (3) Let $\mathbf{p}, \mathbf{q} \in N_\delta(\mathbf{x}_0)$ be distinct and consider

$$\mathbf{x} = t\mathbf{p} + (1-t)\mathbf{q}$$

where $t \in [0,1]$. Since $\mathbf{x} - \mathbf{x}_0 = t(\mathbf{p} - \mathbf{x}_0) + (1-t)(\mathbf{q} - \mathbf{x}_0)$ there follows $\|\mathbf{x} - \mathbf{x}_0\| \leqq \delta$. ■

6.1.3 (1) Property I is immediate. For II, if $\mathbf{x} \in G_1 \cap G_2$ where G_1 and G_2 are open then there are neighborhoods of respective radii δ_1 and δ_2 contained in G_1 and G_2. Consequently $\delta = \min\{\delta_1, \delta_2\}$ defines a neighborhood in $G_1 \cap G_2$. The preceding extends to any finite number of open sets. For III, if $\mathbf{x}$ belongs to the union of any family of open sets then it belongs to at least one member and there will be at least one neighborhood contained in the member, and consequently in the union of the family. ■ (2) Let $A(i) = (-1/i, 1/i)$ for $i = 1, 2, \ldots$ in $\mathfrak{R}$. (3) (I′) The null set and the entire space are closed sets. (II′) The union of any finite number of closed sets is a closed set. (III′) The intersection of any number of closed sets is a closed set. (4) Use complements of sets in (2).

6.1.4 (4) If $N_\delta(\mathbf{x}_0)$ contained only $\mathbf{x}_1, \mathbf{x}_2, \ldots, \mathbf{x}_p$ distinct from $\mathbf{x}_0$ then to

$$2m = \min\{\|\mathbf{x}_0 - \mathbf{x}_i\| : i = 1, 2, \ldots, p\}$$

would correspond $N_m(\mathbf{x}_0)$ containing no other points of A. ■ (9) Suppose $\mathbf{x}$ is a limit point of closed A. If $\mathbf{x} \in A^c$, which is open, then $\mathbf{x}$ has a neighborhood contained in A^c which contains no points of A. Conversely, if A contains all of its limit points then $\mathbf{x} \in A^c$ is not a limit point and has a neighborhood that does not intersect A, that is, it is a subset of A^c. Hence A is closed because A^c is open.

6.1.5 (1) These justify the use of "open" and "closed." Only the finite closed interval $[a, b]$ is compact. (2) Coincides with its interior, boundary is (3), closure is (4), and is open and connected. (3) Interior is $\varnothing$, coincides with its boundary and closure, and is closed and compact. (4) Interior is (2), boundary is (3), coincides with its closure, and is closed and compact.

6.1.6 (1) See 3.4.5 (8). (2) Use (1). (3) Use: the norm of $f(\mathbf{x}_0)$ equals the absolute value of the norm of $f(\mathbf{x}_0)$. (5) Verify $|h/|h|| \to 1$ as $h \to 0$ but $h/|h| \not\to 1$ as $h \to 0$. (If $\mathbf{y}_0 = \mathbf{0}$ then (4) applies.) (7) Use $\|f(\mathbf{x}_0 + \mathbf{h}) - \mathbf{y}_0\| < \varepsilon$ whenever $0 < \|\mathbf{h}\| < \delta$.

6.1.7 (2) Statement: if f is continuous at $\mathbf{x}_0$ then the restrictions of f

$$x_i \mapsto g(t(x_i))$$

where $t(x_i)=[x_{10},x_{20},\ldots,x_i,\ldots,x_{n0}]^T$, are continuous at $x_{i0}\in\mathcal{R}$ for $i=1,2,\ldots,n$. (3) Consider $\mathbf{x}\mapsto x_1x_2(x_1^2+x_2^2)^{-1}$ and $\mathbf{x}_0=\mathbf{0}$ in $\mathcal{R}^2$. (5) Use (6.9) and complements. (6) Use (6.9) and the open sets $(-\infty,a)$ and (a,∞) in $\mathcal{R}$.

6.1.8 (1) For example, use $A\subset \operatorname{clos} A$ and 6.12 (3); or use (6.13). (2) Suppose A is compact and $\mathcal{F}$ has an empty intersection. Then A has an open cover $\mathcal{G}$ of sets

$$G(x)=A-F(x)$$

and if a finite subcover corresponds to $x_1,x_2,\ldots,x_m$ we obtain the contradiction

$$F(x_1)\cap F(x_2)\cap\cdots\cap F(x_m)=\varnothing$$

Conversely, let $\mathcal{G}$ be an open cover of A and define

$$H(y)=A-G(y)$$

Now the collection $\mathcal{H}$ has the finite intersection property but an empty intersection, so necessarily

$$H(y_1)\cap H(y_2)\cap\cdots\cap H(y_p)=\varnothing$$

for some finite set and the corresponding G's form a finite subcover of A. ∎ (5) Use (6.19). (6) Use 6.1.7 (5).

6.1.9 (1) Fill in details between (3.54) and (3.55). (2) Compare (3.45). First, show that the given formulas correctly give the achieved suprema in question. Second, appeal to (6.29). (5) Let $\mathbf{A}$ be $m\times n$. Then $\mathbf{A}^H\mathbf{A}$ is Hermitian and positive semidefinite of order n; it consequently has n nonnegative eigenvalues $\lambda_1,\lambda_2,\ldots,\lambda_n$ with, say, λ_1 largest, and by (5.9) it is diagonalizable by a unitary transformation with matrix

$$\mathbf{U}=[\mathbf{u}_1,\mathbf{u}_2,\ldots,\mathbf{u}_n]$$

where $\mathbf{u}_i$ is a normalized eigenvector corresponding to λ_i for $i=1,2,\ldots,n$. Let $\mathbf{x}\in\mathcal{C}^n$ be nonzero with $\|\mathbf{x}\|=1$ (which is possible without loss of generality). Then since $\mathbf{U}$ is the matrix of an orthonormal basis for $\mathcal{C}^n$

$$\mathbf{x}=c_1\mathbf{u}_1+c_2\mathbf{u}_2+\cdots+c_n\mathbf{u}_n$$

and because $\|\mathbf{x}\|=1$

$$|c_1|^2+|c_2|^2+\cdots+|c_n|^2=1$$

By an easy calculation $\langle \mathbf{Ax}, \mathbf{Ax}\rangle$ equals

$$\langle \mathbf{x}, \mathbf{A}^H\mathbf{Ax}\rangle=|c_1|^2\lambda_1+|c_2|^2\lambda_2+\cdots+|c_n|^2\lambda_n$$

The RHS is a convex combination of the set of eigenvalues and therefore achieves its maximum λ_1 for $c_1=1$ and

$$c_2=c_3=\cdots=c_n=0$$

By (6.26) the standard natural norm of $\mathbf{A}$ is $\lambda_1^{1/2}$. ■ (6) Certainly $\alpha \leqq \|\mathbf{a}_j\|$ holds for each column of $\mathbf{A}$. Since $\|\mathbf{e}_j\|=1$, $\mathbf{Ae}_j=\mathbf{a}_j$ implies

$$\|\mathbf{a}_j\| \leqq \|\mathbf{A}\|$$

by (6.26) and establishes the lower bound. Next use (2.15), (3.44), and 3.4.5 (8) to write

$$\|\mathbf{Ax}\| \leqq (\|\mathbf{a}_1\|+\|\mathbf{a}_2\|+\cdots+\|\mathbf{a}_n\|)\|\mathbf{x}\|$$

Since $\|\mathbf{a}_j\|^2 \leqq m\alpha^2$, also $\|\mathbf{Ax}\| \leqq mn\alpha^2\|\mathbf{x}\|$ and normalizing $\mathbf{x}$ shows that the upper bound also follows from (6.26). ■

6.1.10 (1) Using (6.26) and the Cauchy–Schwarz inequality of (3.45), $\|\mathbf{a}^T\| \leqq \|\mathbf{a}\|$. Considering $\mathbf{a}\neq\mathbf{0}$ and $\mathbf{x}=\mathbf{a}/\|\mathbf{a}\|$ shows that the upper bound is achieved. When $\mathbf{a}=\mathbf{0}$ the statement clearly holds. ■(2) From $\mathbf{x}=\mathbf{A}^{-1}\mathbf{Ax}$ there follows

$$\|\mathbf{x}\|=\|\mathbf{A}^{-1}(\mathbf{Ax})\| \leqq \|\mathbf{A}^{-1}\|\cdot\|\mathbf{Ax}\|$$

using 6.1.9(1). ■

6.1.11 (1) (ii) Use Cauchy–Schwarz inequality. (2) Use complements of sets in 6.1.7(6).

6.1.12 (2) If $\operatorname{int} K=\varnothing$ the conclusion follows. Otherwise, let $\mathbf{p}, \mathbf{q} \in \operatorname{int} K$ be distinct and consider

$$\mathbf{x}=t\mathbf{p}+(1-t)\mathbf{q}$$

where $t\in(0,1)$. Show $\mathbf{x}\in\operatorname{int} K$ as follows. Any point in

$$N_\delta(\mathbf{x})=\{\mathbf{y}:\mathbf{y}=\mathbf{x}+\mathbf{z} \text{ for some } \mathbf{z},\ \|\mathbf{z}\|<\delta\}$$

can be written

$$\mathbf{y}=t(\mathbf{p}+t^{-1}\mathbf{z})+(1-t)\mathbf{q}$$

For $\delta>0$ and sufficiently small, $\mathbf{p}+t^{-1}\mathbf{z}$ belongs to a neighborhood of $\mathbf{p}$ that lies within K and consequently $N_\delta(\mathbf{x})\subset K$. ■ (7) Dimension must be n so neighborhoods can be subsets of the simplex.

6.1.13 (1) Establish equivalence of the set of all affine combinations and the parametric form of a line (4.7).

6.1.14 (3) If C has only one element then the affine hull is the subspace determined by the element and C coincides with its relative interior. In general, if C is of dimension $k>1$ then its affine hull contains a k-simplex that can be used to prove the result for C.

SECTION 6.2

6.2.1 (1) If both F_1 and F_2 satisfied (6.31) then $G=F_1-F_2$ would be the zero transformation since for any $\mathbf{h}\in\mathcal{R}^n$

$$G\mathbf{h}=\|\mathbf{h}\|\lim_{t\to 0}\frac{Gt\mathbf{h}}{\|t\mathbf{h}\|}=\mathbf{0}$$

(6) Use A is a linear transformation so

$$A\mathbf{h}=A(h_1\mathbf{e}_1+\cdots+h_n\mathbf{e}_n)=[A\mathbf{e}_1,\ldots,A\mathbf{e}_n]\mathbf{h}$$

and then use (2.15). (7) Norm for numerator is always standard norm (or natural norm) for vector space where images $F\mathbf{h}, A\mathbf{h}, B\mathbf{h}, C\mathbf{h}$, $G\mathbf{h}$, respectively, lie and norm for denominator is standard norm on $\mathcal{R}^n$ in each case. (8) Counterpart of (6.39). (9) Adapt the proof of (6.44) allowing for the fact that (6.43) does not have an exact counterpart for transformations. Show $\langle\mathbf{v},f(\mathbf{x})-f(\mathbf{y})\rangle=0$ for all $\mathbf{v}\in\mathcal{R}^m$.

6.2.2 and **6.2.3** Drill.

6.2.4 (1) If $f\in C^{(1)}$ then f is differentiable by (6.45) so the transformation $\mathbf{x}\mapsto f'(\mathbf{x})$ is defined on D. If $\mathbf{x}\in D$ and $\varepsilon>0$ is fixed then by continuity of the partial derivatives there exists $\delta>0$ such that

$$|\partial_j f_i(\mathbf{x})-\partial_j f_i(\mathbf{y})|<\varepsilon(mn)^{-\frac{1}{2}}$$

whenever $\|\mathbf{x}-\mathbf{y}\|<\delta$. Consequently, the natural norm satisfies

$$\|f'(\mathbf{x})-f'(\mathbf{y})\|<\varepsilon$$

by 6.1.9 (6). Conversely, let f be continuously differentiable. If $\mathbf{x}\in D$ and $\varepsilon>0$ is fixed then by continuity of f' there exists $\delta'>0$ such that the last displayed inequality holds whenever $\|\mathbf{x}-\mathbf{y}\|<\delta'$. Then the first inequality in 6.1.9 (6) establishes the continuity of all partial derivatives of coordinate functions of f so that $f\in C^{(1)}$. ■

6.2.5 (3) The standard matrix of the derivative of the transformation

$$\mathbf{x}\mapsto \nabla f(\mathbf{x})\in\mathcal{R}^n$$

coincides with the Hessian of f. The transformation

$$\mathbf{x}\mapsto \mathbf{f}'(\mathbf{x}_0)\in\mathcal{M}(1,n)$$

can be identified with the element in $\mathcal{L}(\mathcal{R}^n,\mathcal{R})$ whose derivative is also represented by the Hessian of f.

6.2.6 (1) Denote $\partial_j f(\mathbf{x}_0)=\phi_j(\mathbf{x}_0)$ for $j=1,2,\ldots,n$ so successive rows of Hessian are $\boldsymbol{\phi}_1'(\mathbf{x}_0),\ldots,\boldsymbol{\phi}_n'(\mathbf{x}_0)$. Guided by (6.62), select standard tensor

$$\mathbf{f}'''(\mathbf{x}_0)=\begin{bmatrix}\boldsymbol{\phi}_1''(\mathbf{x}_0)\\ \boldsymbol{\phi}_2''(\mathbf{x}_0)\\ \vdots\\ \boldsymbol{\phi}_n''(\mathbf{x}_0)\end{bmatrix}$$

handle $G\mathbf{h}$ same as $C\mathbf{h}$ and follow pattern for $B\mathbf{k}$ to obtain

$$\left[\mathbf{k}^T\boldsymbol{\phi}_1''(\mathbf{x}_0)\mathbf{h},\ldots,\mathbf{k}^T\boldsymbol{\phi}_n''(\mathbf{x}_0)\mathbf{h}\right]$$

as standard representation for $G\mathbf{hk}$.

6.2.7 The expression "respective Classes $C^{(s)}$" is appropriate because different classes are associated with different pairs of domains and ranges. Use the subject theorem to exhibit the partial derivatives and establish their continuity for $m=1$. Extend to transformations via the coordinate functions.

SECTION 6.3

6.3.1 (1) Use definition (6.69). (2) Use induction for (i) and (ii).

6.3.2 (1) Use Table 6.1(5) here and also in following parts.

6.3.3 (1) Use preceding problem.

6.3.4 and **6.3.5** Drill.

6.3.6 (4) Use Table 6.4(4). (5) Use Table 6.4(8). (6) (i) Differentiate $\mathbf{A}^{-1}\mathbf{A}=\mathbf{I}$. (ii) Substitute $\mathbf{A}$ for $\mathbf{A}^{-1}$. (7) Use $c\mathbf{a}_j^T$ and $(c\mathbf{I})\mathbf{A}$.

6.3.7 (5) Use (6.73) and present (4). (7) Use (6.74) and present (6).

6.3.8 Drill; however, notice analogy in (2) with (2.14) and Table 6.2 (5).

6.3.9 (1) (ii) Can use results for Example 6.4(3) to verify for $h(\mathbf{x})=(f(\mathbf{x}))^{-1}$ and $g(y)=y^{-1}$

$$\mathbf{h}'(\mathbf{x})=-\left[f(\mathbf{x})\right]^{-2}\mathbf{f}'(\mathbf{x})$$

$$\mathbf{h}''(\mathbf{x})=-\left[f(\mathbf{x})\right]^{-2}\mathbf{f}''(\mathbf{x})+2\left[f(\mathbf{x})\right]^{-3}(\mathbf{f}'(\mathbf{x}))^T\mathbf{f}'(\mathbf{x})$$

of correct orders $1\times n$ and $n\times n$, respectively. (2) (i) $-2\langle\mathbf{x},\mathbf{x}\rangle^{-2}\mathbf{x}^T$ (ii) $-2(\mathbf{x}^T\mathbf{A}\mathbf{x})^{-2}\mathbf{x}^T\mathbf{A}$. (iv) $(-2(\mathbf{x}^T\mathbf{B}\mathbf{x})^{-2}\mathbf{x}^T\mathbf{A}\mathbf{x})\mathbf{x}^T\mathbf{B}+2(\mathbf{x}^T\mathbf{B}\mathbf{x})^{-1}\mathbf{x}^T\mathbf{A}$. (4) See, for example, Rektorys (1969, 450–452).

6.3.10 (1) Use first and second derivatives from 6.3.9 (1) (ii). The latter simplifies because here $\mathbf{f}''(\mathbf{x})=\mathbf{O}$. Differentiate second derivative of $(f(\mathbf{x}))^{-1}$ using (1) and (7) in Table 6.4. (3) Note that this problem includes (1) as a special case.

SECTION 6.4

6.4.1 (2) If $\mathbf{x}\mapsto\mathbf{y}$ is the transformation then

$$y_1^2+y_2^2=e^{2x_1},\ y_2\cos x_2=y_1\sin x_2$$

shows that the region between $x_2=0$ and $x_2=2\pi$ is mapped into the entire y_1y_2-plane less the origin. Since the given transformation is periodic in x_2 with period 2π, any region between $x_2=b$ and $x_2=b+2\pi$ is transformed in similar fashion. By (1.1) there can be no global inverse. However, (6.90) does apply and yields local inverses that clearly cannot be extended parallel to the x_1-axis to any distance greater than 2π.

6.4.2 Consider $\mathbf{y}_0\in f(A)$ where $A\subset D$ is open and some one $\mathbf{x}_0\in A$ such that $\mathbf{y}_0=f(\mathbf{x}_0)$. Apply (6.90) to restriction of f to A and find neighborhood N of $\mathbf{x}_0$ such that $f(N)\subset f(A)$ is open. Since $f(N)$ contains a neighborhood of $\mathbf{y}_0$ so also does $f(A)$. ■

6.4.3, **6.4.4**, and **6.4.5** Drill in applying theorems and translating them into different forms.

6.4.6 (1) Start with $m=1$ and relate $f(\mathbf{x})-z=0$ and $u(\mathbf{y})=0$. Then use (6.93) for $m>1$. (2) Use (4.4). (3) The set P is a k-plane specified in the form of the affine subspace resulting from translating the subspace T by $\mathbf{x}_0$ so P and T are parallel k-planes. Substituting $\mathbf{h}=\mathbf{x}-\mathbf{x}_0$ the defining equation for P becomes

$$\mathbf{u}'(\mathbf{x}_0)(\mathbf{x}-\mathbf{x}_0)=\mathbf{0}$$

which is essentially the hyperplane form in (4.11).

SECTION 6.5

6.5.1 (3) Let $\mathbf{p}_i=[\mathbf{x}_i^T,z_i]^T$ denote points in E. Suppose f is convex. All points $\mathbf{p}\in E$ on the line segment connecting distinct $\mathbf{p}_1,\mathbf{p}_2\in E$ correspond to

$$\mathbf{x}=t\mathbf{x}_1+(1-t)\mathbf{x}_2 \quad \text{and} \quad z=tz_1+(1-t)z_2$$

where $t\in[0,1]$. Consequently, $f(\mathbf{x}_1)\leqq z_1$ and $f(\mathbf{x}_2)\leqq z_2$ imply

$$f(\mathbf{x})\leqq tf(\mathbf{x}_1)+(1-t)f(\mathbf{x}_2)\leqq z$$

so $\mathbf{p}\in E$. Conversely, if E is convex and f were not then (6.97) would fail for some $\mathbf{x}_1,\mathbf{x}_2\in D$ and $t\in[0,1]$. This would imply that the corresponding point $\mathbf{p}$ on the line segment connecting $\mathbf{p}_1$ and $\mathbf{p}_2$ would not belong to E. ■ (4) (ii) Use $x\mapsto x^3$ on $\mathfrak{R}$.

6.5.2 Use $f(x)=x^{-1}$ on (0, 1) and $g(x)=x$ on (0, 1] but $g(0)=1$.

6.5.3 (1) Combine the strict version of (6.97) and 6.103 (1) which holds because strict convexity implies convexity.

6.5.4 (1) Use (6.97) which holds as an equality if f is both convex and concave. Show that g is linear where $g(\mathbf{x})=f(\mathbf{x})-f(\mathbf{0})$. (2) Use points in $\mathfrak{R}$. (4) Consider $\mathbf{A}\mathbf{x}_i=\mathbf{y}_i$ for $i=1,2$.

6.5.5 (3) For $n=4$ find second derivative and require $36a_3^2-96a_2\leqq 0$. None are convex for odd degrees $n\geqq 3$. (7) Use derivatives.

6.5.6, **6.5.7**, and **6.5.8** See, for example, Roberts and Varberg (1973).

SECTION 7.1

7.1.1 Drill.

7.1.2 (2) A counterexample is provided by the point (0,0) which does not belong to the interior of the set

$$K=\{(\rho,\theta):0\leqq\rho\leqq r(\theta)\}$$

where (ρ, θ) are the familiar polar coordinates in the real plane and where

$$r(\theta)=2\pi, \qquad \theta=0$$

$$=\theta, \qquad 0<\theta<2\pi$$

That is, the boundary includes a *spiral of Archimedes*.

7.1.3 (2) Minimum at (1, 1). (3) Saddle point at (0,0). (9) See 6.3.9 (2). Objective function is called the *Rayleigh quotient*. (10) Use eigenvalues of $\mathbf{AB}^{-1}$. The case $\mathbf{A}=\mathbf{B}=\mathbf{I}$ shows there are difficulties with repeated eigenvalues.

7.1.4 Drill.

7.1.5 (1) Changing constraint to $\|\mathbf{h}\|_1=1$ yields direction $\pm\mathbf{e}_i$ where ith component of gradient has maximum absolute value and $\pm$ sign, respectively.

7.1.6 For example, use $f(\mathbf{x})=\|\mathbf{x}\|$ and $\mathbf{x}_0=\mathbf{0}$. Another is (7.6).

7.1.7 Drill.

7.1.8 (1) If K is open then by (7.12) there are no solution points. If K is compact then by (6.19) there is at least one solution point to be found by searching the boundary. If K is neither open nor compact then there may or may not be a solution point. Possibly the methods of Section 7.2 can be used for the boundary. (2) The Hessian $\mathbf{f}''$ is symmetric by (6.60) and its definiteness can be resolved at each critical point using Procedure (5.28). Positive definiteness establishes a strict local solution by (7.15). If $\mathbf{f}''$ is positive (or negative) semidefinite-singular then the tests of this section are inconclusive; in this case the behavior of f might be investigated in a neighborhood of the critical point and *level sets* $\{\mathbf{x}:f(\mathbf{x})=k\}$ may help. If $\mathbf{f}''$ is indefinite there is a saddle point. In the remaining two cases—negative definite and negative semidefinite-singular—maxima rather than minima are involved.

SECTION 7.2

7.2.1 (1) $\theta''(t)=[\phi'(t)]^T\mathbf{f}''(\phi(t))\phi'(t)+\mathbf{f}'(\phi(t))\phi''(t)$ (2) Object equals $\boldsymbol{\lambda}^T\otimes \mathbf{y}^T$.

7.2.2 (1) Work with $\lambda=1,4$. (4) Two local minima. (6) Reduce to $x_3=0$ or $\lambda=2$. (7) Maximize $\log f(\mathbf{x})$. (8) Solution point is (1,0) but Lagrange multipliers are not applicable because rank $\mathbf{h}'(\mathbf{x}_0)\neq k$. Example is due to *R.* Courant.

7.2.3 (1) Minimize $\|\cdot\|^2$.

7.2.4 (2) Use an advantageous representation for the line.

7.2.5 (1) $\mathbf{x} = \langle \mathbf{a}, \mathbf{a} \rangle^{-1}\mathbf{a}$. (2) $\mathbf{x} = \pm \|\mathbf{a}\|^{-1}\mathbf{a}$.

7.2.6 Use eigenvalues and diagonalized form of $\mathbf{A}$.

7.2.7 $\mathbf{x}_0 = \mathbf{C}^{-1}\mathbf{A}^T\boldsymbol{\lambda}_0$ where $\boldsymbol{\lambda}_0 = \mathbf{A}\mathbf{C}^{-1}\mathbf{A}^T\mathbf{b}$.

7.2.8 and **7.2.9** Drill.

SECTION 7.3

7.3.1 (1) Use (6.8). (2) Use $g(x) = 0$ on $\mathfrak{R}$. (3) Adapt proof of (7.25). (4) Use $f(x) = x^2$ and $g(x) = -x$ on $\mathfrak{R}$.

7.3.2 and **7.3.3** Drill.

7.3.4 Not in general. Consider $k = 1$ and show that (7.23) holds with multipliers 0, 1, 1 for any differentiable f and g at any $\mathbf{x} \in D$.

7.3.5 (2) All binding constraints are nondegenerate.

7.3.6 (1) Notice that if (7.18) holds with $\mathbf{x}_0$ and $\boldsymbol{\lambda}_0$ it also holds for maximum problem with $\mathbf{x}_0$ and $\boldsymbol{\mu}_0 = -\boldsymbol{\lambda}_0$. (2) Show that pairs $\mathbf{x}_0, \mathbf{u}_0$ and $\mathbf{x}_0, \mathbf{v}_0$ for the respective problems cannot in general exist under (7.24).

7.3.7 Follow pattern.

SECTION 7.4

7.4.1 (1) The result to be proved is

John's Theorem. Suppose f, g, h belong to their respective Classes $C^{(1)}$. If $\mathbf{x}_0$ is a local solution point of Problem (7.4) then there exist $\bar{u} \in \mathfrak{R}, \mathbf{u}_0 \in \mathfrak{R}^m, \boldsymbol{\lambda}_0 \in \mathfrak{R}^k$ such that

(i) $\bar{u} \geqq 0$, $\mathbf{u}_0 \geqq \mathbf{0}$, and $\begin{bmatrix} \bar{u} \\ \mathbf{u}_0 \\ \boldsymbol{\lambda}_0 \end{bmatrix} \neq \mathbf{0}$.

(ii) $\mathbf{u}_0^T g(\mathbf{x}_0) = 0$.

(iii) $\bar{u}\mathbf{f}'(\mathbf{x}_0) + \mathbf{u}_0^T \mathbf{g}'(\mathbf{x}_0) + \boldsymbol{\lambda}_0^T \mathbf{h}'(\mathbf{x}_0) = \mathbf{O}$.

(3) Use $\{\mathbf{y} : \mathbf{h}'(\mathbf{x}_0)\mathbf{y} = \mathbf{0}$ and $\mathbf{g}_j'(\mathbf{x}_0)\mathbf{y} = \mathbf{0}$ for $j \in B^+(\mathbf{x}_0, \mathbf{u}_0)\}$

7.4.2 Drill.

SECTION 7.5

7.5.1 Use 6.1.7 (6) and several other results.

7.5.2 (1) Suppose maximum occurs at $\mathbf{x}_0$, that $\mathbf{x} \in D$ is arbitrary, and write

$$\mathbf{x}_0 = t\mathbf{x} + (1-t)\mathbf{u}$$

for some $\mathbf{u} \in D$. Then use (6.97) to conclude $f(\mathbf{x}) = f(\mathbf{x}_0)$.

7.5.3 (2) Use theorems of the alternative to construct an example where neither primal nor dual has a feasible point. (5) Recall (7.33).

7.5.4 (1) The set of m rows of $\mathbf{A}$ is a subset of the LI set of rows of the partitioned matrix. (2) There is a vertex iff $\mathbf{Pq} = \mathbf{r}$ for some $\mathbf{q} \in \mathcal{R}^m$ where $\mathbf{P}$ is a nonsingular submatrix of $\mathbf{A}^T$ and $\mathbf{r}^T$ is the corresponding submatrix of $\mathbf{c}^T$.

7.5.5 (9) Use (7.29).

7.5.6 Recall Example 7.3.

7.5.7 (1) Use 7.5.4 (2). (2) Assume that a computer program for the simplex method is available and supply details omitted in Example 7.4.

SECTION 7.6

7.6.1 (1) Quadrants I and II. (2) Interior of unit disc and its exterior.

7.6.2 (1) $\{1/4\}, [1/8, 5/8], [1/12, 9/12], \ldots, [1/4n, (4n-3)/4n], \ldots$.

7.6.3 (1) Use (7.33). (2) Use (7.9). (3) Use (6.19). (4) Use indirect proof and expand inner products. (5) Use $\mathbf{p} \in \operatorname{bdry} C$ or $\mathbf{p} \notin \operatorname{bdry} C$. (6) Not essential that $\mathbf{q}_0 \in C$. (7) Write the hyperplanes for $\mathbf{q}_0 \neq \mathbf{0}$. (8) Use line segments as stated. (9) Use (6.16). (10) Go from one set of hypotheses to the other. (11) Show $C_n \subset \operatorname{int} C$. (12) Use 1.2.4 and (6.4). (13) Use C closed. (14) Here D is necessarily a nonempty k-plane where $k < n$, every point in D is a boundary point, and all hypotheses of (7.44) are readily verified.

7.6.4 (2) Drop compactness in (7.39). For example, let C be the set of points on or above $x_1x_2 = 1$ in the interior of the first quadrant and let those of D lie on or below $x_1x_2 = -1$ in the interior of the fourth. Then the x_1-axis is the only supporting hyperplane possessed by C and the same is true for D.

7.6.5 Various statements can be made; for example, a hyperplane through the origin can be identified with subspaces of dimension $n-1$.

SECTION 8.1

8.1.1 Recall from Example 2.1 that $\dim \mathcal{C} = 1$. If $T: \mathcal{C} \to \mathcal{C}$ is linear then for some $a, b \in \mathcal{R}$

$$Tz = (a + ib)(x + iy)$$

Consequently for the linear transformation $A \in \mathcal{L}(\mathcal{R}^2)$, the associated function in $\mathcal{C}$ will be linear iff $a_{11}=a_{22}$ and $a_{12}=-a_{21}$. Now use the standard matrix for the derivative of (8.3) to deduce the Cauchy–Riemann conditions.

8.1.2 (1) Consider $z^{-1}+(z-2)^{-1}$ for $z_0=1$.

SECTION 8.2

8.2.1 Drill.

8.2.2 (1) $\operatorname{res}(p/q, z_0)=[6p'(z_0)q''(z_0)-2p(z_0)q'''(z_0)]/3[q''(z_0)]^2$. (3) (ix) 3 at $z=1$.

8.2.3 (1) Use terms $-(1+k+s)z$ and z^2+k.

8.2.4 Use az^n and $-e^z$.

SECTION 9.1

9.1.1 (1) Show that there always exists an appropriate linear transformation. (2) Can use -1 for LH, and 1 for RH, endpoints whenever endpoints are present.

9.1.2 and **9.1.3** Drill.

9.1.4 (1) Use

$$\mathbf{A}=\begin{bmatrix} a & b \\ c & d \end{bmatrix}$$

and eliminate y_2 from equations to obtain

$$y_1''-(a+d)y_1'+(ad-bc)y_1=0$$

9.1.5 (1) $y'=ay$.

SECTION 9.2

9.2.1 and **9.2.2** Drill.

9.2.3 (2) Use y^{-3} and find $y^4+t=cy^2$.

9.2.4 Drill.

9.2.5 No, maybe there is some other kind of formula.

9.2.6 (1) $3e^t-3-2t$.

SECTION 9.3

9.3.1 (1) Use t^2 and $t|t|$ on $[-1,1]$.

9.3.2 (1) Use $y'=p$

9.3.3 (1) Use $e^{\lambda t}=0$ for no $t\in\mathcal{R}$.

9.3.4 and **9.3.5** Drill.

9.3.6 (6) $4(3)^{-1/2}\exp(-t/2)\sin(3^{1/2}t/2)+t^2-2t$.

9.3.7 (1) Iff real parts of all roots of (9.9) are negative iff $\alpha_0>0$; $\alpha_0,\alpha_1>0$; $\alpha_0,\alpha_1,\alpha_2>0$, and $\alpha_2\alpha_1>\alpha_0$.

9.3.8 and **9.3.9** Drill.

SECTION 9.4

9.4.1 (4) Show that

$$(\mathbf{A}+\mathbf{B})^2=\mathbf{A}^2+2\mathbf{AB}+\mathbf{B}^2$$

and so on, to establish equality of power series.

9.4.2–9.4.5 Drill.

9.4.6 (1) Show that all values of $g(\mathbf{A})$ are determined by the values $g(\lambda)$ where λ ranges over the set of eigenvalues of $\mathbf{A}$. See, for example, Lancaster (1969, Chapter 5).

SECTION 9.5

9.5.1 (6) (viii). $(1/6)t^3+2t^2+10t+22+((2/3)t^3-(5/2)t^2+12t-22)e^t$.

9.5.2–9.5.3 Drill.

9.5.4 Yes, just treat initial values as constants to be determined later.

SECTION 10.1

10.1.1 (1) 2^k is counterpart to e^x. (8) Establish one expression and then change summation variable. (9) Use (10.21); this is a check for manual work with difference tables.

10.1.2 (1) (i) $f(n)=n^3-9n+11$.

10.1.3 (6) Psi function; see Abramowitz and Stegun (1964). (7) (i) Use differentiation with respect to a to find

$$a(a-1)^3\left[n^2a^{n+2}-(2n^2+2n-1)a^{n+1}+(n+1)^2a^n-(a+1)\right]$$

if $a\neq1$, $n(n+1)(2n+1)/6$ if $a=1$. (ii)–(x) Use Table 10.3 and, for trigonometric functions, use (10.25) and (8.5). See, for example, Jolley (1961) or Jordan (1965).

(8) Expansions are

$$u = u^{(1)}$$

$$u^2 = u^{(1)} + u^{(2)}$$

$$u^3 = u^{(1)} + 3u^{(2)} + u^{(3)}$$

$$u^4 = u^{(1)} + 7u^{(2)} + 6u^{(3)} + u^{(4)}$$

(9) Values are

$$\sum_{u=0}^{n} u = \frac{n(n+1)}{2}$$

$$\sum_{u=0}^{n} u^2 = \frac{n(n+1)(2n+1)}{6}$$

$$\sum_{u=0}^{n} u^3 = \left[\frac{n(n+1)}{2}\right]^2$$

$$\sum_{u=0}^{n} u^4 = \frac{n(n+1)(2n+1)(3n^2+3n-1)}{30}$$

(10) Can use sum from 0 to q less that for 0 to $p-1$.

10.1.4 Compare Example 10.3 below.

10.1.5 (1) Use closed form expression for partial sum and take limits.

10.1.6 (3) Find first and second differences by operating on summations and show: a is nonnegative, nondecreasing, and convex; b is nonnegative with $b(0) = m$, nonincreasing, and convex; L is nonnegative with $L(0) = mB$, and convex. (5) 11, 10, 11, 9, 7, respectively. (6) Negative penalties are "premiums" and it may happen that no finite value of n is optimal.

SECTION 10.2

10.2.1 (1) Write out equations and deduce actual domains required. (2) Use suitable recurrence.

10.2.2 and **10.2.3** Drill.

10.2.4 Compare 9.1.4; only new aspect is that there are no special (differentiability) conditions.

10.2.5 (3) $y(k+1) = (k+1)(k-r+1)^{-1}y(k)$. (4) Equate a fourth-order determinant to zero.

10.2.6 Drill.

10.2.7 (1) Unit periodics can be added to solutions of difference equations with no more effect than from adding constants. The same cannot be said for solutions of functional equations as is clear from consideration, for example, of

$$\omega(x) = \sin 2\pi x$$

Whenever difference equations arise from periodic "sampling" of values from a data source governed by a functional equation then it is necessary to take unit periodics into consideration. (3) See Abramowitz and Stegun (1964) or Milne-Thomson (1933).

SECTION 10.3

10.3.1 and **10.3.2** Drill.

10.3.3 (1) Specify $y(r)$ and suppose $a(\cdot)$ is nonzero. (2) $k! + k!\Sigma_{u=0}^{k-1} 1/(u+1)!$, or equivalent form $\Sigma_{u=0}^{k} k^{(u)}$, or others.

10.3.4 Drill.

10.3.5 (1) Set up difference table and proceed as in 10.1.2(2). (3) See 10.1.4.

SECTION 10.4

10.4.1–10.4.3 Drill.

10.4.4 Use same approach as for differential equations. If RHS term is one of

$$k^m, a^k, e^{bk}, a^k k^m, \sin ck, \cos ck$$

then the respective LC used in trial solution $v(k)$ has the form

$$d_0 + d_1 k + \cdots + d_m k^m, d_0 a^k, d_0 e^{bk},$$

$$a^k(d_0 + d_1 k + \cdots + d_m k^m), d_0 \cos ck + d_1 \sin ck, d_0 \cos ck + d_1 \sin ck$$

The corresponding generalizations are used for terms such as $a^k e^{bk}$, $k^m \sin ck$, and so on. Again as for differential equations, there are two cases to distinguish in practice. In the first, none of terms on RHS are solutions to the homogeneous equation (10.64) and above LCs are used unchanged. In the second, solutions appear on RHS and $v(k)$ must be multiplied by k^r where r is the smallest integer such that no terms $k^r v(k)$ are solutions of (10.64). See 10.5.2(3) and 10.5.3(5) for examples of further complications for systems.

10.4.5 Just as is preceding, this is a straightforward counterpart. If $y_1(k)$ and $y_2(k)$ are two LI solutions of (10.64), seek $c_1(k)$ and $c_2(k)$ such that

$$y(k)=c_1(k)y_1(k)+c_2(k)y_2(k)$$

is a particular solution of nonhomogeneous equation. This leads to the particular solution

$$y(k)=\sum_{u=0}^{k} Q(u,k)b(u)$$

where coefficients are quotients of Casorati determinants

$$Q(u,k)=\begin{vmatrix} y_1(u-1) & y_2(u-1) \\ y_1(k) & y_2(k) \end{vmatrix} \div \begin{vmatrix} y_1(u-1) & y_2(u-1) \\ y_1(u) & y_2(u) \end{vmatrix}$$

10.4.6 Drill.

10.4.7 (1) (Following numbers from 10.2.2) (1) Undetermined coefficients method works here to yield $-(9k^2+12k+11)/27$ to which must be added $c_1(2)^k+c_2(-2)^k$. (2) Variation of parameters method works here to yield $(k+2)(k+1)/2$ to which must be added c_1+c_2k. (3) Solution is

$$y(k)=\frac{8}{9}\cos\frac{2k\pi}{3}+\frac{10\sqrt{3}}{9}\sin\frac{2k\pi}{3}+\frac{k^2}{3}-\frac{2k}{3}+\frac{1}{9}$$

(4) $y(k)=[(-4)^k-(-2)^k]/240.$

(5) $y(k)=-\frac{1}{3}(-1)^k+\frac{1}{3}(2)^k=\frac{1}{3}(-1)^{k+1}+\frac{1}{3}(2)^k.$

(6) $y(k)=[(-2)^{k+1}+18k^3-45k^2+21k+2]/162.$

10.4.8 Require moduli of both characteristic roots to be less than unity. For $n=2$ the region is the interior of the triangle in $\alpha_0\alpha_1$-plane determined by $(1,2)$, $(-1,0)$, and $(1,-2)$.

10.4.9 Drill.

10.4.10 One example is 10.2.2(4) whose solution appears just above for 10.4.7.

SECTION 10.5

10.5.1 Drill.

10.5.2 (3) Use $p(k)=(-1)^k\begin{bmatrix} d_0 \\ e_0 \end{bmatrix}+\begin{bmatrix} d_1 \\ e_1 \end{bmatrix}+k\begin{bmatrix} d_2 \\ e_2 \end{bmatrix}+k^2\begin{bmatrix} d_3 \\ e_3 \end{bmatrix}$

10.5.3 (2)(i) Solution is

$$y(k)=\frac{e^k-1}{e-1}\begin{bmatrix}-1\\2\end{bmatrix}+\frac{(-1)^k-e^k}{e+1}\begin{bmatrix}-2\\2\end{bmatrix}+\frac{5}{4}(-1)^k\begin{bmatrix}1\\-1\end{bmatrix}$$

$$+\frac{3}{2}k^2\begin{bmatrix}1\\2\end{bmatrix}+\frac{3}{2}k\begin{bmatrix}-2\\1\end{bmatrix}+\frac{1}{4}\begin{bmatrix}-1\\5\end{bmatrix}$$

(iii) Can use $\mathbf{P}=\begin{bmatrix}2 & 1\\3 & 1\end{bmatrix}$ to find $\mathbf{y}(k)=\begin{bmatrix}2^{k+1}-2k-1\\2^{k+1}-3k\end{bmatrix}$

(3)(i) Set up problem

$$y(k+2)+y(k+1)+y(k)=6$$

$$y(0)=1, \qquad y(1)=2$$

and find $y(k)=2-\cos\frac{2k\pi}{3}-\frac{\sqrt{3}}{3}\sin\frac{2k\pi}{3}$.
(ii) Use $y(k+3)-y(k)=0$. (5) First solve as algebraic system for $y(k+1)$ and $z(k+1)$ and then solve as a difference system.

10.5.4 Require all eigenvalues to be in interior of unit disc in $\mathcal{C}$.

SECTION 10.6

10.6.1 (1) Use geometric series starting with entry (1) and adapt techniques for Laurent's series. (2) Most are straightforward. For entry (3), for example, write out the series

$$y(1)+y(2)z^{-1}+y(3)z^{-2}+\cdots+y(k+1)z^{-k}+\cdots$$

and note that (formally) it equals $z\eta(z)-zy(0)$. (3) (v) Use Table 10.8(6),(7),(8),(11) at least. (7) Check the solutions. (8) Establish existence and uniqueness theorem; check solution.

10.6.2 Drill.

10.6.3 (2)(i) $\frac{\partial Y}{\partial t}=\lambda(s-1)Y$.

SECTION 10.7

10.7.1 (2) $y(k,t)=\Sigma_{u=0}^{k}(at)^u(u!)^{-1}$. (5) Compare obtaining this result from direct solution method on the one hand, and from z-transform method on the other.

10.7.2 (3) Solution is

$$y(k,t)=\sum_{u=0}^{k}\frac{t^u}{u!}\begin{bmatrix}2+3\cdot 2^u-2^{u+2}\\3+3\cdot 2^u-2^{u+2}\end{bmatrix}$$

(4)(i) $y(k,t)=\Sigma_{u=0}^{k}(2t)^u(u!)^{-1}$.
10.7.3 $y(k)=(\lambda/\mu)^k(1-\lambda\mu^{-1})$.
10.7.4 (2) Use summations. (3) Use differential equations. (4) $p(n,t)=\exp(-\mu t)$,

$$p(k,t)=\frac{(\mu t)^{n-k}e^{-\mu t}}{(n-k)!}$$

for $k=n-1,n-2,\ldots,2$ and $p(0,t)$ equals unity minus the sum of all other $p(k,t)$. (5) Use idea of truncation.
10.7.5 Equation is

$$\frac{\partial\Pi(z,t)}{\partial t}=\mu z(1-z)\frac{\partial\Pi(z,t)}{\partial z}+\lambda(z^{-1}-1)\Pi(z,t)$$

10.7.6 (1) Solution is

$$p(k,t)=C(k-1,k-u)e^{-u\lambda t}(1-e^{-\lambda t})^{k-u}$$

(2) $ue^{\lambda t}$ (3) $ue^{\lambda t}(e^{\lambda t}-1)$. (4) zero for all t.
10.7.7 Establish given result as a solution.
10.7.8 Drill.
10.7.9 (2)(ii) $y(t)=\Sigma_{u=0}^{k}\dfrac{(t-u)^u}{u!}$ for $t\in[k,k+1]$.

REFERENCES

Abramowitz, M., and I. A. Stegun (1964). *Handbook of Mathematical Functions.* National Bureau of Standards Applied Mathematics Series, No. 55. U.S. Government Printing Office.

Aczél, J. (1966). *Lectures on Functional Equations and their Applications*. Academic.

Aczél, J. (1969). *On Applications and Theory of Functional Equations.* Academic.

Apostol, T. M. (1974). *Mathematical Analysis*, 2nd ed. Addison-Wesley.

Avriel, M. (1976). *Nonlinear Programming: Analysis and Methods.* Prentice-Hall.

Bellman, R., and K. L. Cooke (1963). *Differential–Difference Equations.* Academic.

Botts, T. (1942). Convex sets. *Amer. Math. Monthly*, **49**, 527–535.

Bourbaki, N. (1953). *Espaces Vectoriels Topologiques*. Actualités Scientifiques et Industrielles No. 1189. Hermann.

Cadzow, J. A. (1973). *Discrete-Time Systems. An Introduction with Interdisciplinary Applications.* Prentice-Hall.

Churchill, R. V. (1972). *Operational Mathematics*, 3rd ed. McGraw-Hill.

Churchill, R. V., J. W. Brown, and R. F. Verhey (1974). *Complex Variables and Applications*, 3rd ed. McGraw-Hill.

Dantzig, G. B. (1963). *Linear Programming and Extensions*. Princeton University Press.

Eggleston, H. G. (1958). *Convexity*. Cambridge University Press.

El'sgol'ts, L. E., and S. B. Norkin (1973). *Introduction to the Theory and Application of Differential Equations with Deviating Arguments* (trans. J. L. Casti). Academic.

Falk, J. E. (1967). Lagrange multipliers and nonlinear programming. *J. Math. Anal. Appl.*, **19**, 141–159.

Falk, J. E. (1969), Lagrange multipliers and nonconvex problems. *SIAM J. Control.* **7**, 534–545.

Farkas, J. (1902). Über die Theorie der einfachen Ungleichungen. *J. Reine Angew. Math.*, **124**, 1–24.

Feller, W. (1968). *An Introduction to Probability Theory and Its Applications*, Vol. 1, 3rd ed. Wiley.

Fenchel, W. (1953). *Convex Cones, Sets, and Functions*. Lecture notes by D. W. Blackett. Department of Mathematics, Princeton University.

Fiacco, A. V. (1976). Sensitivity analysis for nonlinear programming using penalty methods. *Math. Programming*, **10**, 287–311.

Fiacco, A. V., and G. P. McCormick (1968). *Nonlinear Programming: Sequential Unconstrained Minimization Techniques*. Wiley.

Fleming, W. H. (1965). *Functions of Several Variables*. Addison-Wesley.

Gale, D. (1960). *The Theory of Linear Economic Models*. McGraw-Hill.

Gaver, D. P., and G. L. Thompson (1973). *Programming and Probability Models in Operations Research.* Brooke/Cole.

Goldman, A., and A. W. Tucker (1956). Polyhedral convex cones. In H. W. Kuhn and A. W. Tucker (Eds.), *Linear Inequalities and Related Systems*, pp. 19–40. Annals of Mathematics Studies, No. 38. Princeton University Press.

Gordan, P. (1873). Über die Auflösungen linearer Gleichungen mit reelen Coefficienten. *Math. Ann.*, **6**, 23–28.

Gould, H. W. (1972). *Combinatorial Identities*. Henry W. Gould, 1239 College Avenue, Morgantown, WV 26505.

Gross, D., and C. M. Harris (1974). *Fundamentals of Queueing Theory*. Wiley-Interscience.

Henrici, P. (1974). *Applied and Computational Complex Analysis*, Vol. 1. Wiley-Interscience.

Hillier, F. S. and G. J. Lieberman (1974). *Operations Research*, 2nd ed. Holden-Day.

Hogg, R. V., and E. A. Tanis (1977). *Probability and Statistical Inference*. Macmillan.

John, F. (1948). Extremum problems with inequalities as subsidiary conditions. In K. O. Friedrichs, O. E. Neugebauer, and J. J. Stoker (Eds.), *Courant Anniversary Volume*, pp. 187–204. Wiley Interscience.

Jolley, L. B. W. (1961). *Summation of Series*, 2nd ed. Dover.

Jordan, C. (1965). *Calculus of Finite Differences*, 3rd ed. Chelsea.

Kantorovich, L. V. (1939). *Mathematical Methods of Organizing and Planning Production.* Leningrad State University. Reprinted (1959) in *The Application of Mathematics in Economic Analyses* (V. S. Nemchinov, Ed.), pp. 251–309. Publishing House for Socio-Economic Literature, Moscow. Translated (1960) by R. W. Campbell and W. H. Marlow in *Management Sci.*, **6**, 366–422.

Kuczma, M. (1968). *Functional Equations in a Single Variable*. Polish Scientific Publishers, Warsaw.

Kuhn, H. W. and A. W. Tucker (1951). Nonlinear programming. In J. Neyman (Ed.), *Proceedings of the Second Berkeley Symposium on Mathematical Statistics and Probability*, pp. 481–492. University of California Press.

Lancaster, P. (1969). *Theory of Matrices*. Academic.

McCormick, G. P. (1972). Attempts to calculate global solutions of problems that may have local minima. In F. A. Lootsma (Ed.). *Numerical Methods for Non-linear Optimization*, pp. 209–221. Academic.

Mangasarian, O. L. (1969). *Nonlinear Programming*. McGraw-Hill.

Mann, N. R., R. E. Schafer, and N. D. Singpurwalla (1974). *Methods for Statistical Analysis of Reliability and Life Data*. Wiley-Interscience.

Marlow, W. H. (Ed.) (1976). *Modern Trends in Logistics Research*. Massachusetts Institute of Technology Press.

Milne-Thomson, L. M. (1933). *The Calculus of Finite Differences*. Macmillan.

Minkowski, H. (1896) and (1910). *Geometrie der Zahlen*, Parts I and II. Teubner, Leipzig.

von Neumann, J., and O. Morgenstern (1944). *Theory of Games and Economic Behavior*. Princeton University Press.

Phillips, D. T., A. Ravindran, and J. J. Solberg (1976). *Operations Research: Principles and Practice*. Wiley.

Pinney, E. (1959). *Ordinary Difference–Differential Equations*. University of California Press.

Rektorys, K. (Ed.) (1969). *Survey of Applicable Mathematics* (trans. R. Výborný et al.). Massachusetts Institute of Technology Press.

Roberts, A. W., and D. E. Varberg (1973). *Convex Functions*. Academic.

Rockafeller, R. T. (1970). *Convex Analysis*. Princeton University Press.

Stiemke, E. (1915). Über positive Lösungen homogener linearer Gleichungen. *Math. Ann.*, **76**, 340–342.

Stoer, J., and C. Witzgall (1970). *Convexity and Optimization in Finite Dimensions*, Vol. 1. Springer.

Taylor, A. E., and W. R. Manne (1972). *Advanced Calculus*, 2nd ed. Xerox College Publishing.

Tucker, A. W. (1956). Dual systems of homogeneous linear relations. In H. W. Kuhn and A. W. Tucker (Eds.), *Linear Inequalities and Related Systems*, pp. 3–18. Annals of Mathematics Studies, No. 38. Princeton University Press.

Uzawa, H. (1958). A theorem on convex polyhedral cones. In K. J. Arrow, L. Hurwicz, and H. Uzawa (Eds.), *Studies in Linear and Non-Linear Programming*, pp. 23–31. Stanford University Press.

Valentine, F. A. (1964). *Convex Sets*. McGraw-Hill.

Wagner, H. M. (1975). *Principles of Operations Research with Applications to Managerial Decisions*, 2nd ed. Prentice-Hall.

Weyl, H. (1935). Elementare Theorie der konvexen Polyeder. *Comment. Math. Helv.*, **7**, 290–306. Translated (1950) by H. W. Kuhn in *Contributions to the Theory of Games*, Vol. I, pp. 3–18. Annals of Mathematics Studies, No. 24. Princeton University Press.

LIST OF SPECIAL NOTATION

$(x.y)$	Display y in Chapter x, xii
$a.b.c$	Problem set c in Section $a.b$, xii
iff	if and only if, 1
RHS and RH	right-hand side and right-hand, 1
LHS and LH	left-hand side and left-hand, 1
$\blacksquare$	this completes the proof, 1
$a \in A$	a is an element of A, 2
$/$	negation, 2
$A \subset B$	A is a subset of B, 2
$\varnothing$	empty set, 2
A^c	complement of A, 2
$B - A$	relative difference of B and A, 2
$A \cap B$	intersection of A and B, 2
$A \cup B$	union of A and B, 2
$\{x : p(x)\}$	set of all x for which statements $p(x)$ are true, 2
$\{x, y, \dots\}$	set consisting of $x, y, \dots$, 2
$X \times Y$	Cartesian product of X and Y, 2
$f : X \to Y$	function f on X to Y, 3
$x \mapsto x^2$	symbol-free functional notation, 3
$f(A)$	set of images of A, 3
$f^{-1}(B)$	set of inverse images of B, 3
$x \leftrightarrow y$	one-to-one correspondence, 3
f^{-1}	inverse of f, 4
$\mathcal{R}$	real number field or real line, 6, 7
$\mathcal{C}$	complex number field or complex plane, 6, 9
$\lvert x \rvert$	absolute value of x, 7, 8
sup A	supremum of A, 7
max A	maximum of A, 7
inf A	infimum of A, 7
min A	minimum of A, 7

$[a,b]$, (a,b), $[a,b)$,...	real intervals, 8
$\mathrm{Re}\, z$	real part of z, 8
$\mathrm{Im}\, z$	imaginary part of z, 8
$\bar{z}$	complex conjugate of z, 8
$f(x_1-)$	LH limit of f at x_1, 11
$f(x_1+)$	RH limit of f at x_1, 11
$O(v)$	large O of v, 12
$o(v^p)$	small o of v^p, 12
$\approx$	is approximated by, 12
$[x]$	greatest integer function, 13
$\mathcal{V}$	vector space, 15
$\mathbf{a}, \mathbf{b}, \ldots, \mathbf{x}, \mathbf{y}, \ldots$	vectors, 16
$a, b, \ldots, x, y, \ldots$	scalars, 16
$\mathbf{0}$	zero vector, 16
$\mathcal{V}^n$	column vector space, 16
$\mathbf{x}^T$	transpose of $\mathbf{x} \in \mathcal{V}^n$, 16
$\mathcal{R}^n$	Euclidean space, 17, 63
$\mathcal{C}^n$	unitary space, 17, 56
LC	linear combination, 18
LD	linearly dependent, 18
LI	linearly independent, 18
$\dim \mathcal{V}$	dimension of $\mathcal{V}$, 19
$\mathbf{e}_1, \mathbf{e}_{2, \ldots}$	standard basis vectors for $\mathcal{V}^n$, 19
$(\mathbf{v})_A$	column vector of coordinates of $\mathbf{v}$ with respect to ordered basis A, 20
$S_1 + S_2$	linear sum of subspaces S_1 and S_2, 22
$S_1 \oplus S_2$	direct sum of subspaces S_1 and S_2, 22
$\mathbf{A}, \mathbf{B}, \ldots$	matrices, and same lowercase letter usually denotes columns, 23, 25
$m \times n$	order m by n, 23
$\mathbf{A}^T$	transpose of $\mathbf{A}$, 24
$\mathbf{diag}(d_{11}, d_{22}, \ldots, d_{nn})$	diagonal matrix, 24
$\mathbf{O}, \mathbf{O}_{m,n}, \mathbf{O}_n$	zero matrices, 24
$\mathbf{I}, \mathbf{I}_n$	identity matrices, 24
$\mathbf{e}$	$[1, 1, \ldots, 1]^T$, 24
$[\mathbf{a}_1, \mathbf{a}_2, \ldots, \mathbf{a}_n]$	partitioned matrix, 25
$\mathbf{a}_1, \mathbf{a}_2, \ldots$	vectors, and same letter with two subscripts usually denotes components, 25
$[\mathbf{A}, \mathbf{b}]$	partitioned matrix, 26
$\mathbf{A}^{-1}$	inverse matrix, 26
$T\mathbf{x}, T\mathbf{y}, \ldots$	LH notation for transformations, 30
$\mathcal{L}(\mathcal{V}, \mathcal{W}), \mathcal{L}(\mathcal{V})$	vector spaces of linear transformations, 31

diam X	diameter of X, 186
$d(\mathbf{u},X)$	distance from $\mathbf{u}$ to X, 186
$\partial_j f(\mathbf{x}_0)$	partial derivative of f with respect to x_j evaluated at $\mathbf{x}_0$, 189
$\mathbf{f}'(\mathbf{x}_0)$	standard matrix of $f'(\mathbf{x}_0)$, 189, 195
$\nabla f(\mathbf{x}_0)$	gradient vector of f at $\mathbf{x}_0$, 190
$o(\|\mathbf{h}\|)$	small o of $\|\mathbf{h}\|$, 191
$f\in C^{(s)}$	f belongs to Class $C^{(s)}$, 200
$\mathbf{f}''(\mathbf{x}_0)$	standard representation of $f''(\mathbf{x}_0)$, 201
$\partial_i\partial_j f$	mixed partial derivative of f, 199
$\nabla^2 f(\mathbf{x})$	del squared of f at $\mathbf{x}$, 207
$\dfrac{dz}{d\mathbf{x}}, \dfrac{d\mathbf{y}}{d\mathbf{x}}, \dfrac{df}{d\mathbf{x}}$	derivatives with respect to $\mathbf{x}\in\mathfrak{R}^n$, 208, 209
$\mathbf{A}\otimes\mathbf{B}$	Kronecker product of $\mathbf{A}$ and $\mathbf{B}$, 209
$K\mathbf{x}$	$[x_{m+1},x_{m+2},\ldots,x_n]^T$, where $m+k=n$, 232
s.t.	subject to, 248
$\mathbf{A}^L$	least squares inverse of $\mathbf{A}$, 255
$\mathbf{A}^N$	minimum norm inverse of $\mathbf{A}$, 263
$\mathbf{A}^G$	generalized inverse of $\mathbf{A}$, 263
$\mathbf{A}^C$	conditional inverse of $\mathbf{A}$, 267
$B(\mathbf{x}_0)$	binding index set at $\mathbf{x}_0$, 268
$B^+(\mathbf{x}_0,\mathbf{u}_0)$	nondegenerate binding index set corresponding to $\mathbf{u}_0$ at $\mathbf{x}_0$, 273
res (f,z_0)	residue of f at z_0, 305
$\mathcal{I}$	real interval serving as domain for solutions of differential equations, 316
$\mathbf{y}(t),\mathbf{f}(t),\ldots$	values in $\mathfrak{R}^n$ of transformations, 318, 373
$e^{t\mathbf{A}}=\exp t\mathbf{A}$	matrix exponential of $\mathbf{A}$, 333, 334
$\int_o^t \mathbf{h}(u)\,du$	$\left[\int_o^t h_1(u)\,du,\ldots,\int_o^t h_n(u)\,du\right]^T$, 337
$\mathfrak{L}\{y(t)\}$	Laplace transform of y, 344
$\mathfrak{L}^{-1}\{\eta(s)\}$	inverse Laplace transform of η, 344
$y_c(t)$	translation of y by c, 346
$\Delta f(x_k)$	first difference of f at x_k, 353
$Ef(x_k)$	first displacement of f at x_k, 354
$\mathfrak{D}$	set of consecutive integers serving as domain for solutions of difference equations, 355
$x^{(r)}$	rth factorial power of x, 355, 356
$C(x,r)$	binomial coefficient for $x\in\mathfrak{R}$ and integer r, 357
$\mathfrak{S}_n^{(m)}$	Stirling number of the second kind, 361
$\Sigma f(k)$	summation operation, 380
$\underset{u=p}{\overset{q}{\mathbf{S}}} f(u)$	definite summation of f from p to q, 381

$\sum_{u \in \mathfrak{D}'} \mathbf{h}(u)$	$\left[\sum_{u \in \mathfrak{D}'} h_1(u), \ldots, \sum_{u \in \mathfrak{D}'} h_n(u) \right]^T$, 388
$\mathfrak{Z}\{y(k)\}$	z-transform of y, 394
$\mathfrak{Z}^{-1}\{\eta(s)\}$	inverse z-transform of η, 394
$\mathcal{G}\{y(k)\}$	generating function of y, 408

Index

A CATALOG OF SELECTED

DOVER BOOKS

IN SCIENCE AND MATHEMATICS

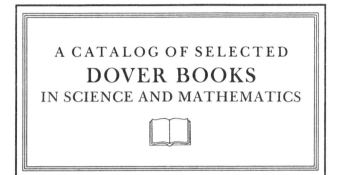

A CATALOG OF SELECTED
DOVER BOOKS
IN SCIENCE AND MATHEMATICS

QUALITATIVE THEORY OF DIFFERENTIAL EQUATIONS, V.V. Nemytskii and V.V. Stepanov. Classic graduate-level text by two prominent Soviet mathematicians covers classical differential equations as well as topological dynamics and ergodic theory. Bibliographies. 523pp. 5⅜ x 8½. 65954-2 Pa. $14.95

MATRICES AND LINEAR ALGEBRA, Hans Schneider and George Phillip Barker. Basic textbook covers theory of matrices and its applications to systems of linear equations and related topics such as determinants, eigenvalues and differential equations. Numerous exercises. 432pp. 5⅜ x 8½. 66014-1 Pa. $12.95

QUANTUM THEORY, David Bohm. This advanced undergraduate-level text presents the quantum theory in terms of qualitative and imaginative concepts, followed by specific applications worked out in mathematical detail. Preface. Index. 655pp. 5⅜ x 8½. 65969-0 Pa. $15.95

ATOMIC PHYSICS (8th edition), Max Born. Nobel laureate's lucid treatment of kinetic theory of gases, elementary particles, nuclear atom, wave-corpuscles, atomic structure and spectral lines, much more. Over 40 appendices, bibliography. 495pp. 5⅜ x 8½. 65984-4 Pa. $13.95

ELECTRONIC STRUCTURE AND THE PROPERTIES OF SOLIDS: The Physics of the Chemical Bond, Walter A. Harrison. Innovative text offers basic understanding of the electronic structure of covalent and ionic solids, simple metals, transition metals and their compounds. Problems. 1980 edition. 582pp. 6⅛ x 9¼. 66021-4 Pa. $19.95

BOUNDARY VALUE PROBLEMS OF HEAT CONDUCTION, M. Necati Özisik. Systematic, comprehensive treatment of modern mathematical methods of solving problems in heat conduction and diffusion. Numerous examples and problems. Selected references. Appendices. 505pp. 5⅜ x 8½. 65990-9 Pa. $12.95

A SHORT HISTORY OF CHEMISTRY (3rd edition), J.R. Partington. Classic exposition explores origins of chemistry, alchemy, early medical chemistry, nature of atmosphere, theory of valency, laws and structure of atomic theory, much more. 428pp. 5⅜ x 8½. (Available in U.S. only) 65977-1 Pa. $12.95

A HISTORY OF ASTRONOMY, A. Pannekoek. Well-balanced, carefully reasoned study covers such topics as Ptolemaic theory, work of Copernicus, Kepler, Newton, Eddington's work on stars, much more. Illustrated. References. 521pp. 5⅜ x 8½. 65994-1 Pa. $15.95

PRINCIPLES OF METEOROLOGICAL ANALYSIS, Walter J. Saucier. Highly respected, abundantly illustrated classic reviews atmospheric variables, hydrostatics, static stability, various analyses (scalar, cross-section, isobaric, isentropic, more). For intermediate meteorology students. 454pp. 6⅛ x 9¼. 65979-8 Pa. $14.95

CHALLENGING MATHEMATICAL PROBLEMS WITH ELEMENTARY SOLUTIONS, A.M. Yaglom and I.M. Yaglom. Over 170 challenging problems on probability theory, combinatorial analysis, points and lines, topology, convex polygons, many other topics. Solutions. Total of 445pp. 5⅜ x 8½. Two-vol. set.
Vol. I: 65536-9 Pa. $8.95
Vol. II: 65537-7 Pa. $7.95

FIFTY CHALLENGING PROBLEMS IN PROBABILITY WITH SOLUTIONS, Frederick Mosteller. Remarkable puzzlers, graded in difficulty, illustrate elementary and advanced aspects of probability. Detailed solutions. 88pp. 5⅜ x 8½.
65355-2 Pa. $4.95

EXPERIMENTS IN TOPOLOGY, Stephen Barr. Classic, lively explanation of one of the byways of mathematics. Klein bottles, Moebius strips, projective planes, map coloring, problem of the Koenigsberg bridges, much more, described with clarity and wit. 43 figures. 210pp. 5⅜ x 8½. 25933-1 Pa. $8.95

RELATIVITY IN ILLUSTRATIONS, Jacob T. Schwartz. Clear nontechnical treatment makes relativity more accessible than ever before. Over 60 drawings illustrate concepts more clearly than text alone. Only high school geometry needed. Bibliography. 128pp. 6⅛ x 9¼. 25965-X Pa. $7.95

AN INTRODUCTION TO ORDINARY DIFFERENTIAL EQUATIONS, Earl A. Coddington. A thorough and systematic first course in elementary differential equations for undergraduates in mathematics and science, with many exercises and problems (with answers). Index. 304pp. 5⅜ x 8½. 65942-9 Pa. $9.95

FOURIER SERIES AND ORTHOGONAL FUNCTIONS, Harry F. Davis. An incisive text combining theory and practical example to introduce Fourier series, orthogonal functions and applications of the Fourier method to boundary-value problems. 570 exercises. Answers and notes. 416pp. 5⅜ x 8½. 65973-9 Pa. $13.95

AN INTRODUCTION TO ALGEBRAIC STRUCTURES, Joseph Landin. Superb self-contained text covers "abstract algebra": sets and numbers, theory of groups, theory of rings, much more. Numerous well-chosen examples, exercises. 247pp. 5⅜ x 8½.
65940-2 Pa. $8.95

STARS AND RELATIVITY, Ya. B. Zel'dovich and I. D. Novikov. Vol. 1 of *Relativistic Astrophysics* by famed Russian scientists. General relativity, properties of matter under astrophysical conditions, stars and stellar systems. Deep physical insights, clear presentation. 1971 edition. References. 544pp. 5⅜ x 8½.
69424-0 Pa. $14.95

Prices subject to change without notice.

Available at your book dealer or write for free Mathematics and Science Catalog to Dept. GI, Dover Publications, Inc., 31 East 2nd St., Mineola, N.Y. 11501. Dover publishes more than 250 books each year on science, elementary and advanced mathematics, biology, music, art, literature, history, social sciences and other areas.